Vibrational Optical Activity

Vibrational Optical Activity

Principles and Applications

LAURENCE A. NAFIE

Department of Chemistry, Syracuse University
Syracuse, New York, 13244-4100, USA

WILEY

A John Wiley & Sons, Ltd., Publication

This edition first published 2011

Registered office
John Wiley & Sons Ltd, The Atrium, Southern Gate, Chichester, West Sussex, PO19 8SQ, United Kingdom

For details of our global editorial offices, for customer services and for information about how to apply for permission to reuse the copyright material in this book please see our website at www.wiley.com.

Library of Congress Cataloging-in-Publication Data

Nafie, Laurence A.
Vibrational optical activity : principles and applications / Laurence A. Nafie.
p. cm.
Includes bibliographical references and index.
ISBN 978-0-470-03248-0 (cloth)
1. Vibrational spectra. I. Title.
QC454.V5N34 2011
539.6–dc22

2011012255

A catalogue record for this book is available from the British Library.

Print ISBN: 9780470032480
ePDF ISBN: 9781119976509
oBook ISBN: 9781119976516
ePub ISBN: 9781119977537
Mobi: 9781119977544

Set in 9/11pt Times by Thomson Digital, Noida, India

This book is dedicated to the loving, nurturing, and inspiring support of both my parents, Marvin Daniel and Edith Fletcher Nafie and my mother's parents Frederic Stark and Edith Webster Fletcher, and to my loving wife Rina Dukor who, for the last 15 years, has been my business and scientific partner in helping me to bring vibrational optical activity to the world, and who recently became, as well, my life's partner in marriage.

Contents

Preface

During the years surrounding the new millennium, the field of vibrational optical activity (VOA), comprised principally of vibrational circular dichroism (VCD) and vibrational Raman optical activity (ROA), underwent a transition from a specialized area of research that had been practiced by a handful of pioneers into an important new field of spectroscopy practiced by an increasing number of scientists worldwide. This transition was made possible by the development of commercial instrumentation and software for the routine measurement and quantum chemical calculation of VOA. This development in turn was fueled by the growing focus among chemists for controlling and characterizing molecular chirality in synthesis, dynamics, analysis, and natural product isolation. The emphasis on chirality was particularly important in the pharmaceutical industry, where the most effective new drugs were single enantiomers and where new federal regulations required specifying proof of absolute configuration and enantiomeric purity for each new drug molecule developed. Today, more than a decade beyond the start of this renaissance, chemists and spectroscopists are discovering the power of VOA to provide, directly, the stereo-specific information needed to further enhance the ongoing revolution in the application of chirality across all fields of molecular science.

The impact of VOA has not been restricted to applications centered on molecular chirality. A concurrent revolution is currently taking place in the field of biotechnology. All biological molecules are chiral, where the chirality is specified by the homochirality of our biosphere, for example L-amino acids and D-sugars. The role of chirality here is not with the specification of absolute configuration but with the specification of the solution-state conformation of biological molecules in native environments. VOA has been found to be hypersensitive to the conformational state in all classes of biological molecules, including amino acids, peptides, proteins, sugars, nucleic acids, glycoprotiens, in addition to fibrils, viruses, and bacteria. Now that the human genome has been coded, emphasis has shifted to understanding what proteins and related molecules are specified in the genetic code. What is their structure and function? Thus VOA is particularly useful as a sensitive new probe of the solution structure of these new protein molecules by classification of their folding family in solution.

What is it about VOA that allows it to determine absolute configuration and molecular conformation in new ways? It is simply that the field of VOA is fulfilling its promise of combining the detailed structural sensitivity of vibrational spectroscopy with the three-dimensional stereo-sensitivity of traditional forms of optical activity. The actual realization of the foreseen potential of VOA has been delivered by sweeping advances in the last two decades of both instrumentation for the measurement of VOA, and software for its calculation and accurate spectral simulation. As will be seen in the chapters of this book, VOA spectra are accompanied by their parent normal vibrational spectra, vibrational absorption, and Raman scattering, and the additional VOA spectrum, linked to a traditional spectrum, is what confers the specific new spectral information.

Beyond the practical benefits to those needing information about the stereochemical structure of chiral molecules, VOA is also providing deep insights into our understanding of the theoretical and computational basis of chemistry. At the theoretical level, VOA intensities require contributions from the interaction of radiation with matter that lie beyond the normal electric-dipole interaction, which by itself is blind to chirality. The new interactions manifested in VOA spectra are the interference of the electric-dipole mechanism with the magnetic-dipole mechanism, and in the case of ROA, the electric-quadrupole mechanism, as well. In addition, VCD in particular requires a theoretical description that lies beyond the Born–Oppenheimer approximation and gives new information about the correlation of the nuclear velocities with molecular electron current density. This is new terrain that lies beyond the traditional Born–Oppenheimer base view of conceptualizing molecules in terms of correlations between nuclear positions and electron probability density. VOA spectra are also proving to be delicate points of reference for quantum chemists who are seeking to improve the accuracy of descriptions of molecules from small organics to proteins and nucleic acids with increasingly realistic models of solvent and intermolecular interactions.

Although VCD and ROA were discovered about the same time in the early to mid-1970s, they have evolved along distinctly different paths in terms of instrumentation and theoretical description. VCD progressed dramatically by taking advantage of Fourier transform infrared spectrometers while ROA gained enormously in efficiency by using advanced solid-state lasers and multi-channel charge-coupled device detectors. ROA theory emerged early and directly from within the Born–Oppenheimer approximation, while VCD theory had to await a deeper understanding of the theory beyond the Born–Oppenheimer approximation for its complete formulation. On the other hand, VCD is simpler and more efficient to calculate whereas ROA is more challenging and requires more intensive calculations. Owing to differences in the relative advantages of infrared absorption and Raman scattering, VCD and ROA tend to be applied to different types of molecules in different types of sampling environments. As a result, papers on VOA, with a few recent exceptions, tend to involve either VCD or ROA, but not both. Nevertheless, despite these relatively separate lines of development, VCD and ROA have a great deal in common, and taken together contain complementary and reinforcing spectral information.

The goal of this book is to bring together, in one place, a comprehensive description of the fundamental principles and applications of both VCD and ROA. An effort has been made to describe these two fields using a unified theoretical description so that the similarities and differences between VCD and ROA can most easily be seen. Both of these fields rest on the foundations of vibrational spectroscopy and the science of describing the vibrational motion of molecules, and both are forms of molecular optical activity sensitive to chirality in molecules. After a basic and somewhat historical introduction to VOA in Chapter 1, the fundamentals of vibrational spectroscopy are presented in Chapter 2 where the formalism of the complete adiabatic approximation, needed for the theoretical description of VCD and a refined description of ROA, is provided. Chapter 3 contains the fundamentals of molecular chirality and the mathematical formalism needed for understanding the theory of both VCD as given in Chapter 4 and ROA as given in Chapter 5. Having completed the necessary theoretical basis of VOA, the focus of the book shifts to instrumentation. The language of describing optical instrumentation and measured VOA intensities, including interfering intensities from birefringence, is the Stokes–Mueller formalism. This is introduced in Chapter 6 for a description of fundamental and advanced methods of VCD instrumentation and is continued in Chapter 7 as a basis for describing ROA instrumentation. The focus of Chapter 8 is the measurement of VOA spectra followed by a description of the methods used for calculating VOA spectra in Chapter 9. In Chapter 10, the final chapter of the book, highlights and selected examples of VOA applications are described. Here VCD and ROA applications are interwoven to better gain an appreciation for both the differences and features in common between these two areas of VOA.

As can be seen from this description of the contents of the book, the material flows from basic principles through theoretical and experimental methods to applications. An effort has been made with the book as a whole, as well as with the individual chapters, to begin with an overview of contents. Thus, Chapter 1 gives a bird's eye view of the entire book and each chapter begins with a descriptive overview at an elementary level of the contents of that chapter. Continued reading in the book or in each chapter carries the reader deeper into the subject with the most advanced material presented usually in last parts of each chapter.

The intended readership for the book is the complete range from beginner to expert in the field of VOA. The book attempts to bridge the gap between the fundamentals of vibrational spectroscopy, chirality, and optical activity and the frontier of research and applications of VOA. The book could serve both as a textbook for graduate courses in chemistry or biophysics as well as a reference for the experienced researcher or scientist. A basic understanding of spectroscopy and quantum mechanics is assumed, but beyond that, nothing further is needed besides patience and a desire to learn new concepts and ideas. Hopefully, the book can serve as a foundation for the continued advancement and development of the exciting new field of VOA.

The book contains many equations, and as a result, alas, it won't ever make the New York Times Bestseller's List. In fact, at the theoretical level, the book is essentially a carefully crafted set of explained equations. Equations are numbered by chapter. When an equation is presented that is based on a previously presented equation, even if it is the same equation, reference to the earlier equation is given to allow the reader to go back and see in more detail the equation's origin in the book. References are provided in the text in a format that identifies authors and years of publication. In the electronic version of the book these are, where possible, live HTML links that take the reader to the source of electro-nic publication. For the most part, chapters are written to be self-consistent and thus can be read individually in any order depending on the particular interests and background knowledge of the reader.

As with any book requiring years of preparation, the author is deeply grateful for the help, collaboration and support of many individuals without whom this book could not have been written. Gratitude begins with my Ph.D. advisor Warner L. Peticolas, who sadly passed away in 2009, and my postdoctoral advisor Philip J. Stephens who started me off on the road to VCD. Warner taught me the excitement of scientific discovery and opened the doors for me to the world of Raman spectroscopy, and Philip taught me the importance of precision and discipline in the way science is practiced and gave me the opportunity to explore and discover the world of infrared vibrational optical activity. I am also grateful to Gershon Vincow, Chairman of the Chemistry Department at Syracuse University who in 1975 hired me as a new Assistant Professor and supported the beginning and growth of my research program in VCD and ROA, and to then Assistant Professor William (Woody) Woodruff who welcomed me to the department and shared his facilities with me to help jump start the construction of my first ROA spectrometer.

I owe endless gratitude to my many graduate students and postdoctoral associates who have worked with me over the years at Syracuse University. Of particular importance are my first postdoctoral associates, Max Diem and Prasad Polavarapu, both of whom went on to distinguished academic careers. I also give very special acknowledgment to Teresa (Tess) Freedman who, as a Research Professor at Syracuse University, collaborated with me on VOA for nearly three decades and helped guide my research program from 1984 to 2000, when I was busy as Chair of the Chemistry Department. Her talent for planning VOA experiments, writing papers, advising students, and carrying out calculations complemented my own love of developing VOA theory and new methods of VOA instrumentation. Without her daily support over those many years, my research in VOA could not have progressed as broadly as it did. Special thanks also go to my former postdoctoral associate, Xiaolin Cao, now a research scientist at Amgen, Inc., who contributed significantly to the optimization of the first dual-PEM, dual-source FT-VCD spectrometer at Syracuse University.

I would like to thank Dr. Rina K. Dukor for being my partner in founding BioTools, Inc., starting in 1996, with the central goal of commercializing VCD and ROA instrumentation. This was achieved in stages, first with VCD in 1997 and then with ROA in 2003. With Rina, my focus on VOA changed from Syracuse University to the world, from pure academic pursuit to facilitating the measurement and calculation of VOA by anyone who wanted to explore this new field of spectroscopy. For the birth of commercial VCD instrumentation, special thanks go Henry Buijs, Gary Vail, Jean-René Roy, Allan Rilling, and many others at Bomem for helping to bring dedicated VCD instrumentation to commercial availability, and again to Philip Stephens for purchasing this first VCD instrument and helping to refine its testing and performance. For ROA instrumentation, special thanks go to Werner Hug for his unfailing encouragement and providing, with help from Gilbert Hangartner, the details of his revolutionary new design for the measurement of ROA. I would also like to thank Omar Rahim and David Rice of Critical Link, LLC for working with BioTools to design and build the first generation of commercial ROA spectrometers, and to Laurence Barron of Glasgow University for purchasing the first of these spectrometers and assisting with Lutz Hecht in the improvement of its performance.

I owe a debt of gratitude to all the employees and close customers of BioTools, Inc. who helped advance the cause of VOA, with special thanks to Oliver McConnell, Doug Minick, Anders Holman, Hiroshi Izumi, Don Pivonka, Ewan Blanch, and Salim Abdali. I would also like to thank those at Gaussian Inc., specifically Mike Frisch and Jim Cheeseman, for being the first to bring VCD and ROA software to commercial availability.

Finally, I would like to thank all other colleagues and collaborators not yet mentioned, who have joined with me in helping to explore and extend the frontiers of VCD and ROA.

Palm Beach Gardens, Florida, USA
February, 2011

1

Overview of Vibrational Optical Activity

1.1 Introduction to Vibrational Optical Activity

Vibrational optical activity (VOA) is a new form of natural optical activity whose early history dates back to the nineteenth century. We now know that the original observations of optical activity, the rotation of the plane of linearly polarized radiation, termed optical rotation (OR), or the differential absorption of left and right circularly polarized light, circular dichroism (CD), have their origins in electronic transitions in molecules. Not until after the establishment of quantum mechanics and molecular spectroscopy in the twentieth century was the physical basis of natural optical activity revealed for the first time.

1.1.1 Field of Vibrational Optical Activity

Vibrational optical activity, as the name implies, is the area of spectroscopy that results from the introduction of optical activity into the field of vibrational spectroscopy. VOA can be broadly defined as the difference in the interaction of left and right circularly polarized radiation with a molecule or molecular assembly undergoing a vibrational transition. This definition allows for a wide variety of spectroscopies, as will be discussed below, but the most important of these are the forms of VOA associated with infrared (IR) absorption and Raman scattering. The infrared form is known as vibrational circular dichroism, or VCD, while the Raman form is known as vibrational Raman optical activity, VROA, or usually just ROA (Raman optical activity). VCD and ROA were discovered experimentally in the early 1970s and have since blossomed independently into two important new fields of spectroscopy for probing the structure and conformation of all classes of chiral molecules and supramolecular assemblies.

VCD has been measured from approximately $600\,\mathrm{cm}^{-1}$ in the mid-infrared region, into the hydrogen stretching region and through the near-infrared region to almost the visible region of the spectrum at $14\,000\,\mathrm{cm}^{-1}$. The infrared frequency range of up to $4000\,\mathrm{cm}^{-1}$ is comprised mainly of fundamental transitions, while higher frequency transitions in the near-infrared are dominated by

Vibrational Optical Activity: Principles and Applications, First Edition. Laurence A. Nafie.

overtone and combination band transitions. ROA has been measured to as low as $50\,cm^{-1}$, a distinct difference compared with VCD, but ROA is more difficult to measure beyond the range of fundamental transitions and is typically only measured for vibrational transitions below $2000\,cm^{-1}$. VCD and ROA can both be measured as electronic optical activity in molecules possessing low-lying electronic states, although in the case of VCD it is appropriate to refer to these phenomena as infrared electronic circular dichroism, IR-ECD or IRCD, and electronic ROA, or EROA.

VCD and ROA are typically measured for liquid or solution-state samples. VCD has been measured in the gas phase and in the solid phase as mulls, KBr pellets and films of various types. When sampling solids, distortions of the VCD spectra due to birefringence and particle scattering need to be avoided. To date, ROA has not been measured in gases or diffuse solids, but nothing precludes this sampling option, although technical issues may arise, such as sufficient Raman intensity for gases and competing particle scattering for diffuse solids.

At present, there is only one form of VCD, namely the one-photon differential absorption form, although recently, a second manifestation of VCD, the differential refractive index, termed the called vibrational circular birefringence (VCB), has been measured. A VCB spectrum is the Kramers–Kronig transform of a VCD spectrum and is also known as vibrational optical rotatory dispersion (VORD). As we shall see, ORD is the oldest form of optical activity and the form of VOA that was sought in the 1950s and 1960s before the discovery of VCD. By comparison, ROA is much richer in experimental possibilities. Because one can consider circular (or linear) polarization differences in Raman scattering intensity associated with the incident or scattered radiation, or both, in-phase and out-of-phase, there are four (eight) distinct forms of ROA. Further, for ROA there are choices of scattering geometry and the frequency of the incident radiation, both of which give rise to different ROA spectra. As a result, there is in principle a continuum of different types of VOA measurements that can be envisioned for a given choice of sample molecule.

Beyond this, many other forms of VOA are possible. One form is reflection vibrational optical activity, which would include VCD measured as specular reflection, diffuse reflection or attenuated total reflection (ATR). In principle, VCD could also be measured in fluorescence. Because fluorescence depends on the third power of the exciting frequency, infrared fluorescence VOA would be very weak relative to VCD and thus very difficult to measure. As with fluorescence in the visible and ultraviolet regions of the spectrum, fluorescence VCD could be measured in two forms, fluorescence detected VCD or circularly polarized emission VCD. In the former, one would measure all the fluorescence intensity resulting from the differential absorbance of left and right circularly polarized infrared radiation (VCD) or measure the difference in left and right circularly polarized infrared emission from unpolarized exciting infrared radiation. Finally, we note the various manifestations of nonlinear or multi-photon VCD, such as two-photon infrared absorption VCD.

In the case of ROA there are a variety of different forms of VOA yet to be measured. One recently reported for the first time is near-infrared excited ROA. Other forms of ROA yet to be measured are ultraviolet resonance Raman ROA, surface-enhanced ROA, coherent anti-Stokes ROA, and hyper-ROA in which two laser photons generate an ROA spectrum in the region of twice the laser frequency. Second harmonic generation (SHG) ROA at two-dimensional interfaces has been measured, and attempts have been made to measure sum frequency generation (SFG) VOA, which is an interesting form of optical activity that depends on transition moments which arise in both VCD and ROA.

Another class of optical activity that has VOA content is vibronic optical activity. Here the source of optical activity is a combination of electronic optical activity (EOA) and VOA when changes to both electronic and vibrational states occur in a transition. This form of EOA–VOA arises in ECD whenever vibronic detail is observed. The analogous form of ROA is either vibronically resolved electronic ROA or ROA arising from strong resonance with particular vibronic states of a molecule.

Finally, we consider other forms of radiation that may affect vibrational transitions in molecules. In particular, it is possible to create beams of neutrons that are circular polarized either to the left or to the right. This phenomenon has been considered theoretically, but experimental attempts at measurement have not been reported. Another common form of vibrational spectroscopy that does not involve photons as the source of radiation interaction is electron energy loss spectroscopy. This is essentially Raman scattering using electrons. If modulation between left and right circularly polarized electrons could be realized, then this could become a new form of VOA in the future.

1.1.2 Definition of Vibrational Circular Dichroism

VCD is defined as the difference in the absorbance of left minus right circularly polarized light for a molecule undergoing a vibrational transition. For VCD to be non-zero, the molecule must be chiral or else be in a chiral molecular environment, such as a non-chiral molecule in a chiral molecular crystal or bound to a chiral molecule. The definition of VCD is illustrated in Figure 1.1 for a molecule undergoing a transition from the zeroth (0) to the first (1) vibrational level of the ground electronic state (g) of a molecule.

More generally, we can define VCD for a transition between any two vibrational sublevels ev and ev' of an electronic state e as:

$$\text{VCD} \qquad (\Delta A)^a_{ev',ev} = (A_L)^a_{ev',ev} - (A_R)^a_{ev',ev} \tag{1.1}$$

where A_L is the absorbance for left circularly polarized light and A_R is the absorbance for right circularly polarized light. The superscript a refers to the vibrational mode, or modes, associated with the vibrational transition. The sense of the definition of VCD is left minus right circularly polarization in conformity with the definition used for electronic circular dichroism (ECD). The parent ordinary infrared absorption intensity associated with VCD, also referred to as vibrational absorbance (VA), is defined as the average of the individual absorbance intensities for left and right circularly polarized radiation, namely:

$$\text{VA} \qquad (A)^a_{ev',ev} = \frac{1}{2}\left[(A_L)^a_{ev',ev} + (A_R)^a_{ev',ev}\right] \tag{1.2}$$

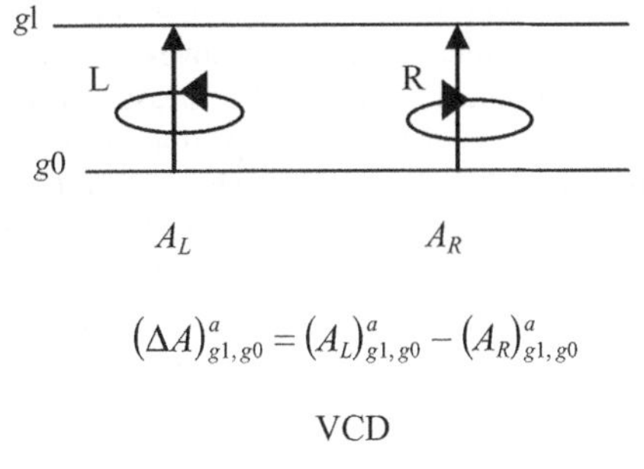

Figure 1.1 Energy-level diagram illustrating the definition of VCD for a molecule undergoing a transition from the zeroth to the first vibrational level of the ground electronic state

These definitions of VCD and VA represent the total intensity associated with a given vibrational transition with the label a. Experimentally, one measures VCD and VA spectra as bands in the spectrum that have a shape or distribution as a function of radiation frequency ν, which is expressed as $f_a'(\nu)$ for each vibrational transition. The reason for the prime will be explained in Chapter 3. An experimentally measured VCD or VA spectrum is therefore related to the defined quantities in Equations (1.1) and (1.2) by sums over all the vibrational transitions a in the spectrum as:

$$\Delta A(\nu) = \sum_a (\Delta A)^a_{ev',ev} f_a'(\nu) \tag{1.3}$$

$$A(\nu) = \sum_a (A)^a_{ev',ev} f_a'(\nu) \tag{1.4}$$

From these expressions it can also be seen that the original definitions of VCD and VA in Equations (1.1) and (1.2) represent integrated intensities over the measured VCD, or VA, band of vibrational transition a by writing for example:

$$\Delta A^a_{ev',ev} = \int_a \Delta A(\nu) \mathrm{d}\,\nu = \int_a \Delta A^a_{ev',ev} f_a'(\nu) \mathrm{d}\,\nu = \Delta A^a_{ev',ev} \int_a f_a'(\nu) \mathrm{d}\,\nu \tag{1.5}$$

where the last integral on the right-hand side of this expression is equal to 1 when a normalized bandshape of unit area is used as:

$$\int_a f_a'(\nu) \mathrm{d}\,\nu = 1 \tag{1.6}$$

Experimentally, the VA intensities are defined by the relationship:

$$A(\nu) = -\log_{10}[I(\nu)/I_0(\nu)] = \varepsilon(\nu) bC \tag{1.7}$$

where $I(\nu)$ is the IR transmission intensity of the sample, which is divided by the reference transmission spectrum of the instrument, $I_0(\nu)$, usually without the sample in place. Normalization of the sample transmission by the reference spectrum removes the dependence of the measurement on the characteristics of the instrument used for the measurement of the spectrum, namely throughput and spectral profile. The second part of Equation (1.7) assumes Beer–Lambert's law and defines the molar absorptivity of the sample, $\varepsilon(\nu)$, where b and C are the pathlength and molar concentration in the case of solution-phase samples, respectively. The experimental measurement of VCD is similar, but more complex than the definition of VA in Equation (1.7), and we defer description of this definition until Chapter 6, when the measurement of VCD is described in detail. The definition of the molar absorptivity in Equation (1.7) yields a molecular-level definition of VCD intensity, $\Delta\varepsilon(\nu)$, which is free of the choice of the sampling variables pathlength and concentration. This is given by:

$$\Delta\varepsilon(\nu) = \Delta A(\nu)/(ee)bC \tag{1.8}$$

where (ee) is the enantiomeric excess of the sample. The (ee) can be defined as the concentration of the major enantiomer, C_M, minus that of the minor enantiomer, C_m, divided by the sum of their concentrations, which is also the total concentration.

$$(ee) = \frac{C_M - C_m}{C_M + C_m} = \frac{C_M - C_m}{C} \tag{1.9}$$

The value of (ee) can vary from unity for a sample of only a single enantiomer to zero for a racemic mixture of both enantiomers, such that neither enantiomer is in excess. Thus we can write:

$$\Delta\varepsilon(\nu) = \Delta A(\nu)/b(C_M - C_m) \tag{1.10}$$

This definition of VCD represents a molecular-level quantity that has been corrected for the pathlength and concentrations of both enantiomers. The intensity expressed as molar absorptivity of a VCD band for vibrational transition a, $(\Delta\varepsilon)^a_{ev',ev}$, can be extracted from the experimentally measured molar absorptivity VCD spectrum by integration over the VCD band of transition a, as:

$$(\Delta\varepsilon)^a_{ev',ev} = \int_a \Delta\varepsilon(\nu)\mathrm{d}\nu \tag{1.11}$$

The quantity $(\Delta\varepsilon)^a_{ev',ev}$ can be compared directly with theoretical expressions of VCD intensity.

A transition between vibrational levels separated by a single quantum of vibrational energy corresponds to a fundamental transition and is described by the superscript a for a particular vibrational mode in the definitions above. In the case of higher level vibrational transitions, more than one vibrational quantum number is needed, such as ab for a a combination band of mode a and mode b, or $2a$ for the first overtone of mode a. All fundamental transitions occur in the IR region below a frequency of $4000\,\mathrm{cm}^{-1}$ and all vibrational transitions above that frequency in the near-infrared region involve only overtones and combination bands.

1.1.3 Definition of Vibrational Raman Optical Activity

ROA is defined as the difference in Raman scattering intensity for right minus left circularly polarized incident and/or scattered radiation. There are four forms of circular polarization ROA. Energy-level diagrams are given in Figure 1.2 for a molecule undergoing a transition from the zeroth to the first vibrational level of the ground electronic state. The left-hand vertical upward-pointing arrows represent the incident laser radiation, and the right-hand downward-pointing arrows represent the scattered Raman radiation. A Stokes Raman scattering process is assumed such that the molecule gains vibrational energy while the scattering Raman radiation is red-shifted from the incident laser radiation by the same energy. The initial and final states of the Raman-ROA transitions, $g0$ and $g1$, are the same as those in Figure 1.1 for VA-VCD transitions. The excited vibrational–electronic (ev) states of the molecule are represented by energy levels above the energy of the incident laser radiation, which applies for the common case in which the incident radiation has lower energy than any of the allowed electronic states of the molecule.

The original form of ROA is now called incident circular polarization (ICP) ROA. Here the incident laser is modulated between right and left circular polarization states, and the Raman intensity is measured at a fixed linear or unpolarized radiation state. The second form of ROA is called scattered circular polarization (SCP) ROA. In this form, fixed linear or unpolarized incident laser radiation is used and the difference in the right and left circularly polarized Raman scattered light is measured. The third form of ROA is in-phase dual circular polarization ($\mathrm{DCP_I}$) ROA. Here the

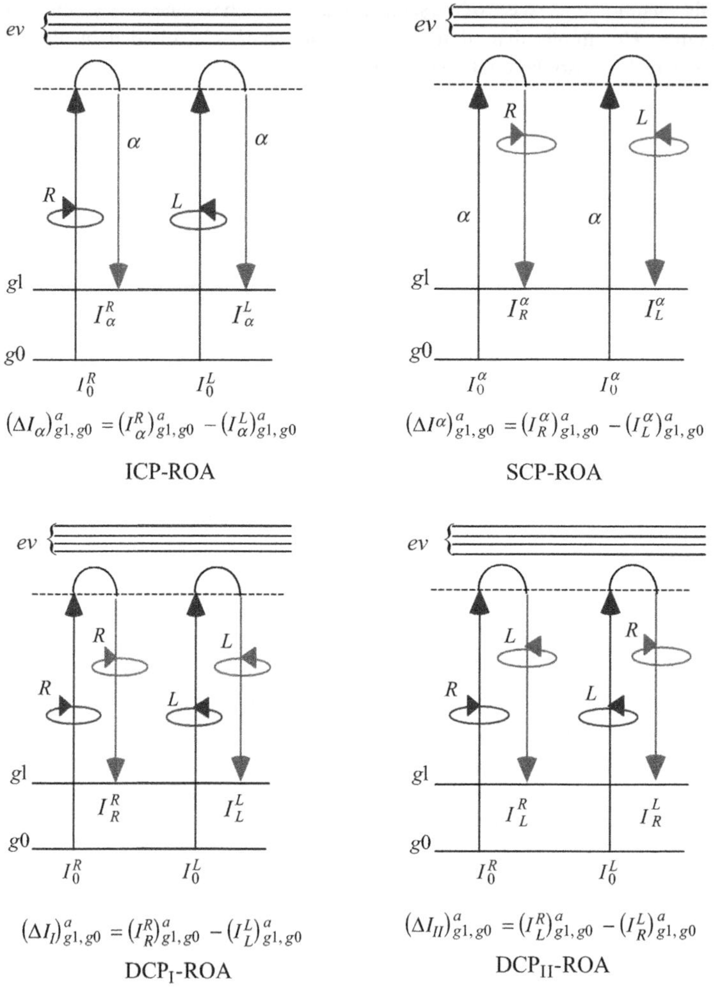

Figure 1.2 Energy-level diagrams illustrating the definition of ROA for a molecule undergoing a transition from the zeroth (g0) to the first (g1) vibrational level of the ground electronic state, where the excited intermediate states of the Raman transition are represented by electronic–vibrational levels (ev)

polarization states of both the incident and scattered radiation are switched synchronously between right and left circular states. The last form of ROA is called out-of-phase dual circular polarization (DCP_{II}) ROA, where the polarization states of both the incident and scattered radiation are switched oppositely between left and right circular states. The definitions of these forms of ROA for any vibrational transition involving normal mode a between states ev and ev' are given by the following expressions.

$$\text{ICP ROA} \quad (\Delta I_\alpha)^a_{ev',ev} = \left(I^R_\alpha\right)^a_{ev',ev} - \left(I^L_\alpha\right)^a_{ev',ev} \tag{1.12a}$$

$$\text{SCP ROA} \quad (\Delta I^\alpha)^a_{ev',ev} = \left(I^\alpha_R\right)^a_{ev',ev} - \left(I^\alpha_L\right)^a_{ev',ev} \tag{1.12b}$$

$$\text{DCP}_\text{I}\ \text{ROA}\quad (\Delta I_I)^a_{ev',ev} = \left(I^R_R\right)^a_{ev',ev} - \left(I^L_L\right)^a_{ev',ev} \tag{1.12c}$$

$$\text{DCP}_\text{II}\ \text{ROA}\quad (\Delta I_{II})^a_{ev',ev} = \left(I^R_L\right)^a_{ev',ev} - \left(I^L_R\right)^a_{ev',ev} \tag{1.12d}$$

The definition of the corresponding total Raman intensity is given as the sum, not the average, of the intensities for right and left circularly polarized radiation.

$$\text{ICP-Raman}\quad (I_\alpha)^a_{ev',ev} = \left(I^R_\alpha\right)^a_{ev',ev} + \left(I^L_\alpha\right)^a_{ev',ev} \tag{1.13a}$$

$$\text{SCP-Raman}\quad (I^\alpha)^a_{ev',ev} = \left(I^\alpha_R\right)^a_{ev',ev} + \left(I^\alpha_L\right)^a_{ev',ev} \tag{1.13b}$$

$$\text{DCP}_\text{I}\text{-Raman}\quad (I_I)^a_{ev',ev} = \left(I^R_R\right)^a_{ev',ev} + \left(I^L_L\right)^a_{ev',ev} \tag{1.13c}$$

$$\text{DCP}_\text{II}\text{-Raman}\quad (I_{II})^a_{ev',ev} = \left(I^R_L\right)^a_{ev',ev} + \left(I^L_R\right)^a_{ev',ev} \tag{1.13d}$$

The intensity of Raman scattering per unit solid angle Ω collected from a cone of angle θ and an illumination volume V of sample varies linearly with the incident laser intensity I_0 and the molar concentration C. Hence, an effective molecular DCP_I Raman differential scattering cross-section $[\mathrm{d}\sigma_\text{I}(\theta)/\mathrm{d}\Omega]^a_{ev',ev}$ can be defined by the expression

$$(I_\text{I})^a_{ev',ev} = I_0 NCV[\mathrm{d}\sigma_\text{I}(\theta)/\mathrm{d}\Omega]^a_{ev',ev} \tag{1.14}$$

where N is Avagadro's number. In an analogous manner, the DCP_I ROA molecular cross-section $[\Delta\,\mathrm{d}\sigma_\text{I}(\theta)/\mathrm{d}\Omega]^a_{ev',ev}$ can be defined as:

$$[\Delta\,\mathrm{d}\sigma_\text{I}(\theta)/\mathrm{d}\Omega]^a_{ev',ev} = \frac{1}{I_0 NCV(ee)}(\Delta I_\text{I})^a_{ev',ev} = \frac{1}{I_0 NV(C_M - C_m)}(\Delta I_\text{I})^a_{ev',ev} \tag{1.15}$$

and where (ee), the enantiomeric excess, is defined in Equation (1.9). Using the definitions of the lineshape functions for individual Raman transitions for modes labeled a, we can express the measured ROA and Raman spectra as sums over individual transitions multiplied by their lineshape functions as:

$$\Delta I_\text{I}(\nu) = \sum_a (\Delta I_\text{I})^a_{ev',ev} f'_a(\nu) \tag{1.16}$$

$$I_\text{I}(\nu) = \sum_a (I_\text{I})^a_{ev',ev} f'_a(\nu) \tag{1.17}$$

1.1.4 Unique Attributes of Vibrational Optical Activity

Vibrational optical activity possesses many unique properties that distinguish it from other forms of spectroscopy. As such it will have an enduring place in the set of available spectroscopic probes of molecular properties. These unique attributes are discussed below.

1.1.4.1 VOA is the Richest Structural Probe of Molecular Chirality

Chirality is arguably one of the most subtle and important properties of our world of three spatial dimensions. Similarly, molecular chirality is one of the most subtle and important characteristics of molecular structure. Of all the available spectroscopic probes of molecular chirality, such as optical rotation and electronic circular dichroism, VOA is by far the richest in structural detail. The IR and VCD spectra, or Raman and ROA spectra, of a chiral molecule sample contain sufficient stereochemical detail to be consistent with only a single absolute configuration and a unique solution-state conformation, or distribution of conformations, of the molecule. In addition, the magnitude of a VOA spectrum relative to its parent IR or Raman spectrum is proportional to the enantiomeric excess of the sample.

1.1.4.2 VOA is the Most Structurally Sensitive Form of Vibrational Spectroscopy

VCD and ROA spectra add a new dimension of stereo-sensitivity to their parent IR and Raman spectra, which are already the most structurally rich forms of solution-state optical spectroscopy. VOA spectra possess a hypersensitivity to the three-dimensional structures of chiral molecules that surpasses ordinary IR and Raman spectroscopy. This is most evident in the VOA spectra of complex biological molecules, such as peptides, proteins, carbohydrates, and nucleic acids, in addition to biological assemblies such as membranes, protein fibrils, viruses, and bacteria. In many cases, VOA spectra exhibit distinct differences in the conformations of biological molecules that are only apparent in the IR and Raman spectra as minor, non-specific changes in frequency or bandshape.

1.1.4.3 VOA Can be Used to Determine Unambiguously the Absolute Configuration of a Chiral Molecule

VOA measurements compared with the results of quantum chemistry calculations of VOA spectra can determine the absolute configuration of a chiral molecule from a solution or liquid state measurement without reference to any prior determination of absolute configuration, modification of the molecule, or reference to a chirality rule or approximate model. Samples need not be enantiomerically pure and minor amounts of impurities can be tolerated. By contrast, the determination of absolute configuration using X-ray crystallography requires single crystals of the sample molecules in enantiomerically pure form. VOA provides either a supplemental check or a viable alternative to X-ray crystallography for the determination of the absolute configuration of chiral molecules. As a bonus, the solution- or liquid-state conformational state of the molecule is also specified when the absolute conformation is determined.

1.1.4.4 VOA Spectra Can be Used to Determine the Solution-State Conformer Populations

Vibrational spectroscopy, as well as electronic spectroscopy, is sensitive to superpositions of conformer populations as conformers interconvert on a time scale slower than vibrational frequencies. VOA spectra of samples containing more than one contributing conformer can be simulated by calculating the VOA of each contributing conformer and combining the conformer spectra with a population distribution of the conformers. When a close match between measured and theoretical simulated VOA and parent IR or Raman spectra is achieved, the solution-state population of conformers used in the simulation is a close representation of the actual solution-state conformer distribution. By contrast, NMR spectra represent only averages of conformer populations interconverting faster than the microsecond timescale. As a result, for such conformers, VOA is currently the only spectroscopic method capable of determining the major solution-state conformers of chiral molecules with more than one contributing conformer.

1.1.4.5 VOA Can be Used to Determine the ee of Multiple Chiral Species of Changing Absolute and Relative Concentration

VCD and ROA are the only forms of optical activity with true simultaneity of spectral measurement at multiple frequencies. For VCD this is achieved with Fourier transform spectroscopy and ROA uses multi-channel array detectors called charge-coupled device (CCD) detectors. All other forms of optical activity are either single-frequency measurements or scanned multi-frequency measurements.

The structural richness of IR or Raman spectra permits the determination of the concentration of multiple species present in solution as a function of time for a single non-repeating kinetic process. The corresponding VCD and ROA spectra depend on both the concentrations and the *ee* values of the multiple chiral species present. The *ee* of multiple species as a function of time can be extracted from VOA spectra by first eliminating their dependence on the concentration of the species present. As a result, VOA has the potential to be used as a unique *in situ* monitor of species concentration and *ee* for reactions of chiral molecules.

While VOA has many unique advantages and capabilities, as highlighted above, most problems of molecular structure are best approached by a combination of techniques. In addition, VOA cannot presently be used in all cases, such as low concentration or rapid timescales, where other methods, such as electronic circular dichroism or femtosecond spectroscopy, have been successfully used. Nevertheless, VOA does have a unique place among the many powerful spectroscopic methods available for molecular structure determination in diverse environments. It should be mentioned that recently VCD has been measured with sub-picosecond laser pulses raising the prospect that the limitation of VCD measurement with rapid time evolution may be overcome in the near future.

1.2 Origin and Discovery of Vibrational Optical Activity

The emergence of VOA in the early 1970s was preceded by many earlier efforts to uncover the effects of vibrational transitions in optical activity spectra, primarily optical rotation measurements in the near-infrared and infrared regions. Tracing the origins and subsequent development of ROA and VCD can only be done at a relatively superficial level. What follows in this and subsequent sections is an attempt to capture the highlights of this story, but leaving out many closely related developments that cannot be included by virtue of limited space. A more complete description of the history and development of VOA requires its own dedicated treatment in order to arrive at a more thorough account of all the key events.

1.2.1 Early Attempts to Measure VOA

The discovery of optical activity in electronic transitions pre-dates the discovery of vibrational optical by more than a century. The measurement of optical rotation (OR) dates back to early nineteenth century (Arago, 1811) when the rotation of the plane of polarized light passing through quartz was first measured. Subsequently, the same phenomenon in simple chiral organic liquids was observed for the first time (Biot, 1815). The first measurements of circular dichroism (CD), the differential absorption of opposite circular polarization states, were not achieved until much later (Haidinger, 1847) and were made in the amethyst form of quartz. CD in liquids was not measured until nearly 50 years later (Cotton, 1895) for solutions of chiral tartrate metal complexes. For those interested in further details of the origins of natural optical activity, several excellent reviews have been written of the history and development of optical activity and the origins of circular polarization of radiation and molecular chirality (Lowry, 1935; Mason, 1973; Barron, 2004). As will be shown in detail in Chapter 3, OR and CD are closely related phenomena. The presence of OR at any wavelength in the spectrum of a sample requires the presence of CD at the same or a different region of the spectrum, and vice versa. Because OR is a dispersive phenomena related to the index of refraction, it appears virtually throughout the spectrum at some level. As such, it is always accessible for measurement, whereas CD is restricted to those regions of the spectrum where absorption bands occur.

The search for vibrational optical activity followed a path similar to that of electronic optical activity just discussed. Early attempts to measure vibrational optical activity consisted of measurements of OR extending to longer wavelengths towards the infrared spectral region. The earliest such measurements (Lowry, 1935) yielded no indications that new sources of CD might lie in the vibrational region of the spectrum. Anomalous OR in α-quartz (Gutowsky, 1951) was reported for the infrared region, but this was challenged and not contested (West, 1954), and was attributed to an instrumental artifact. Similarly, reports of anomalies in the OR of chiral organic liquids were published (Hediger and Gunthard, 1954), but later these observations were also concluded to be instrumental artifacts (Wyss and Gunthard, 1966).

The earliest indication of VCD was the measurement of OR in the near-infrared (near-IR) region (Katzin, 1964) where the monotonic behavior of the OR curve with wavelength (also known as ORD) in α-quartz, indicated a source of CD further into the IR region. Similar conclusions were reached a few years later (Chirgadze *et al.*, 1971) regarding samples of chiral polymers. These two reports refer only to indirect measurements of VOA using OR, and not of the VOA in the region of the originating vibrational transition, called a Cotton effect. Beyond this point in the history of VOA, no further OR measurements, either in the near-IR or the IR region, were reported until recently, as mentioned above and discussed further in Chapter 3 (Lombardi and Nafie, 2009). This absence of VORD occurred because instrumental artifacts are difficult to control for very small OR measurements, and because OR curves are difficult to translate into quantities of direct quantum mechanical significance.

1.2.2 Theoretical Predictions of VCD

The discovery of CD in vibrational transitions from isolated molecules was guided by early theoretical studies. These efforts described VCD intensities through a blend of simple models of CD with those for vibrational absorption intensities. Two theoretical predictions were particularly important in that they predicted VCD intensities that appeared to be within the range of measurable magnitude. The need to resort directly to simple models of VCD, rather than full quantum formulations of VCD stemmed from the fact, as we shall see in detail in Chapter 4, that a complete theoretical description of VCD is not possible within the Born–Oppenheimer approximation. This failure occurs because the electronic contribution to the magnetic-dipole transition moment vanishes for a vibrational transition taking place within a single electronic state of the molecule. This failure yielded a physical inconsistency, as the unscreened nuclear contribution to VCD intensity proved to be no problem whatsoever. Thus, the theory of VCD appeared to possess an internal enigma, and it was not at all clear prior to its experimental discovery whether VCD would be an observable phenomenon. As a result, the publication of simple model calculations was vital for the advancement of the field beyond the level of intellectual speculation. It was not until the early 1980s that the theory of VCD was understood in depth for the first time.

The first model formulation of VCD that could be applied to simple chiral molecules (Holzwarth and Chabay, 1972) was based on the coupled oscillator model of electronic CD (See Appendix A for theoretical description). The problem of the vanishing electronic contribution to the magnetic-dipole transition moment in the Born–Oppenheimer approximation was avoided by developing an expression for VCD based on a pair of chirally-disposed electric-dipole transition moments. If two coupled electric-dipole vibrational transition moments in a molecule are separated in space and twisted with respect to one another, their vibrational motion supports both VA and VCD. The pair of transition moments can be thought of as a coupled-dimer pair of vibrations and their transitions as the action of a coupled-oscillator pair of transitions. This model of CD is known as the coupled oscillator (CO) or exciton coupling model and is described theoretically in Appendix A. The two coupled oscillators result in two vibrational transitions that are slightly separated in frequency and have vibrational motions that are in- and out-of-phase leading to a characteristic VCD couplet that is either positive–negative from high to low frequency in the spectrum or the reverse depending on the twist angle of the two oscillators. In the mirror-image (enantiomer) of the chiral molecule, the structure of the pair of oscillators is identical but the twist angle, and hence the sense of the VCD couplet, is the opposite. The predicted ratio of VCD to VA intensities for typical values of the electric-dipole transition moments of the dimer pair were reported to be in the range of from 10^{-4} to 10^{-5}, which was just within reach of infrared CD instrumentation available at the time.

A year later, a paper was published (Schellman, 1973) that gave further impetus to the search for VCD spectra. In this paper, VA and VCD intensities were modeled by assigning a charge to each nucleus that represents the nuclear charge minus a fixed electronic screening of the nuclear charge. The motion of these fixed partial charges located at the nuclei of a chiral molecule provided sufficient physics for the determination of the electric- and magnetic-dipole transition moments, and hence VA and VCD intensities, for any vibrational mode in the molecule. The problem of the vanishing contribution of the electrons to the magnetic-dipole transition moment was avoided by transferring that contribution, as static quantities, to the nuclear contribution where no such problem was present. Here again, predicted intensities were in the range of from 10^{-4} to 10^{-5} for the ratio of VCD to VA for particular transitions. This method of calculating VCD intensities, although somewhat crude, is important because of its generality and absence of assumptions on the nature of the chiral molecule or its vibrational modes. The model subsequently became known as the fixed partial charge (FPC) model remains important today for its conceptual significance. A brief theoretical description of the FPC model is given in Appendix A.

1.2.3 Theoretical Predictions of ROA

As with VCD, the discovery of ROA was preceded by theoretical prediction. In this case, a single paper (Barron and Buckingham, 1971) established the theoretical foundation for ROA, both experimentally and theoretically. For ROA, no fundamental enigma is present at the level of the Born–Oppenheimer approximation, and hence there was no impediment to writing down a complete and internally self-consistent theoretical formation. This paper was preceded by a description (Atkins and Barron, 1969) of the Rayleigh scattering of left and right circularly polarized light by chiral molecules. These two papers, taken together, established a completely new form of natural optical activity, namely optical activity in light scattering, which became the theoretical basis for both Rayleigh and Raman optical activity.

The experimental focus of the first ROA paper was a form of ROA that today is known as incident circular polarization (ICP) ROA, defined above in Figure 1.2 and Equation (1.12a), although the SCP form of ROA was also described by means of a quantity termed the degree of circular polarization of the scattered beam. The scattering geometry assumed for ROA measurements was the classical right-angle scattering that was predominant at the time. The ratio of the intensity of ROA to Raman was predicted to be in the range of from 10^{-3} to 10^{-4}, although model calculations were not carried out until after ROA was discovered experimentally. These estimates, as in the case of VCD, were sufficiently encouraging, relative to the sensitivity of existing experimental Raman instrumentation, that several research groups undertook the attempt to measure ROA for the first time.

1.2.4 Discovery and Confirmation of ROA

The discovery of the first genuine ROA spectra was reported in 1973, a year before the discovery of VCD was published. Three papers were published in that year from the laboratory of A.D. Buckingham at the University of Cambridge, which were co-authored with postdoctoral associate Laurence D. Barron and graduate student M.P. Bogaard (Barron *et al.*, 1973a; Barron *et al.*, 1973b; Barron *et al.*, 1973c). One of the molecules exhibiting ROA was α-phenylethylamine in the spectral region of low frequency vibrational modes between roughly 250 and 400 cm^{-1}. This first ROA spectrum is reproduced in Figure 1.3 (left) for both enantiomers of α-phenylethylamine (Barron *et al.*, 1973a). The reported ROA spectra from all three of these papers remained unconfirmed

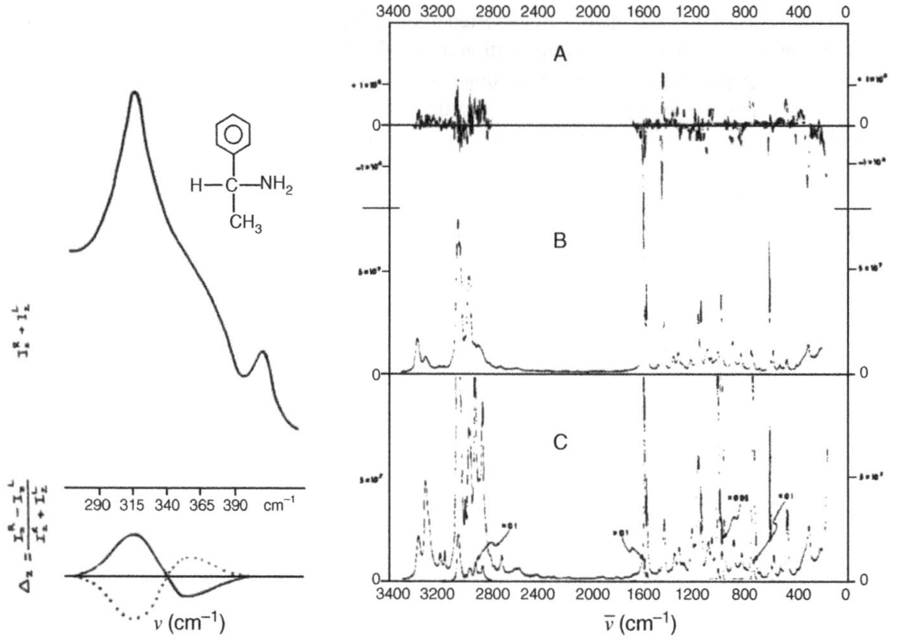

Figure 1.3 Discovery (left) and confirmation (right A) of the ROA spectrum of a neat liquid sample of (+)-α-phenylethylamine with the depolarized Raman spectra (left, and right B) and the polarized Raman spectrum C. Reproduced with permission from the American Chemical Society (Left: Barron et al., 1973c; Right: Hug et al., 1975)

until 1975 when Werner Hug, working the laboratory of James Scherer at the University of California, confirmed the ROA measurement of neat α-pinene and α-phenylethylamine (Hug *et al.*, 1975). This work also extended the spectral range of measurement to include fundamental normal modes from approximately 200 to 3400 cm^{-1}. This confirmation spectrum is shown in Figure 1.3 (right).

It should be noted that in 1972 a research group from the University of Toronto (Bosnick *et al.*, 1972) and then in early 1973 from the University of Toledo (Diem *et al.*, 1973) published papers reporting ROA (termed Raman circular dichroism in the first case and circularly differential Raman in the second) from simple chiral liquids. In both cases, samples of mirror-image pairs of molecules gave equal and oppositely signed ROA spectra, although all the bands in the ROA spectra of each enantiomer were the same sign. Both of these ROA spectra, measured in polarization perpendicular to the scattering plane, where polarized Raman scattering is present, were eventually shown to be the result of instrumental polarization artifacts sensitive to the optical alignment and possibly the optical rotation of the chiral sample molecules. The genuine ROA spectra reported from Cambridge were measured in parallel polarization as depolarized Raman scattering, and were approximately an order of magnitude smaller, with the signs of the ROA varying across the spectrum between positive and negative values for different vibrational modes of the same molecule. These early erroneous reports threw an air of caution into the search for the first genuine VCD spectra, which was underway at the same time that the discovery and verification of ROA was taking place.

1.2.5 Discovery and Confirmation of VCD

The first measurement of VCD for a vibrational mode of an individual molecule was published in 1974 from the laboratory of George Holzwarth at the University of Chicago (Holzwarth *et al.*, 1974). The sample was 2,2,2-trifluoromethyl-1-phenylethanol as a neat liquid. This paper was co-authored by postdoctoral associate E.C. Hsu with collaborators Albert Moscowitz and John Overend of the University of Minnesota, who provided theoretical support, and Harry Mosher from Stanford University, who provided the sample. The vibrational mode was the lone methine C–H stretching mode of the hydrogen on the asymmetric carbon center of this chiral molecule. This first published spectrum of VCD is reproduced in Figure 1.4 (left). It consists of the VCD spectra of the (+)-enantiomer, the (–)-enantiomer and the racemic mixture. It is clear from this figure that this first VCD spectrum is just barely discernable above the noise of the VCD spectrometer. Because of the difficulties encountered with the discovery of ROA, discussed above, this result stood for more than a year as an unconfirmed report until this first spectrum was measured and confirmed independently by a different research group using an IR-CD instrument of a different design.

The confirmation of Holzwarth's measurement was published from the laboratory of Philip J. Stephens at the University of Southern California in 1975 and was co-authored by postdoctoral associates Laurence A. Nafie and Jack Cheng (Nafie *et al.*, 1975). As with ROA, the original measurement was not only confirmed but was improved and extended to other vibrational modes. The VCD spectrum of 2,2,2-trifluoromethyl-1-phenylethanol confirming the discovery of VCD is presented in Figure 1.4 (right).

In addition to the VCD spectrum in the C–H stretching mode, strong VCD spectra were also recorded in the free and hydrogen bonded OH stretching region. In 1976, the first full paper on VCD was published by Nafie, Keiderling, and Stephens, extending the measurement of VCD to dozens of otherwise non-exceptional chiral molecules in the hydrogen stretching region (Nafie *et al.*, 1976). This paper showed that instrumentation could be constructed for the

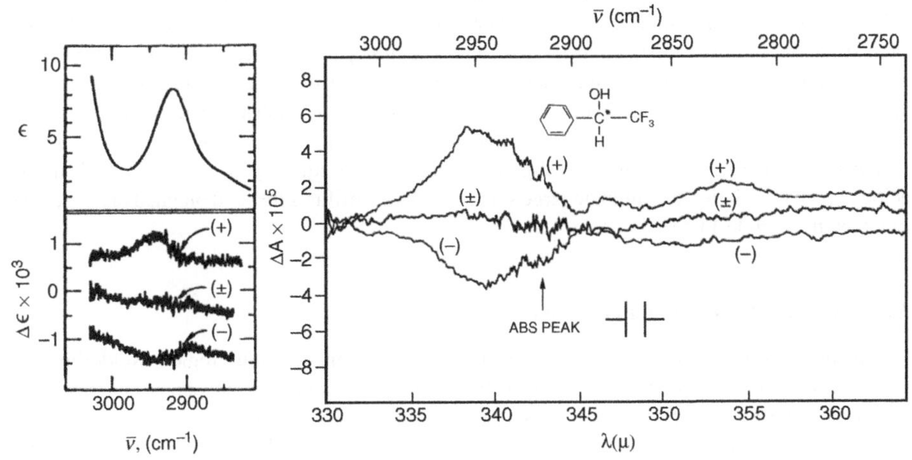

Figure 1.4 *Discovery (left) and confirmation (right) of VCD spectra from individual molecules for a sample of neat 2,2,2-trifluoro-1-phenylethanol. Reproduced with permission from the American Chemical Society (Left: Holzwarth et al., 1974; Right: Nafie et al., 1975)*

routine measurement of the VCD spectra of ordinary chiral molecules, including metal complexes and polymers.

1.3 VCD Instrumentation Development

Instrumentation for the measurement of VCD spectra has undergone several stages of significant advancement, which we briefly discuss here. More complete descriptions of these advances will be provided in Chapter 6 where VCD instrumentation is considered in detail.

1.3.1 First VCD Measurements – Dispersive, Hydrogen-Stretching Region

The first instruments constructed for measuring VCD spectra were dispersive scanning instruments extended from operation in the visible and near-IR regions (Osborne *et al.*, 1973) to cover the wavelength regions of the hydrogen-stretching and mid-IR vibrational frequencies. The hydrogen-stretching region extends from 4000 to 2000 cm^{-1}, or 2.5 to 5.0 microns (μm). It consists primarily of fundamentals of O–H, N–H, and C–H stretching modes, as well as their deuterium analogs, O–D, N–D, and C–D stretching modes. The higher-frequency region from 4000 to 14 000 cm^{-1} is widely regarded as the near-IR region populated by overtone and combination bands of fundamental vibrational modes, as well as low-lying electronic states of some metal complexes. The region from 2000 cm^{-1} to approximately 400 cm^{-1} is usually called the mid-IR region.

The first VCD measurements, which took place in the hydrogen-stretching region, were made possible by several important technological advances that permitted measurements of absorbance intensities in the IR to the level of 10^{-5} absorbance units. These advances were: (i) liquid-nitrogen cooled semiconductor detectors with low noise and response times in the region of 1 μs, (ii) infrared photoelastic modulators (PEMs) with high modulation frequencies and large optical apertures of IR-transparent materials, and (iii) solid-state lock-in amplifiers with high-stability, low-noise and high gain. The components used for the first VCD measurement at the University of Chicago were a Nernst glower source, a Ge PEM, and an InSb photovoltaic detector, whereas the group at the University of Southern California (USC) employed a quartz-halogen lamp, two ZnSe PEMs and an InSb detector. The InSb detector has a low-frequency cut-off near 2000 cm^{-1} and responds electrically to individual photon strikes as opposed to standard IR detectors, which depend on the slower thermal diffusion process. The fast response of the semiconductor detectors was critical for following the modulation frequency of the PEMs, which are in the frequency region of tens of kilohertz. The two PEMs used at USC represented a significant optical advance known as polarization scrambling (Cheng *et al.*, 1975), in which the second PEM is operated at a slightly different PEM frequency from the first PEM and at a specific optical retardation to eliminate large sources of VCD artifacts that had plagued measurement attempts with a single PEM.

1.3.2 Near-IR VCD Measurements

With capability established for measuring VCD in the hydrogen-stretching region, Keiderling and Stephens carried out the first near-IR VCD measurement of combination bands and overtones (Keiderling and Stephens, 1976). Essentially the same instrumentation employed in the hydrogen-stretching region was used except that an InAs detector, with a cutoff of close to 3000^{-1} was substituted for the InSb detector. More than a decade passed before Abbate undertook additional near-IR VCD measurements (Abbate *et al.*, 1989) by converting a visible electronic CD spectrometer into near-IR operation to about 6000 cm^{-1}. The emphasis of this work is the study of the VCD of second and higher overtones of fundamental CH-stretching vibrational modes.

1.3.3 Mid-IR VCD Measurements

The low-frequency limit of VCD observation in the first few years following the discovery of VCD was about 2800 cm^{-1}, the low-energy limit of CH stretching vibrations. This limit was extended considerably in 1978 to 1600 cm^{-1} using a PbSnTe detector on the USC dispersive VCD spectrometer (Stephens and Clark, 1979). In 1980, Keiderling, at the University of Illinois, Chicago, extended VCD measurements through much of the mid-IR to approximately 1200 cm^{-1} using an HgCdTe detector (Su *et al.*, 1980). Further extension of VCD in the mid-IR did not occur until Fourier transform (FT) VCD spectrometers were developed and optimized by Nafie and co-workers at Syracuse University where vibrational transition to as low of 800 cm^{-1} were measured (see the next section). Following the advent of FT-VCD instrumentation, the low-frequency limit of VCD measurement was extended to 650 cm^{-1} (Devlin and Stephens, 1987) using dispersive instrumentation and a silicon detector with a low frequency cut-off. Attempts to extend this limit to 300 cm^{-1} using a Fourier transform polarization division interferometer were reported (Polavarapu and Deng, 1996) but currently remain uncertain due to the small size of the possible VCD features relative to the noise level.

1.3.4 Fourier Transform VCD Instrumentation

The theory of rapid-scan double modulation Fourier transform difference spectroscopy, with application to both circular and linear dichroism, or any other high-frequency modulation of the IR beam, was published in 1979 (Nafie and Diem, 1979). The concept of Fourier transform CD measurement had been pursued and demonstrated in France in the 1960s using step-scan instrumentation in the visible region of the spectrum (Russel *et al.*, 1972). The double modulation rapid-scan approach used the idea of inserting a PEM, operating in the tens of kilohertz region, prior to the sample in an FT-IR spectrometer and separating the double-modulated high-frequency VCD interferogram from the much lower frequency ordinary IR interferogram using electronic filters. The output of a lock-in amplifier tuned to the PEM frequency with a sub-millisecond time constant produces a VCD interferogram that can be Fourier transformed and further processed to produce a final VCD spectrum.

Based on this methodology, the first FT-VCD measurements were carried out by Nafie and Vidrine at the headquarters of the Nicolet Instrument Corporation in Madison, Wisconsin, USA, in the summer of 1978 (Nafie *et al.*, 1979). These measurements employed an InSb detector and yielded FT-VCD in the medium-IR region for the CH-stretching modes of camphor. Subsequently, at Syracuse University in 1981, FT-VCD measurements were extended to the mid-IR region where VCD spectra of high quality and high spectral resolution were measured from 1600 to 900 cm^{-1} for a variety of chiral organic molecules, thus demonstrating the generality of FT-VCD instrumentation (Lipp *et al.*, 1982). The low-frequency limit of these measurements was imposed principally by the cut-off of the type-A HgCdTe (MCT) detector with a low-frequency limit of detection of approximately 800 cm^{-1}. This advance not only ushered in a new era of VCD measurements in terms of simultaneous quality, resolution and spectral range of VCD measurement, but it opened the mid-IR region to routine measurement of VCD and considerably eased the task of constructing a VCD spectrometer. One could now begin construction of a VCD spectrometer starting from a sophisticated computer-controlled FT-IR spectrometer. Only the VCD accessory bench and auxiliary electronics needed to be assembled. This brought VCD instrumentation within the reach of manufacturers of FT-IR spectrometers and ultimately led to the commercialization of VCD spectrometers in the 1990s.

1.3.5 Commercially Available VCD Instrumentation

The commercialization of instrumentation for VCD measurements took place gradually. Both BioRad (Digilab) and Nicolet (now Thermo Nicolet) helped interested customers equip their FT-IR spectrometers with VCD accessory benches in the mid- to late 1980s, but neither company actively advertised

VCD in their product literature. Throughout the early 1990s, both Nicolet and BioRad were helping VCD researchers (Nicolet with Nafie at Syracuse University and BioRad with Keiderling at the University of Illinois, Chicago) in various ways and were learning from them how to best equip an FT-IR spectrometer for VCD operation. With the advent of step-scan FT-IR instrumentation in the mid-1990s, interest in the possibility of commercially available VCD increased further, and Bruker started working with Nafie at Syracuse University to develop a commercial VCD accessory.

A major breakthrough in commercial VCD instrumentation occurred when Rina Dukor, alongwith Nafie, formed BioTools Inc., and convinced Henry Buijs and Garry Vail at Bomem Inc., an FT-IR manufacturer in Quebec City, Canada, to team with BioTools to build a dedicated FT-VCD spectrometer optimized with hardware and software for VCD measurement. In 1997, Bomem and BioTools, in a joint venture, introduced the Chiral*IR* FT-VCD spectrometer to the market place as the first stand-alone, fully-dedicated instrument for VCD operation. This instrument revolutionized the field of VCD. For the first time, one could watch VCD spectra being collected and improved second-by-second with each FT-scan, as the various steps of phase correction of interferograms, division of VCD transmission by IR transmission spectra, and intensity calibration were automatically incorporated into the spectral collection and output displays.

Following this advance, other FT-IR manufactures, such as Nicolet, BioRad (now Varian), and Bruker offered improved VCD accessory benches with dedicated software for VCD measurements. However, the Bomem–BioTools VCD instrument (now assembled solely by BioTools) remained the only single platform VCD spectrometer with factory pre-aligned optics available commercially. More recently, Jasco has commercialized a single-platform VCD spectrometer in Japan. In 2009 BioTools offered a second-generation FT-VCD spectrometer called the Chiral*IR-2X* in which all electronics processing is reduced to a single PC card in the VCD computer, and up to three interferograms can be collected simultaneously, one for the normal IR spectrum and two for the VCD and the VCD baseline, the latter two of which are dynamically subtracted with each interferogram scan.

Over the past several years the number of research groups involved in measuring or calculating VCD has increased from approximately four to well over 50, more than an order of magnitude, with the level of interest and activity increasing each year.

1.4 ROA Instrumentation Development

The construction of an ROA instrument that is relatively free of optical artifacts is much more difficult to achieve than the corresponding instrumentation for VCD. As a result, prior to the development of commercially available ROA instrumentation, the prevalence of ROA instrumentation world-wide was limited, except for brief efforts, to essentially three research groups, namely those of Laurence Barron in Glasgow, UK, Werner Hug in Fribourg Switzerland, and Laurence Nafie in Syracuse, New York, USA, and where only the instruments at Glasgow and Syracuse have maintained activity over the 30 years since the discovery of ROA. In this section, the historical development of the major advances in ROA instrumentation is briefly described, with more technical details provided in Chapter 7 on ROA instrumentation.

1.4.1 First ROA Measurements – Single Channel ICP-ROA

The first ROA measurements were carried out with dispersive scanning monochromators equipped with single-channel photomultiplier tube detectors and photo-counting electronics. The optical layout was right-angle scattering and the polarization modulation scheme was depolarized ICP-ROA. In particular, the incident laser radiation was square-wave modulated between right and left circular polarization states and the scattered light was passed through a linear polarizer that was set to be parallel to the plane of scattering, thus producing depolarized Raman scattering intensities and eliminating polarized Raman scattering. The earliest measurements used either the 448 or the 514 nm lines of an argon ion laser. An advance using this basic instrumentation included the first measurements of anti-Stokes ROA in Glasgow, which confirmed the theoretical prediction of the signs and intensities for this type of ROA (Barron, 1976).

1.4.2 Multi-Channel ROA Measurements

Multi-channel ICP-ROA measurements were reported from the laboratory of Werner Hug at the University of Fribourg in 1979 (Hug and Surbeck, 1979) and shortly thereafter from the laboratory of Martin Moskovits at the University of Toronto (Brocki *et al.*, 1980). Multi-channel detection overcomes one of the most serious drawbacks of ROA measurement, namely the length of time required for the detection of a spectrum with sufficient signal-to-noise ratio. Multi-channel detection reduces the time required by nearly two orders of magnitude, thus dramatically increasing the access of ROA spectra to experimental measurement. The measurements of Hug also featured a dual arm collection scheme that permitted the first reported measurements of ROA in perpendicular polarization (polarized ICP-ROA) where artifacts were reduced by cancellation of artifacts of opposite sign in the two collection arms.

1.4.3 Backscattering ROA Measurements

In 1982, Hug reported the design of an ICP-ROA instrument for measurements in backscattering geometry (Hug, 1982). Unfortunately, this instrument, as well as Hug's entire ROA laboratory, was destroyed by a fire in the Chemistry Department at the University of Fribourg. With insufficient funds to reconstruct his laboratory, Hug turned his attention for several years to theoretical calculations of ROA; however, in 1989, with encouragement from Laurence Barron, he collaborated with Barron by providing parts recovered from his 1982 instrument to build a new backscattering ROA spectrometer in Glasgow (Barron *et al.*, 1989). This work included the first measurements of ROA with a charged-coupled device (CCD) detector and revolutionized the measurement of ICP-ROA by reaching unprecedented levels of speed of collection and spectral quality.

1.4.4 SCP-ROA Measurements

The theoretical basis for SCP-ROA was established by Barron and Buckingham when they first considering Rayleigh and Raman optical activity in 1971. It was referred to as P_c, the degree of circularity of the scattered beam, if linear polarized light was incident on the sample. However, it was regarded more as a theoretical curiosity rather than an ROA intensity that could be readily measured with available technology. The conceptual barrier was that the definition of P_c was couched more in terms of a single property of scattered light, the small degree to which left or right circularly polarized was in excess for the various Raman scattering intensities.

However, in 1988 Nafie reasoned that the entire scattered beam could be thought of as consisting of only left and right circularly polarized contributions. The contributions could be measured separately by using a zeroth-order quarter-wave plate such that one circular component would be converted into vertically-polarized intensity and the other circular component would be horizontally-polarized intensity. The two circular polarization components of the scattered beam could be then be separately measured by using a linear polarizer to select either the right or left circularly polarized component of the scattered radiation. If these two intensities were added, the ordinary polarized or depolarized Raman spectrum would be obtained, and if subtracted, the P_c could be measured, not as a single measurement but as the difference of two Raman spectra, in the same way that ICP-ROA was measured by switching the circular polarization states of the incident light back and forth and measuring the difference in Raman intensities. To draw attention to this analogy, the measured P_c spectrum was referred to as scattered circular polarization ROA or SCP-ROA (Spencer *et al.*, 1988). Subsequently, with the help of Lutz Hecht and Diping Che, this instrument was redesigned and upgraded (Hecht *et al.*, 1991) to allow detailed comparisons of ICP-ROA and SCP-ROA of the same sample under the same instrument conditions (Hecht *et al.*, 1992).

1.4.5 DCP-ROA Measurements

The demonstration of the feasibility of the measurement of ICP and SCP forms of ROA led to the theoretical prediction of two new dual circular polarization forms of ROA (Nafie and Freedman, 1989), designated DCP_I-ROA and DCP_{II}-ROA and defined in Section 1.1.3. Backscattering DCP_I-ROA is a purely depolarized form of ROA, and is the most efficient form of ROA relative to the intensity of the parent Raman spectrum, DCP_I-Raman, that can be performed. On the other hand, DCP_{II}-ROA is a very weak effect that vanishes in the far-from-resonance (FFR) approximation. The first measurements of DCP_I-ROA were published in 1991 and confirmed the theoretical prediction of the effect made two years earlier (Che *et al.*, 1991). In general, the sum of DCP_I-ROA and DCP_{II}-ROA equals ICP_u-ROA, that is ICP-ROA measured with no analyzer or polarization discrimination in the scattered beam. In 1994, DCP_{II}-ROA was measured for the first time as the difference in between backscattering ICP_u-ROA and DCP_I-ROA in a series of four molecules with increasing double-bond character, starting from *trans*-pinane, which has no functionality (Yu and Nafie, 1994). The corresponding increase in the intensity of the DCP_{II}-ROA spectrum with increasing organic functionality signaled the breakdown of the FFR approximation and the onset of pre-resonance Raman intensity.

1.4.6 Commercially Available ROA Instruments

The first, and to date the only, commercially available instrumentation for the measurement of ROA was introduced by BioTools Inc., in 2003. The Chiral*RAMAN* spectrometer is an SCP-ROA spectrometer with laser excitation at 532 nm designed along the lines of an SCP-ROA spectrometer built by Hug at the University of Zurich. The details of the optical design of the Hug instrument were described first (Hug and Hangartner, 1999) and its artifact reduction features were published a few years later (Hug, 2003). Hug's design embodies a number of novel features that make the measurement of ROA routine and more efficient than any ROA or Raman spectrometer previously constructed. The most significant of these is that the right and left circularly polarized scattered radiation are measured simultaneously on the upper and lower halves of the CCD detector, thereby eliminating the effects of laser intensity variations and sample flicker noise in the measured SCP-ROA spectrum. The second novel feature of this instrument is the use of electronically-controlled half-wave plates to eliminate linear polarization components and equalize over time any bias in the instrument for detection of left and right circularly polarized scattered light. The result is an instrument that measures SCP-ROA spectra of high quality in a routine fashion thereby freeing the user to focus on the measurement and interpretation of ROA spectra rather than optimizing and adjusting the performance of the instrument.

1.5 Development of VCD Theory and Calculations

Understanding the origin of VCD intensities and developing software for calculating VCD intensities has passed through a number of stages, beginning with simple conceptual models and culminating with the current status of sophisticated quantum calculations that closely simulate experimentally measured spectra. The two conceptual models that emerged from considerations of electronic optical activity in the decades preceding the development of VCD were the coupled oscillator model and the one-electron model for a charge following a helical path. Not surprisingly, both of these conceptual models can be found in the development of descriptions of VCD intensity.

1.5.1 Models of VCD Spectra

Models of VCD spectra were important for the development of this field because, as mentioned above, the complete quantum mechanical description of VCD lies beyond the Born–Oppenheimer approximation. A review of the various models developed for the description of VCD has been published that connects these models to the formal theory of VCD (Freedman and Nafie, 1994). The first theoretical model of VCD intensity was published in two papers by Deutche and Moscowitz for vibrational modes in polymers (Deutsch and Moscowitz, 1968; Deutsche and Moscowitz, 1970). The basic idea was that of moving charges on the nuclei that comprised the structure of the polymer. The mathematical formalism used in these papers was complex and simple expressions were not offered that could be generally applied to other situations.

1.5.1.1 Coupled Oscillator Model

The first publication of a model of the VCD having wide applicability was on the coupled oscillator model (Holzwarth and Chabay, 1972) (Appendix A). As noted above, VCD was shown to arise from two identical oscillators in a molecule that were separated by a fixed distance and were skewed relative to one another. These two oscillators were assumed to couple such that two new vibrational modes ensued from their coupling. One was the mode where the two oscillators moved in-phase with respect to each other and the other where they moved out-of-phase. The model predicts VCDs of equal magnitude and opposite sign for the two coupled modes. Estimates of the magnitudes of expected VCD intensities were provided that served as an incentive for the experimental search for VCD spectra. This model was failry general but nevertheless was restricted to cases where two near-identical coupled vibrational motions could be found in a molecule.

1.5.1.2 Fixed Partial Charge Model

The second early model of VCD intensity was the fixed partial charge model of VCD intensity (Schellman, 1973) (Appendix A). The model had its conceptual roots in the papers by Deutche and Moscowitz, but more specifically used theoretically derived estimates of the excess positive or negative fractional charge found on the individual nuclei of a molecule. This model shows that IR and VCD intensities can be calculated for any chiral molecule for which partial charges are assigned to each atomic nucleus and for which the relative displacements of each atom in the normal modes of the molecule are known. Predictions of expected VCD intensities were given for selected vibrational modes in a series methyl pyrrolidones, which were encouraging for the observability of VCD.

1.5.1.3 Localized Molecular Orbital Model

In 1977 Nafie and Walnut published the first model of VCD that used a quantum mechanical description of the electronic motion (Nafie and Walnut, 1977) (Appendix A). In this approach, the molecular orbitals of the molecule are first localized (LMOs) by one of several available methods, resulting in orbitals with pairs of electrons corresponding to inner shell atomic orbitals, bonding orbitals, and lone pairs. The motion of the centroid (average position) of charge of each LMO is combined with the motion of each nucleus during a normal mode to predict the VA and VCD intensity.

1.5.1.4 Charge Flow Model

A generalization of the FPC model was published in 1981 by Abbate, Laux, Overend, and Moscowitz (Abbate *et al.*, 1981) and a similar version was published soon thereafter (Moskovits and Gohin, 1982). The idea is to permit the fixed charges on the nuclei to vary with nuclear motion rather than to remain fixed, as in the FPC model, leading to charge fluxes at the nuclei and charge currents between nuclei along connecting bond lines. This flexibility overcame a major limitation to the FPC model but was never implemented to any significant extent. Instead, this model was important conceptually as it presaged the solution to the problem of the Born–Oppenheimer approximation.

1.5.1.5 Ring Current Model

The ring current (RC) model was first proposed in 1983 by Nafie, Oboodi, and Freedman to explain the anomalously large positive VCD associated with the lone methane (C_α–H) stretching mode of all the L-amino acids (Nafie *et al.*, 1983; Nafie and Freedman, 1986). All previously proposed empirical models of VCD were conservative in the sense that the sum of the VCD over all coupled modes yielded zero net VCD intensity, and thereby could not explain this excess VCD associated with the C_α–H stretching modes of the L-amino acids, and subsequently other such modes. The basic idea of the RC model is that a single oscillator generates a ring of vibrationally induced current that is not accompanied by a corresponding motion of the nuclei. The current in the ring generates its own unshielded oscillating magnetic dipole moment that combines with the electric-dipole moment of the current-generating oscillator, such as a CH bond stretch, to produce large monosignate VCD intensity. After many years of successful application, cases were found that did not conform to the prediction of the RC model (Bursi *et al.*, 1990) and its active use was discontinued in favor of *ab initio* calculations of VCD. Further discussion of the ring current and other current models of VCD is given in Appendix A.

Other empirical or molecular orbital models of VCD were proposed over time (Freedman and Nafie, 1994), but only the coupled oscillator model in a more general form has persisted past the development of the rigorous formulation of VCD intensity and its subsequent implementation using modern quantum chemistry methods, as described below.

1.5.2 Vibronic Coupling Theory of VCD

As mentioned previously, the Born–Oppenheimer (BO) approximation does not include the dynamical response of the electron density in a molecule to nuclear velocity. This problem was solved exactly for the first time (Nafie and Freedman, 1983) where it was demonstrated that the lowest-order correction to the BO approximation contains the missing vibronic coupling term that gives formally equivalent electronic contributions to VA intensities using either the position or the velocity dipole moment operators. By extension, a formally correct and complete description is thereby obtained for the electronic contribution to the magnetic-dipole moment for a vibrational transition within a single non-degenerate electronic state. Subsequently, it was shown (Nafie, 1983a) that the identified BO correction term could be made imaginary by converting a quantum mechanical momentum operator into a classical momentum coordinate, thus producing a missing correlation between electronic current density in molecules correlated with associated classical nuclear velocities. This theory supplements the normal BO correlation between static electron density and classical nuclear positions. The new adiabatic wavefunction having parametric dependence on both classical nuclear positions and velocities is termed the complete adiabatic (CA) wavefunction. The complete vibronic coupling theory (VCT) by necessity includes a summation over all the excited states (SOS) of the molecule, a fact that temporarily impeded the implementation of the theory of VCD for practical calculations for several more years (Dutler and Rauk, 1989).

1.5.3 Magnetic Field Perturbation Formulation of VCD

Two years after the VCT theory was published, a magnetic field perturbation (MFP) formulation of VCD was proposed (Stephens, 1985) based on the VCT theory of VCD published by Nafie. This was followed a year later by the implementation of the MFP theory for two simple rigid molecules that were chiral by deuterium substitution (Lowe *et al.*, 1986a; Lowe *et al.*, 1986b). The results were encouraging and represented the first VCD spectra calculated from first principles, using *ab initio* quantum mechanical algorithms, without underlying approximations in the theoretical expression used. This represented a major advance to the field of VCD. In the MFP formulation of VCD, it was shown that the explicit non-Born–Oppenheimer sum over electronic excited states in VCT theory could be circumvented by replacing the summation with the algebraically equivalent perturbation of the electronic wave function with an applied magnetic field, even though a magnetic field is not present during a VCD measurement. This step also creates an imaginary component to the electronic wavefunction, which supports the magnetic field generated electronic current density. The MFP formalism was independently developed in the Ph.D. thesis of Galwas at the University of Cambridge under the supervision of Buckingham (Galwas, 1983). This work was subsequently published but calculated VCD intensities were not reported (Buckingham *et al.*, 1987).

The SOS and MFP formulations of the VCT theory of VCD intensity are formally the same, but offer different computation routes to the same result. Both formulations sample the physics of all the excited states defined within the finite basis set used for the quantum chemistry calculations. In the MFP formulations the sampling of the excited electronic states takes place within the solution of coupled perturbed Hartree–Fock (CPHF) equations, whereas in the SOS formulation, the excited states are used directly within the second-order perturbation theory with explicit excited-state energies without perturbing the electronic wavefunction in a self-consistent way. In 1989 the SOS formulation of VCT was first implemented for VCD calculations (Dutler and Rauk, 1989).

1.5.4 Nuclear Velocity Perturbation Formulation of VCD

In 1992 the third formulation of VCD amenable for use with *ab initio* quantum chemistry programs was published (Nafie, 1992). In the nuclear velocity perturbation (NVP) formulation of VCT theory, the non-BO correction term is parameterized using the classical momenta of the nuclei and is used as an energy perturbation to carry out CPHF theory of the electronic wavefunction. Nuclear velocity dependence of the electronic wavefunction was included by the use of an exponential gauge factor on the atomic orbitals, similar to the gauge factors used to describe so-called London orbitals (London, 1937) or gauge-invariant atomic orbitals (GIAOs) (Ditchfield, 1974). GIAOs are currently used in the calculation of magnetic properties of molecules, where, in the absence of such orbitals, the choice of origin of the magnetic moment operator leads to undesired variation, origin dependence, in the calculated results. The NVP gauge factor carries an explicit dependence of the nuclear velocity of the basis-function orbitals on which the orbitals are centered. As the NVP formulation of VCD is a CPHF approach to VCT theory, it represents an origin-independent alternative to avoid the explicit SOS states of VCT theory. To date the NVP formulation of VCD has not yet been implemented. In the NVP paper of Nafie, it was also demonstrated for the first time that GIAOs in the MFP formation of VCD also produce an origin-independent formulation of VCD intensities. In 1993, the first calculations of VCD using London orbitals were published (Bak *et al.*, 1993).

1.5.5 *Ab Initio* Calculations of VCD Spectra

Since the first *ab initio* calculations of VCD reported in 1985, a number of significant advances have been made, mostly pioneered by Philip Stephens in collaboration with Michael Frisch at Gaussian Inc. These include overcoming the problem of origin dependence by implementing a distributed origin treatment (Jalkanen and Stephens, 1988) and then implementing GIAOs, testing of basis sets for relative accuracy, testing post-Hartree–Fock programs, such as second order Moller–Plesset (MP2) (Stephens *et al.*, 1994) and implementing density functional theory (DFT) and exploring available functionals (Devlin *et al.*, 1996). The result of this work is a minimal recommendation for the calculation of VCD intensities as a GIAO basis set at the level of 6-31G(d) or higher and DFT with hybrid functionals such as B3LYP or B3PW94. These minimal options for VCD calculations are available in several quantum chemistry programs.

1.5.6 Commercially Available Software for VCD Calculations

With the breakthroughs in the practical formulation and *ab initio* calculations of VCD intensities, it was not long before quantum chemistry programs included VCD as an option in the menu of available calculated molecular properties. Several quantum chemistry packages have become available for the calculation of VCD intensities. All use the MFP formulation of VCD intensities and offer a variety of choices of basis sets and approaches to intensity calculations. The first of these was the Cambridge Analytical Derivative Program Package (CADPAC, http://www-theor.ch.cam.ac.uk/software/cadpac.html) from the University of Cambridge. The Dalton Program (http://www-theor.ch.cam.ac.uk/software/cadpac.html) from the University of Olso, Norway, and Gaussian 98, 03 and 09 from Gaussian Inc., (www.gaussian.com) in Wallingford, CT, USA soon followed with available VCD subroutines. More recently the ADF programs from the Amsterdam Density Functional (ADF) software packages from Scientific Computing and Modeling (http://www.scm.com/) offer subroutines for calculating VCD and ROA based density function theory (DFT) as opposed to a wider spectrum of quantum chemistry methods. Of these, the Gaussian program package is currently the longest, most established commercially available software for VOA calculations. With the advent of well-maintained software for the calculation of VCD spectra, the power of VCD to elucidate the structure of chiral molecules became significantly enhanced and is now widely available to all who wish to compare the measured and calculated VCD spectra.

1.6 Development of ROA Theory and Calculations

The theory of ICP-ROA in the far-from-resonance approximation has been known since 1971, more than a decade before VCD reached an equivalent level of complete theory. Due in part, however, to its greater theoretical complexity as a second-order perturbation phenomena with respect to the interaction of light and matter, ROA theory passed through a long period of intensity models and spectra–structure empirical correlation before the first ROA intensities were calculated using *ab initio* quantum chemistry programs (Polavarapu, 1990). In parallel with this, a more complete elucidation of the theory of ROA ensued where a more general theory was established (Hecht and Nafie, 1991) and limiting cases, such as resonance ROA (Nafie, 1996), were examined. In this section we highlight some of the key developments between the initial statement of the theory and our present day understanding. A complete treatment of ROA theory is given in Chapter 5.

1.6.1 Original Theory of ROA

For want of a better term, the original theory of ROA as published by Barron and Buckingham in 1971 described ICP-ROA for right-angle scattering using a theoretical treatment that avoided vibronic detail in the sum over excited electronic states. In this limit, the so-called FFR limit, the resonance response of the molecule is taken to be equivalent for both the incident and scattered radiation, the Raman tensor is symmetric, there are only two Raman invariants and three ROA invariants. This theory was the guiding light of ROA experiments for over a decade before a number of improvements were introduced. The first of these was by Hug where backscattering ICP-ROA was first described (Hug, 1982). Here it was demonstrated that ROA could be measured with several advantages in backscattering compared with the traditional right-angle scattering geometry.

1.6.2 Models of ROA Spectra

During the early years of the exploration of ROA, simple models played an important role in understanding the possible origin of significant ROA spectral features. In the absence of *ab initio* calculations, models and empirical correlation between spectra and structure were the only means to understand ROA spectra. Prominent among these models is the perturbed degenerate mode model used to explain bisignate ROA arising from near-degenerate vibrations of locally symmetric groups, such as the degenerate anti-symmetric methyl deformation modes near 1450 cm^{-1}. Another important model is the two-group model that predicts bisignate ROA from pairs of coupled locally symmetric polarizability groups. This model is the analog of the degenerate coupled oscillator (DCO) model of VCD, for which many examples have been found experimentally. For ROA, however, few examples of the two-group model of ROA have been identified experimentally, perhaps due to differences in the origins of ROA and VCD intensities. Finally, there is the torsion mode model of ROA intensities that is applicable to low-frequency torsional modes in molecules. These models are extensively described by Barron (2004) in his book on light scattering and optical activity.

1.6.3 General Unrestricted Theory of Circular Polarization ROA

In contrast to the original theory, the general unrestricted (GU) theory of ROA describes the full extent of the theory of the various forms and scattering geometries of circular and linear polarization ROA measurements (Nafie and Che, 1994). The forms of circular polarization (CP) ROA have been described from the experimental viewpoint in Section 1.4, but beyond this there are four forms of linear polarization (LP) ROA, yet to be discovered, as described in the next section. Behind these developments lies a rich theory of ROA that has not yet been implemented in commercially available computational software programs. Nevertheless, we describe here what this full theory entails and the key advances in the theory of ROA that have led to its present stage of development.

The first steps toward the full unrestricted theory of ROA were taken by 1985 in a paper describing the asymmetry that arises between Stokes and anti-Stokes ICP-ROA when the symmetry of the Raman and ROA polarizability and optical activity tensors is not assumed (Barron and Escribano, 1985). A closely related symmetry breakdown was also noted between what is now called the ICP and SCP forms of ROA, although the latter form of ROA was called the degree of circular polarization in advance of its experimental discovery and renaming in 1987. Following the experimental discovery of SCP and the theoretical prediction of DCP forms of ROA, the distinct theory of all four forms of CP ROA was described by Hecht and Nafie for all possible scattering angles and typical linear polarization schemes. No assumptions were made that would limit the applicability of the theory (Hecht and Nafie, 1991). The aim of the paper was to compare and clarify the relative advantages and disadvantages of the various ways of measuring ROA. The analysis confirmed that ordinary Raman scattering is described by three Raman invariants, and ten ROA invariants. Experimental schemes can be devised to isolate all three Raman invariants, but only six linearly independent combinations of the ten ROA can be measured.

1.6.4 Linear Polarization ROA

While working on the GU theory of ROA, a new form of ROA was discovered in backscattering or forward scattering, known as linear polarization ROA (Hecht and Nafie, 1990). Here linear polarization incident on the sample is scattered with the linear polarization state rotated to the left or the right depending on the sign of the LP ROA. In LP ROA, one uses the imaginary part of the electric-quadrupole ROA tensor and the real part of the magnetic-dipole ROA tensor. It can be shown that the LP ROA tensors are only non-zero in the limit of resonance with one or more particular excited electronic states of the molecule.

1.6.5 Theory of Resonance ROA in the SES Limit

The theory of ROA for a molecule in resonance with a single electronic state (SES) is the simplest theoretical expression for ROA scattering (Nafie, 1996). This theory, like the FFR theory, represents a limiting case of the theory of ROA. In this limit, only A-term resonance Raman scattering is present, and it is shown that the resulting ROA borrows all of its sense of chirality from the electronic circular dichroism (ECD) associated with the single resonant electronic state. In fact, it is demonstrated that the ratio of the ECD intensity of this band to the intensity of its parent absorption band, the anisotropy ratio, is equal and opposite in sign to the corresponding ratio of the resonance ROA (RROA) to its parent resonance Raman (RR) scattering intensity for all bands in the spectrum. The opposite sign occurs because of the reverse definition of positive intensity in ROA (right minus left) as debated in the literature (Barron and Vrbancich, 1983; Nafie, 1983b). The direct correlation between SES-RROA and the ECD of the resonant state means that the RROA, in this limit, is monosignate and the same shape as its parent Raman spectrum, only smaller by the anisotropy ratio of the resonant electronic state. Subsequently, the theory was confirmed for a pair of chiral molecules, each with a single electronic state in resonance with the incident laser radiation (Vargek *et al.*, 1998). This limit breaks down as resonance with other states becomes important or when the resonance state is vibronically coupled to nearby electronic states. If these other states have ECD of the opposite sign to the resonant state, then the ROA spectrum can begin to have ROA bands of both positive and negative signs. The theory of resonance ROA in the SES limit shows the most primitive association between the occurrence of ROA and the ECD of all the excited electronic states responsible for the Raman polarizability and ROA tensors.

1.6.6 Near Resonance Theory of ROA

More recently, a new theoretical limit of ROA has been defined called the near resonance (NR) theory (Nafie *et al.*, 2007), which connects the original far-from-resonance (FFR) theory to the SES theory. In the FFR theory, as mentioned above, the Raman tensor is symmetric and there are only two Raman tensor invariants and three ROA tensor invariants. Symmetry in the Raman tensor signifies that the tensor response is equivalent in degree of resonance for the incident and scattered Raman radiation. As resonance is approached, this equivalence of resonance degree breaks down. Once the Raman tensor becomes even slightly non-symmetric, the number of tensor invariants increases from two to three for Raman scattering and three to ten for ROA scattering. In the NR limit differences can been seen theoretically between ICP and SCP ROA and DCP_{II} ROA becomes non-zero. Unlike the FFR limit, the NR limit, although almost as simple in form as the FFR theory, carries the full richness of theoretical structure as the GU theory of Raman and ROA.

1.6.7 *Ab Initio* Calculations of ROA Spectra

As mentioned above, the first *ab initio* calculations of ROA were carried out by Polavarapu in 1990. The calculations were carried out at the Hartree–Fock level in the zero-frequency static limit for the incident radiation without correction for origin dependence. Improvement in this methodology was advanced by the use of London (GIAO) orbitals (discussed above for VCD theory) and linear response theory (Bak *et al.*, 1994). Using this methodology, calculations of ROA were carried out that were independent of molecular origin and the frequency of the incident radiation was specified rather than set to zero value. More recently, Hug has reported ROA calculations using rarefied basis sets optimized for the calculation of Raman and ROA intensities (Zuber and Hug, 2004). For the Raman and ROA intensities he employed Hartree–Fock linear response theory using London orbitals as implemented in the Dalton Program with a basis set that includes ample diffuse functions but limited polarization functions. The calculation of the optimized geometry, force field, and normal mode displacements were carried with more traditional basis sets, such as 6-311G**. This dual basis-set approach yielded ROA intensities at the same level of accuracy as an aug-cc-pVDZ basis set, but at a fraction of the cost.

1.6.8 Quantum Chemistry Programs for ROA Calculations

As mentioned in the previous section, the Dalton Program has included ROA calculations as an option since 1997. Similarly to their VCD program package, the ROA calculations are implemented at a variety of levels of theory, including self-consistent field (SCF), multiple-configuration SCF (MCSCF), second-order Moller–Plesset (MP2) perturbation theory, coupled-cluster theory, and density functional theory (DFT). The Gaussian program package included ROA calculations starting with Gaussian 03 (Frisch *et al.*, 2003) and with further improvements in Gaussian 09 (Frisch, 2009). The ROA subroutines run more slowly than the corresponding VCD subroutines for the same molecule for two reasons. The first is that Raman and ROA intensities place more demanding requirements on the basis set used due to the need to calculate the response of the polarizability and optical activity tensors with the nuclear motion, compared with electric and magnetic dipole moments for IR and VCD. The second reason is that the Gaussian 03 programs for ROA tensor derivatives were executed by finite difference which is very time-consuming, but the Gaussian 09 version has implemented analytic derivatives for the ROA tensor derivatives, which now eliminates this second shortcoming.

1.7 Applications of Vibrational Optical Activity

The areas of application of VOA range almost as broadly as those of infrared and Raman spectroscopy with the restriction that the samples be chiral molecules with a measurable amount of enantiomeric excess. There are many ways to classify the applications of VOA. Most commonly, this is done either by the type of sample, the type of measurement, or the type of information obtained. In this section we provide an overview of applications of VOA without considering VCD and ROA separately. In Section 1.8 we will compare and contrast these two pillars of VOA spectroscopy.

1.7.1 Biological Applications of VOA

Virtually all biological molecules of significance are chiral. Most of these biological molecules display conformational preferences that are important for their roles as biological agents of structure or dynamics. The structure of complex biological molecules that have well-defined conformations can be determined by X-ray crystallography if the molecule can be isolated, purified, and crystallized. Alternatively, solution-phase determinations at the level of atomic resolution can also be determined by nuclear magnetic resonance (NMR) spectroscopy. Relative to these high-resolution techniques, VOA offers some unique insights and contributions to our understanding of the structure of biological molecules. Firstly, VOA is sensitive to chirality, as is X-ray crystallography, but without the restriction of a single crystalline sample and a single conformational state. Secondly, VOA is measured in solution under a variety of sampling conditions, similar to those of NMR, but NMR is blind to chirality. Further, as mentioned above, VOA is not restricted to the size of the biological molecule for which information can be extracted, as occurs with NMR above a certain molecular weight.

VOA spectra have been measured for amino acids, peptides, proteins, sugars, carbohydrates, oligosaccharides, nucleotides, DNA, RNA, glycoproteins, viruses, and bacteria. In short, all major classes of biological molecules have been investigated. Most studies of biomolecules with VOA have been aimed at correlating spectra with structural changes in the biomolecules on an empirical basis or interpreted in terms of some model of VOA intensity. More recently progress has been made toward the *ab initio* DFT calculation of VCD intensity for oligopeptides with up to as many as 20 residues for particular secondary structure motifs (Bour and Keiderling, 2005). Parallel with these developments, a sophisticated form of the coupled oscillator model has been used that simultaneously models IR, polarized Raman, depolarized Raman, and VCD for a large number of tripeptides (Schweitzer-Stenner *et al.*, 2007). It has been shown that these molecules adopt a regular well-defined solution-state conformation. Similar progress in ROA is possible in principle, but is currently hampered by lower molecular size constraints for carrying out ROA *ab initio* calculations, although some progress in this direction has been made recently (Herrmann and Reiher, 2007).

1.7.2 Absolute Configuration Determination

With the availability over the past decade of commercial instrumentation for the routine measurement of VOA spectra, coupled with software packages for the accurate *ab initio* calculation of VOA spectra over almost the same period, a new important area of application has rapidly emerged, namely the determination of the absolute configuration of chiral molecules. Most of this activity has occurred for VCD as both instrumentation for measurement and commercial software for calculation of ROA have been available only since 2004. Nevertheless, everything that has been achieved with VCD in this area can in principle be performed with ROA, albeit not as simply nor as quickly.

Part of the rapid growth in VCD over the past decade, and ROA more recently, has been due to the commercial availability of instrumentation combined with a growing need for proof of absolute configuration in the pharmaceutical industry. More than half of all the new pharmaceutical drugs in discovery, development, and testing are chiral. Before a new pharmaceutical molecule can be presented to a regulatory agency for approval of sale to the general public, the absolute configuration must be known beyond any doubt. The gold standard for the determination of the absolute configuration of a molecule has been single-crystal X-ray diffraction using the Bijovet method. In this method, the absolute stereochemistry of the single crystal sample can be determined from its anomalous X-ray diffraction pattern by using a heavy atom to specify the phase of the diffraction. However, obtaining single crystals of sufficient quality for the determination of absolute configuration can be difficult, and sometimes impossible, to achieve. In the vast majority of such cases, it has been shown that VCD can determine the absolute configuration of such chiral pharmaceutical molecules without ambiguity.

The VCD method of absolute configuration determination is carried out by comparing the experimentally measured solution-phase VCD spectrum with the *ab initio* calculated VCD spectrum (Freedman *et al.*, 2003; Stephens and Devlin, 2000). In general, there are several tens of bands to compare in sign and relative magnitude. When a good match is obtained between measured and calculated IR and VCD spectra, the absolute configuration of the physical sample is known because the absolute configuration of the molecular structure used for the theoretically calculated spectrum is known. If the calculated VCD spectrum agrees in relative intensities but with opposite signs, then the wrong enantiomer was chosen for the theoretical calculation, and the problem is easily corrected by multiplying the theoretical spectrum by minus one.

1.7.3 Solution-State Conformation Determination

The IR absorption and VCD spectrum of a molecule is sufficiently sensitive that a good match between measured and calculated intensities is not possible unless the correct conformation, or distribution of conformations, of the molecule in solution is first identified. In the case of small to medium sized organic molecules or inorganic complexes, if more than one solution-state conformer is present in significant population, the interchange between conformers occurs on the order of picoseconds, much faster than the NMR timescale. As a result, NMR conformation analysis yields only an average of the conformers present whereas vibrational spectra in general, and VOA in particular, consists of linear superpositions of the contribution of each conformer present. In most cases, as the result of the large number of vibrational transitions in the spectrum, individual bands can often be found in the measured IR and VCD spectra that are specific to a particular conformer. When a distribution of conformers is identified by *ab initio* calculation and a close match with measured IR and VCD intensities is achieved, the resulting information obtained about the solution-state conformers present is unique. No other technique currently available has the sensitivity to identify solution-state conformer populations of chiral molecules, in addition to the identification of their absolute configuration.

1.7.4 Enantiomeric Excess and Reaction Monitoring

One of the simplest applications of any form of optical activity is the measurement of enantiomeric excess (*ee*), also called optical purity, defined earlier in the chapter. A common form of the measurement of optical purity is optical rotation, or specific rotation at a particular wavelength, where the rotation value compared with a known standard gives the *ee* of the sample. Measurement of *ee* was carried out for the first time in 1990 with VCD (Spencer *et al.*, 1990) and in 1995 with ROA (Hecht *et al.*, 1995). An advantage of using VOA for *ee* measurements is that the molecule is identified by its VOA and parent vibrational spectrum at the same time that its *ee* is measured. This ensures that unknown impurities are not present and that the parent IR or Raman spectrum provides a secondary check on the identity and quantity of sample being measured. In optical rotation, for example, there is no independent check if the concentration of the sample is correct for a known optical pathlength. This is one reason that the temperature must be specified for optical rotation measurements, but no such precaution is normally required for VOA determinations of *ee*.

In VOA, each band in the VOA spectrum represents an independent measure of the *ee* when ratioed to its parent vibrational band or compared with the corresponding VOA intensity of a known standard of the same sample molecule. Owing to the large number of such bands that are unique to each molecular species, and because these bands are well resolved from one another, the possibility exists of simultaneous measurement of the *ee* of more than one molecule at a time. Such a determination is not possible with optical rotation, and is difficult for electronic optical activity because electronic absorption or CD bands are not well resolved or sufficiently distinct from one molecule to another.

Recently it was demonstrated that VCD could be employed to monitor the conversion of one chiral molecule into another by continuous flow-cell sampling where, for a sequence of mid-IR FT-VCD measurements, the mole fractions and the *ee* s of the two molecules were changing as a function of time (Guo *et al.*, 2004). In so doing, the ability of VCD to monitor the progress of a reaction of one chiral molecule to another in terms of mole fraction and *ee* of both species was demonstrated. A similar demonstration using near-IR FT-VCD showed that chiral reaction monitoring could be carried out in either the mid-IR and the near-IR as desired with an accuracy of approximately 2% *ee* (Guo *et al.*, 2005). FT-VCD reaction monitoring of chiral purity of multiple species is possible not only because of the structural richness of the IR and VCD spectra collected, but because all spectral frequencies are collected simultaneously in Fourier transform spectroscopy, thus eliminating spectral time biasing, which is unavoidable using a scanning spectrometer as required for ECD.

Presently, there are no spectroscopic techniques that combine sensitivity to chirality with traditional forms of kinetics measurements to study the reaction mechanisms of chiral molecules. Moreover, in the industrial sector, there is no process monitoring technique that combines traditional spectroscopic probes, for example near-IR sensors, with chiral sensitivity. VCD offers a unique opportunity to probe the dynamics of reactions of chiral molecules that will become increasingly important as the capabilities of FT-VCD reaction monitoring are realized and implemented.

1.7.5 Applications with Solid-Phase Sampling

A new area of VCD application is solid-phase sampling. This has been made possible by the reduction in birefringence artifacts associated with the dual-polarization modulation methodology. Solid-phase sampling has not been reported for ROA. This may be due to the higher susceptibility of ROA to interference from particle scattering because of the shorter wavelengths in ROA compared with VCD. However, VCD spectra have recently been reported for films of proteins where it was demonstrated that the absence of solvent absorption greatly facilitates the acquisition of spectra and reduces the amount of sample required for measurements (Shanmugam and Polavarapu, 2005). By careful film preparation, VCD spectra almost identical with those of solution can be obtained. Other laboratories have reported VCD in mulls, KBr pellets, and spray dried films, and it has been shown that VCD is very sensitive to the particle-size distribution of the sample and the crystal morphology above the size of the unit cell (Nafie and Dukor, 2007). Solid-phase sampling also permits direct comparison of VCD structure determination with that of X-ray crystallography, however, calculations of VCD for molecular solids has not yet advanced to the state where an accurate VCD spectrum can be calculated for an arbitrary solid. Even without the connection to calculations, structural characterization of solids and solid formulations of pharmaceutical products has become a sensitive new area of application of VCD.

1.8 Comparison of Infrared and Raman Vibrational Optical Activity

It is natural to want to compare VCD and ROA. Both are forms of VOA. It is possible to display IR, VCD, Raman, and ROA of the same sample over the same range of vibrational frequencies (Qu *et al.*, 1996). When this is done, one finds that VCD and ROA are as different from one another as the parent IR and Raman are from each other, and perhaps even more so because of the added dimension of the sign of each VOA band. It might be suspected that the sign of a VCD band and its ROA companion are related by the common motion of the molecule in that vibrational mode. If there were such a correlation, or anti-correlation given the opposite definition of the sign conventions in VCD and ROA, then the concept of vibrational chirality would have merit. However, no such correlation has been discerned, and one is left to conclude that the sign of a VCD or ROA band has as much to do with the electronic mechanism, dipole moment derivatives with respect to nuclear motion for VCD versus the corresponding polarizability derivatives for ROA, as it does with the chirality of the nuclear motion itself. In this section, a brief comparison of VCD and ROA is provided to give the reader a better appreciation of the relative strengths and weaknesses of these two complementary forms of vibrational optical activity.

1.8.1 Frequency Ranges and Structural Sensitivities

ROA was discovered in the low-frequency range below 500 cm^{-1} where, even after almost four decades of development, VCD has yet to be observed. The frontiers of ROA were advanced from low to high frequency, and, even today, ROA intensities in the high-frequency range of hydrogen stretching motions are weak and difficult to measure. By contrast, the frontiers of VCD have been advanced from high to low frequency down to about 600 cm^{-1} but no further. The effective spectral range of VCD is from 600 to 14 000 cm^{-1} when overtone and combination band transitions are included. By contrast, the range of ROA coverage is from 50 to 4000 cm^{-1}, although most ROA spectra are restricted to below 2000 cm^{-1}. A great advantage of ROA is the coverage of low-frequency vibrational modes that may be particularly sensitive to the conformation and dynamics of biological molecules. VCD on the other hand has coverage to very high frequencies where near-IR process monitoring applications have the clear potential to become very important.

Both VCD and ROA are sensitive to vibrational coupling in molecules. Their intensities derive from the stereo-specific way in which vibrational motion, including the electronic response, is distributed over the framework of the molecule. There are perceived differences, however, in the sensitivities of VCD and ROA to this motion. Because electric dipole moments can couple through space by a radiative mechanism that differs from through bond mechanical coupling, VCD appears to exhibit through space coupling over distances larger than those seen in ROA. For example, in proteins, VCD shows large intensities in the amide I region that change dramatically with secondary structure through dipole–dipole coupling of the carbonyl stretching motions, whereas ROA is much more sensitive in the amide III region, that arises from localized hydrogen bending motions near the chiral alpha carbon center of peptides residues, which also vary with secondary structure. These different sensitivities of ROA and VCD make these techniques even more complementary than they might otherwise be. This means that adding an ROA spectrum to a VCD spectrum does more than offer the same information in a different form. Different structural information is emphasized and these two techniques are not redundant in a way that is even more striking than how ordinary IR and Raman are complementary and non-redundant.

1.8.2 Instrumental Advantages and Disadvantages

ROA instrumentation is approximately twice as expensive as VCD instrumentation. Some of the added cost derives from the fact that a Raman spectrometer equipped with a multi-element CCD camera is more expensive than an FT-IR spectrometer of similar quality and spectroscopic resolution, and the rest of the cost difference is due to the additional expense associated with the laser. Beyond the cost difference, ROA instruments are more complex than the equivalent VCD instrument. ROA is more difficult to measure than VCD due to the additional sensitivity of ROA to imbalances to circular polarization modulation. The additional sensitivity comes from the variation of a Raman spectrum with different states of relative polarization between the incident and scattered light beams, otherwise known as the depolarization ratio. This ratio is huge relative to ROA intensities, and, as a result, any leakage of the depolarization ratio into an ROA measurement introduces a large instrumental artifact in the spectrum. The effect is most easily noticed in highly polarized bands that have the largest differences between the polarization of the Raman scattering that is parallel versus that perpendicular to the scattering plane. Commercial ROA instrumentation overcomes this sensitivity and other imbalances through a series of additional optical elements and timing circuitry, as described under the development of ROA instrumentation in Section 1.4, and also in Chapter 7.

1.8.3 Sampling Methods and Solvents

The most significant difference between sampling for VCD and ROA is the nature of the solvents that are most easily and most commonly used. For VCD, the best solvents are relatively non-polar. If hydrogen is present in the solvent, the deuterated analog is preferred as all hydrogen vibrational bands are shifted to lower frequencies away from solute bands of interest. Polar solvents typically have high absorbances and fewer windows for viewing the VCD spectra of solutes. In the case of water, absorption bands are so broad and intense that VCD spectra can only be obtained for samples in high concentrations using very short pathlengths, usually less the 10 microns (μm). Water, on the other hand, is a weak Raman scatterer, and as a result, aqueous solutions pose no difficulties in the measurement of ROA. This is a significant advantage when studying biological molecules. However, polar solvents have very high Raman scattering intensities that can interfere with the collection of ROA spectra of molecules dissolved in such solvents.

Sampling techniques for VCD and ROA follow closely those associated with their parent IR and Raman spectroscopies. Because the laser beam can be focused to spot sizes measured in microns, sample volumes for ROA can be in the microliter range without loss of light collection efficiency. The volumes of sample required for VCD spectral collection can also be reduced to sub-milliliter amounts by a suitable choice of sample cell that has little if any dead space. In general, it can be concluded that VCD and ROA have differences in sampling methods but neither can be said to be superior to the other.

In the area of sample preparation there is a significant difference between ROA and VCD, particularly for samples of a biological origin. This difference is associated with fluorescence, which can be very detrimental to the measurement of ROA spectra. In order to measure high quality ROA spectra, samples must be chemically purified to eliminate as many impurities as possible, then filtered with a micron hole size filter, kept free of dust and particulate matter, and if necessary filtered again with activated charcoal. Only distilled, de-ionized water should be used in the preparation of aqueous solutions for ROA measurements. The time invested in sample preparation is more than worth its tradeoff in total measurement time in obtaining a high quality ROA spectrum. No such precautions are needed for solution-phase VCD measurements.

1.8.4 Computational Advantages and Disadvantages

Aside from the issue of the non-Born–Oppenheimer contributions necessary for the formulation of VCD intensities, the calculation of VCD spectra is much easier and faster than for ROA spectra. To obtain similar levels of accuracy, lower demands are placed on VCD calculations compared with ROA calculations in terms of basis sets and time required for the completion of runs. ROA requires more sophisticated basis sets with ample diffuse functions to describe the more tenuous behavior off the polarizability of a molecule in its response to incident radiation. Owing to its inherently more complex theoretical expression with higher-order tensors for ROA compared with VCD, and its resulting higher requirements for basis set quality, it appears as though ROA will always be more demanding computationally than VCD.

Neither VCD nor ROA place as many demands on calculations as those required for the calculation of an electronic CD spectrum of comparable quality, as VCD and ROA spectra do not typically depend critically on individual excited electronic states, the properties of which are intrinsically more difficult to calculate than those of the ground state of a molecule. Furthermore, in order to obtain the proper bandshape of an ECD spectrum, the entire vibronic manifold must be calculated, which is similar to calculating the entire VCD spectrum for each band in an ECD spectrum. Based on these considerations, it appears that VCD enjoys a permanent advantage over all other forms of natural optical activity in terms of its structural content and relative ease of accurate calculation. As computers improve and become more powerful in the future, the advantage of VCD and ROA will continue to be present as calculated spectra move closer to their measured counterparts.

1.9 Conclusions

In this chapter, a brief outline has been given of the origins, development and current state of the art of VOA. The aim has been to provide an overview of the topics that will be dealt with in more detail in the following chapters. Space does not permit a more detailed description of the history of VOA. By necessity some important developments and contributions have not been included, but could be in a more expanded treatment of this subject. If desired, this chapter could be read in isolation from the rest of the book, not as a comprehensive review, but more of an impressionistic overview of what VOA is, where did it come from, and how did it get to its present state of development. In the following chapters, details are provided in all key areas of VOA for the reader to investigate as desired.

References

Abbate, S., Laux, L., Overend, J., and Moscowitz, A. (1981) A charge flow model for vibrational rotational strengths. *J. Chem. Phys.*, **75**, 3161–3164.

Abbate, S., Longhi, G., Ricard, L. *et al.* (1989) Vibrational circular dichroism as a criterion for local mode versus normal mode behavior. Near-infrared circular dichroism of some monoterpenes. *J. Am. Chem. Soc.*, **111**, 836–840.

Arago, D.F. (1811) *Mem. de L'Inst.*, **12**, part 1, 93.

Atkins, P.W., and Barron, L.D. (1969) Rayleigh scattering of polarized photons by molecules. *Mol. Phys.*, **16**, 453–466.

Bak, K.L., Jorgensen, P., Helgaker, T. *et al.* (1993) Gauge-origin independent multiconfigurational self-consistent-field theory for vibrational circular dichroism. *J. Chem. Phys.*, **98**, 8873–8887.

Bak, K.L., Jorgensen, P., Helgaker, T., and Ruud, K. (1994) Basis-set convergence and correlation-effects in vibrational circular-dichroism calculations using London atomic orbitals. *Faraday Discuss. R. Soc. Chem.*, **99**, 121–129.

Barron, L.D. (1976) Anti-Stokes Raman optical activity. *Mol. Phys.*, **31**, 1929–1931.

Barron, L.D. (2004) *Molecular Light Scattering and Optical Activity*, 2nd edn, Cambridge University Press, Cambridge.

Barron, L.D., and Buckingham, A.D. (1971) Rayleigh and Raman scattering from optically active molecules. *Mol. Phys.*, **20**, 1111–1119.

Barron, L.D., Bogaard, M.P., and Buckingham, A.D. (1973a) Raman scattering of circularly polarized light by optically active molecules. *J. Am. Chem. Soc.*, **95**, 603–605.

Barron, L.D., Bogaard, M.P., and Buckingham, A.D. (1973b) Differential Raman scattering of right and left circularly polarized light by asymmetric molecules. *Nature*, **241**, 113–114.

Barron, L.D., Bogaard, M.P., and Buckingham, A.D. (1973c) Raman circular intensity differential observations on some monoterpenes. *J. Chem. Soc., Chem. Commun.*, 152–153.

Barron, L.D., and Vrbancich, J. (1983) On the sign convention for Raman optical activity. *Chem. Phys. Lett.*, **102**, 285–286.

Barron, L.D., and Escribano, J.R. (1985) Stokes Antistokes asymmetry in natural Raman optical activity. *Chem. Phys.*, **98**, 437–446.

Barron, L.D., Hecht, L., Hug, W., and MacIntosh, M.J. (1989) Backscattered Raman optical activity with CCD detector. *J. Am. Chem. Soc.*, **111**, 8731–8732.

Biot, J.B. (1815) *Bull. Soc. Philomath.*, 190.

Bosnick, B., Moskovits, M., and Ozin, G.A. (1972) Raman circular dichroism. Its observation in á-phenylethylamine. *J. Am. Chem. Soc.*, **94**, 4750–4751.

Bour, P., and Keiderling, T.A. (2005) Vibrational spectral simulation for peptides of mixed secondary structure: Method comparisons with the trpzip model hairpin. *J. Phys. Chem. B*, **109**, 23687–23697.

Brocki, T., Moskovits, M., and Bosnich, B. (1980) Vibrational optical activity. Circular differential Raman scattering from a series of chiral terpenes. *J. Am. Chem. Soc.*, **102**, 495–450.

Buckingham, A.D., Fowler, P.W., and Galwas, P.A. (1987) Velocity-dependent property surfaces and the theory of vibrational circular dichroism. *Chem. Phys.*, **112**, 1–14.

Bursi, R., Devlin, F.J., and Stephens, P.J. (1990) Vibrationally induced ring currents? The vibrational circular dichroism of methyl lactate. *J. Am. Chem. Soc.*, **112**, 9430–9432.

Che, D., Hecht, L., and Nafie, L.A. (1991) Dual and incident circular polarization Raman optical activity backscattering of (−)-*trans*-pinane. *Chem. Phys. Lett.*, **180**, 182–190.

Cheng, J.C., Nafie, L.A., and Stephens, P.J. (1975) Polarization scrambling using a photoelastic modulator: Application to circular dichroism measurement. *J. Opt. Soc. Am.*, **65**, 1031–1035.

Chirgadze, Y.N., Venyaminov, S.Y., and Lobachev, V.M. (1971) Optical rotatory dispersion of polypeptides in the near-infrared region. *Biopolymers*, **10**, 809–820.

Cotton, A. (1895) *Compt. Rend.*, **120**, 989.

Deutsch, C.W., and Moscowitz, A. (1968) Optical activity of vibrational origin. I. A model helical polymer. *J. Chem. Phys.*, **49**, 3257–3272.

Deutsche, C.W., and Moscowitz, A. (1970) Optical activity of vibrational origin. II. Consequences of polymer conformation. *J. Chem. Phys.*, **53**, 2630–2644.

Devlin, F.J., Stephens, P.J., Cheeseman, J.R., and Frisch, M.J. (1996) Prediction of vibrational circular dichroism spectra using density functional theory: Camphor and fenchone. *J. Am. Chem. Soc.*, **118**, 6327–6328.

Devlin, R., and Stephens, P.J. (1987) Vibrational circular dichroism measurement in the frequency range of 800 to 650 cm^{-1}. *Appl. Spectrosc.*, **41**, 1142–1144.

Diem, M., Fry, J.L., and Burow, D.F. (1973) Circular differential Raman spectra of carvone. *J. Am. Chem. Soc.*, **95**, 253–255.

Ditchfield, R. (1974) A gauge-invariant LCAO method for N. M. R. chemical shifts. *Mol. Phys.*, **27**, 798–807.

Dutler, R., and Rauk, A. (1989) Calculated infrared absorption and vibrational circular dichroism intensities of oxirane and its deuterated analogues. *J. Am. Chem. Soc.*, **111**, 6957–6966.

Freedman, T.B., and Nafie, L.A. (1994) Theoretical formalism and models for vibrational circular dichroism intensity. In: *Modern Nonlinear Optics, Part 3* (eds M. Evans and S. Kielich), John Wiley & Sons, Inc., New York, pp. 207–263.

Freedman, T.B., Cao, X., Dukor, R.K., and Nafie, L.A. (2003) Absolute configuration determination of chiral molecules in the solution state using vibrational circular dichroism. *Chirality*, **15**, 743–758.

Frisch, M.J., Trucks, G.W., Schlegel, H.B. *et al.* (2003) Gaussian 03. Revision B. 03 ed. Gaussian, Inc., Pittsburgh, PA.

Frisch, M.J., Trucks, G.W., Schlegel, H.B. *et al.* (2009) Gaussian 09. Revision A. 1 ed. Gaussian, Inc., Wallingford, CT.

Galwas, P.A. (1983) Ph.D. thesis.University of Cambridge.

Guo, C., Shah, R.D., Dukor, R.K. *et al.* (2004) Determination of enantiomeric excess in samples of chiral molecules using Fourier transform vibrational circular dichroism spectroscopy: Simulation of real-time reaction monitoring. *Anal. Chem.*, **76**, 6956–6966.

Guo, C., Shah, R.D., Cao, X. *et al.* (2005) Enantiomeric excess determination by Fourier transform near-infrared vibrational circular dichroism spectroscopy: Simulation of real-time process monitoring. *Appl. Spectrosc.*, **59**, 1114–1124.

Gutowsky, H.S. (1951) Optical rotation of quartz in the infrared to 9.7 μ. *J. Chem. Phys.*, **19**, 438–441.

Haidinger, W. (1847) *Ann. Phys.*, **70**, 531.

Hecht, L., and Nafie, L.A. (1990) Linear polarization Raman optical activity: a new form of natural optical activity. *Chem. Phys. Lett.*, **174**, 575–582.

Hecht, L., Che, D., and Nafie, L.A. (1991) A new scattered circular polarization Raman optical activity instrument equipped with a charged coupled device detector. *Appl. Spectrosc.*, **45**, 18–25.

Hecht, L., and Nafie, L.A. (1991) Theory of natural Raman optical activity I. Complete circular polarization formalism. *Mol. Phys.*, **72**, 441–469.

Hecht, L., Che, D., and Nafie, L.A. (1992) Experimental comparison of scattered and incident circular polarization Raman optical activity in pinanes and pinenes. *J. Phys. Chem.*, **96**, 4266–4270.

Hecht, L., Phillips, A.L., and Barron, L.D. (1995) Determination of enantiomeric excess using Raman optical-activity. *J. Raman Spectrosc.*, **26**, 727–732.

Hediger, H.J., and Gunthard, H.H. (1954) Optical rotatory power of organic substances in the infrared. *Helv. Chim. Acta*, **37**, 1125–1133.

Herrmann, C., and Reiher, M. (2007) First principles approach to vibrational spectroscopy of biomolecules. *Top. Curr. Chem.*, **268**, 85–132.

Holzwarth, G., and Chabay, I. (1972) Optical activity of vibrational transitions. Coupled oscillator model. *J. Chem. Phys.*, **57**, 1632–1635.

Holzwarth, G., Hsu, E.C., Mosher, H.S. *et al.* (1974) Infrared circular dichroism of carbon–hydrogen and carbon–deuterium stretching modes. Observations. *J. Am. Chem. Soc.*, **96**, 251–252.

Hug, W. (1982) Instrumental and theoretical advances in Raman optical activity. In: *Raman Spectroscopy* (ed. J. Lascombe), John Wiley & Sons, Ltd, Chichester, pp. 3–12.

Hug, W. (2003) Virtual enantiomers as the solution of optical activity's deterministic offset problem. *Appl. Spectrosc.*, **57**, 1–13.

Hug, W., Kint, S., Bailey, G.F., and Scherer, J.R. (1975) Raman circular intensity differential spectroscopy. The spectra of (−)-a-pinene and (+)-a-phenylethylamine. *J. Am. Chem. Soc.*, **97**, 5589–5590.

Hug, W., and Surbeck, H. (1979) Vibrational Raman optical activity spectra recorded in perpendicular polarization. *Chem. Phys. Lett.*, **60**, 186–192.

Hug, W., and Hangartner, G. (1999) A very high throughput Raman and Raman optical activity spectrometer. *J. Raman Spectrosc.*, **30**, 841–852.

Jalkanen, K.J., and Stephens, P.J. (1988) Gauge dependence of vibrational rotational strengths: NHDT. *J. Phys. Chem.*, **92**, 1781–1785.

Katzin, L.I. (1964) Rotatory dispersion of quartz. *J. Phys. Chem.*, **68**, 2367–2370.

Keiderling, T.A., and Stephens, P.J. (1976) Vibrational circular dichroism in overtone and combination bands. *Chem. Phys. Lett.*, **41**, 46–48.

Lipp, E.D., Zimba, C.G., and Nafie, L.A. (1982) Vibrational circular dichroism in the mid-infrared using Fourier transform spectroscopy. *Chem. Phys. Lett.*, **90**, 1–5.

Lombardi, R.A., and Nafie, L.A. (2009) Observation and calculation of a new form of vibrational optical activity: Vibrational circular birefringence. *Chirality*, **21**, E277–E286.

London, F. (1937) Théorie quantique des courants interatomiques dans les combinaisons aromatiques. *J. Phys. Radium Paris*, **8**, 397–409.

Lowe, M.A., Segal, G.A., and Stephens, P.J. (1986a) The theory of vibrational circular dichroism: *trans*-1, 2-dideuteriocyclopropane. *J. Am. Chem. Soc.*, **108**, 248–256.

Lowe, M.A., Stephens, P.J., and Segal, G.A. (1986b) The theory of vibrational circular dichroism: *trans*-1, 2-dideuteriocyclobutane and propylene oxide. *Chem. Phys. Lett.*, **123**, 108–116.

Lowry, T.M. (1935) *Optical Rotatory Power*, Longmans, Green and Co., London.

Mason, S.F. (1973) The development of theories of optical activity and their applications. In: *Optical Rotatory Dispersion and Circular Dichroism* (eds F. Ciardelli and P. Salvadori), Heydon & Son, London, pp. 27–40.

Moskovits, M., and Gohin, A. (1982) Vibrational circular dichroism: Effect of charge fluxes and bond currents. *J. Phys. Chem.*, **86**, 3947–3950.

Nafie, L.A. (1983a) Adiabatic behavior beyond the Born–Oppenheimer approximation. Complete adiabatic wavefunctions and vibrationally induced electronic current density. *J. Chem. Phys.*, **79**, 4950–4957.

Nafie, L.A. (1983b) An alternative view on the sign convention of Raman optical activity. *Chem. Phys. Lett.*, **102**, 287–288.

Nafie, L.A. (1992) Velocity-gauge formalism in the theory of vibrational circular dichroism and infrared absorption. *J. Chem. Phys.*, **96**, 5687–5702.

Nafie, L.A. (1996) Theory of resonance Raman optical activity – the single electronic-state limit. *Chem. Phys.*, **205**, 309–322.

Nafie, L.A., Cheng, J.C., and Stephens, P.J. (1975) Vibrational circular dichroism of 2,2, 2-trifluoro-1-phenylethanol. *J. Am. Chem. Soc.*, **97**, 3842–3843.

Nafie, L.A., Keiderling, T.A., and Stephens, P.J. (1976) Vibrational circular dichroism. *J. Am. Chem. Soc.*, **98**, 2715–2723.

Nafie, L.A., and Walnut, T.H. (1977) Vibrational circular dichroism theory: A localized molecular orbital model. *Chem. Phys. Lett.*, **49**, 441–446.

Nafie, L.A., and Diem, M. (1979) Theory of high frequency differential interferometry: Application to infrared circular and linear dichroism via Fourier transform spectroscopy. *Appl. Spectrosc.*, **33**, 130–135.

Nafie, L.A., Diem, M., and Vidrine, D.W. (1979) Fourier transform infrared vibrational circular dichroism. *J. Am. Chem. Soc.*, **101**, 496–498.

Nafie, L.A., and Freedman, T.B. (1983) Vibronic coupling theory of infrared vibrational intensities. *J. Chem. Phys.*, **78**, 7108–7116.

Nafie, L.A., Oboodi, M.R., and Freedman, T.B. (1983) Vibrational circular dichroism in amino acids and peptides. 8. A chirality rule for the methine C*H stretching mode. *J. Am. Chem. Soc.*, **105**, 7449–7450.

Nafie, L.A., and Freedman, T.B. (1986) The ring current mechanism of vibrational circular dichroism. *J. Phys. Chem.*, **90**, 763–767.

Nafie, L.A., and Freedman, T.B. (1989) Dual circular polarization Raman optical activity. *Chem. Phys. Lett.*, **154**, 260–266.

Nafie, L.A., and Che, D. (1994) Theory and measurement of Raman optical activity. In: *Modern Nonlinear Optics, Part 3* (eds M. Evans, and S. Kielich), John Wiley & Sons, Inc., New York, pp. 105–149.

Nafie, L.A., Brinson, B.E., Cao, X. *et al.* (2007) Near-infrared excited Raman optical activity. *Appl. Spectrosc.*, **61**, 1103–1106.

Nafie, L.A., and Dukor, R.K. (2007) Pharmaceutical applications of vibrational optical activity. In: *Applications of Vibrational Spectroscopy in Pharmaceutical Research and Development* (eds. D. Pivonka, P.R. Griffiths, and J.M. Chalmers), John Wiley & Sons, Ltd., Chichester, pp. 129–154.

Osborne, G.A., Cheng, J.C., and Stephens, P.J. (1973) Near-infrared circular dichroism and magnetic circular dichroism instrument. *Rev. Sci. Instrum.*, **44**, 10.

Polavarapu, P.L. (1990) *Ab initio* Raman and Raman optical activity spectra. *J. Phys. Chem.*, **94**, 8106–8112.

Polavarapu, P.L., and Deng, Z.Y. (1996) Measurement of vibrational circular-dichroism below $\sim$600 cm^{-1}: Progress towards meeting the challenge. *Appl. Spectrosc.*, **50**, 686–692.

Qu, X., Lee, E., Yu, G.S. *et al.* (1996) Quantitative comparison of experimental infrared and Raman optical-activity spectra. *Appl. Spectrosc.*, **50**, 649–657.

Russel, M.I., Billardon, M., and Badoz, J.P. (1972) Circular and linear dichrometer for the near infrared. *Appl. Opt.*, **11**, 2375–2378.

Schellman, J.A. (1973) Vibrational optical activity. *J. Chem. Phys.*, **58**, 2882–2886.

Schweitzer-Stenner, R., Measey, T., Kakalis, L. *et al.* (2007) Conformations of alanine-based peptides in water probed by FTIR, Raman, vibrational circular dichroism, electronic circular dichroism, and NMR spectroscopy. *Biochemistry*, **46**, 1587–1596.

Shanmugam, G., and Polavarapu, P.L. (2005) Film techniques for vibrational circular dichroism measurements. *Appl. Spectrosc.*, **59**, 673–681.

Spencer, K.M., Freedman, T.B., and Nafie, L.A. (1988) Scattered circular polarization Raman optical activity. *Chem. Phys. Lett.*, **149**, 367–374.

Spencer, K.M., Cianciosi, S.J., Baldwin, J.E. *et al.* (1990) Determination of enantiomeric excess in deuterated chiral hydrocarbons by vibrational circular dichroism spectroscopy. *Appl. Spectrosc.*, **44**, 235–238.

Stephens, P.J., and Clark, R. (1979) Vibrational circular dichroism: The experimental viewpoint. In: *Optical Activity and Chiral Discrimination* (ed. S. F. Mason), D. Reidel, Dordrecht, pp. 263–287.

Stephens, P.J. (1985) Theory of vibrational circular dichroism. *J. Phys. Chem.*, **89**, 748–752.

Stephens, P.J., Chabalowski, C.F., Devlin, F.J., and Jalkanen, K.J. (1994) *Ab initio* calculation of vibrational circular-dichroism spectra using large basis-set MP2 force-fields. *Chem. Phys. Lett.*, **225**, 247–257.

Stephens, P.J., and Devlin, F.J. (2000) Determination of the structure of chiral molecules using *ab initio* vibrational circular dichroism spectroscopy. *Chirality*, **12**, 172–179.

Su, C.N., Heintz, V.J., and Keiderling, T.A. (1980) Vibrational circular dichroism in the mid-infrared. *Chem. Phys. Lett.*, **73**, 157–159.

Vargek, M., Freedman, T.B., Lee, E., and Nafie, L.A. (1998) Experimental observation of resonance Raman optical activity. *Chem. Phys. Lett.*, **287**, 359–364.

West, C.D. (1954) Anomalous rotatory dispersion of quartz above 3.7 μ. *J. Chem. Phys.*, **22**, 749–750.

Wyss, H.R., and Gunthard, H.H. (1966) Diphenyl-dinaphto-(2′,1′:1,2;1″, 2″:3,4)-5,8-diaza-cyclooctatetraen (I) und 3′,6″-Dimethyl-1,2:3,4-dibenz-1,3-cycloheptadien-6-on (II) im Infrarot. *Helv. Chim. Acta*, **49**, 660–663.

Yu, G.-S., and Nafie, L.A. (1994) Isolation of preresonance and out-of-phase dual circular polarization Raman optical activity. *Chem. Phys. Lett.*, **222**, 403–410.

Zuber, G., and Hug, W. (2004) Rarefied basis sets for the calculation of optical tensors. 1. The importance of gradients on hydrogen atoms for the Raman scattering tensor. *J. Phys. Chem. A*, **108**, 2108–2118.

2

Vibrational Frequencies and Intensities

2.1 Separation of Electronic and Vibrational Motion

Understanding vibrational spectroscopy begins with understanding the nature of the motion of electrons and nuclei in molecules during vibrational oscillations. In this chapter we will see that electronic motion determines the equilibrium positions of nuclei in electronic states as well as defining the way nuclei vibrate about these equilibrium positions. Infrared absorption intensities arise from the combined contributions of the motion of the positively charged nuclei followed closely, but not perfectly, by the negatively charged electrons. Raman scattering intensity, on the other hand, is a result of only the electrons as they respond to the influence of the electromagnetic field of incident laser radiation while being modulated by the underlying vibrational motion of the nuclei. The story of how this is described mathematically begins with the separation in the quantum mechanical description of the motions of the electrons and the nuclei in molecules.

2.1.1 Born–Oppenheimer Approximation

We begin by defining the Schrödinger equation of a molecule in terms of the total Hamiltonian $\mathcal{H}$ as:

$$\mathcal{H}\Psi_m(\boldsymbol{r}, \boldsymbol{R}) = \mathcal{E}_m \Psi_m(\boldsymbol{r}, \boldsymbol{R}) \tag{2.1}$$

where $\Psi_m(\boldsymbol{r}, \boldsymbol{R})$ is the wavefunction of the molecule in the mth state with energy $\mathcal{E}_m$. The wavefunction depends on the sets of Cartesian coordinates for the electrons $\boldsymbol{r}$ and nuclei $\boldsymbol{R}$. In writing Equation (2.1), the focus is on the electronic and nuclear motion, and we have excluded terms for electron and nuclear spins as well as any external fields. We next write the total Hamiltonian in terms of the kinetic energy of the electrons and the nuclei, the electron–electron repulsion energy of the electrons, the electron–nuclear attraction energy, and the nuclear–nuclear repulsion energy.

$$\mathcal{H} = T_E + T_N + V_{EE} + V_{EN} + V_{NN} = H_E + T_N \tag{2.2}$$

All the these energy terms, except for the nuclear kinetic energy, T_N, can be gathered together to form the Hamiltonian for the motion of the electrons, H_E. The nuclear repulsion energy is included

Vibrational Optical Activity: Principles and Applications, First Edition. Laurence A. Nafie.

with the electronic Hamiltonian as a constant value of energy for a given set of nuclear positions, $\boldsymbol{R}$. In the Born–Oppenheimer (BO) approximation, the electronic motion is separated from the nuclear motion by first solving the electronic problem with the nuclei fixed at the nuclear position $\boldsymbol{R}$. This is accomplished by assigning two quantum numbers, ev, to the total wavefunction, $\Psi_{ev}^{A}(\boldsymbol{r},\boldsymbol{R})$, and then writing this wavefunction as a product of the electronic and nuclear wavefunctions, $\psi_{e}^{A}(\boldsymbol{r},\boldsymbol{R})\phi_{ev}(\boldsymbol{R})$, where the superscript A refers to adiabatic as the wavefunction changes gradually and reversibly with changes in nuclear position. With these changes, Equation (2.1) becomes:

$$\mathcal{H}\Psi_{ev}^{A}(\boldsymbol{r},\boldsymbol{R}) = (H_E + T_N)\psi_{e}^{A}(\boldsymbol{r},\boldsymbol{R})\phi_{ev}(\boldsymbol{R}) = E_{ev}^{A}\psi_{e}^{A}(\boldsymbol{r},\boldsymbol{R})\phi_{ev}(\boldsymbol{R}) \tag{2.3}$$

where $\psi_{e}^{A}(\boldsymbol{r},\boldsymbol{R})$ is the electronic wavefunction and $\phi_{ev}(\boldsymbol{R})$ is the nuclear wavefunction responsible for describing the vibrational states (and rotational states, if any) v in the electronic state e of the molecule. The operation of the electronic Hamiltonian on the electronic wavefunction is described by the electronic Schrödinger equation, written as:

$$H_E\psi_{e}^{A}(\boldsymbol{r},\boldsymbol{R}) = E_{e}^{A}(\boldsymbol{R})\psi_{e}^{A}(\boldsymbol{r},\boldsymbol{R}) \tag{2.4}$$

The BO approximation continues by assuming that terms arising from the operation of the nuclear kinetic energy operator on the *electronic* wavefunction can be ignored. With this assumption, the last two parts of Equation (2.3) can be written as:

$$\psi_{e}^{A}(\boldsymbol{r},\boldsymbol{R})[E_{e}^{A}(\boldsymbol{R}) + T_N]\phi_{ev}(\boldsymbol{R}) = \psi_{e}^{A}(\boldsymbol{r},\boldsymbol{R})E_{ev}^{A}\phi_{ev}(\boldsymbol{R}) \tag{2.5}$$

This equation now reduces to the Schrödinger equation of the nuclear motion as the electronic wavefunction can be cancelled on both sides of the equation because there are no further operators acting on it, and we can write:

$$H_N\phi_{ev}(\boldsymbol{R}) = [T_N + E_{e}^{A}(\boldsymbol{R})]\phi_{ev}(\boldsymbol{R}) = E_{ev}^{A}\phi_{ev}(\boldsymbol{R}) \tag{2.6}$$

where H_N is the Hamiltonian associated with the total energy of the nuclei. It can be seen that the eigenvalue of the electronic motion $E_{e}^{A}(\boldsymbol{R})$ determined in Equation (2.4), which is a function of the nuclear position, becomes the potential energy for the nuclear motion in electronic state e of the molecule.

2.1.2 Electronic Structure Problem

Within the BO approximation, the description of the electronic structure of the molecule is governed by Equation (2.4). In general, there is a lowest energy electronic state, called the ground electronic state of the molecule. There are also excited electronic states ranging from the lowest excited electronic states up to a continuum of excited states, the most energetic of which correspond to ionization of electrons from the molecule. For most organic molecules, the lowest excited electronic state lies in the visible or near-ultraviolet (near-UV) region of the spectrum and the higher electronic states lie deeper into the UV and X-ray regions of the spectrum.

The motion of individual electrons is too fast to follow, or even to think about. They are true quantum, indistinguishable particles. The best we can do is to describe their probability density in space. The electron probability density, as a function of the nuclear positions, is defined by writing:

$$\rho_e(\boldsymbol{r},\boldsymbol{R}) = \psi_{e}^{A}(\boldsymbol{r},\boldsymbol{R})^{*}\psi_{e}^{A}(\boldsymbol{r},\boldsymbol{R}) = \left|\psi_{e}^{A}(\boldsymbol{r},\boldsymbol{R})\right|^{2} \tag{2.7}$$

The electron probability density for state e is the product of the electronic wavefunction and its complex conjugate. This product is also the absolute square of the electronic wavefunction and is denoted in Equation (2.7) by the square of the wavefunction flanked by vertical lines. The complex conjugate is necessary in case the wavefunction is complex or pure imaginary as the probability density is a real, observable property of the molecule.

The electronic structure of a molecule consists of the set of wavefunctions $\psi_e^A(\boldsymbol{r}, \boldsymbol{R})$ for all the electronic states of the molecule. The probability density of each electronic state can be visualized by plotting its probability density, $\rho_e(\boldsymbol{r}, \boldsymbol{R})$, for some particular set of positions $\boldsymbol{R}$ of the nuclei. The probability density of the ground electronic state usually has the simplest shape with no nodes, or planes of zero electron density. Higher excited electronic states have more constrained electron density with increasing numbers of nodal planes and increasing electron energy relative to the ground electronic state. This is analogous to a constrained oscillating string with energy levels that increases as the number of nodes along the string increases and its oscillating wavelength gets shorter. The electrons in excited electronic states in molecules have a similar set of constrained nodal motions described spatially by the various electron probability densities, $\rho_e(\boldsymbol{r}, \boldsymbol{R})$.

2.1.3 Nuclear Structure Problem

A major goal of this chapter is to understand vibrational motion in molecules in the ground electronic state. In modern quantum chemistry this begins by solving the Schrödinger equation for the motion of the electrons given by Equation (2.4) and then finding sets of nuclear positions, $\boldsymbol{R}$, that correspond to local minima of the multi-dimensional energy surface $E_g^A(\boldsymbol{R})$, where we have used subscript g to denote the ground electronic state. Each local minimum corresponds to a stable conformation of the molecule. It can be shown that the best ground-state electronic wavefunction, for a particular stable conformation, is the one that gives the lowest value for the energy $E_g^A(\boldsymbol{R})$ in its local minimum. Thus, for the ground state, the problem of finding the electronic wavefunction, and hence its energy as a function of nuclear position, can be carried out iteratively as adjustments are made to the wavefunction that reduce the value of the electronic energy to an acceptable local minimum. Higher excited electronic states for each conformation can then be defined relative to the ground electronic state. Once an electronic wavefunction has been determined and its nuclear potential energy found, Equation (2.6) can be used to find the vibrational sublevels of the electronic state.

The motion of the nuclei is governed by a Schrödinger equation in the same way the electronic motion is defined. Furthermore, the allowed vibrational energy levels of the nuclei are quantized, just as the electronic energy levels are quantized, but because the nuclei have masses that are approximately three orders of magnitude greater than those of the electrons, nuclear vibrational motion is much slower and can be thought about and visualized classically. In fact, the BO approximation is the mathematical step that makes this description possible. Nuclei are represented as being located at well-defined spatial positions with zero velocity and their vibrational motion can be pictured using arrows, or vectors, whose magnitude and direction describe the relative motion of the nuclei away from their equilibrium position for a given vibrational motion in the molecule. Simultaneous knowledge of the position and velocity of a particle is forbidden by the uncertainty principle of quantum mechanics, but is allowed for the visual description of nuclear position and displacements or velocities for a molecule undergoing vibrational motion within the BO approximation.

As a result, the vibrational structure problem lies at the very boundary between the world of quantum mechanics and the world of classical physics that govern our perceived everyday experience. It should be appreciated that the BO approximation therefore underlies our conceptualization of molecular structure in terms of nuclei existing at well-defined positions, and vibrational structure as periodic departures of the nuclei from those equilibrium nuclear positions.

2.1.4 Nuclear Potential Energy Surface

Within the past two decades, there has been a revolution in quantum chemistry made possible by ever faster desktop computers and improvements in the quantum mechanical descriptions of molecules, primarily the description embodied by density functional theory (DFT). Before this revolution, vibrational spectroscopists labored to define the potential energy of molecules, and therefore their vibrational motion, in terms of parameterized force fields. Today, such efforts are rarely used as the determination of the potential surface defined by $E_g^A(\boldsymbol{R})$obtained from quantum chemistry calculations exceeds the accuracy obtainable from parameterization methods. In quantum chemistry today, the vibrational problem is solved by finding the values of the derivatives of the Taylor expansion of the electronic energy as a function of the set of nuclear positions $\boldsymbol{R}$ about the equilibrium nuclear position $\boldsymbol{R}_0$ given by:

$$E_g^A(\boldsymbol{R}) = E_g^A(\boldsymbol{R}_0) + \sum_{J=1}^{3N} \left(\frac{\partial E_g^A}{\partial \boldsymbol{R}_J} \right)_{R=R_0} \Delta \boldsymbol{R}_J + \frac{1}{2} \sum_{J,J'=1}^{3N} \left(\frac{\partial^2 E_g^A}{\partial \boldsymbol{R}_J \partial \boldsymbol{R}_{J'}} \right)_{R=R_0} \Delta \boldsymbol{R}_J \Delta \boldsymbol{R}_{J'} + \cdots \tag{2.8}$$

The energy of the molecule at the equilibrium nuclear position is given by $E_g^A(\boldsymbol{R}_0)$ and $\Delta \boldsymbol{R}_J$ is the displacement of the Jth nucleus from its equilibrium position $\boldsymbol{R}_J - \boldsymbol{R}_{J,0}$. The energy $E_g^A(\boldsymbol{R}_0)$ is the value at the local minimum of the multi-dimensional potential energy surface of a particular stable molecular conformation. The slope and hence all first derivatives $[\partial E_g^A(\boldsymbol{R})/\partial \boldsymbol{R}_J]_{R=R_0}$ are equal to zero at $\boldsymbol{R}_0$. The second derivatives $(\partial^2 E_g^A/\partial \boldsymbol{R}_J \partial \boldsymbol{R}_{J'})_{R=R_0}$ are the curvatures at the minimum of the potential energy and also serve as the force constants defining the harmonic nuclear motion. When the second derivatives represent quadratic displacements of the same nuclear coordinates, they are termed diagonal force constants, and when they represent the product of two different coordinates, they are termed off-diagonal force constants. Higher order derivatives describe anharmonic corrections to the harmonic force field.

2.1.5 Transitions Between Electronic States

Descriptions of the intensities of vibrational transitions in molecules rely on how the electrons respond to vibrational motion. These descriptions in turn depend on electronic transitions between, typically, the ground electronic state and higher electronic states. In this section we present theoretical expressions for transitions in a molecule between pairs of electronic states and describe simple visualizations of what happens in a molecule undergoing a transition between such pairs of states. A more detailed, formal derivation of the expressions given in this section including the reduction of the probability density of a multi-electron wavefunction to a probability over a single spatial coordinate, $\boldsymbol{r}$, is given in Nafie (1997) and in Appendix B.

Returning to the probability density for electrons in Equation (2.7), we suppress the dependence on the nuclear position by considering the electron density for the ground electronic state at the equilibrium nuclear position for that state.

$$\rho_g(\boldsymbol{r}, \boldsymbol{R}_0) = \psi_g^A(\boldsymbol{r}, \boldsymbol{R}_0)^* \psi_g^A(\boldsymbol{r}, \boldsymbol{R}_0) = \psi_g^0(\boldsymbol{r}) \psi_g^0(\boldsymbol{r}) = \rho_{g,0}(\boldsymbol{r}) \tag{2.9}$$

For simplicity we have considered the electronic state to be non-degenerate and hence its wavefunction is real.

Similarly, we can define the electron probability density of a molecule in any *excited* electronic state, e, at the same equilibrium nuclear position, namely the equilibrium position of the *ground* electronic state, as:

$$\rho_{e,0}(\boldsymbol{r}) = \psi_e^0(\boldsymbol{r})\psi_e^0(\boldsymbol{r}) \tag{2.10}$$

An electronic transition from the ground electronic state to an excited electronic state occurs on a time scale faster than the nuclei can move, and hence it is useful consider the equilibrium position of the ground electronic state as a convenient reference for describing transitions between the ground state and any excited state.

A quantity called the transition probability density (TPD) is defined by:

$$\Theta_{ge}^0(\boldsymbol{r}) \equiv \psi_g^0(\boldsymbol{r})\psi_e^0(\boldsymbol{r}) \tag{2.11}$$

The TPD describes the overlap probability density between the ground and excited electronic states. It is the simplest spatial representation of the relationship of electronic structure between any two electronic states. During an electronic transition, the TPD oscillates at the transition frequency, given by:

$$\nu_{eg}^0 = (E_e^0 - E_g^0)/h \tag{2.12}$$

where E_g^0 and E_e^0 are the energies of the two electronic states at the ground-state equilibrium position. Pairs of electronic states in a molecule are orthogonal, which means that the integral over all space of the TPD is zero.

$$\int_{-\infty}^{\infty} \Theta_{ge}^0(\boldsymbol{r})\mathrm{d}\boldsymbol{r} = \int_{-\infty}^{\infty} \psi_g^0(\boldsymbol{r})\psi_e^0(\boldsymbol{r})\mathrm{d}\boldsymbol{r} = 0 \tag{2.13}$$

This happens in this case because the wavefunction of the ground electronic state is positive everywhere with no nodal plane or surface where the sign of the wavefunction changes from positive to negative. On the other hand, electronic excited states have one or more nodal surfaces. Summing up the product of the ground and excited state electronic wavefunction, or the product of any two electronic-state wavefunctions in the molecule, over all space yields zero exactly.

In order to describe an electronic transition, integration over all space is required and some operator is needed to make the transition integral, or matrix element, non-zero.

The simplest such operator is the electron position operator given by

$$\int_{-\infty}^{\infty} \psi_e^0(\boldsymbol{r})^* \boldsymbol{r} \psi_g^0(\boldsymbol{r})\mathrm{d}\boldsymbol{r} = \left\langle \psi_e^0 \middle| \boldsymbol{r} \middle| \psi_g^0 \right\rangle \tag{2.14}$$

In writing Equation (2.14), we have adopted the bra and ket notation of quantum mechanics to represent integrals over all space, where g is the initial state and e is the final state. The dipole moment operator is defined as the product of the charge e_j times the position $\boldsymbol{r}_j$ of its constituent particles given by:

$$\boldsymbol{\mu} = \sum_j e_j \boldsymbol{r}_j = -e\sum_j \boldsymbol{r}_j = -e\boldsymbol{r} \tag{2.15}$$

which for electronic states can effectively be just the sum of the contributions of the electrons, as expressed by the second and third equalities. The electric dipole transition moment of a molecule is given by:

$$\langle \boldsymbol{\mu} \rangle_{eg}^{0} = \left\langle \psi_e^0 \middle| \boldsymbol{\mu} \middle| \psi_g^0 \right\rangle \tag{2.16}$$

where, on the left side, we have further simplified the notation describing the integrals over all space. The transition dipole moment is not directly observable, but its absolute square, $D_{r,eg}^0$, the dipole strength, is proportional to the absorption intensity of the transition between the ground and excited electronic states of the molecule. The position form of the dipole strength is given by:

$$D_{r,eg}^{0} = \left\langle \psi_e^0 \middle| \boldsymbol{\mu} \middle| \psi_g^0 \right\rangle^* \cdot \left\langle \psi_e^0 \middle| \boldsymbol{\mu} \middle| \psi_g^0 \right\rangle = \left| \left\langle \psi_e^0 \middle| \boldsymbol{\mu} \middle| \psi_g^0 \right\rangle \right|^2 \tag{2.17}$$

Finally, we define the transition dipole density (TDD) as the electron position-weighted TPD given by Nafie (1997) and Freedman *et al.* (1998):

$$\boldsymbol{M}_{ge}^{0}(\boldsymbol{r}) \equiv \boldsymbol{r}\Theta_{ge}^{0}(\boldsymbol{r}) = \psi_g^0(\boldsymbol{r})\boldsymbol{r}\psi_e^0(\boldsymbol{r}) \tag{2.18}$$

The TDD is the integrand of the position form of the electric dipole transition moment in Equation (2.16) and is a vector field emanating from the origin of coordinates of the molecule and describing the contribution of each point in space to the observed dipole strength. Because the field lines in Equation (2.18) emanate from the origin of coordinates, the field is origin dependent even though the corresponding integral over all space is not. Shifting the origin of coordinates by a constant amount in the transition moment in Equation (2.16) has no effect as the constant is not an operator, and the electronic wavefunctions are orthogonal.

2.1.6 Electronic Transition Current Density

In the section above we described transitions between electronic states in molecules in terms of the probability density and the position operator of the electrons. In this section, we complement that description by considering the corresponding time-dependent description involving electron current density. We will need the current density description of transitions between electronic states in a molecule when we describe electron current density generated by the velocities of nuclei undergoing vibrational motion. This description is accomplished by defining the transition current density (TCD) of a molecule between any pair of pure electronic states, in this case the ground and an electronic excited state (Nafie, 1997; Freedman *et al.*, 1998), as:

$$\boldsymbol{J}_{ge}^{0}(\boldsymbol{r}) \equiv \frac{\hbar}{2m}\left[\psi_g^0(\boldsymbol{r})\nabla\psi_e^0(\boldsymbol{r}) - \psi_e^0(\boldsymbol{r})\nabla\psi_g^0(\boldsymbol{r})\right] \tag{2.19}$$

where the operator ∇ is the vector derivative operator $\partial/\partial\boldsymbol{r}$. The TCD describes the vector field associated with the motion of electron probability density as it oscillators at transition frequency ν_{eg}^0. TCD is the dynamic velocity-based description of the TPD defined above, both of which oscillate at ν_{eg}^0 during an electronic transition between the states e and g of the molecule. This close relationship can be expressed in terms of the continuity equation:

$$-\nabla \cdot \boldsymbol{J}_{ge}^{0}(\boldsymbol{r}) = \omega_{eg}^{0}\Theta_{ge}^{0}(\boldsymbol{r}) \tag{2.20}$$

Here the radial transition frequency ω_{eg}^0 is equal to $2\pi\nu_{eg}^0$. Equation (2.20) states that the gradient of the TCD across the boundary of a volume element in the space of the molecule is equal to the transition frequency times the TPD within the volume element of space. In other words, the oscillation of TPD at the frequency ω_{eg}^0 during an electronic transition is a conserved quantity. It is neither created nor destroyed at any point in space, but rather flows in the form of TCD from one point to another like a continuous fluid.

A velocity version of the dipole strength can be defined that is complementary to the position version given in Equation (2.17) as:

$$D_{v,eg}^0 = \left(\frac{1}{\omega_{eg}^0}\right)^2 \left|\left\langle \psi_e^0 \middle| \dot{\boldsymbol{\mu}} \middle| \psi_g^0 \right\rangle\right|^2 \tag{2.21}$$

where the velocity form of the electric-dipole transition moment, assuming a real electronic wavefunction, can be written as:

$$\begin{aligned}
\left\langle \psi_e^0 \middle| \dot{\boldsymbol{\mu}} \middle| \psi_g^0 \right\rangle &= -e\int_{-\infty}^{\infty} \psi_e^0(\boldsymbol{r})\dot{\boldsymbol{r}}\psi_g^0(\boldsymbol{r})\mathrm{d}\boldsymbol{r} \\
&= \frac{ie\hbar}{m}\int_{-\infty}^{\infty} \psi_e^0(\boldsymbol{r})(\partial/\partial\boldsymbol{r})\psi_g^0(\boldsymbol{r})\mathrm{d}\boldsymbol{r} \\
&= -\frac{ie\hbar}{2m}\int_{-\infty}^{\infty} [\psi_g^0(\boldsymbol{r})\nabla\psi_e^0 - \psi_e^0(\boldsymbol{r})\nabla\psi_g^0(\boldsymbol{r})]\mathrm{d}\boldsymbol{r} \\
&= -ie\int_{-\infty}^{\infty} \boldsymbol{J}_{\mathrm{ge}}^0(\boldsymbol{r})\mathrm{d}\boldsymbol{r}
\end{aligned} \tag{2.22}$$

Here we have used the quantum mechanical definition of the electron velocity operator as its momentum divided by its mass given by:

$$\dot{\boldsymbol{r}} = \boldsymbol{p}/m = -i\hbar(\partial/m\partial\boldsymbol{r}) = -i\hbar\nabla/m \tag{2.23}$$

We can see that the TCD defined in Equation (2.19) is simply the integrand of the velocity form of the electric dipole transition moment defined in Equation (2.22). In contrast to the TDD, $\boldsymbol{M}_{eg}^0(\boldsymbol{r})$, the TCD, $\boldsymbol{J}_{eg}^0(\boldsymbol{r})$, is origin independent and hence is a unique spatial representation of the oscillatory motion of electron charge density during an electronic transition. For exact electronic wavefunctions, the integration of both the TDD and the TCD yields the same value of the dipole strength in Equations (2.17) and (2.22) for the transition between electronic states g and e, because according to the hypervirial theorem of quantum mechanics the position and velocity forms of the transition matrix elements are related by:

$$\left\langle \psi_e^0 \middle| \dot{\boldsymbol{\mu}} \middle| \psi_g^0 \right\rangle = i\omega_{eg}^0 \left\langle \psi_e^0 \middle| \boldsymbol{\mu} \middle| \psi_g^0 \right\rangle \tag{2.24}$$

2.2 Normal Modes of Vibrational Motion

Molecules are assemblies of atomic nuclei held in a potential of electron density that balances the attraction of negatively charged electrons to positively charged nuclei against the mutual repulsions of electrons among themselves and the nuclei among themselves. The nuclei possess nearly all the mass of the molecule. Because the nuclei are so much heavier than electrons, their vibrational motion can be considered classically, even though vibrational energy levels are quantized. This alone is not surprising as many classical systems that are confined by boundary conditions, such as a string tightly held at each end, experience quantized allowed energy levels. We consider first the degrees of freedom of a molecule compared with the degrees of freedom of its individual atomic nuclei.

2.2.1 Vibrational Degrees of Freedom

A molecule comprised of N nuclei has a total of $3N$ degrees of spatial freedom, three Cartesian directions, x, y, and z, for each nucleus. However, because the molecule is an intact structure, three of these degrees of freedom must be reserved for the translational freedom of the molecule as a whole. Furthermore, the molecule has three degrees of rotational freedom, unless it is a perfectly linear molecule, in which case it has only two degrees of rotation freedom. If these degrees of freedom are subtracted from the total, one obtains $3N-6$ internal degrees of spatial freedom, or $3N-5$ in the case of a linear molecule. These internal degrees of freedom correspond to the vibrational degrees of freedom of the molecule for which the center of mass of the molecule does not move and the molecule as a whole does not rotate about any axis.

2.2.2 Normal Modes of Vibrational Motion

Nature partitions the vibrational degrees of freedom of a molecule into normal modes of motion. Each normal mode is independent of all the other normal modes even though, in principle barring symmetry restrictions, every nucleus of the molecule participates to some degree in every normal mode of vibration. Each normal mode also has a well-defined vibrational frequency, which is obtained by starting from the nuclear wave equation given in Equation (2.6). Inserting the expression for the nuclear kinetic energy in Cartesian coordinates into this equation for the ground electronic state we obtain:

$$\left[\sum_{J}^{N}\frac{1}{2}M_J\dot{\boldsymbol{R}}_J^2 + E_g^A(\boldsymbol{R})\right]\phi_{gv}(\boldsymbol{R}) = E_{gv}^A\phi_{gv}(\boldsymbol{R}) \tag{2.25}$$

where the nuclear masses and velocities are M_J and $\dot{\boldsymbol{R}}_J$, respectively. If we combine this equation with the expression for the vibrational potential energy in Cartesian coordinates, $E_g^A(\boldsymbol{R})$, given in Equation (2.8), we obtain the following equation for the vibrational nuclear motion of the ground electronic state:

$$\left[\sum_{J}^{N}\frac{1}{2}M_J\dot{\boldsymbol{R}}_J^2 + \frac{1}{2}\sum_{J,J'=1}^{N}\left(\frac{\partial^2 E_g^A}{\partial\boldsymbol{R}_J\partial\boldsymbol{R}_{J'}}\right)_{R=0}\Delta\boldsymbol{R}_J\Delta\boldsymbol{R}_{J'}\right]\phi_{gv}(\boldsymbol{R}) = E_{gv}^A\phi_{gv}(\boldsymbol{R}) \tag{2.26}$$

Here we have utilized the fact that at a local equilibrium nuclear equilibrium position, $\boldsymbol{R}_0$, the constant term $E_g^A(\boldsymbol{R}_0)$can be taken to be zero for the vibrational problem, and all the first derivatives of the potential energy $E_g^A(\boldsymbol{R})$can also be set equal to zero when evaluated at $\boldsymbol{R} = \boldsymbol{R}_0$. We have kept only the quadratic terms for the potential energy which restricts the solution of Equation (2.25) to the harmonic approximation.

It is important to notice that, unlike the potential energy, the nuclear kinetic energy has no terms that couple the contributions of different nuclear coordinates. As shown below, the solution of

Equation (2.26) in terms of normal coordinates maintains the independence of nuclear contributions in the kinetic energy while creating the same independence for the nuclear potential energy. The transformation from Cartesian coordinates to normal coordinates therefore completely decouples the normal coordinates from one another, reducing an equation with dimensions of $3N$ to a set of $3N-6$ independent one-dimensional problems, each with its own energy eigenvalue and vibrational frequency. The coordinate transformation that connects each Cartesian coordinate to the set of normal coordinate is given by:

$$\Delta\boldsymbol{R}_J = \sum_{a}^{3N-6}\left(\frac{\partial \boldsymbol{R}_J}{\partial Q_a}\right)_{Q=0} Q_a \qquad \Delta\boldsymbol{R}_{J'} = \sum_{b}^{3N-6}\left(\frac{\partial \boldsymbol{R}_{J'}}{\partial Q_b}\right)_{Q=0} Q_b \tag{2.27}$$

Substitution of this transformation into the expression for the potential energy in Equation (2.26) initially gives the expression:

$$E_g^A(\boldsymbol{Q}) = \frac{1}{2}\sum_{J,J'=1}^{N}\sum_{a,b}^{3N-6}\left(\frac{\partial \boldsymbol{R}_J}{\partial Q_a}\right)_{Q=0}\cdot\left(\frac{\partial^2 E_g^A}{\partial \boldsymbol{R}_J \partial \boldsymbol{R}_{J'}}\right)_{R=0}\cdot\left(\frac{\partial \boldsymbol{R}_{J'}}{\partial Q_b}\right)_{Q=0} Q_a Q_b \tag{2.28}$$

Because this transformation eliminates all cross-terms from different normal coordinates, Q_a and Q_b, to the potential energy, after summation over J and J', it reduces to an expression with only squares of individual normal coordinates Q_a:

$$E_g^A(\boldsymbol{Q}) = \frac{1}{2}\sum_{a}^{3N-6}\left(\frac{\partial^2 E_g^A}{\partial Q_a^2}\right)_{Q=0} Q_a^2 = \frac{1}{2}k_{g,a}Q_a^2 \tag{2.29}$$

Here $k_{g,a}$ is the force constant for normal mode a, which is simply the second-derivative of the potential energy, or curvature, associated with the potential well of normal mode a in the ground electronic state. We can now write the vibrational wave equation in Equation (2.25) in terms of normal coordinates as:

$$\left[\frac{1}{2}\sum_{a}^{3N-6}\dot{Q}_a^2 + \frac{1}{2}\sum_{a}^{3N-6}k_{g,a}Q_a^2\right]\phi_{gv}(\boldsymbol{Q}) = E_{gv}^A\phi_{gv}(\boldsymbol{Q}) \tag{2.30}$$

The nuclear masses no longer appear explicitly as the normal coordinates are mass weighted coordinates with dimensions of length times mass to the one-half power ($l\cdot m^{1/2}$). Because there are no terms that depend on products of different normal coordinates, Equation (2.30) can be written as a set of $3N-6$ isolated wave equations for each normal coordinate of vibration:

$$\left[\frac{1}{2}\dot{Q}_a^2 + \frac{1}{2}k_{g,a}Q_a^2\right]\phi_{gv}(Q_a) = E_{gv,a}^A\phi_{gv}(Q_a) \tag{2.31}$$

2.2.3 Visualization of Normal Modes

Once the force constant matrix given by the potential energy expression in Equation (2.26) has been diagonalized to give the new potential energy expression Equation (2.29), the transformation from

Cartesian to normal coordinates is known. This transformation can been inverted to give the expressions for each normal mode in terms of Cartesian displacement vectors:

$$Q_a = \sum_J^N \left(\frac{\partial Q_a}{\partial \boldsymbol{R}_J}\right)_{R=0} \cdot \Delta \boldsymbol{R}_J \tag{2.32}$$

The derivative evaluated at $R = 0$ represents a set of coefficients describing the vector contribution of each nucleus to the normal mode Q_a. This equation can be written explicitly for the x, y and z components of each Cartesian displacement vector, $\Delta \boldsymbol{R}_J$ as:

$$Q_a = \sum_J^N \sum_\alpha^{x,y,z} \left(\frac{\partial Q_a}{\partial R_{J\alpha}}\right)_{R=0} \Delta R_{J\alpha} = \sum_J^N \left(\frac{\partial Q_a}{\partial R_{J\alpha}}\right)_{R=0} \Delta R_{J\alpha} \tag{2.33}$$

In the second part of the equation, we have adopted the Einstein summation convention where repeated Greek subscripts representing the Cartesian components are automatically summed over x, y, and z without an explicit summation sign.

The contribution of each nucleus J to the normal mode Q_a is also contained in the original transformation in Equation (2.27) where we define the so-called S-vector as:

$$\boldsymbol{S}_{J,a} = \left(\frac{\partial \boldsymbol{R}_J}{\partial Q_a}\right)_{Q=0} \quad \text{or} \quad S_{J\alpha,a} = \left(\frac{\partial R_{J\alpha}}{\partial Q_a}\right)_{Q=0} \tag{2.34}$$

and where we have written the S-vector both in vector form and Cartesian component form.

The S-vectors provide a basis for visualizing the nuclear displacements associated with each normal mode. They are the vectors of each nucleus, the magnitude and direction of which depict how each nucleus moves in concert in normal mode Q_a.

2.2.4 Vibrational Energy Levels and States

Equation (2.31) can be solved by using the following relation for the quantum mechanical operator for velocity of the ath normal mode, $\dot{Q}_a = P_a = -i\hbar\partial/\partial Q_a$ to yield:

$$\frac{1}{2}\left[-\frac{\hbar^2\partial^2}{\partial Q_a^2} + k_{g,a}Q_a^2\right]\phi_{gv}^a(Q_a) = E_{gv,a}^A\phi_{gv}^a(Q_a) \tag{2.35}$$

No mass factor appears for the kinetic energy term or between the velocity and momentum, $\dot{Q}_a = P_a$, because, as noted above, the normal mode coordinate Q_a is mass weighted. The solution of Equation (2.35) yields evenly separated energy levels of the harmonic oscillator, namely,

$$E_{gv,a}^A = (v + 1/2)\hbar\omega_a \tag{2.36}$$

where $\omega_a = (k_{g,a})^{1/2}$ and $\phi_{gv}^a(Q_a)$ is the wavefunction of the vth energy level of normal mode a.

2.2.5 Transitions Between Vibrational States

In the harmonic oscillator approximation transitions between vibrational states are governed by selection rules that allow only transitions from state v to state v' between adjacent energy levels, namely, $v' = v \pm 1$, according to which,

$$\left\langle \phi^a_{gv'}(Q_a) \middle| Q_a \middle| \phi^a_{gv}(Q_a) \right\rangle = \left[v + \begin{pmatrix} 1 \\ 0 \end{pmatrix} \right]^{1/2} \left(\frac{\hbar}{2\omega_a} \right)^{1/2} \tag{2.37}$$

where in the column bracket, the upper option, 1, holds for $v' = v + 1$ and the lower option, 0, holds for $v' = v - 1$. Thus, for transitions between the levels 0 and 1, either up (0 to 1 first transition integral below) or down (1 to 0 second transition integral below) we have:

$$\left\langle \phi^a_{g1} \middle| Q_a \middle| \phi^a_{g0} \right\rangle = \left\langle \phi^a_{g0} \middle| Q_a \middle| \phi^a_{g1} \right\rangle = \left(\frac{\hbar}{2\omega_a} \right)^{1/2} \tag{2.38}$$

Here we have simplified the notation for the vibrational wavefunction to be ϕ^a_{gv} instead of $\phi^a_{gv}(Q_a)$. A similar result is obtained for the velocity (or momentum) form of the vibrational transition moment. In this instance the expression analogous to Equation (2.37) is:

$$\left\langle \phi^a_{gv'} \middle| \dot{Q}_a \middle| \phi^a_{gv} \right\rangle = \pm i \left[v + \begin{pmatrix} 1 \\ 0 \end{pmatrix} \right]^{1/2} \left(\frac{\hbar\omega_a}{2} \right)^{1/2} \tag{2.39}$$

and for fundamental transitions between the zeroth and first vibrational levels:

$$\left\langle \phi^a_{g1} \middle| \dot{Q}_a \middle| \phi^a_{g0} \right\rangle = -\left\langle \phi^a_{g0} \middle| \dot{Q}_a \middle| \phi^a_{g1} \right\rangle = i \left(\frac{\hbar\omega_a}{2} \right)^{1/2} \tag{2.40}$$

Notice that this transition integral is pure imaginary, due to the velocity or momentum operator, given just before Equation (2.35), and the integral has *opposite signs* for the transitions 0 to 1 and 1 to 0. Finally, the vibrational position and velocity transition moments obey the hypervirial relationship in the same way that transition moments for pure electronic transitions do in Equation (2.24), namely:

$$\left\langle \phi^a_{g1} \middle| \dot{Q}_a \middle| \phi^a_{g0} \right\rangle = i\omega_a \left\langle \phi^a_{g1} \middle| Q_a \middle| \phi^a_{g0} \right\rangle \tag{2.41}$$

2.2.6 Complete Adiabatic Approximation

We now introduce a generalization of the BO approximation known as the complete adiabatic (CA) approximation, which adds nuclear momentum or velocity dependence of the nucleus to the electronic wavefunction. This generalization is needed for formulating VCD intensities in terms of nuclear-velocity dipole-moment derivatives to be described in Chapter 4. Previously, we wrote the adiabatic wavefunction as:

$$\Psi^A_{ev}(\boldsymbol{r}, \boldsymbol{R}) = \psi^A_e(\boldsymbol{r}, \boldsymbol{R}) \phi_{ev}(\boldsymbol{R}) \tag{2.42}$$

as the form of the solution of the BO Schrödinger equation given in Equation (2.3)

$$(H_E + T_N)\psi_e^A(\boldsymbol{r},\boldsymbol{R})\phi_{ev}(\boldsymbol{R}) = E_{ev}^A\psi_e^A(\boldsymbol{r},\boldsymbol{R})\phi_{ev}(\boldsymbol{R}) \tag{2.43}$$

We evaluated this equation by ignoring terms resulting from the operation of the nuclear kinetic energy on the electronic wavefunction by writing $T_N\psi_e^A(\boldsymbol{r},\boldsymbol{R})\phi_{ev}(\boldsymbol{R})$ as equal to $\psi_e^A(\boldsymbol{r},\boldsymbol{R})T_N\phi_{ev}(\boldsymbol{R})$. Instead, we now include cross-terms responsible for coupling between nuclear and electronic velocities by writing:

$$\begin{aligned} T_N\psi_e^A(\boldsymbol{r},\boldsymbol{R})\phi_{ev}(\boldsymbol{R}) &= \frac{\boldsymbol{P}^2}{2M}\psi_e^A(\boldsymbol{r},\boldsymbol{R})\phi_{ev}(\boldsymbol{R}) = \frac{(-i\hbar\partial/\partial\boldsymbol{R})^2}{2M}\psi_e^A(\boldsymbol{r},\boldsymbol{R})\phi_{ev}(\boldsymbol{R}) \\ &= -i\hbar\frac{\partial\psi_e^A(\boldsymbol{r},\boldsymbol{R})}{\partial\boldsymbol{R}}\cdot\left(\frac{-i\hbar}{M}\frac{\partial}{\partial\boldsymbol{R}}\right)\phi_{ev}(\boldsymbol{R}) = -i\hbar\frac{\partial\psi_e^A(\boldsymbol{r},\boldsymbol{R})}{\partial\boldsymbol{R}}\cdot\dot{\boldsymbol{R}}\phi_{ev}(\boldsymbol{R}) \end{aligned} \tag{2.44}$$

Here $\boldsymbol{P}$ is the general nuclear momentum equal to $M\dot{\boldsymbol{R}}$ where M is the mass of whichever nucleus is referred to by the general nuclear coordinate $\boldsymbol{R}$. The first step in this equation is to write $T_N = M\dot{\boldsymbol{R}}^2/2 = \boldsymbol{P}^2/2M$. Then the quantum mechanical momentum operator $-i\hbar\partial/\partial\boldsymbol{R}$ is substituted for $\boldsymbol{P}$. Next, the cross-term is created by operating with each of the two momentum operators only once on the electronic wavefunction while the other operator, as a velocity operator, passes through to the nuclear wavefunction. In the last step of this equation, the nuclear velocity operator, $\boldsymbol{P}/M$, is converted back into its classical form as a parametric variable of the electronic wavefunction by the substitution

$$-\frac{i\hbar}{M}\frac{\partial}{\partial\boldsymbol{R}} = \dot{\boldsymbol{R}} \tag{2.45}$$

The last part of Equation (2.44) gives the new perturbation operator that is added to the BO electronic wave equation in Equation (2.4) to form the corresponding CA electronic wave equation given below. The classical parametric variable $\dot{\boldsymbol{R}}$ for the nuclear velocity can be converted back into its quantum mechanical operator form whenever integration over the nuclear coordinates is carried out after the CA electronic wavefunction has been determined. The CA electronic wave equation, generalized from Equation (2.4), with the lowest order nuclear kinetic energy perturbation from Equation (2.44) is given by:

$$\left(H_E - i\hbar\frac{\partial}{\partial\boldsymbol{R}}\cdot\dot{\boldsymbol{R}}\right)\tilde{\psi}_e^{CA}(\boldsymbol{r},\boldsymbol{R},\dot{\boldsymbol{R}}) = E_e^A(\boldsymbol{R})\tilde{\psi}_e^{CA}(\boldsymbol{r},\boldsymbol{R},\dot{\boldsymbol{R}}). \tag{2.46}$$

Here, the CA nuclear velocity perturbation operator is pure imaginary, and as such it adds imaginary character to the CA electronic wavefunction, $\tilde{\psi}_e^{CA}(\boldsymbol{r},\boldsymbol{R},\dot{\boldsymbol{R}})$, which now becomes complex, as designated by a superscript tilda. A similar change occurs when a molecule is perturbed by a magnetic field that adds imaginary character to the wavefunction, which in turn permits the description of magnetic-field induced current density in the molecule. As we will show, the imaginary kinetic energy perturbation introduced here leads to a description of nuclear-velocity induced current density within the adiabatic approximation of a factorable wavefunction.

The CA wavefunction, which is an extension of the adiabatic wavefunction given in Equation (2.42), is:

$$\tilde{\Psi}_{ev}^{CA}(\boldsymbol{r},\boldsymbol{R},\dot{\boldsymbol{R}}) = \tilde{\psi}_e^{CA}(\boldsymbol{r},\boldsymbol{R},\dot{\boldsymbol{R}})\phi_{ev}(\boldsymbol{R}) \tag{2.47}$$

where the electronic part of the CA wavefunction can be written in terms of real and imaginary parts as:

$$\begin{aligned}\tilde{\psi}_e^{CA}\left(\boldsymbol{r},\boldsymbol{R},\dot{\boldsymbol{R}}\right) &= \psi_e^A(\boldsymbol{r},\boldsymbol{R}) + i\psi_e^{CA}\left(\boldsymbol{r},\boldsymbol{R},\dot{\boldsymbol{R}}\right)\\ &= \psi_e^0(\boldsymbol{r}) + \sum_J\left(\frac{\partial\psi_e^A(\boldsymbol{r})}{\partial\boldsymbol{R}_J}\right)_{R=0}\boldsymbol{R}_J + i\sum_J\left(\frac{\partial\psi_e^{CA}(\boldsymbol{R})}{\partial\dot{\boldsymbol{R}}_J}\right)_{\substack{\dot{R}=0\\R=0}}\dot{\boldsymbol{R}}_J + \ldots\end{aligned} \tag{2.48}$$

From these expressions it can be seen that the CA wavefunction contains a new imaginary term, $i\psi_e^{CA}\left(r,\boldsymbol{R},\dot{\boldsymbol{R}}\right)$, beyond the usual expression for the BO adiabatic wavefunction.

2.2.7 Vibrational Probability Density and Vibrational Transition Current Density

The BO adiabatic wavefunction describes the correlation between nuclear positions and electron probability density while the new terms in the CA wavefunction describe contributions that provide correlation between nuclear velocities and electron current density. For the standard BO correlation between positions of nuclei and electron density, we set the nuclear velocities to zero and expand the electron density to first order in nuclear position by writing:

$$\begin{aligned}\rho_g^{CA}(r,\boldsymbol{R},0) = \rho_g^A(r,\boldsymbol{R}) &= \psi_g^A(\boldsymbol{r})\psi_g^A(\boldsymbol{r}) \doteq \rho_g^0(\boldsymbol{r}) + \sum_J\left(\frac{\partial\rho_g^A(\boldsymbol{r})}{\partial\boldsymbol{R}_J}\right)_{R=0}\cdot\boldsymbol{R}_J\\ &= \psi_g^0(\boldsymbol{r})^2 + 2\sum_J\psi_g^0(\boldsymbol{r})\left(\frac{\partial\psi_g^A(\boldsymbol{r},\boldsymbol{R})}{\partial\boldsymbol{R}_J}\right)_0\cdot\boldsymbol{R}_J\end{aligned} \tag{2.49}$$

Here the probability density for vibrational motion depends only on the adiabatic part of the CA electronic wavefunction. By contrast, the corresponding nuclear velocity dependence depends only on the imaginary part of the CA wavefunction.

$$\begin{aligned}\boldsymbol{j}_g^{CA}\left(\boldsymbol{r},\boldsymbol{R},\dot{\boldsymbol{R}}\right) &= \frac{\hbar}{2mi}\left[\tilde{\psi}_g^{CA*}(\boldsymbol{r})\nabla\tilde{\psi}_g^{CA}(\boldsymbol{r}) - \tilde{\psi}_g^{CA}(\boldsymbol{r})\nabla\tilde{\psi}_g^{CA*}(\boldsymbol{r})\right] \cong \sum_J\left(\frac{\partial\boldsymbol{j}_g^{CA}(\boldsymbol{r})}{\partial\dot{\boldsymbol{R}}_J}\right)_{\substack{R=0\\\dot{R}=0}}\cdot\dot{\boldsymbol{R}}_J\\ &= -\frac{\hbar}{m}\left[\sum_J\left(\frac{\partial\psi_g^{CA}(\boldsymbol{r})}{\partial\dot{\boldsymbol{R}}_J}\right)_{0,0}\nabla\psi_g^0(\boldsymbol{r}) - \psi_g^0(\boldsymbol{r})\nabla\left(\frac{\partial\psi_g^{CA}(\boldsymbol{r})}{\partial\dot{\boldsymbol{R}}_J}\right)_{0,0}\right]\cdot\dot{\boldsymbol{R}}_J\end{aligned} \tag{2.50}$$

There is no zeroth order term because there is no vibrational current density when the velocities of the nuclei are zero. As in the case of pure electronic transition, it is straightforward to show that the changes in probability density obey the continuity equation (Freedman et al., 1997)

$$-\nabla\cdot\boldsymbol{j}_g^{CA}\left(\boldsymbol{r},0,\dot{\boldsymbol{R}}\right) = \frac{\partial\rho_g^{CA}(\boldsymbol{r},\boldsymbol{R},0)}{\partial t} \tag{2.51}$$

In terms of nuclear coordinate derivatives given in Equations (2.49) and (2.50) the continuity equations becomes

$$-\nabla \cdot \left(\frac{\partial j_g^{CA}(\boldsymbol{r})}{\partial \dot{\boldsymbol{R}}_J} \right)_{\substack{\dot{R}=0 \\ R=0}} = \left(\frac{\partial \rho_g^A(\boldsymbol{r})}{\partial \boldsymbol{R}_J} \right)_{R=0} \tag{2.52}$$

This equation states that the change in electron current density, induced by a nuclear velocity, across a boundary surrounding any point $\boldsymbol{r}$ in the space of the molecule is equal to the change in the probability density induced by a corresponding nuclear displacement. It also balances phenomena that are beyond the BO (non-BO) on the left side of the equation with phenomena that are within the BO approximation on the right side. This demonstrates that the CA approximation restores information missing from the BO approximation because in the BO approximation the nuclei are fixed and unmoving, and the information added by the CA approximation is nothing more than the dynamic equivalent of that contained within the BO approximation. The CA approximation repairs the BO approximation without losing the separation of electronic and nuclear motion that is sought by invoking the BO approximation, that is, the CA wavefunction in Equation (2.47) is still a factored wavefunction and the CA approximation has no effect on the vibrational wavefunctions that arise within the BO approximation.

2.3 Infrared Vibrational Absorption Intensities

In this section we use the formalism developed above for electronic and nuclear motion in molecules to describe infrared (IR) vibrational absorption (VA) intensities. We use the abbreviation IR when referring to the infrared spectral region and the closely related abbreviation VA when referring to the intensities of vibrational absorbance peaks in the spectrum. For example, most VA and VCD intensities are measured in the IR and near-IR regions of the spectrum. It is well known that VA intensities are proportional to the derivative of the electric dipole moment of the molecule with nuclear displacement along the normal coordinate. Continuing our definitions of vibrational transitions in Chapter 1, we write the absorbance A in terms of the experimental quantities of transmission of the sample with I, the transmission of the reference as I_0, the pathlength l and the concentration C, where the molar absorptivity ε is a property of the molecular sample independent of the choice of instrumentation or sampling conditions. For a transition between state ev and ev' for normal mode a, we have the standard relation for VA intensity given previously in Equation (1.7),

$$(A)^a_{ev',ev} = -\log_{10}(I/I_0)^a_{ev',ev} = (\varepsilon)^a_{ev',ev} Cl \tag{2.53}$$

The ratio of the IR transmission intensity of the sample, I, also know as the single beam intensity, and the reference intensity, I_0, provides a normalization of the sample transmission that removes the dependence of the measurement on the instrument spectral intensity and profile. The second part of this equation assumes Beer–Lambert's law and defines the molar absorptivity of the sample, $(\varepsilon)^a_{ev',ev}$, a molecular property.

Measurement of an entire spectrum can be described as a summation of the absorbance for each normal mode a, each with it own frequency location ν and bandshape in the spectrum

$$A(\nu) = \sum_a (A)^a_{ev',ev}(\nu) \tag{2.54}$$

The experimental transmission spectrum across the entire spectrum is given by:

$$I(\nu) = I_0(\nu)10^{-A(\nu)} = I_0(\nu)10^{-\varepsilon(\nu)Cl} \tag{2.55}$$

The molar absorptivity spectrum, $\varepsilon(\nu)$, can be also written as a sum over all normal mode transitions a, each of which has a spectral distribution, or bandshape, $f'_a(\nu)$, usually taken to be a Lorentzian bandshape in the spectrum,

$$\varepsilon(\nu) = \frac{8\pi^3 N\nu}{3000hc\ln(10)}\sum_a D_a f'_a(\nu) = \frac{\nu}{9.184\times 10^{-39}}\sum_a D_a f'_a(\nu) \tag{2.56}$$

Here D_a is the dipole strength, usually given as a number $\times 10^{-40}$ esu^2 cm^2 (electrostatic units) for the most commonly used units of Planck's constant, h, and the speed of light, c, and where N is Avogadro's number. The frequency, ν, is given in wavenumbers, cm^{-1}, and the normalized Lorentzian lineshape is:

$$f'_a(\nu) = \frac{\gamma_a/\pi}{(\nu_a-\nu)^2+\gamma_a^2} \qquad \int_0^\infty f'_a(\nu)\mathrm{d}\nu = 1 \tag{2.57}$$

where γ_a is the half-width of the band at its half-height. The prime on the function $f'_a(\nu)$ refers to the imaginary part of the complex index of refraction and will be described in detail in the next chapter. The dipole strength, D_a, is given by:

$$D_a = \left|\left(\frac{\partial\langle\boldsymbol{\mu}\rangle}{\partial Q_a}\right)_{Q_a=0} Q_a\right|^2 \tag{2.58}$$

where $\langle\boldsymbol{\mu}\rangle$ is the electric dipole moment of the entire molecule, including both electronic and nuclear contributions. This brings us back to the classical view of the origin of VA vibrational intensities, namely they are proportional to the square of the derivative of the electric dipole moment of the molecule with respect to the normal coordinate of vibration, Q_a.

As a prelude to our consideration of the origin of VCD intensities in Chapter 4, we will next provide descriptions of VA intensities both in terms of the traditional nuclear displacement along normal coordinates, Q_a, and in terms of the corresponding normal coordinate velocity $\dot{Q}_a$ or P_a.

2.3.1 Position and Velocity Dipole Strengths

The VA intensities of infrared absorption bands are expressed theoretically in terms of the dipole strength. We have previously defined the position and velocity forms of the dipole strength for the case of pure electronic transitions in Equations (2.17) and (2.21). These definitions can be applied to a vibrational transition between levels gv and gv' of the ath normal mode of the ground electronic state as:

$$D^a_{r,gv',gv} = \left|\left\langle \Psi^A_{gv'}(\boldsymbol{r},Q_a)\middle|\boldsymbol{\mu}\middle|\Psi^A_{gv}(\boldsymbol{r},Q_a)\right\rangle\right|^2 \tag{2.59}$$

$$D^a_{v,gv',gv} = \omega^{-2}_{gv',gv}\left|\left\langle \tilde{\Psi}^{CA}_{gv'}(\boldsymbol{r},Q_a,\dot{Q}_a)\middle|\dot{\boldsymbol{\mu}}\middle|\tilde{\Psi}^{CA}_{gv}(\boldsymbol{r},Q_a,\dot{Q}_a)\right\rangle\right|^2 \tag{2.60}$$

We identify the position form of electric-dipole transition moment as:

$$
\begin{aligned}
(\boldsymbol{\mu})^a_{gv',gv} &= \left\langle \Psi^A_{gv'}(\mathbf{r},Q_a)\middle|\boldsymbol{\mu}\middle|\Psi^A_{gv}(\mathbf{r},Q_a)\right\rangle \\
&= \left\langle \psi^A_g(\mathbf{r},Q_a)\phi^a_{gv'}(Q_a)\middle|\boldsymbol{\mu}\middle|\psi^A_g(\mathbf{r},Q_a)\phi^a_{gv}(Q_a)\right\rangle
\end{aligned} \tag{2.61}
$$

The corresponding velocity-dipole transition moment is:

$$
\begin{aligned}
(\dot{\boldsymbol{\mu}})^a_{gv'gv} &= \left\langle \tilde{\Psi}^{CA}_{gv'}(\mathbf{r},Q_a,\dot{Q}_a)\middle|\dot{\boldsymbol{\mu}}\middle|\tilde{\Psi}^{CA}_{gv}(\mathbf{r},Q_a,\dot{Q}_a)\right\rangle \\
&= \left\langle \tilde{\psi}^{CA}_g(\mathbf{r},Q_a,\dot{Q}_a)\phi^a_{gv'}(Q_a)\middle|\dot{\boldsymbol{\mu}}\middle|\tilde{\psi}^{CA}_g(\mathbf{r},Q_a,\dot{Q}_a)\phi^a_{gv}(Q_a)\right\rangle
\end{aligned} \tag{2.62}
$$

First we develop the position form of the dipole transition moment and then summarize the corresponding development of the velocity form. Because the dipole strength of a vibrational transition involves contributions from both the electronic and nuclear motion, the dipole moment operator above is written as sums over the positions times the charges of all the electrons, $\boldsymbol{\mu}_E$, and nuclei, $\boldsymbol{\mu}_N$, in the molecule as:

$$
\boldsymbol{\mu} = \boldsymbol{\mu}_E + \boldsymbol{\mu}_N = -\sum_j e\mathbf{r}_j + \sum_J Z_J e\mathbf{R}_J \tag{2.63}
$$

The position form of the transition moment can be developed further as:

$$
(\boldsymbol{\mu})^a_{gv'gv} = \left\langle \phi^a_{gv'}(Q_a)\middle|\left\langle \psi^A_g(\mathbf{r},Q_a)\middle|\boldsymbol{\mu}_E+\boldsymbol{\mu}_N\middle|\psi^A_g(\mathbf{r},Q_a)\right\rangle\middle|\phi^a_{gv}(Q_a)\right\rangle \tag{2.64a}
$$

$$
= \left\langle \phi^a_{gv'}(Q_a)\middle|\left[\left\langle \psi^A_g(\mathbf{r},Q_a)\middle|\boldsymbol{\mu}_E\middle|\psi^A_g(\mathbf{r},Q_a)\right\rangle + \boldsymbol{\mu}_N\right]\middle|\phi^a_{gv'}(Q_a)\right\rangle \tag{2.64b}
$$

$$
= (\boldsymbol{\mu}_E)^a_{gv',gv} + (\boldsymbol{\mu}_N)^a_{gv',gv} \tag{2.64c}
$$

where we have used the fact that the nuclear position operator in Equation (2.64a) does not operate on the electronic wavefunction and hence, for that term, the electronic wavefunctions can be integrated over all space to unity, as the wavefunction is taken to be normalized.

To evaluate the electronic contribution to Equation (2.64b) further, we expand the nuclear motion dependence of the electronic wavefunction to first order in the normal mode coordinate, Q_a, about the equilibrium position $Q_a = 0$:

$$
\psi^A_g(\mathbf{r},Q_a) = \psi^A_g(\mathbf{r},0) + \left(\partial\psi^A_g(\mathbf{r},Q_a)/\partial Q_a\right)_{Q_a=0} Q_a \tag{2.65}
$$

Inserting this expression into the electronic part of the transition moment in Equation (264) gives:

$$
\begin{aligned}
(\boldsymbol{\mu}_E)^a_{gv',gv} = &\left[\left\langle \psi^A_g(\mathbf{r},0)\middle|\boldsymbol{\mu}_E\middle|\left(\partial\psi^A_g(\mathbf{r},Q_a)/\partial Q_a\right)_{Q_a=0}\right\rangle\right. \\
&\left.+\left\langle \left(\partial\psi^A_g(\mathbf{r},Q_a)/\partial Q_a\right)_{Q_a=0}\middle|\boldsymbol{\mu}_E\middle|\psi^A_g(\mathbf{r},0)\right\rangle\right] \\
&\times\left\langle \phi^a_{gv'}(Q_a)\middle|Q_a\middle|\phi^a_{gv}(Q_a)\right\rangle
\end{aligned} \tag{2.66}
$$

This expression can be recognized to be the electronic contribution to the derivative of the electric dipole moment of the molecule with respect to Q_a.

$$(\boldsymbol{\mu}_E)^a_{gv',gv} = \left(\frac{\partial \left\langle \psi^A_g(\mathbf{r}, Q_a) \middle| \boldsymbol{\mu}_E \middle| \psi^A_g(\mathbf{r}, Q_a) \right\rangle}{\partial Q_a} \right)_{Q_a=0} \left\langle \phi^a_{gv'} \middle| Q_a \middle| \phi^a_{gv} \right\rangle \tag{2.67}$$

The analogous expression for the nuclear contribution to the dipole transition moment in Equation (2.64) can be written as:

$$(\boldsymbol{\mu}_N)^a_{gv',gv} = \left(\frac{\partial \boldsymbol{\mu}_N}{\partial Q_a} \right)_{Q_a=0} \left\langle \phi^a_{gv'} \middle| Q_a \middle| \phi^a_{gv} \right\rangle \tag{2.68}$$

Thus the total transition moment in Equation (2.64) can be written in terms of the derivative of the total dipole moment of the molecule with respect to displacement along normal mode Q_a as:

$$(\boldsymbol{\mu})^a_{gv',gv} = \left[\left(\frac{\partial \langle \boldsymbol{\mu}_E(Q_a) \rangle_g}{\partial Q_a} \right)_{Q_a=0} + \left(\frac{\partial \boldsymbol{\mu}_N}{\partial Q_a} \right)_{Q_a=0} \right] \left\langle \phi^a_{gv'} \middle| Q_a \middle| \phi^a_{gv} \right\rangle \tag{2.69a}$$

$$= \left(\frac{\partial \langle \boldsymbol{\mu} \rangle}{\partial Q_a} \right)_{Q_a=0} \left\langle \phi_{gv'} \middle| Q_a \middle| \phi_{gv} \right\rangle \tag{2.69b}$$

where we have used the equalities

$$\langle \boldsymbol{\mu}_E(Q_a) \rangle_g = \left\langle \psi^A_g(\mathbf{r}, Q_a) \middle| \boldsymbol{\mu}_E \middle| \psi^A_g(\mathbf{r}, Q_a) \right\rangle \tag{2.70a}$$

$$\langle \boldsymbol{\mu} \rangle = \langle \boldsymbol{\mu}_E(Q_a) \rangle_g + \boldsymbol{\mu}_N \tag{2.70b}$$

The absolute square of Equation (2.69b) is the full quantum mechanical expression for the dipole strength given at the beginning of this section in Equation (2.58).

The corresponding velocity-dipole expression from Equation (2.62) is easily written, where the electronic contribution involves derivatives of the CA electronic wavefunction with respect to $\dot{Q}_a$, as:

$$(\dot{\boldsymbol{\mu}})^a_{gv',gv} = \left[\left(\frac{\partial \langle \dot{\boldsymbol{\mu}}_E(\dot{Q}_a) \rangle_g}{\partial \dot{Q}_a} \right)_{\dot{Q}_a=0} + \left(\frac{\partial \dot{\boldsymbol{\mu}}_N}{\partial \dot{Q}_a} \right)_{\dot{Q}_a=0} \right] \left\langle \phi^a_{gv'} \middle| \dot{Q}_a \middle| \phi^a_{gv} \right\rangle \tag{2.71a}$$

$$= \left(\frac{\partial \dot{\boldsymbol{\mu}}}{\partial \dot{Q}_a} \right)_{\dot{Q}_a=0} \left\langle \phi^a_{gv'} \middle| \dot{Q}_a \middle| \phi^a_{gv'} \right\rangle \tag{2.71b}$$

with

$$\langle \dot{\boldsymbol{\mu}}_E(\dot{Q}_a) \rangle = \left\langle \tilde{\psi}^{CA}_g(\mathbf{r}, 0, \dot{Q}_a) | \dot{\boldsymbol{\mu}}_E \middle| \tilde{\psi}^{CA}_g(\mathbf{r}, 0, \dot{Q}_a) \right\rangle \tag{2.72}$$

2.3.2 Atomic Polar Tensors

It convenient to express the derivatives of the dipole moments with respect to normal coordinates in terms of components of Cartesian displacement or velocity vectors. For VA intensities, these are termed atomic polar tensors (APTs) and the position form of the APT is defined by:

$$P^J_{r,\alpha\beta} = E^J_{r,\alpha\beta} + N^J_{r,\alpha\beta} \tag{2.73}$$

$$E^J_{r,\alpha\beta} = \left[\frac{\partial}{\partial R_{J,\alpha}}\left\langle \psi^A_g(\boldsymbol{r},\boldsymbol{R})|\mu_{E,\beta}\middle|\psi^A_g(\boldsymbol{r},\boldsymbol{R})\right\rangle\right]_{R=0} \tag{2.74}$$

where $\mu_{E,\beta}$ is the β-Cartesian component of the electronic contribution to the electric dipole moment operator and $R_{J,\alpha}$ is the α-Cartesian component of the position of the Jth nucleus. The nuclear contribution to the APT is written as:

$$N^J_{r,\alpha\beta} = \left(\frac{\partial\mu_{N,\beta}}{\partial R_{J\alpha}}\right)_{R=0} = \sum_I \left(\frac{\partial Z_I e R_{I\beta}}{\partial R_{J\alpha}}\right)_{R=0} = Z_J e \delta_{\alpha\beta}. \tag{2.75}$$

Here $\delta_{\alpha\beta}$ is the Kronecker delta equal to 1 when the Cartesian directions, $\alpha, \beta = x, y, z$ are equal and 0 otherwise. The expressions for the β-component of the electric-dipole transition moment, given in Equation (2.61) can be written in terms of polar tensors as:

$$(\mu_\beta)^a_{gv',gv} = \left\langle \Psi^A_{gv'}(\boldsymbol{r},Q_a)\middle|\mu_\beta\middle|\Psi^A_{gv}(\boldsymbol{r},Q_a)\right\rangle = \sum_J P^J_{r,\alpha\beta} S_{J\alpha,a}\left\langle \phi^a_{gv'}\middle|Q_a\middle|\phi^a_{gv}\right\rangle \tag{2.76}$$

The corresponding velocity-dipole transition moment from Equation (2.62) is given by:

$$(\dot{\mu}_\beta)^a_{gv',gv} = \left\langle \tilde{\Psi}^{CA}_{gv'}(\boldsymbol{r},Q_a,\dot{Q}_a)\middle|\dot{\mu}_\beta\middle|\tilde{\Psi}^{CA}_{gv'}(\boldsymbol{r},Q_a,\dot{Q}_a)\right\rangle = \sum_J P^J_{v,\alpha\beta} S_{J\alpha,a}\left\langle \phi^a_{gv'}|\dot{Q}_a\middle|\phi^a_{gv}\right\rangle \tag{2.77}$$

where the velocity form of the APT is expressed as:

$$P^J_{v,\alpha\beta} = E^J_{v,\alpha\beta} + N^J_{v,\alpha\beta} \tag{2.78}$$

and the velocity form of the electronic contribution to the APT requires the CA electronic wavefunction, in contrast to the standard adiabatic wavefunction required for Equation (2.74).

$$E^J_{v,\alpha\beta} = \left[\frac{\partial}{\partial \dot{R}_{J\alpha}}\left\langle \tilde{\psi}^{CA}_g(\boldsymbol{r},\boldsymbol{R},\dot{\boldsymbol{R}})\middle|\dot{\mu}_{E,\beta}\middle|\tilde{\psi}^{CA}_g(\boldsymbol{r},\boldsymbol{R},\dot{\boldsymbol{R}})\right\rangle\right]_{\substack{R=0\\ \dot{R}=0}} \tag{2.79}$$

The nuclear velocity contribution, $N^J_{v,\alpha\beta}$, is the same for both the position and velocity forms of the APT as:

$$N^J_{v,\alpha\beta} = \left(\frac{\partial\dot{\mu}_{N,\beta}}{\partial \dot{R}_{J\alpha}}\right)_{\dot{R}_J=0} = \sum_I \left(\frac{\partial Z_I e \dot{R}_{I\beta}}{\partial \dot{R}_{J\alpha}}\right)_{\dot{R}_J=0} = Z_J e \delta_{\alpha\beta} \tag{2.80}$$

This can be compared with Equation (2.75). In addition, the S-vectors, $S_{J\alpha,a}$, describing the transformation between Cartesian and normal coordinates, defined in Equation (2.34) and used here in Equations (2.76) and (2.77), are also the same in both the position and velocity formulations of the APT, namely,

$$S_{J\alpha,a} = \left(\frac{\partial R_{J\alpha}}{\partial Q_a}\right)_{Q_a=0} = \left(\frac{\partial \dot{R}_{J\alpha}}{\partial \dot{Q}_a}\right)_{\dot{Q}_a=0} \tag{2.81}$$

2.3.3 Nuclear Dependence of the Electronic Wavefunction

The dependence of the ground-state electronic wavefunction on nuclear motion can be expressed in terms of coupling with the complete set of excited electronic states of the molecule evaluated at the equilibrium nuclear position of the ground electronic state. To see how this is accomplished, we write the CA electronic wavefunction, given previously for any electronic state e in Equation (2.48), for the ground electronic state as:

$$\tilde{\psi}_g^{CA}(\boldsymbol{r},\boldsymbol{R},\dot{\boldsymbol{R}}) = \psi_g^0(\boldsymbol{r}) + \sum_J \left(\frac{\partial \psi_g^A(\boldsymbol{r})}{\partial R_{J\alpha}}\right)_{R=0} R_{J\alpha} + i\sum_J \left(\frac{\partial \psi_g^{CA}(\boldsymbol{r})}{\partial \dot{R}_{J\alpha}}\right)_{\substack{R=0\\ \dot{R}=0}} \dot{R}_{J\alpha} + \ldots \tag{2.82}$$

where we again use the Einstein summation convention for repeated Greek subscripts. The derivatives of the electronic wavefunction with respect to nuclear position can be written as a sum over all electronic states e of the molecule as:

$$\left(\frac{\partial \psi_g^A(\boldsymbol{r})}{\partial R_{J\alpha}}\right)_{R=0} = \sum_{e\neq g} \left\langle \psi_e^0 \middle| \left(\frac{\partial \psi_g^A}{\partial R_{J\alpha}}\right)_{R=0} \right\rangle \psi_e^0(\boldsymbol{r}) = \sum_{e\neq g} C_{eg,\alpha}^{J,0} \psi_e^0(\boldsymbol{r}) \tag{2.83}$$

where $C_{eg,\alpha}^{J,0}$ represents the vibronic coupling matrix element and is the coefficient of each excited electronic state in the series expansion that describes the changes in the ground electronic state with displacement along $\boldsymbol{R}_J$. The left-hand side of this equation can be recovered by carrying out the sum over e of the complete set of wavefunctions to closure, which is formally expressed in quantum mechanics by:

$$\sum_e |\psi_e^0\rangle\langle\psi_e^0| = 1 \tag{2.84}$$

and where $C_{eg,\alpha}^{J,0}$ vanishes for $e = g$ because a wavefunction and its first derivative with respect to nuclear displacement are orthogonal. An analogous expression to Equation (2.83) can be obtained for the derivative of the electronic wavefunction with respect to nuclear velocity from first-order perturbation theory using the perturbation operator $-i\hbar\dot{\boldsymbol{R}}\cdot\partial/\partial\boldsymbol{R}$ introduced previously in Equation (2.46) (Nafie, 1997).

$$\left(\frac{\partial \psi_g^{CA}(\boldsymbol{r})}{\partial \dot{R}_{J\alpha}}\right)_{R=0} = \sum_{e\neq g} \left\langle \psi_e^0 \middle| \left(\frac{\partial \psi_g^A}{\partial R_{J\alpha}}\right)_{R=0} \right\rangle \psi_e^0(\boldsymbol{r}) \left(\frac{\hbar}{E_{eg}^0}\right) \tag{2.85}$$

The summation over e to closure cannot be carried out because the energy denominator $E_{eg}^0 = E_e^0 - E_g^0$ is also included in the summation giving different weightings to each of the excited electronic states. Using these expressions, the CA electronic wavefunction can now be written as:

$$\tilde{\psi}_g^{CA}(\boldsymbol{r}, \boldsymbol{R}, \dot{\boldsymbol{R}}) = \psi_g^0(\boldsymbol{r}) + \sum_J \sum_{e \neq g} \left\langle \psi_e^0 \middle| \left(\frac{\partial \psi_g^A}{\partial R_{J\alpha}} \right)_{R=0} \right\rangle \psi_e^0(\boldsymbol{r}) \cdot \left(R_{J\alpha} + \frac{i\hbar}{E_{eg}^0} \dot{R}_{J\alpha} \right) \tag{2.86}$$

In more compact notation we can write this equation as:

$$\tilde{\psi}_g^{CA}(\boldsymbol{r}, \boldsymbol{R}, \dot{\boldsymbol{R}}) = \psi_g^0(\boldsymbol{r}) + \sum_{J, e \neq g} C_{eg,\alpha}^{J,0} \psi_e^0(\boldsymbol{r}) \left(R_{J\alpha} + i(\omega_{eg}^0)^{-1} \dot{R}_{J\alpha} \right) \tag{2.87}$$

Equation (2.86) or (2.87) represents a unified expression of the CA electronic wavefunction that can be used to describe the dependence of molecular electronic properties on both nuclear positions and nuclear velocities.

2.3.4 Vibronic Coupling Formulation of VA Intensities

We can now write expressions for the transition moment in Equation (2.61) in terms of vibronic coupling. Only electronic contributions to the transition moments need be considered as vibronic coupling does not affect the description of the nuclear contributions. From Equation (2.67) we can write a series of expressions based on notation established above for a transition from vibrational state $g0$ to $g1$ in the ground electronic state:

$$(\mu_{E,\beta})_{g1,g0}^a = \sum_J \left(\frac{\partial \left\langle \psi_g^A \middle| \mu_{E,\beta} \middle| \psi_g^A \right\rangle}{\partial R_{J,\alpha}} \right)_{R=0} \left(\frac{\partial R_{J,\alpha}}{\partial Q_a} \right)_{Q_a=0} \left\langle \phi_{g1}^a \middle| Q_a \middle| \phi_{g0}^a \right\rangle \tag{2.88a}$$

$$= 2 \sum_J \left\langle \psi_g^A \middle| \mu_{E,\beta} \middle| \frac{\partial \psi_g^A}{\partial R_{J\alpha}} \right\rangle_{R=0} \cdot S_{J\alpha,a} \left\langle \phi_{g1}^a \middle| Q_a \middle| \phi_{g0}^a \right\rangle \tag{2.88b}$$

$$= 2 \sum_J \sum_{e \neq g} \left\langle \psi_g^0 \middle| \mu_{E,\beta} \middle| \psi_e^0 \right\rangle \left\langle \psi_e^0 \middle| \left(\frac{\partial \psi_g^A}{\partial R_{J\alpha}} \right)_{R=0} \right\rangle S_{J\alpha,a} \left\langle \phi_{g1}^a \middle| Q_a \middle| \phi_{g0}^a \right\rangle \tag{2.88c}$$

$$= 2 \sum_J \sum_{e \neq g} \left\langle \mu_{E,\beta} \right\rangle_{eg}^0 C_{eg,\alpha}^{J,0} S_{J\alpha,a} \left\langle \phi_{g1}^a \middle| Q_a \middle| \phi_{g0}^a \right\rangle \tag{2.88d}$$

$$= \sum_J E_{r,\alpha\beta}^J S_{J\alpha,a} \left\langle \phi_{g1}^a \middle| Q_a \middle| \phi_{g0}^a \right\rangle \tag{2.88e}$$

Equation (2.88b) is obtained by differentiation of the matrix element of the electronic contribution to the electric dipole moment of the ground electronic state and the definition of the S-vector in Equation (2.81). Equation (2.88c) is obtained by introduction of a summation over all electronic states followed by the definitions of the vibronic coupling matrix element in Equation (2.88d) and the electronic contribution to the position form of the APT in Equation (2.88e). The corresponding expressions for the velocity formulation of the electric dipole transition moment, starting from

Equation (2.62) are

$$(\dot{\mu}_{E,\beta})^a_{g1,g0} = \sum_J \left(\frac{\partial \left\langle \tilde{\psi}_g^{CA} \middle| \dot{\mu}_{E,\beta} \middle| \tilde{\psi}_g^{CA} \right\rangle}{\partial \dot{R}_{J,\alpha}} \right)_{R=0} \left(\frac{\partial \dot{R}_{J,\alpha}}{\partial \dot{Q}_a} \right)_{Q_a=0} \left\langle \phi^a_{g1} \middle| \dot{Q}_a \middle| \phi^a_{g0} \right\rangle \tag{2.89a}$$

$$= 2 \sum_J \left\langle \psi_g^A \middle| \dot{\mu}_{E,\beta} \middle| \frac{\partial \tilde{\psi}_g^{CA}}{\partial \dot{R}_{J\alpha}} \right\rangle_{R=0} S_{J\alpha,a} \left\langle \phi^a_{g1} \middle| \dot{Q}_a \middle| \phi^a_{g0} \right\rangle \tag{2.89b}$$

$$= 2i \sum_J \sum_{e \neq g} \frac{\left\langle \psi_g^0 | \dot{\mu}_{E,\beta} | \psi_e^0 \right\rangle}{\omega^0_{eg}} \left\langle \psi_e^0 \middle| \left(\frac{\partial \psi_g^A}{\partial R_{J\alpha}} \right)_{R=0} \right\rangle S_{J\alpha,a} \left\langle \phi^a_{g1} \middle| \dot{Q}_a \middle| \phi^a_{g0} \right\rangle \tag{2.89c}$$

$$= 2i \sum_J \sum_{e \neq g} \left\langle \dot{\mu}_{E,\beta} \right\rangle^0_{ge} (\omega^0_{eg})^{-1} C^{J,0}_{eg,\alpha} S_{J\alpha,a} \left\langle \phi^a_{g1} \middle| \dot{Q}_a \middle| \phi^a_{g0} \right\rangle \tag{2.89d}$$

$$= \sum_J E^J_{v,\alpha\beta} S_{J\alpha,a} \left\langle \phi^a_{g1} \middle| \dot{Q}_a \middle| \phi^a_{g0} \right\rangle \tag{2.89e}$$

The velocity formulation of the electric-dipole transition moment uses the CA electronic wavefunction, the velocity-dipole operator, and the nuclear velocity coordinate.

2.3.5 Equivalence Relationships

The formal equivalence of these two formulations of infrared intensity can be demonstrated by starting with the velocity-dipole moment expression in Equation (2.89d) and using Equation (2.24) for converting between the velocity and position forms of the electric dipole transition moment between pure electronic states e and g, and the corresponding conversion for normal coordinate transition moments given in Equation (2.41) yielding:

$$(\dot{\mu}_{E,\beta})^a_{g1,g0} = 2i \sum_J \sum_{e \neq g} i\omega^0_{ge} \left\langle \mu_{E,\beta} \right\rangle^0_{ge} (\omega^0_{eg})^{-1} C^{J,0}_{eg,\alpha} S_{J\alpha,a} i\omega_a \left\langle \phi^a_{g1} \middle| Q_a \middle| \phi^a_{g0} \right\rangle \tag{2.90a}$$

$$= 2i\omega_a \sum_J \sum_{e \neq g} \left\langle \mu_{E,\beta} \right\rangle^0_{ge} C^{J,0}_{eg,\alpha} S_{J\alpha,a} \left\langle \phi^a_{g1} \middle| Q_a \middle| \phi^a_{g0} \right\rangle \tag{2.90b}$$

$$= i\omega_a (\mu_{E,\beta})^a_{g1,g0} \tag{2.90c}$$

where the last equation is obtained from Equation (2.90b) by using the expression for $(\mu_{E,\beta})^a_{g1,g0}$ given in Equation (2.88d). Equation (2.90c) is the normal conversion between the velocity and position forms of a transition moment with a separation in energy of $E = \hbar\omega_a$. These relationships also demonstrate the equivalence of the electronic contribution to the APT, namely:

$$E^J_{v,\alpha\beta} = 2i \sum_{e \neq g} \left\langle \dot{\mu}_{E,\beta} \right\rangle^0_{ge} C^{J,0}_{eg,\alpha} S_{J\alpha,a} (\omega^0_{eg})^{-1} \tag{2.91a}$$

$$= 2i \sum_{e \neq g} i\omega^0_{ge} \left\langle \mu_{E,\beta} \right\rangle^0_{ge} C^{J,0}_{eg,\alpha} S_{J\alpha,a} (\omega^0_{eg})^{-1} \tag{2.91b}$$

$$= 2 \sum_{e \neq g} \left\langle \mu_{E,\beta} \right\rangle^0_{ge} C^{J,0}_{eg,\alpha} S_{J\alpha,a} = E^J_{r,\alpha\beta} \tag{2.91c}$$

Finally, the equivalence of the expressions for the position and velocity forms of the dipole strength given at the beginning of this section in Equations. (2.58) and (2.59)

$$D^{a}_{g1,g0} = \left|\left(\mu_{\beta}\right)^{a}_{g1,g0}\right|^{2} = \omega_{a}^{-2}\left|\left(\dot{\mu}_{\beta}\right)^{a}_{g1,g0}\right|^{2} \tag{2.92}$$

can be seen easily from Equation (2.90c) and inclusion of the corresponding equivalence relationship for the nuclear contribution given by:

$$\begin{aligned}(\dot{\mu}_{N,\beta})^{a}_{g1,g0} &= \sum_{J} N^{J}_{v,\alpha\beta} S_{J\alpha,a}\left\langle \phi^{a}_{g1}\middle|\dot{Q}_{a}\middle|\phi^{a}_{g0}\right\rangle \\ &= \sum_{J} N^{J}_{r,\alpha\beta} S_{J\alpha,a} i\omega_{a}\left\langle \phi^{a}_{g1}\middle|Q_{a}\middle|\phi^{a}_{g0}\right\rangle = i\omega_{a}(\mu_{N,\beta})^{a}_{g1,g0}\end{aligned} \tag{2.93}$$

2.4 Vibrational Raman Scattering Intensities

Raman scattering in vibrational transitions can be described using much of the same formalism developed above for infrared vibrational absorption. As described in Chapter 1, the Raman scattering process involves a two-photon interaction between the molecule and the radiation field in which incident radiation is scattered with loss or gain of a quantum of vibrational energy. At the classical level this involves the response of the polarizability of the molecule $\alpha_{\alpha\beta}(t)$ to the incident electric field $E^{(0)}_{\beta}(t)$ of the incoming radiation with angular frequency $\omega_0 = 2\pi\nu_0$. At the simplest level, this response takes the form of an induced dipole electric moment $\mu^{I}_{\alpha}(t)$, which radiates at the frequency ω_s of the Raman scattered light with intensity $I(\omega_0, \omega_s)$ according to the expression:

$$I(\omega_0, \omega_s) = k'_{\omega}\omega_s^4(\mu^{I}_{\alpha})^2\sin^2\theta \tag{2.94}$$

where the angle θ is the direction of the radiation with respect to the axis of the induced electric dipole moment and k'_{ω} is given by:

$$k'_{\omega} = \frac{1}{32\pi^2\varepsilon_0 c^3} \tag{2.95}$$

where ε_0 is the dielectric constant of free space and c is the speed of light. In the classical theory the time dependence of the induced dipole moment is expressed as:

$$\mu^{I}_{\alpha}(t) = \sum_{\beta}\alpha_{\alpha\beta}(t)E^{(0)}_{\beta}(t) = \alpha_{\alpha\beta}(t)E^{(0)}_{\beta}(t) \tag{2.96}$$

The Greek subscripts refer to Cartesian components, and in the second part of the equation we again introduce the Einstein convention for implied summation over repeated Greek subscripts. Thus, in terms of Cartesian components, the incident radiation polarized along the β-direction is converted by the polarizability tensor $\alpha_{\alpha\beta}$ into an induced dipole polarized along the α-direction. The time dependences of the incident radiation oscillating at ω_0 and that of the molecular polarizability, which includes oscillation at the frequency ω_a of normal mode Q_a, are given by:

$$E^{(0)}_{\beta}(t) = E^{(0)}_{\beta,0}\cos\omega_0 t \tag{2.97}$$

$$\alpha_{\alpha\beta}(t) = \alpha_{\alpha\beta,0} + \left(\frac{\partial \alpha_{\alpha\beta}}{\partial Q_a}\right)_{Q_a=0} Q_a \cos \omega_a t \tag{2.98}$$

$$\mu_\alpha^I(t) = \alpha_{\alpha\beta,0} E_{\beta,0}^{(0)} \cos(\omega_0 t) + \left(\frac{\partial \alpha_{\alpha\beta}}{\partial Q_a}\right)_0 Q_a E_{\beta,0}^{(0)} \cos(\omega_0 t) \cos(\omega_a t) \tag{2.99}$$

Using a standard trigonometric identity, we can write Equation (2.99) as:

$$\mu_\alpha^I(t) = \alpha_{\alpha\beta,0} E_{\beta,0}^{(0)} \cos(\omega_0 t) + \frac{1}{2}\left(\frac{\partial \alpha_{\alpha\beta}}{\partial Q_a}\right)_0 Q_a E_{\beta,0}^{(0)} [\cos(\omega_0 - \omega_a)t + \cos(\omega_0 + \omega_a)t] \tag{2.100}$$

The first term can be identified as Rayleigh scattering at the incident radiation frequency while the second and third terms correspond to Raman scattering intensity at the Stokes ($\omega_s = \omega_0 - \omega_a$) and anti-Stokes ($\omega_s = \omega_0 + \omega_a$) frequencies, respectively. The classical expression takes no account of vibrational energy levels or their relative populations and therefore predicts incorrectly that Stokes and anti-Stokes Raman scattering have the same intensities. Nevertheless, the classical description does reveal that Raman intensity is associated with the derivative of the polarizability of the molecule with respect to the normal coordinate of vibration.

2.4.1 General Unrestricted (GU) Theory of Raman Scattering

The general unrestricted (GU) quantum mechanical theory of the intensity of light scattering from a molecule with complex polarizability, $\tilde{\alpha}_{\alpha\beta}$, where again we denote a complex quantity with a tilda, interacting with the unit complex polarization vectors of the incident electric field $\tilde{e}_\beta^i$ and scattered electric field $\tilde{e}_\alpha^d$, is given by:

$$I(\tilde{e}^d, \tilde{e}^i) = 90K\left\langle \left| \tilde{e}_\alpha^{d*} \tilde{\alpha}_{\alpha\beta} \tilde{e}_\beta^i \right|^2 \right\rangle \tag{2.101}$$

where the angled brackets represent the average of all orientations of the polarizability in the molecule frame of reference relative to the polarization vectors that are fixed in the laboratory frame of reference. The Einstein convention used above applies to all repeated Greek subscripts. The constant K includes the dependence of the scattering on the incident electric field intensity $\tilde{E}^{(0)}$, the scattered frequency of the radiation ω_R, the distance from the scattering molecule to the detector R, and the magnetic permeability for free space μ_0.

$$K = \frac{1}{90}\left(\frac{\omega_R^2 \mu_0 \tilde{E}^{(0)}}{4\pi R}\right)^2 \tag{2.102}$$

The transition polarizability of the molecule from state n to state m is expressed using time-dependent perturbation theory as:

$$(\tilde{\alpha}_{\alpha\beta})_{mn} = \frac{1}{\hbar}\sum_{j \neq m,n}\left[\frac{\langle \tilde{\Psi}_m | \hat{\mu}_\alpha | \tilde{\Psi}_j \rangle \langle \tilde{\Psi}_j | \hat{\mu}_\beta | \tilde{\Psi}_n \rangle}{\omega_{jn} - \omega_0 - i\Gamma_j} + \frac{\langle \tilde{\Psi}_m | \hat{\mu}_\beta | \tilde{\Psi}_j \rangle \langle \tilde{\Psi}_j | \hat{\mu}_\alpha | \tilde{\Psi}_n \rangle}{\omega_{jn} + \omega_R + i\Gamma_j}\right] \tag{2.103}$$

where $\omega_{jn} = \omega_j - \omega_n$ represents the frequency (energy) difference between the initial state n and the intermediate state j. The summation is over all excited states of the molecule, and $i\Gamma_j$ is the imaginary damping term inversely proportional to the lifetime for state j. We make no assumptions about the wavefunctions, whether they are real or complex, adiabatic, complete adiabatic, or non-adiabatic. In this equation, the first term from the left is the resonance term. It represents a time-ordered sequence, read in the numerator from right to left, in which the incident photon with Cartesian component β interacts with the molecule as described by the first matrix element, and this is followed by the interaction of the scattered radiation with Cartesian component α in the second matrix element. The real part of the denominator becomes zero when the frequency of the incident photon ω_0 matches the transition frequency, ω_{jn}, from state n to the virtual intermediate state j. The non-resonance term is written next. Here the time-order of the interactions is reversed, the scattered photon is emitted followed by the uptake of the incident photon, and the frequency denominator never approaches zero in the absence of the rare occurrence that the intermediate state j is lower in energy than the initial state n. An alternative form of the energy denominator of the non-resonance term is $\omega_{jm} + \omega_0$, which equals $\omega_{jn} + \omega_R$ as the frequency of the final state m differs from the initial state n by the same amount as the Raman scattered frequency ω_R differs from the incident frequency ω_0. These alternatives become identical in the case of Rayleigh scattering where states n and m are the same and the incident and scattered radiation frequencies are equal. The energy denominators in the resonance and non-resonance terms are simply statements of the energy balance of the molecule and the radiation field after the interaction with the first photon represented by the first matrix element on the right in the numerator of each term. The resonance and non-resonance Raman terms in Equation (2.103) as described here can be represented by time-ordered diagrams given in Figure 2.1.

2.4.2 Vibronic Theory of Raman Intensities

We next develop the expression for the Raman transition polarizability in Equation (2.103) within the BO approximation where we separate the wavefunction into a product of electronic and vibrational parts. The formalism can be extended beyond the BO approximation to include the CA wavefunctions described above for IR absorption intensities (Nafie, 2008), but as, unlike VCD, ROA does not require wavefunctions beyond the BO approximation, we do not include this feature here. To introduce separate electronic and vibrational wavefunctions, we use the notation for the state of the molecule developed for IR intensities, except for the last step where we reduce the notation further for economy of style.

$$\left|\tilde{\Psi}_j(\boldsymbol{r}, \boldsymbol{R})\right\rangle = \left|\Psi^A_{ev}(\boldsymbol{r}, \boldsymbol{R})\right\rangle = \left|\psi^A_e(\boldsymbol{r}, \boldsymbol{R})\phi_{ev}(\boldsymbol{R})\right\rangle = \left|e\right\rangle\left|\phi_{ev}\right\rangle \tag{2.104}$$

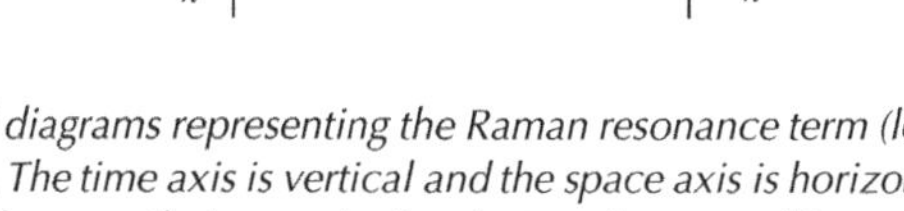

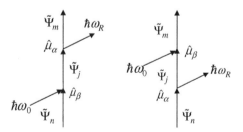

Figure 2.1 *Time-ordered diagrams representing the Raman resonance term (left) and non-resonance term (right) of Equation (2.103). The time axis is vertical and the space axis is horizontal. Molecular state vectors are represented by vertical arrows that move in time but not in space. Diagonal arrows represent photons traveling both in time and in space. The interaction vertices labeled with Greek letters represent the polarization direction of the photon interaction that changes the state vectors. In the resonance term, the molecule first destroys the incident photon and then creates the scattered photon, whereas in the non-resonance term the time-order of the photon interactions is reversed*

Using this compact notation, the Raman transition polarizability from vibrational levels $g0$ to $g1$ of the ground electronic state for normal mode a is given by:

$$(\tilde{\alpha}_{\alpha\beta})^a_{g1,g0} = \frac{1}{\hbar}\sum_{ev}\left[\frac{\langle\phi^a_{g1}|\langle g|\hat{\mu}_\alpha|e\rangle|\phi_{ev}\rangle\langle\phi_{ev}|\langle e|\hat{\mu}_\beta|g\rangle|\phi^a_{g0}\rangle}{\omega_{ev,g0}-\omega_0-i\Gamma_{ev}} + \frac{\langle\phi^a_{g1}|\langle g|\hat{\mu}_\beta|e\rangle|\phi_{ev}\rangle\langle\phi_{ev}|\langle e|\hat{\mu}_\alpha|g\rangle|\phi^a_{g0}\rangle}{\omega_{ev,g0}+\omega_R+i\Gamma_{ev}}\right] \tag{2.105}$$

We next describe the dependence of the electronic wavefunction on nuclear motion by writing the adiabatic electronic wavefunction to first order in normal coordinate dependence,

$$\psi^A_g(Q_a) = \psi_{g0} + \left(\frac{\partial\psi_g}{\partial Q_a}\right)_0 Q_a \tag{2.106}$$

Using the notation from above for the first derivative of the entire electronic matrix element $\langle g|\hat{\mu}_\alpha|e\rangle^{Q_a}_0$ can be written as:

$$\langle g|\hat{\mu}_\alpha|e\rangle^{Q_a}_0 = \left(\frac{\partial\langle g|\hat{\mu}_\alpha|e\rangle}{\partial Q_a}\right)_0 = \left\langle\left(\frac{\partial\psi_g}{\partial Q_a}\right)_0\left|\hat{\mu}_\alpha\right|\psi_{e0}\right\rangle + \left\langle\psi_{g0}\left|\hat{\mu}_\alpha\right|\left(\frac{\partial\psi_e}{\partial Q_a}\right)_0\right\rangle \tag{2.107}$$

which enables us to write

$$\begin{aligned}(\tilde{\alpha}_{\alpha\beta})^{Q_a}_{g1,g0} = \frac{1}{\hbar}\sum_{\substack{ev\\ e\neq g}}\Bigg\{&\left(\frac{\langle g|\hat{\mu}_\alpha|e\rangle_0\langle e|\hat{\mu}_\beta|g\rangle_0}{\omega_{ev,g0}-\omega_0-i\Gamma_{ev}} + \frac{\langle g|\hat{\mu}_\beta|e\rangle_0\langle e|\hat{\mu}_\alpha|g\rangle_0}{\omega_{ev,g0}+\omega_R+i\Gamma_{ev}}\right)\langle\phi^a_{g1}|\phi_{ev}\rangle\langle\phi_{ev}|\phi^a_{g0}\rangle \\ &+\left(\frac{\langle g|\hat{\mu}_\alpha|e\rangle^{Q_a}_0\langle e|\hat{\mu}_\beta|g\rangle_0}{\omega_{ev,g0}-\omega_0-i\Gamma_{ev}} + \frac{\langle g|\hat{\mu}_\beta|e\rangle^{Q_a}_0\langle e|\hat{\mu}_\alpha|g\rangle_0}{\omega_{ev,g0}+\omega_R+i\Gamma_{ev}}\right)\langle\phi^a_{g1}|Q_a|\phi_{ev}\rangle\langle\phi_{ev}|\phi^a_{g0}\rangle \\ &+\left(\frac{\langle g|\hat{\mu}_\alpha|e\rangle_0\langle e|\hat{\mu}_\beta|g\rangle^{Q_a}_0}{\omega_{ev,g0}-\omega_0-i\Gamma_{ev}} + \frac{\langle g|\hat{\mu}_\beta|e\rangle_0\langle e|\hat{\mu}_\alpha|g\rangle^{Q_a}_0}{\omega_{ev,g0}+\omega_R+i\Gamma_{ev}}\right)\langle\phi^a_{g1}|\phi_{ev}\rangle\langle\phi_{ev}|Q_a|\phi^a_{g0}\rangle\Bigg\}\end{aligned} \tag{2.108}$$

For each term, we have factored the transition tensor into electronic and vibrational parts. This expression is general in that it can be used for both resonance and non-resonance limits, the latter being the far-from-resonance (FFR) approximation that will be considered in detail below. The first term in Equation (2.108) is the so-called Frank–Condon A-term that is important only in the case of resonance Raman scattering as it depends on the Frank–Condon overlap integrals between ground- and excited-state vibrational wavefunctions. This term vanishes in the FFR approximation. The second and third terms contain the so-called B- and C-terms to be defined below. The second term describes vibronic coupling with the second photonic interaction, while the third term describes vibronic coupling with the first photonic interaction.

2.4.3 Raman Scattering Tensors and Invariants

The elements of the Raman polarizability tensor, $(\tilde{\alpha}_{\alpha\beta})_{g1,g0}^{Q_a}$, depend on the orientation of the molecule with respect to an *xyz*-reference frame. However, certain linear combinations of tensor elements are invariant to changes in the orientation of the molecule in the reference frame. These combinations are called invariants. For the Raman tensor, there are three such combinations called the isotropic invariant, α^2, the square of the average of the diagonal terms of the tensor $(\tilde{\alpha}_{xx}+\tilde{\alpha}_{yy}+\tilde{\alpha}_{zz})/3$, the symmetric anisotropy, $\beta_S(\tilde{\alpha})^2$, sensitive to differences in the diagonal elements and the symmetric average of corresponding off-diagonal elements, and the anti-symmetric anisotropy, $\beta_A(\tilde{\alpha})^2$, sensitive to differences between corresponding off-diagonal elements. These invariants are given by:

$$\alpha^2 = \frac{1}{9}\mathrm{Re}\left[(\tilde{\alpha}_{\alpha\alpha})^S(\tilde{\alpha}_{\beta\beta})^{S*}\right] \tag{2.109}$$

$$\beta_S(\tilde{\alpha})^2 = \frac{1}{2}\mathrm{Re}\left[3(\tilde{\alpha}_{\alpha\beta})^S(\tilde{\alpha}_{\alpha\beta})^{S*} - (\tilde{\alpha}_{\alpha\alpha})^S(\tilde{\alpha}_{\beta\beta})^{S*}\right] \tag{2.110}$$

$$\beta_A(\tilde{\alpha})^2 = \frac{1}{2}\mathrm{Re}\left[3(\tilde{\alpha}_{\alpha\beta})^A(\tilde{\alpha}_{\alpha\beta})^{A*}\right] \tag{2.111}$$

where the symmetric and anti-symmetric combinations of tensor elements are defined as:

$$(\tilde{T}_{\alpha\beta})^S = \frac{1}{2}\left[(\tilde{T}_{\alpha\beta}) + (\tilde{T}_{\beta\alpha})\right] \tag{2.112a}$$

$$(\tilde{T}_{\alpha\beta})^A = \frac{1}{2}\left[(\tilde{T}_{\alpha\beta}) - (\tilde{T}_{\beta\alpha})\right] \tag{2.112b}$$

Raman tensor invariants are important when considering samples of liquids and solutions where the molecules are randomly oriented relative to the laboratory reference frame. In order to obtain expressions for Raman scattering experiments in terms of the Raman tensor elements of the scattering molecules, the Raman tensor must be averaged over all possible orientations of the molecule relative to the laboratory reference frame. This can be accomplished with direction cosines, $\Phi_{A\alpha}$, which specify the projection of, or cosine of the angle between, the laboratory A-axis and the molecule α-direction. Using direction cosines Equation (2.101) can be re-written such that the polarization vectors are in the fixed laboratory frame of measurement, while the Raman tensors of the molecules can be averaged over all possible values of the direction cosines. Thus we have:

$$I(\tilde{e}^d,\tilde{e}^i) = 90K\left|\tilde{e}_A^{d*}\langle\Phi_{A\alpha}\tilde{\alpha}_{\alpha\beta}\Phi_{B\beta}\rangle\tilde{e}_B^i\right|^2 \tag{2.113}$$

where A an B specify fixed laboratory Cartesian directions for a particular measurement setup, while the expression in brackets indicate that the molecular Raman tensor must be averaged over all possible orientation of the molecule relative to the laboratory reference frame. The only combinations of Raman tensor elements to survive the averaging process are the invariants in Equations (2.109), (2.110) and (2.111), and hence only these Raman tensor invariants appear in expressions for particular Raman and ROA measurement setups.

2.4.4 Polarization Experiments and Scattering Geometries

The intensities for Raman scattering from various polarization setups and scattering geometries can be written in terms of the Raman tensor invariants defined above. While there are many such expressions that can be defined, we select here the ones that are most often used to perform Raman scattering measurements. To describe these setups and intensities, we first specify the scattering geometry in the laboratory frame (upper case Cartesian letters). For right-angle or 90° scattering, the incident light propagates along the Z-direction and scatters along the Y-direction forming a ZY-scattering plane. We first consider the polarization setup where the incident light is X-polarized and the scattered light is Z-polarized. The polarization vectors are orthogonal, and this type of scattering is referred to as depolarized right-angle Raman scattering and is expressed as:

$$I_Z^X(90^\circ) = 2K\left[3\beta_S(\tilde{\alpha})^2 + 5\beta_A(\tilde{\alpha})^2\right] \tag{2.114}$$

where the upper subscript designates the polarization state of the incident radiation and the lower subscript that of the scattered radiation. If the analyzing polarizer for the scattered light is changed from Z-polarized in the scattering plane to X-polarized perpendicular to the scattering plane, then the polarization vectors of the incident and scattered radiation are now parallel, and this is referred to as right-angle polarized Raman scattering.

$$I_X^X(90^\circ) = 2K\left[45\alpha^2 + 4\beta_S(\tilde{\alpha})^2\right] \tag{2.115}$$

We note that the isotropic invariant contributes only to polarized scattering whereas both anisotropic invariants contribute to depolarized scattering. Next we consider the same scattering geometry but with the incident radiation unpolarized. The unpolarized state may be considered to be a random superposition of orthogonal polarization states, and so we can write

$$I_Z^U(90^\circ) = I_Z^X(90^\circ) + I_Z^Y(90^\circ) = 2K\left[6\beta_S(\tilde{\alpha})^2 + 10\beta_A(\tilde{\alpha})^2\right] \tag{2.116}$$

$$I_X^U(90^\circ) = I_X^X(90^\circ) + I_X^Y(90^\circ) = 2K\left[45\alpha^2 + 7\beta_S(\tilde{\alpha})^2 + 5\beta_A(\tilde{\alpha})^2\right] \tag{2.117}$$

The depolarized intensity $I_Z^U(90^\circ)$ is twice the depolarized intensity of Equation (2.114) because it is the sum of two intensities with orthogonal polarization states, and similarly $I_X^U(90^\circ)$ is simply the polarized intensity of Equation (2.115) plus an additional depolarized contribution. It can be shown that all right angle scattering intensities are simply linear combinations of the depolarized and polarized intensities of Equations (2.114) and (2.115), and yet there are three invariants in these two sets of intensities. In order to isolate all three invariants, a new scattering geometry involving circularly polarized radiation is needed. We consider two that are the circular polarization analogs of the polarized and depolarized intensities given above. These are co-rotating and contra-rotating circular polarization backscattering intensities given by:

$$I_R^R(180^\circ) = I_L^L(180^\circ) = 2K\left[6\beta_S(\tilde{\alpha})^2\right] \tag{2.118}$$

$$I_L^R(180^\circ) = I_R^L(180^\circ) = 2K\left[45\alpha^2 + \beta_S(\tilde{\alpha})^2 + 5\beta_A(\tilde{\alpha})^2\right] \tag{2.119}$$

It can be seen that the co-rotating backscattering intensity is depolarized and depends on only one invariant. Although $I_R^R(180^\circ)$ is called co-rotating because only one sense of circularly polarized light is used for incident and scattered radiation beams, the sense of circular polarization actually reverses along the direction of propagation on going from incident to scattered radiation, and this is the reason why the scattering is depolarized. By contrast, the contra-rotating backscattered Raman intensity is a strongly polarized intensity. Again, the reason here is that a beam of circularly polarized light that is back-reflected from a mirror changes sense of polarization from right to left or vice versa, only because of a reversal in direction, and not because a reversal of the angular direction of the circular polarization vector.

2.4.5 Depolarization and Reversal Ratios

Depolarization ratios are a well-known method of analyzing Raman spectra to aid in vibrational assignments. For right-angle scattering using vertically polarized incident light, the depolarization ratio is the ratio of the depolarized to polarized scattering intensities given above:

$$\rho_l(90^\circ) = \frac{I_Z^X(90^\circ)}{I_X^X(90^\circ)} = \frac{3\beta_S(\tilde{\alpha})^2 + 5\beta_A(\tilde{\alpha})^2}{45\alpha^2 + 4\beta_S(\tilde{\alpha})^2} \tag{2.120}$$

Far from resonance when the anti-symmetric anisotropy $\beta_A(\tilde{\alpha})^2$ is zero, this depolarization ratio varies from zero only when α^2 is non-zero, as in the case of a spherically symmetric transition polarizability tensor (only equal diagonal elements), to 3/4 when α^2 is zero, as is the case for non-totally symmetric modes. In the rare cases in resonance Raman scattering when $\beta_A(\tilde{\alpha})^2$ is the only non-zero invariant, the depolarization ratio is essentially infinite. Such bands are called inverse polarized Raman bands. The corresponding depolarization ratio for unpolarized incident radiation is given by:

$$\rho_n(90^\circ) = \frac{I_Z^U(90^\circ)}{I_X^U(90^\circ)} = \frac{6\beta_S(\tilde{\alpha})^2 + 10\beta_A(\tilde{\alpha})^2}{45\alpha^2 + 7\beta_S(\tilde{\alpha})^2 + 5\beta_A(\tilde{\alpha})^2} \tag{2.121}$$

This ratio varies between 0 and 6/7 when $\beta_A(\tilde{\alpha})^2$ is zero for the same limits as discussed above. When only the anti-symmetric anisotropic invariant is non-zero, the maximum value of this depolarization ratio is 2.

When circularly polarized radiation is used, the relevant depolarization ratio is called the reversal ratio, the ratio of co-rotating to contra-rotating circular polarization intensities, or in terms of the Raman intensity associated with DCP scattering, it is the ratio of $\mathrm{DCP_I}$ divided by $\mathrm{DCP_{II}}$ Raman intensity:

$$R(180^\circ) = \frac{I_R^R(180^\circ)}{I_L^R(180^\circ)} = \frac{I_L^L(180^\circ)}{I_R^L(180^\circ)} = \frac{6\beta_S(\tilde{\alpha})^2}{45\alpha^2 + \beta_S(\tilde{\alpha})^2 + 5\beta_A(\tilde{\alpha})^2} \tag{2.122}$$

This ratio is defined for backscattering, but other scattering angles are possible as long as right-angle scattering is not chosen, because at right angles there is no difference in intensity between co- and contra-rotating circular polarization intensities. At 90° looking back at the sample at the moment of scattering, there is no projection visible of whether the incident light was left or right circularly polarized, and hence the reversal ratio is unity for all bands in the spectrum. In the FFR approximation, $R(180^\circ)$ varies between 0 and 6 in going from a spherically symmetric Raman polarizability tensor to a non-totally symmetric tensor with α^2 equal to zero. Another dimensionless quantity that is useful for symmetry analysis when circularly polarized light is involved is the degree of circularity defined for backscattering geometry as:

$$^{R}C(180^\circ) = \frac{I_R^R(180^\circ) - I_L^R(180^\circ)}{I_R^R(180^\circ) + I_L^R(180^\circ)} = \frac{5\beta_S(\tilde{\alpha})^2 - 45\alpha^2 - 5\beta_A(\tilde{\alpha})^2}{45\alpha^2 + 7\beta_S(\tilde{\alpha})^2 + 5\beta_A(\tilde{\alpha})^2} \tag{2.123}$$

Here the degree of circularity is defined for right incident circular polarization, but the convention that is important is that it is that the intensity difference in the numerator is always co-rotating minus contra-rotating (i.e., DCP_I minus DCP_{II} Raman) intensity. This ratio has a sign depending on whether the co-rotating or contra-rotating scattering is more intense. In the FFR approximation, the degree of circularity varies from -1 for spherically polarized scattering, when only α^2 is non-zero, to 5/7 for depolarized bands when only $\beta_S(\tilde{\alpha})^2$ is non-zero.

2.4.6 Isolation of Raman Scattering Invariants

As mentioned above, three linearly independent Raman intensity measurements are needed to isolate all three Raman invariants. Three such measurements, selected from the intensities given above, are:

$$\alpha^2 = \frac{1}{90K}\left[I_X^X(90°) - \frac{2}{3}I_R^R(180°)\right] \tag{2.124}$$

$$\beta_S(\tilde{\alpha})^2 = \frac{1}{12K}I_R^R(180°) \tag{2.125}$$

$$\beta_A(\tilde{\alpha})^2 = -\frac{1}{20K}\left[I_R^R(180°) - 2I_Z^X(90°)\right] \tag{2.126}$$

It is important to ensure that the Raman intensities for different scattering geometries are properly related. One way is first to measure the Raman intensity of a sample which is far from any electronic resonance. In this case, $\beta_A(\tilde{\alpha})^2 = 0$, and hence from Equation (2.126), or Equations (2.114) and (2.118), the relationship $I_R^R(180°) = 2I_Z^X(90°)$ for two different scattering geometries must be satisfied.

2.4.7 Far-From-Resonance Approximation

We have referred several times to the so-called far-from-resonance (FFR) approximation. If the lowest excited electronic state in a molecule is much higher in energy than the incident or scattered photon energies, then the vibronic detail in the energy (frequency) denominators, including the imaginary damping terms, of Equation (2.105) can be dropped by writing

$$\omega_{ev,g0} - \omega_0 - i\Gamma_{ev} \approx \omega_{eg}^0 - \omega_0 \tag{2.127}$$

$$\omega_{ev,g0} + \omega_R + i\Gamma_{ev} = \omega_{ev,g1} + \omega_0 + i\Gamma_{ev} \approx \omega_{eg}^0 + \omega_0 \tag{2.128}$$

$$(\tilde{\alpha}_{\alpha\beta})_{g1,g0}^a = \frac{1}{\hbar}\sum_{ev}\left[\frac{\langle\phi_{g1}^a|\langle g|\hat{\mu}_\alpha|e\rangle|\phi_{ev}\rangle\langle\phi_{ev}|\langle e|\hat{\mu}_\beta|g\rangle|\phi_{g0}^a\rangle}{\omega_{eg}^0 - \omega_0}\right.$$
$$\left.+\frac{\langle\phi_{g1}^a|\langle g|\hat{\mu}_\beta|e\rangle|\phi_{ev}\rangle\langle\phi_{ev}|\langle e|\hat{\mu}_\alpha|g\rangle|\phi_{g0}^a\rangle}{\omega_{eg}^0 + \omega_0}\right] \tag{2.129}$$

Essentially this means that any vibronic detail in the frequency denominators, or any vibrational difference between the incident and scattered radiation frequencies, is not significant and can be dropped, leaving frequency denominators that depend only on pure electronic state frequency differences and then only in the incident laser radiation. This simplification permits summation to unity, similar to Eq. (2.84), over all the vibrational sublevels of the excited electronic states in Equation (2.129) leaving the following expression involving only an electronic summation e and ground electronic state vibrational wavefunctions.

$$\left(\tilde{\alpha}_{\alpha\beta}\right)^{a}_{g1,g0} = \left\langle \phi^{a}_{g1} \middle| \frac{1}{\hbar}\sum_{e}\left[\frac{\langle g|\hat{\mu}_{\alpha}|e\rangle\langle e|\hat{\mu}_{\beta}|g\rangle}{\omega^{0}_{eg}-\omega_{0}} + \frac{\langle g|\hat{\mu}_{\beta}|e\rangle\langle e|\hat{\mu}_{\alpha}|g\rangle}{\omega^{0}_{eg}+\omega_{0}}\right] \middle| \phi^{a}_{g0}\right\rangle \tag{2.130}$$

If we assume real wavefunctions and the Hermitian property of the matrix elements, this expression can be simplified further by first noting that with respect to interchange of Greek subscripts

$$\langle g|\hat{\mu}_{\beta}|e\rangle\langle e|\hat{\mu}_{\alpha}|g\rangle = \langle g|\hat{\mu}_{\alpha}|e\rangle^{*}\langle e|\hat{\mu}_{\beta}|g\rangle^{*} = \langle g|\hat{\mu}_{\alpha}|e\rangle\langle e|\hat{\mu}_{\beta}|g\rangle \tag{2.131}$$

This equality equates the numerators in Equation (2.130) and ensures the symmetry of $\left(\tilde{\alpha}_{\alpha\beta}\right)^{a}_{g1,g0}$ in the FFR approximation. Combining the two terms in Equation (2.130) over a common denominator, we can write:

$$\left(\alpha_{\alpha\beta}\right)^{a}_{g1,g0} = \left\langle \phi^{a}_{g1}|\alpha_{\alpha\beta}|\phi^{a}_{g0}\right\rangle = \left\langle \phi^{a}_{g1} \middle| \frac{2}{\hbar}\sum_{e\neq g}\frac{\omega^{0}_{eg}\langle g|\hat{\mu}_{\alpha}|e\rangle\langle e|\hat{\mu}_{\beta}|g\rangle}{\left(\omega^{0}_{eg}\right)^{2}-\omega_{0}^{2}} \middle| \phi^{a}_{g0}\right\rangle \tag{2.132}$$

The FFR approximation allows complete isolation of the expression for the polarizability of a molecule from its vibrational transition moment where we can write the expression:

$$\alpha_{\alpha\beta} = \frac{2}{\hbar}\sum_{e\neq g}\frac{\omega^{0}_{eg}}{\left(\omega^{0}_{eg}\right)^{2}-\omega_{0}^{2}}\langle g|\hat{\mu}_{\alpha}|e\rangle\langle e|\hat{\mu}_{\beta}|g\rangle \tag{2.133}$$

and where $\alpha_{\alpha\beta} = \alpha_{\beta\alpha}$ and therefore $\beta_{A}(\tilde{\alpha})^{2} = 0$. As a result, the three Raman scattering invariants reduce to two invariants in the FFR approximation and are given from Equations (2.109) and (2.110) as:

$$\alpha^{2} = \frac{1}{9}\alpha_{\alpha\alpha}\alpha_{\beta\beta} \tag{2.134}$$

$$\beta(\alpha)^{2} = \frac{1}{2}\left(3\alpha_{\alpha\beta}\alpha_{\alpha\beta} - \alpha_{\alpha\alpha}\alpha_{\beta\beta}\right) \tag{2.135}$$

where the superscripts denoting symmetric and anti-symmetric parts are no longer needed as the Raman polarizability is symmetric in the FFR approximation. The depolarization ratios in Equations (2.120)–(2.123) now become:

$$\rho_{l}(90^{\circ}) = \frac{I^{X}_{Z}(90^{\circ})}{I^{X}_{X}(90^{\circ})} = \frac{3\beta(\alpha)^{2}}{45\alpha^{2}+4\beta(\alpha)^{2}} \tag{2.136}$$

$$\rho_{n}(90^{\circ}) = \frac{I^{U}_{Z}(90^{\circ})}{I^{U}_{X}(90^{\circ})} = \frac{6\beta(\alpha)^{2}}{45\alpha^{2}+7\beta(\alpha)^{2}} \tag{2.137}$$

$$R(180^{\circ}) = \frac{I^{R}_{R}(180^{\circ})}{I^{R}_{L}(180^{\circ})} = \frac{I^{L}_{L}(180^{\circ})}{I^{L}_{R}(180^{\circ})} = \frac{6\beta(\alpha)^{2}}{45\alpha^{2}+\beta(\alpha)^{2}} \tag{2.138}$$

$${}^{R}C(180^{\circ}) = \frac{I^{R}_{R}(180^{\circ})-I^{R}_{L}(180^{\circ})}{I^{R}_{R}(180^{\circ})+I^{R}_{L}(180^{\circ})} = \frac{5\beta(\alpha)^{2}-45\alpha^{2}}{45\alpha^{2}+7\beta(\alpha)^{2}} \tag{2.139}$$

Finally, simple relationships now exist between all these dimensionless ratios. In particular, the depolarization ratios, the reversal ratio, and the degree of circularity are related by:

$$\rho_n(90^\circ) = \frac{2\rho_l(90^\circ)}{1+\rho_l(90^\circ)} \tag{2.140}$$

$$R(180^\circ) = \frac{2\rho_l(90^\circ)}{1-\rho_l(90^\circ)} \tag{2.141}$$

$$^{R}C(180^\circ) = \frac{3\rho_l(90^\circ)-1}{1+\rho_l(90^\circ)} \tag{2.142}$$

The non-vanishing of the Raman transition tensor in Equation (2.130) requires a first-order dependence of $\alpha_{\alpha\beta}$ on Q_a to be expressed as:

$$\left(\alpha_{\alpha\beta}\right)^{Q_a}_{g1,g0} = \frac{1}{\hbar}\sum_e \left[\frac{\langle g|\hat{\mu}_\alpha|e\rangle_0^{Q_a}\langle e|\hat{\mu}_\beta|g\rangle_0 + \langle e|\hat{\mu}_\alpha|g\rangle_0\langle g|\hat{\mu}_\beta|e\rangle_0^{Q_a}}{\omega_{eg}^0 - \omega_0} + \frac{\langle g|\hat{\mu}_\beta|e\rangle_0^{Q_a}\langle e|\hat{\mu}_\alpha|g\rangle_0 + \langle e|\hat{\mu}_\beta|g\rangle_0\langle g|\hat{\mu}_\alpha|e\rangle_0^{Q_a}}{\omega_{eg}^0 + \omega_0}\right]\left\langle \phi_{g1}^a|Q_\alpha|\phi_{g0}^a\right\rangle \tag{2.143}$$

where the notation of Equation (2.107) has been used to denote the first derivatives of the electronic matrix elements with respect to the coordinate of *a*th normal mode. The form of Equation (2.143) can be recognized as the equivalent to the simple classical expression for Raman intensity, as proportional to the derivative of the polarizability with respect to the normal mode of vibration, first presented in Equation (2.98), namely,

$$\left(\alpha_{\alpha\beta}\right)^{Q_a}_{g1,g0} = \left(\frac{\partial\alpha_{\alpha\beta}}{\partial Q_a}\right)_{Q=0}\left\langle \phi_{g1}^a\middle|Q_\alpha\middle|\phi_{g0}^a\right\rangle \tag{2.144}$$

2.4.8 Near Resonance Theory of Raman Scattering

Before we discuss resonance Raman scattering, we present a level of the theory of Raman scattering that we call the near resonance (NR) theory, which is not as severe as the FFR theory. (Nafie, 2008) and allows description of Raman scattering in the regime of pre-resonance between the FFR theory and various forms of resonance Raman theory to be described below. In the FFR theory, the incident and scattered photons are regarded as having the same (FFR) degree of resonance enhancement, all vibrational sublevel details are removed from the frequency denominators of the polarizability tensor terms. In the NR theory, the simplest level of vibrational sublevel detail is retained so that a difference develops between the resonance responses of the incident and Raman scattered radiation. As we shall see in Chapter 5, the NR level of theory is the simplest level of theory for which the full richness of the various forms of ROA can be seen, where there are differences between ICP- and SCP-ROA, and where DCP_{II}-ROA is non-vanishing. For ordinary Raman scattering the NR theory is very close in form to the FFR theory, but with an important distinction that breaks the symmetry of the Raman tensor and allows the anti-symmetric anisotropic invariant to be non-zero. We begin by a different modification of Equation (2.105) than was used for the FFR approximation. Here we retain the vibronic details in the denominator, but make the simplifying assumption that excited vibrational states are the same form and energy as the ground vibrational states. The imaginary damping terms are also dropped as these are only important close to resonance with excited states. With these simplifications we have:

$$\left(\tilde{\alpha}_{\alpha\beta}\right)^{a}_{g1,g0} = \frac{1}{\hbar}\sum_{ev}\left[\frac{\langle\phi^{a}_{g1}|\langle g|\hat{\mu}_{\alpha}|e\rangle|\phi_{gv}\rangle\langle\phi_{gv}|\langle e|\hat{\mu}_{\beta}|g\rangle|\phi^{a}_{g0}\rangle}{\omega_{ev,g0}-\omega_{0}} + \frac{\langle\phi^{a}_{g1}|\langle g|\hat{\mu}_{\beta}|e\rangle|\phi_{gv}\rangle\langle\phi_{gv}|\langle e|\hat{\mu}_{\alpha}|g\rangle|\phi^{a}_{g0}\rangle}{\omega_{ev,g0}+\omega_{R}}\right] \tag{2.145}$$

If we next expand the matrix elements to first order in the normal coordinate, Q_a, we can write the following expression which is a modification of Equation (2.108),

$$\left(\tilde{\alpha}_{\alpha\beta}\right)^{a}_{g1,g0} = \frac{1}{\hbar}\sum_{ev}\left\{\left(\frac{\langle g|\hat{\mu}_{\alpha}|e\rangle^{Q_a}_{0}\langle e|\hat{\mu}_{\beta}|g\rangle_{0}}{\omega_{ev,g0}-\omega_{0}} + \frac{\langle g|\hat{\mu}_{\beta}|e\rangle^{Q_a}_{0}\langle e|\hat{\mu}_{\alpha}|g\rangle_{0}}{\omega_{ev,g0}+\omega_{R}}\right)\langle\phi^{a}_{g1}|Q_a|\phi_{gv}\rangle\langle\phi_{gv}|\phi^{a}_{g0}\rangle + \left(\frac{\langle g|\hat{\mu}_{\alpha}|e\rangle_{0}\langle e|\hat{\mu}_{\beta}|g\rangle^{Q_a}_{0}}{\omega_{ev,g0}-\omega_{0}} + \frac{\langle g|\hat{\mu}_{\beta}|e\rangle_{0}\langle e|\hat{\mu}_{\alpha}|g\rangle^{Q_a}_{0}}{\omega_{ev,g0}+\omega_{R}}\right)\langle\phi^{a}_{g1}|\phi_{ev}\rangle\langle\phi_{ev}|Q_a|\phi^{a}_{g0}\rangle\right\} \tag{2.146}$$

In this equation, the group of the first two terms will have the following term in the frequency denominator,

$$\omega_{ev,g0} = \omega_{e0,g0} = \omega^{0}_{eg} \tag{2.147}$$

because the vibrational matrix element $\langle\phi_{gv}|\phi^{a}_{g0}\rangle$ requires that $ev = e0$. For the second group of terms, the vibrational matrix element $\langle\phi_{ev}|Q_a|\phi^{a}_{g0}\rangle$ requires that $ev = e1$ so that the following expression occurs in the frequency denominator of the third and fourth terms, respectively,

$$\omega_{ev,g0}-\omega_{0} = \omega_{e1,g0}-\omega_{0} = \omega_{e0,g0}-\omega_{R} = \omega^{0}_{eg}-\omega_{R} \tag{2.148a}$$

$$\omega_{ev,g0}+\omega_{R} = \omega_{e1,g0}+\omega_{R} = \omega_{e0,g0}+\omega_{0} = \omega^{0}_{eg}+\omega_{0} \tag{2.148b}$$

Using the normalization relationships that $\langle\phi^{a}_{g0}|\phi^{a}_{g0}\rangle$ and $\langle\phi^{a}_{g1}|\phi^{a}_{g1}\rangle$ equal unity, the following expression for the NR theory can be written as:

$$\left(\alpha_{\alpha\beta}\right)^{Q_a}_{g1,g0} = \frac{1}{\hbar}\sum_{e}\left[\frac{\langle g|\hat{\mu}_{\alpha}|e\rangle^{Q_a}_{0}\langle e|\hat{\mu}_{\beta}|g\rangle_{0}}{\omega^{0}_{eg}-\omega_{0}} + \frac{\langle e|\hat{\mu}_{\beta}|g\rangle_{0}\langle g|\hat{\mu}_{\alpha}|e\rangle^{Q_a}_{0}}{\omega^{0}_{eg}+\omega_{0}} + \frac{\langle e|\hat{\mu}_{\alpha}|g\rangle_{0}\langle g|\hat{\mu}_{\beta}|e\rangle^{Q_a}_{0}}{\omega^{0}_{eg}-\omega_{R}} + \frac{\langle g|\hat{\mu}_{\beta}|e\rangle^{Q_a}_{0}\langle e|\hat{\mu}_{\alpha}|g\rangle_{0}}{\omega^{0}_{eg}+\omega_{R}}\right]\langle\phi^{a}_{g1}|Q_{\alpha}|\phi^{a}_{g0}\rangle \tag{2.149}$$

This equation is analogous to Equation (2.143) except that the photon energy is ω_R instead of ω_0 for the matrix element derivatives involving the incident radiation with the operator μ_β. This expression can be further consolidated to be analogous to Equation (2.132) of the FFR approximation

$$\left(\alpha_{\alpha\beta}\right)^{a}_{g1,g0} = \frac{2}{\hbar}\sum_{e\neq g}\omega^{0}_{eg}\left[\frac{\langle g|\hat{\mu}_\alpha|e\rangle^{Q_a}_{0}\langle e|\hat{\mu}_\beta|g\rangle_0}{\left(\omega^{0}_{eg}\right)^2-\omega_0^2}+\frac{\langle g|\hat{\mu}_\alpha|e\rangle_0\langle e|\hat{\mu}_\beta|g\rangle^{Q_a}_{0}}{\left(\omega^{0}_{eg}\right)^2-\omega_R^2}\right]\left\langle \phi^{a}_{g1}\middle|Q_a\middle|\phi^{a}_{g0}\right\rangle \tag{2.150}$$

It can easily be seen that setting ω_R equal to ω_0 in Euations (2.149) and (2.150) reduces these NR expressions for the Raman polarizability tensor to the corresponding FFT expressions. The symmetry of the Raman tensor is clearly broken for interchange of operator subscripts α and β, and hence all three Raman invariants found at the GU level of Raman theory are restored at the NR level of theory.

2.4.9 Resonance Raman Scattering

We return again to the general vibronic expression for the Raman transition tensor in Equation (2.105) and consider the case where the photon frequency closely matches a particular electronic state e, such that $\omega_{ev,g0} \approx \omega_0$. In this limit of strong resonance, the non-resonant terms can be dropped and only the resonance term needs to be considered for the expression of the Raman transition tensor. Furthermore, for this resonance term only a summation over vibronic sublevels v of the resonant state e remains, namely,

$$\left(\tilde{\alpha}_{\alpha\beta}\right)^{a}_{g1,g0} = \frac{1}{\hbar}\sum_{v}\frac{\langle\phi^{a}_{g1}|\langle g|\hat{\mu}_\alpha|e\rangle|\phi_{ev}\rangle\langle\phi_{ev}|\langle e|\hat{\mu}_\beta|g\rangle|\phi^{a}_{g0}\rangle}{\omega_{ev,g0}-\omega_0-i\Gamma_{ev}} \tag{2.151}$$

If we consider the nuclear coordinate dependence of the electronic states, as was carried out to reach the vibronic coupling expression for the Raman polarizability in Equation (2.108), the transition tensor can be written as the summation of three well-known terms as:

$$\left(\tilde{\alpha}_{\alpha\beta}\right)^{Q_a}_{g1,g0} = \left(A_{\alpha\beta}\right)^{Q_a}_{g1,g0} + \left(B_{\alpha\beta}\right)^{Q_a}_{g1,g0} + \left(C_{\alpha\beta}\right)^{Q_a}_{g1,g0} \tag{2.152}$$

$$\left(A_{\alpha\beta}\right)^{Q_a}_{g1,g0} = \frac{1}{\hbar}\langle g|\hat{\mu}_\alpha|e\rangle_0\langle e|\hat{\mu}_\beta|g\rangle_0\sum_{v}\frac{\langle\phi^{a}_{g1}|\phi_{ev}\rangle\langle\phi_{ev}|\phi^{a}_{g0}\rangle}{\omega_{ev,g0}-\omega_0-i\Gamma_{ev}} \tag{2.153}$$

$$\begin{aligned}\left(B_{\alpha\beta}\right)^{Q_a}_{g1,g0} = \frac{1}{\hbar}\Bigg[&\langle g|\hat{\mu}_\alpha|(e)^{Q_a}\rangle_0\langle e|\hat{\mu}_\beta|g\rangle_0\sum_{v}\frac{\langle\phi^{a}_{g1}|Q_a|\phi_{ev}\rangle\langle\phi_{ev}|\phi^{a}_{g0}\rangle}{\omega_{ev,g0}-\omega_0-i\Gamma_{ev}}\\ &+\langle g|\hat{\mu}_\alpha|e\rangle_0\langle(e)^{Q_a}|\hat{\mu}_\alpha|g\rangle_0\sum_{v}\frac{\langle\phi^{a}_{g1}|\phi_{ev}\rangle\langle\phi_{ev}|Q_a|\phi^{a}_{g0}\rangle}{\omega_{ev,g0}-\omega_0-i\Gamma_{ev}}\Bigg]\end{aligned} \tag{2.154}$$

$$\begin{aligned}\left(C_{\alpha\beta}\right)^{Q_a}_{g1,g0} = \frac{1}{\hbar}\Bigg[&\left\langle(g)^{Q_a}\middle|\hat{\mu}_\alpha\middle|e\right\rangle_0\left\langle e\middle|\hat{\mu}_\beta\middle|g\right\rangle_0\sum_{v}\frac{\langle\phi^{a}_{g1}|Q_a|\phi_{ev}\rangle\langle\phi_{ev}|\phi^{a}_{g0}\rangle}{\omega_{ev,g0}-\omega_0-i\Gamma_{ev}}\\ &+\langle g|\hat{\mu}_\alpha|e\rangle_0\langle e|\hat{\mu}_\alpha|(g)^{Q_a}\rangle_0\sum_{v}\frac{\langle\phi^{a}_{g1}|\phi_{ev}\rangle\langle\phi_{ev}|Q_a|\phi^{a}_{g0}\rangle}{\omega_{ev,g0}-\omega_0-i\Gamma_{ev}}\Bigg]\end{aligned} \tag{2.155}$$

where we have used the following notation for the first derivative of the electronic wavefunctions:

$$\langle g|\hat{\mu}_\alpha|e\rangle_0^{Q_a} = \left\langle \left(\frac{\partial \psi_g}{\partial Q_a}\right)_0 |\hat{\mu}_\alpha|\psi_{e,0}\right\rangle + \left\langle \psi_{g,0}|\hat{\mu}_\alpha|\left(\frac{\partial \psi_e}{\partial Q_a}\right)_0\right\rangle = \left\langle (g)^{Q_a}|\hat{\mu}_\alpha|e\right\rangle_0 + \left\langle g|\hat{\mu}_\alpha|(e)^{Q_a}\right\rangle_0 \tag{2.156}$$

The A-term is the Franck–Condon term, the B-term represents vibronic coupling of the resonant excited states with other electronic states, whereas the C-term represents vibronic coupling of the ground electronic state with other electronic states. To gain insight into the relative importance of the B- and C-terms, we write the nuclear coordinate dependence of the electronic wavefunctions as a quantum mechanical perturbation expression known as the Herzberg–Teller (HT) expansion,

$$\psi_e(Q_a) = \psi_e^0 + \left(\frac{\partial \psi_e}{\partial Q_a}\right)_0 Q_a + \ldots = \psi_e^0 + \sum_{s\neq e} \frac{\langle \psi_s^0|(\partial H_E/\partial Q_a)_0|\psi_e^0\rangle}{E_e^0 - E_s^0}\psi_s^0 Q_a + \ldots \tag{2.157}$$

where in more compact notation the HT expansion coefficient is written as:

$$\frac{\langle \psi_s^0|(\partial H_E/\partial Q_a)_0|\psi_e^0\rangle}{E_e^0 - E_s^0} = \frac{h_{se,0}^a}{\hbar\omega_{es}^0} \tag{2.158}$$

where the energy dependence of coupling to other states, s, is clearly seen to favor nearby states, s, rather that states that are energetically far removed from the resonant excited state. If vibronic coupling of the ground electronic state is considered, it can also be seen that a large energy denominator will be present for all molecules except those with low-lying excited electronic states. As a result, the C-term is usually ignored in the theory of resonance Raman scattering, and vibronic coupling need only be considered for the B-term. Using the notation of Equation (2.158) the expression for the B-term becomes:

$$(B_{\alpha\beta})_{g1,g0}^{Q_a} = \frac{1}{\hbar^2}\left[\sum_{s\neq e} \langle g|\hat{\mu}_\alpha|s\rangle_0 \frac{h_{se,0}^a}{\omega_{es}^0}\langle e|\hat{\mu}_\beta|g\rangle_0 \sum_v \frac{\langle \phi_{g1}^a|Q_a|\phi_{ev}\rangle\langle \phi_{ev}|\phi_{g0}^a\rangle}{\omega_{ev,g0} - \omega_0 - i\Gamma_{ev}}\right.$$
$$\left. + \langle g|\hat{\mu}_\alpha|e\rangle_0 \frac{h_{es,0}^a}{\omega_{es}^0}\langle s|\hat{\mu}_\beta|g\rangle_0 \sum_v \frac{\langle \phi_{g1}^a|\phi_{ev}\rangle\langle \phi_{ev}|Q_a|\phi_{g0}^a\rangle}{\omega_{ev,g0} - \omega_0 - i\Gamma_{ev}}\right] \tag{2.159}$$

In many cases only one or two nearby electronic states, s, need to be considered to understand the origin and intensity of resonance Raman bands.

2.4.10 Single Electronic State Resonance Approximation

As a final topic in this chapter, we consider the simplest of all expressions for Raman scattering intensity, the case of strong resonance with a single electronic state (SES) where no nearby excited electronic state need be considered for the vibrational mode in question (Nafie, 1996). The Raman transition tensor in this case is written simply as:

$$(\tilde{\alpha}_{zz})_{g1,g0}^{Q_a} = \frac{1}{\hbar}\langle g|\hat{\mu}_z|e\rangle_0 \langle e|\hat{\mu}_z|g\rangle_0 \sum_v \frac{\langle \phi_{g1}^a|\phi_{ev}\rangle\langle \phi_{ev}|\phi_{g0}^a\rangle}{\omega_{ev,g0} - \omega_0 - i\Gamma_{ev}} \tag{2.160}$$

where only the A-term is responsible for the observed intensity, and the contributions of different excited-state vibrational sublevels is governed by products of Franck–Condon vibrational overlap integrals between the ground and resonant excited electronic states. The molecular reference frame can be chosen such that electronic transition moment of the resonant state lies along the molecular z-axis, leaving only one non-zero Raman tensor element $(\tilde{\alpha}_{zz})^{Q_a}_{g1,g0}$. In this case, the isotropic invariant and the symmetric anisotropic invariant are the same except for a factor of nine in the anisotropic invariant. We can now write:

$$\alpha^2 = \frac{1}{9}\left|(\tilde{\alpha}_{zz})^{Q_a}_{g1,g0}\right|^2 = \frac{1}{9\hbar^2}\left|\langle g|\hat{\mu}_z|e\rangle_0\langle e|\hat{\mu}_z|g\rangle_0\sum_v \frac{\langle\phi^a_{g1}|\phi_{ev}\rangle\langle\phi_{ev}|\phi^a_{g0}\rangle}{\omega_{ev,g0}-\omega_0-i\Gamma_{ev}}\right|^2$$

$$= \frac{1}{9\hbar^2}\left|\langle e|\hat{\mu}_z|g\rangle_0\right|^4 U^a_{g1,g0}(\omega_0) = \frac{1}{9\hbar^2}\left(D^0_{eg}\right)^2 U^a_{g1,g0}(\omega_0) \tag{2.161}$$

$$\beta_S(\tilde{\alpha})^2 = \left|(\tilde{\alpha}_{zz})^{Q_a}_{g1,g0}\right|^2 = \frac{1}{\hbar^2}\left(D^0_{eg}\right)^2 U^a_{g1,g0}(\omega_0) \tag{2.162}$$

where $U^a_{g1,g0}(\omega_0)$ is a lineshape factor sensitive to the incident laser frequency for the particular normal mode a. It can be seen that the lone SES Raman invariant is proportional to the square of the dipole strength D^0_{eg} of the resonant electronic state. The relative values of the invariants in Equations (2.161) and (2.162) yield a value of 1/3 for the depolarization ratio $\rho_l(90^\circ)$ in Equation (2.120). This depolarization ratio is a signature of resonance Raman scattering in the SES limit.

References

Freedman, T.B., Shih, M.-L., Lee, E., and Nafie, L.A. (1997) Electron transition current density in molecules. 3. *Ab initio* calculations of vibrational transitions in ethylene and formaldehyde. *J. Am. Chem. Soc.*, **119**, 10620–10626.

Freedman, T.B., Gao, X., Shih, M.-L., and Nafie, L.A. (1998) Electron transition current density in molecules. 2. *Ab initio* calculations for electronic transitions in ethylene and formaldehyde. *J. Phys. Chem., A* **102**, 3352–3357.

Nafie, L.A. (1996) Theory of resonance Raman optical activity: The single-electronic state limit. *Chem. Phys.*, **205**, 309–322.

Nafie, L.A. (1997) Electron transition current density in molecules. 1. Non-Born–Oppenheimer theory of vibronic and vibrational transition. *J. Phys. Chem. A*, **101**, 7826–7833.

Nafie, L.A. (2008) Theory of Raman scattering and Raman optical activity. Near resonance theory and levels of approximation. *Theor. Chem. Acc.*, **119**, 39–55.

3

Molecular Chirality and Optical Activity

The field of vibrational optical activity rests on the foundations of two major pillars of molecular science, vibrational spectroscopy and optical activity. Having just completed an introduction to the basic principles of vibrational spectroscopy, we next turn our attention to the field of optical activity and its physical basis in molecular chirality. Optical activity refers to all phenomena associated with the differential interaction of left versus right circularly polarized radiation for a sample possessing chirality, or handedness, at the molecular level. Optical activity is a measurable molecular property resulting from the interaction of radiation and matter, whereas molecular chirality is a more fundamental property, with a basis in pure geometry, which is associated with the molecule independent of its interaction with radiation.

Many excellent descriptions of molecular chirality and optical activity can be found in existing books and publications (Lowry, 1935; Salvadori and Ciardelli, 1973; Mason, 1973; Barron, 2004). We will not try to cover the whole of that material here. Our goal instead is to provide a basic conceptual foundation of optical activity and at the same time introduce any formalism needed for the comprehensive descriptions of VCD and ROA (Nafie and Freedman, 2000) to be developed in the subsequent chapters.

3.1 Definition of Molecular Chirality

Molecular chirality can be defined on several levels. The most common is the lay definition that a molecule is chiral if its mirror image cannot be superimposed on itself, or in other words, that the molecule has a handedness. Molecular chirality can also be defined in terms of group theory by means of the point group properties of individual molecules. It can also be defined in terms of the space groups of molecular solids in which a solid assembly of either chiral or achiral molecules can be a chiral sample. Molecular chirality can be further defined in liquid crystals, disordered solids, polymers, and other types of molecular assemblies. Finally, there is the definition of molecular chirality at the level of elementary particle physics, which requires the true enantiomer of a chiral molecule to be not only its

Vibrational Optical Activity: Principles and Applications, First Edition. Laurence A. Nafie.

spatial mirror image, but also it must consist of the anti-particles of all the original particles of the molecule, namely anti-protons for protons, anti-neutrons for neutrons, and anti-electrons (positrons) for electrons. Without the anti-particle substitutions there are very small energy differences between a molecule and its spatial mirror image arising from the presence of the weak force between the particles (Barron, 1994). The weak force is the only known fundamental force in the universe that is sensitive to spatial chirality.

3.1.1 Historical Origins

The origins of molecular chirality emerged gradually over the much of the nineteenth century based on observations of optical activity embodied as the rotation of a plane of polarized light by an optically active medium (Barron, 2004; Mauskopf, 2006) This was first published 1811 by Arago who reported observing the rotation of the plane of linearly polarized light passing through quartz crystals (Arago, 1811). Fresnel in 1824 demonstrated that a plane of polarized light is rotated by circular birefringence, the difference in the index of refraction of left and right circularly polarized light in a medium, $n_L - n_R$, and proposed that the molecules comprising an optically active medium possess a right-handed or a left-handed helical form (Fresnel, 1824). Pasteur in 1848 was the first to connect optical activity in mirror-image pairs of crystals with the optical activity of their corresponding solutions (Pasteur, 1848). He did this by noting that for mirror-image crystals of tartaric acid, isolated by hand from a racemic mixture of such crystals, a plane of polarization of light passing through their corresponding solutions is rotated in opposite directions. These observations marked the birth of our understanding of optical activity in nature and its association with mirror symmetry at the molecular level. Although tartaric acid crystals are comprised of chiral molecules, Pasteur and others only speculated on the helical nature of the constituent molecules, and did not suggest any specific geometric connection between the structure of individual molecules and their associated chiral crystal morphology. The significant first step from crystal morphology to the structure of individual molecules as the source of optical rotation came with the introduction of the three-dimensional structure of the asymmetric carbon atom located at the center of an irregular tetrahedron (Le Bel, 1874; van't Hoff, 1874). This three-dimensional molecular structure clearly came in two distinct forms that were interchanged by mirror symmetry.

The term chirality, which comes from the Greek word for hand, *kheir*, was first brought into general use by Lord Kelvin, when in 1904 he stated that 'I call any geometrical figure, or any group of points, chiral, and say that it has chirality, if its image in a plane mirror, ideally realized, cannot be brought to coincide with itself' (Kelvin, 1904).

3.1.2 Molecular Symmetry Definition of Chirality

The symmetry properties of molecules consist of axes of rotation of angle, $2\pi/n$, given the symbol C_n, symmetry planes, σ, inversion centers, i, and improper axes of rotations, S_n, also called reflection–rotation axes. An improper axis S_n is defined as a rotation through $2\pi/n$ followed by a reflection through a plane perpendicular to the axis of rotation. The special case of an improper rotation axis S_1 is the same as any plane of symmetry, σ, and an improper axis S_2 is the same as an inversion center, i. Hence, there are actually only three classes of symmetry elements needed to describe the symmetry properties of a molecule, namely rotation axes, reflection planes, and improper rotation axes, which are a particular combination of rotation axes and reflection planes. It is sufficient to say a molecule is chiral if, and only if, it lacks an improper rotation axis, S_n, of any type. Chiral molecules that lack all symmetry elements, and hence are without any symmetry, are termed *asymmetric*. On the other hand, it is possible for a molecule to possess only a rotation axis without possessing an inversion center or a plane of symmetry. Such chiral molecules with some symmetry elements are termed *dissymmetric*.

Point groups are the classifications of symmetry elements of any object that does not include the translation of the object, namely at least a single point remains fixed during all symmetry operations. Any molecule can be assigned to a particular point group. If the molecule possesses no symmetry elements, its point group is C_1. Molecules in this point group are asymmetric and are chiral. The point groups C_n for $n > 1$ have only rotation axes of angle $2\pi/n$. Molecules in these point groups are dissymmetric and chiral. The point groups D_n contain molecules with rotation axes of angle $2\pi/n$ and n C_2 axes perpendicular to the C_n axis, but no plane of symmetry. Molecules in these point groups are also dissymmetric and chiral. In addition, molecules that belong to the symmetry classifications of tetrahedral, octahedral and icosahedral, *but possess no planes of symmetry*, belong to the point groups *T*, *O*, and *I*, and are also dissymmetric and chiral. Stable chiral molecules that belong to these rare, highly symmetric point groups have only recently been identified.

3.1.3 Absolute Configuration of Chiral Molecules

In the previous section, we presented the group theoretical definition of chirality. Here we ask the questions: what structural elements are present in molecules that allow them to be chiral and how does one specify the absolute sense of a chiral structural element, and hence the absolute configuration of a chiral molecule (Mislow, 1966)? The structural elements that comprise the sources of molecular chirality are the chiral center, the helix, the chiral axis, and the chiral plane.

3.1.3.1 Chiral Center

The classical structural element is the chiral center. Typically, the chiral center is a tetrahedral carbon atom possessing sp^3 atomic orbital symmetry with four unequal groups attached, but other elements such as nitrogen, where the lone pair is considered a group, can also form a chiral center. For example, the molecule 2-bromo-2-hydroxyethane is a chiral molecule with a single chiral center. The chirality of

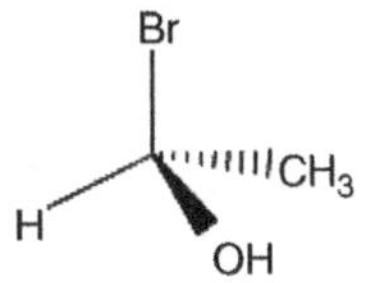

chiral centers can be specified by the letters *R* for rectus (right) or *S* for sinister (left) based on the Cahn–Ingold–Prelog system of absolute configuration assignment (Cahn, 1964; Cahn *et al.*, 1966). In some cases, such as sugars, amino acids, peptides or carbohydrates, the symbols L or D, based on an older, less universal, and somewhat inadequate nomenclature based on the method of Fischer projections, are used for the same purpose. The nomenclature based on *R* or *S* is more general and can be applied to any chiral center, including, for example, sugars and amino acids. The chirality, or absolute configuration, of any chiral center can be specified by first placing the group with lowest priority away from the viewer. The priority order of the groups can be assigned based on a hierarchy of atomic masses, for which well-defined rules have been established. For chiral centers with hydrogen, ^1H, as one of its attached groups, and no lone pairs, the hydrogen is the group with lowest mass priority. Lone pairs have a lower priority than hydrogen, and deuterium, ^2H, has a higher priority than hydrogen. The remaining three groups of the chiral center are now pointing toward the viewer and can then be arranged from highest to lowest priority in either a clockwise or a counterclockwise rotational pattern. If clockwise, the designation is *R*, and if counterclockwise, the designation is *S*. For the example, as illustrated above, the absolute configuration is specified as *S*-2-bromo-2-hydroxyethane.

3.1.3.2 Helix

A helix is one of the simplest embodiments of a chiral structure. One molecule that possesses a simple helix shape is hexahelicene, illustrated below. The sense of chirality here is that of positive or *P*-helicity, or a right-handed helix.

As one traces out an arc from the closest to the furthest part of this structure, the sense of motion is in the clockwise direction. The mirror-image structure embodies the shape of a left-handed helix and carries the designation *M*-helicity for a minus helix.

3.1.3.3 Chiral Axis

A molecule with a chiral axis can be thought of as a structure that starts from a regular tetrahedron with four groups, and two of the groups are stretched away from the other two where the center of chirality now becomes a line of chirality. For this structure the four groups no longer need be different from one another, as in the allene and biphenyl chiral structures shown below.

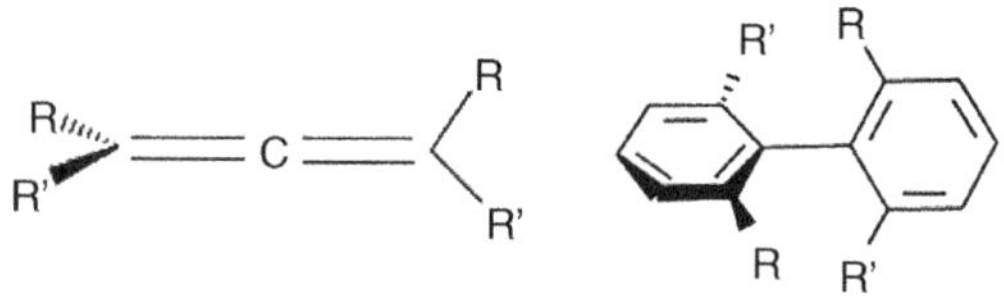

All that is required is that the two groups at each end of the chiral axis not be the same. Molecules with chiral axes possess rotational symmetry and can be assigned to point groups with C_n or D_n symmetry.

The designation of molecules with chiral axes can follow either the *R* or *S* nomenclature or the helicity nomenclature of *P* or *M*. For the *R*–*S* system, one sets relative priority of the two groups at each end of the chiral axis. Next one sets the priority of all groups nearest the viewer when looking down the chiral axis above those at the far end of the axis. Finally, one disregards the lowest priority (fourth level of priority) group and traces an arc from near to far from the highest to the third highest priority. If the arc is clockwise, the structure is designated *R*, while counterclockwise is *S*. For the structures shown above, one can arbitrarily chose one group (R or R′) as having the higher priority, and one finds the allene structure has an *S* chiral axis, and hence *S* absolute configuration, while the biphenyl structure has an *R* chiral axis and *R* absolute configuration.

Alternatively, the absolute configuration of molecules with chiral axes can be assigned using the sense of a helix or propeller formed by the molecule, connecting like groups at each end of the chiral axis with a line that forms a surface akin to the blade of a propeller. In the cases above, the propellers are each twofold. Looking down the chiral axis and tracing the connecting line between the like groups from front to back leads to either a clockwise or counterclockwise arc. The clockwise arc is designated *P* and the counterclockwise arc is designated *M*. This system is most often used for molecules with threefold or higher symmetry. For the molecules shown above, the allene has a *P* chiral axis and the biphenyl has an *M* chiral axis. The *P–M* system of assignment is used more often than the *R–S* system since chiral axis are more easily identified with helices than with tetrahedra.

3.1.3.4 Chiral Plane

Finally, we consider the chiral plane as a source of structural chirality. A chiral plane is a structural plane in a molecule with a group substituted in the plane that destroys a symmetry plane perpendicular to the structural plane. The cyclophane molecule shown below is an example of such a molecule. The carboxyl group destroys the plane bisecting the six-membered ring. The absolute configuration can be assigned based on the *R–S* system. One first chooses a reporter atom above or below the plane and then defines a curved path, using mass priority if choices are present along the path, from the reporter atom to the substituent group that creates the chirality. In the case below, if the methylene chain lies above the plane of the molecule, then the set of atoms C(reporter)–O–C–C–C(carboxylate) forms a clockwise arc when viewed from above, and the chiral plane has *R* absolute configuration.

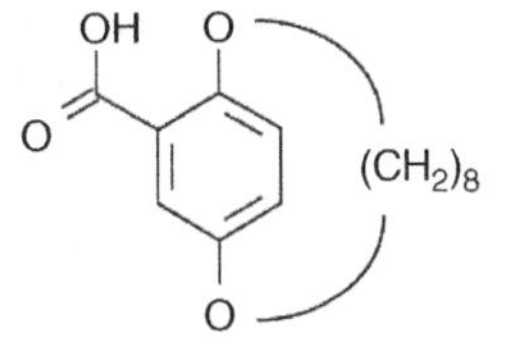

3.1.4 True and False Chirality

While the definitions of molecular and structural chirality given above are conceptually straightforward, there has been confusion over the years in the literature concerning what constitutes a chiral object or a chiral medium. The confusion originates with the role of magnetic fields and time with respect to the definition of chirality. For example, an object possessing infinite rotational symmetry and directionality, such as a cone, might be considered chiral if it is spinning one way or the other about its rotational axis. However, as discussed in detail by Barron, such an object displays what is called false chirality (Barron, 2004). A true chiral object, such as a chiral molecule, is converted into its mirror image by the parity operator, P (x to $-x$, y to $-y$, and z to $-z$), and is called parity odd, but is unchanged by the time reversal operation T (t to $-t$) and hence is even with respect to time reversal. Such objects support observables that are time-even pseudo-scalars. On the other hand, a spinning cone is even under the parity operation and is odd under time reversal as the sense of spinning is reversed. Such an object can support time-odd scalars, but not time-even pseudo-scalars. These concepts will be discussed further when magnetic optical activity is considered later in this chapter.

3.1.5 Enantiomers, Diastereomers, and Racemic Mixtures

The word *enantiomer* (from the Greek word *enantio* meaning opposite) refers to one member of a pair of chiral molecules that are mirror images of one another, but otherwise equal. Two molecules are therefore enantiomers of one another if they differ in structure only by mirror symmetry and have opposite senses of chirality.

The term diastereomer applies to a molecule with more than one structural chirality element. Each unique combination of absolute configurations of these structural elements is called a diastereomer. For example, if a molecule has two chiral centers, then the molecule labeled *R,S* is a diastereomer of the molecule labeled *R,R*. Diastereomers are distinctly different molecules and have different physical properties. On the other hand, the molecules *R,R* and *S,S*, and the molecules *R,S* and *S,R*, are pairs of enantiomers of one another and are distinguished only by their structural mirror symmetry or any physical property that can distinguish the structure of a chiral molecule from its enantiomer.

A racemic mixture refers not to a single molecule, but rather to a collection of molecules that consists of equal numbers of both enantiomers of a chiral molecule. Any appreciable excess of one enantiomer over the other is called the enantiomeric excess, *ee*, or sometimes the per cent enantiomeric excess or % *ee*. The % *ee* is defined as the difference in the amount (moles, concentration, or number of molecules) of each enantiomer divided by the sum of the amounts multiplied by 100. A pure sample consisting of only one enantiomer has 100% *ee*, whereas a racemic mixture has 0% *ee*.

Actually, a *perfect* racemic mixture cannot in general be realized, even for a macroscopic racemic sample with measured 0% *ee*. A typical Gaussian error-function spread about a racemic mixture containing on the order of 10^{20} molecules is the square root of the number, or 10^{10}. Thus for a sample that has no measurable % *ee*, even to eight decimal places of accuracy, the actual excess of one enantiomer over the other is typically of the order of 10 billion molecules.

Another interesting fact, as mentioned above, is that a chiral molecule and its enantiomer do not have exactly the same energy. The difference is due to the effect of the weak force in the molecule, otherwise known as parity violation, which causes an energy difference of the order of one part in 10^{17} of the total energy of the molecule (Quack and Stohner, 2005). This difference in energy for enantiomers is removed if one considers a chiral molecule and its mirror image (parity operator) where all the particles are replaced by their corresponding anti-particles (charge conjugation operator). Thus, a true pair of equal-energy enantiomers consists of two molecules that are both mirror images and anti-particle interchanges of one another.

3.2 Fundamental Principles of Natural Optical Activity

Natural optical activity, as distinct from magnetic optical activity, can be defined as the differential interaction of a chiral molecule with left versus right circularly polarized radiation. This is a broad definition covering all types of radiation and all levels of interactions between radiation and matter. We begin by developing a precise definition of the polarization states of electromagnetic radiation.

3.2.1 Polarization States of Radiation

Electromagnetic radiation is a simultaneous transverse oscillating wave of electric and magnetic fields that are in phase and orthogonal to each other. The classical description of radiation emerges from Maxwell's equations. As the interaction of electromagnetic radiation with molecules is usually stronger for the electric field, rather than the magnetic field, it suffices in most cases to describe radiation in terms of the spatial and temporal dependence of an oscillating electric field:

$$\tilde{\boldsymbol{E}}(\boldsymbol{r},t) = E_0\tilde{\boldsymbol{e}}\exp[i(\tilde{\boldsymbol{k}}\cdot\boldsymbol{r} - \omega t)] \tag{3.1}$$

where $\tilde{\boldsymbol{E}}(\boldsymbol{r},t)$ is a complex vector representing the electric field, E_0 is a scalar representing the maximum magnitude of the field, $\tilde{\boldsymbol{e}}$ is the complex polarization vector of unit magnitude representing the polarization state of the radiation that is perpendicular to the direction of propagation of the transverse wave. The argument of the exponential specifies the phase of the wave for any point $\boldsymbol{r}$ in space or any moment t in time. The wave vector $\tilde{\boldsymbol{k}}$ and the angular frequency ω are given by the relationships

$$\tilde{\boldsymbol{k}} = 2\pi\tilde{\boldsymbol{n}}/\lambda \qquad \omega = 2\pi\nu \tag{3.2}$$

The vector $\tilde{\boldsymbol{n}}$ is a vector in the direction of the propagation of the radiation with a magnitude equal to the complex refractive index of the medium. In a medium without losses, the refractive index vector $\boldsymbol{n}$ is real, and its magnitude, n, is equal to the index of refraction. In a vacuum, n is equal to unity. The magnitude of $\boldsymbol{k}$ is 2π times the spatial frequency of the light, where λ/n is the wavelength of the light. Similarly, ω is equal to 2π times the frequency of the radiation, ν. The speed of propagation of the radiation in a medium of refractive index n is c/n equal to $\lambda\nu/n$.

The spatial and temporal contributions to the phase of the wave in Equation (3.1) are opposite in sign, because traveling along points in space in the positive Z-direction from Z_1 to Z_2, for example, corresponds to phase points further in the past (negative time direction), as the wave at Z_2 passed through Z_1 earlier in time by the amount $(Z_2 - Z_1)/c$.

The polarization vector $\tilde{e}$ carries all the information needed to describe the polarization state of a beam of radiation. To simplify the discussion, we assume that the radiation is traveling in the laboratory Z-direction, which then confines the states of polarization to the XY-plane as the wave is transverse. If the vertical direction is taken to be the X-axis, then vertically polarized radiation will have a polarization vector equal to $\tilde{e} = \hat{e}_X$ and horizontally polarized radiation will have a polarization vector given by $\tilde{e} = \hat{e}_Y$, where $\hat{e}_X$ and $\hat{e}_Y$ are unit vectors in the X- and Y- directions, respectively. For radiation propagation under these coordinate specifications, Equation (3.1) becomes

$$\tilde{\boldsymbol{E}}(\boldsymbol{r},t) = (\tilde{\boldsymbol{E}}_X + \tilde{\boldsymbol{E}}_Y)\exp[i(kz - \omega t)] = E_0(\tilde{e}_X + \tilde{e}_Y)\exp[i(kz - \omega t)] \tag{3.3}$$

Circularly polarized radiation can be described using the complex representation of the polarization vectors. By convention among chemists, the polarization vector associated with right circular polarization (RCP) traces out a right-handed helix in space. When the radiation is viewed propagating toward an observer, the polarization vector of RCP radiation rotates clockwise as the wave (helix) passes through any XY-plane at a particular location along the Z-axis. We can write the following expressions for the complex polarization vectors representing left circular polarization (LCP) and RCP radiation

$$\tilde{e}_L = \frac{1}{\sqrt{2}}(\widehat{e}_X + \mathrm{i}\widehat{e}_Y) \qquad \tilde{e}_R = \frac{1}{\sqrt{2}}(\widehat{e}_X - \mathrm{i}\widehat{e}_Y) \tag{3.4}$$

The validity of these expressions relative to the verbal definition of LCP and RCP radiation is demonstrated by the following relationships for RCP radiation:

$$\begin{aligned}\tilde{\boldsymbol{E}}_R(\boldsymbol{r},t) &= \frac{E_0}{\sqrt{2}}(\widehat{e}_X - \mathrm{i}\widehat{e}_Y)\exp[i(kz - \omega t)] \\ &= \frac{E_0}{\sqrt{2}}\{\widehat{e}_X \exp[i(kz - \omega t)] - \widehat{e}_Y \exp[i(kz - \omega t + \pi/2)]\}\end{aligned} \tag{3.5}$$

where we have used the relationship that $i = \exp(i\pi/2)$ to create a phase shift of $\pi/2$ in the Y-component of the wave relative to the X-component. If we focus on the real part of the electric field vector we can write more simply

$$\begin{aligned}\mathrm{Re}\left[\tilde{\boldsymbol{E}}_R(\boldsymbol{r},t)\right] &= \frac{E_0}{\sqrt{2}}[\widehat{e}_X \cos(kz - \omega t) - \widehat{e}_Y \cos(kz - \omega t + \pi/2)] \\ &= \frac{E_0}{\sqrt{2}}[\widehat{e}_X \cos(kz - \omega t) + \widehat{e}_Y \sin(kz - \omega t)]\end{aligned} \tag{3.6}$$

Equation (3.6) describes an electric field wave polarized in the *X*-direction accompanied by a second wave polarized in the *Y*-direction and retarded in phase by $\pi/2$, or one quarter of a wavelength. The sum of these two waves is a right-circularly polarized light wave. The spatial dependence at $t=0$ traces out a right-handed helix in space that is pure *X*-polarized at $z=0$, at then pure *Y*-polarized at $z=\lambda/4$, and so forth. Alternatively, we can observe the polarization in the plane defined by $z=0$ from some point along the positive Z-axis, and watch the polarization vector rotate clockwise as time advances and the wave as a whole translates along the positive *Z*-axis.

The polarization state of any light beam can be characterized by a two-dimensional Jones vector defined as:

$$\tilde{s}_J = E_0 \begin{pmatrix} \tilde{e}_X \\ \tilde{e}_Y \end{pmatrix} = \begin{pmatrix} \tilde{E}_X \\ \tilde{E}_Y \end{pmatrix} \tag{3.7}$$

where $\tilde{e}_X$ and $\tilde{e}_Y$ are the complex numbers reflecting the amplitudes and phases of the polarization states of the light in the *X*- and *Y*-directions, respectively. The transformation of Jones vectors between different polarization states as a result of passing through various optical elements or material media can be described by two-dimensional complex matrices. The focus of the Jones calculus is on the two spatial degrees of freedom of the polarization states of light as it propagates in space. An alternative, more flexible and somewhat simpler description of the polarization states of light beams is the Stoke–Mueller method, which we describe in the next section.

3.2.2 Mueller Matrices and Stokes Vectors

The Stokes–Mueller formalism is a four-dimensional system of Stokes vectors and Mueller matrices where all the elements refer to radiation intensities, not complex field vectors as in the Jones formalism, and are therefore real. A Stokes vector is defined as:

$$\boldsymbol{S} = \begin{pmatrix} S_0 \\ S_1 \\ S_2 \\ S_3 \end{pmatrix} \tag{3.8}$$

where S_0 represents the total intensity, or absolute square of the electric field vector, of the light beam according the relationships

$$S_0 = I_0 = \tilde{\boldsymbol{E}}^* \cdot \tilde{\boldsymbol{E}} = \tilde{E}_X^* \tilde{E}_X + \tilde{E}_Y^* \tilde{E}_Y = I_X + I_Y \tag{3.9}$$

The Stokes vectors, S_1, S_2, and S_3, are the intensities of *X*-polarized (0°) minus *Y*-polarized (90°) radiation, (45°)-degree polarized minus (− 45°)-degree polarized radiation (clockwise positive rotation when viewed toward the source of radiation), and right minus left circularly polarized radiation, respectively (Barron, 2004).

$$S_1 = \tilde{E}_X^* \tilde{E}_X - \tilde{E}_Y^* \tilde{E}_Y = I_X - I_Y \tag{3.10}$$

$$S_2 = -(\tilde{E}_X^* \tilde{E}_Y + \tilde{E}_Y^* \tilde{E}_X) = I_{45^\circ} - I_{-45^\circ} \tag{3.11}$$

$$S_3 = -i(\tilde{E}_X^* \tilde{E}_Y - \tilde{E}_Y^* \tilde{E}_X) = I_R - I_L \tag{3.12}$$

Transformations between different Stokes vectors, such as $\boldsymbol{S}_A$ to $\boldsymbol{S}_B$, as a result of radiation passing through, or interacting with, an optical element or sample is represented by a Mueller matrix $\boldsymbol{M}$ given by:

$$\boldsymbol{S}_B = \begin{pmatrix} S_{B,0} \\ S_{B,1} \\ S_{B,2} \\ S_{B,3} \end{pmatrix} = \boldsymbol{M}_B \cdot \boldsymbol{S}_A = \begin{pmatrix} M_{00} & M_{01} & M_{02} & M_{03} \\ M_{10} & M_{11} & M_{12} & M_{13} \\ M_{20} & M_{21} & M_{22} & M_{23} \\ M_{30} & M_{31} & M_{32} & M_{33} \end{pmatrix} \begin{pmatrix} S_{A,0} \\ S_{A,1} \\ S_{A,2} \\ S_{A,3} \end{pmatrix} \tag{3.13}$$

We shall return to Mueller matrices and Stokes vectors in Chapters 6, 7 and 8 when describing optical setups used for VCD and ROA instrumentation in which polarized radiation passes through various optical elements on its way from the source to detector.

3.2.3 Definition of Optical Activity

Optical activity involves the combination of the concepts of molecular chirality with those of the circular polarization states of radiation. As we have seen, a chiral molecule exists as either one structural antipode or its exact three-dimensional mirror image (ignoring the miniscule energy difference of parity violation due to the weak force). Similarly, left and right circularly polarization states of radiation are spatial mirror images of one another when considering their pattern of the electric or magnetic field vectors in space at any instant in time, that is, the field vectors are arrayed in left and right helical patterns, respectively, as a function of location along the path of radiation propagation, as described above.

A general definition of molecular optical activity can be stated as follows. *Natural optical activity is the difference in the interaction of a chiral molecule with left versus right circularly polarized radiation.* These differences in interaction exist because they are diastereomeric in form. The interaction of right circularly polarized radiation with a ‘right-handed’ enantiomer of a chiral molecule (R_{rad}, R_{mol}) is different from the interaction of left circularly polarized radiation with the same ‘right-handed’ chiral molecule (L_{rad}, R_{mol}). This is the same conceptual difference as the diastereomers of a molecule with two chiral centers, as discussed in Section 3.1.5. The (R_{rad}, R_{mol}) ‘diastereomeric interaction’ has different physical properties than the corresponding (L_{rad}, R_{mol}) ‘diastereomeric interaction’. The interaction between molecules and radiation typically leads to transitions between quantum energy states in molecules. The nature of these transitions, and the nature of the radiation, provides a means of classifying different types of molecular optical activity. For example, interactions involving electronic state transitions are classified as electronic optical activity, whereas those involving vibrational transitions are classified as vibrational optical activity.

In addition to electromagnetic radiation, neutrons can be polarized in left and right circular polarization states, and hence neutron optical activity can exist and has been described theoretically (Cox and Richardson, 1977). In general, quantum mechanical particles have spin polarization. Fermions, such as quarks and leptons, have half-integer spin while bosons have zero or integral spin. The spin quantum number, and hence polarization state, of an individual photon, a boson, is either $+1$ corresponding to right circular polarization or -1 for left circular polarization. Neutrons are fermions and can have spin polarization quantum numbers of $+\frac{1}{2}$ and $-\frac{1}{2}$, for example, for right and left polarization states, respectively.

3.2.4 Optical Activity Observables

We have seen earlier that the first measurements of optical activity involved observations of rotation of the plane of polarized light as radiation passed through a chiral medium. The classical forms of optical activity observables are optical rotation and circular dichroism, both of which were originally expressed in terms of angles associated with the polarization state of elliptically polarized light as it passed through the optically active medium. As we shall see, optical rotation is the sign and magnitude of angle of rotation induced on linearly polarized radiation or the major axis of elliptically polarized radiation, and circular dichroism is the sign and angle of ellipticity imparted on the initially linearly polarized radiation. These measured optical properties reflect the nature of the medium, its index of refraction, its absorption, its density, and the wavelength of the radiation. In this section we identify the classical optical activity variables and develop a connection between these material observables and the molecular properties from which they arise.

3.2.4.1 Complex Index of Refraction

We begin by considering the effects of the complex refractive index $\tilde{\boldsymbol{n}}$ introduced earlier in connection with the description of the electric field of electromagnetic radiation given in Equations (3.1) and (3.2). We can combine these equations with Equation (3.9) for the Stokes vector definition of the total intensity of radiation, S_0, in terms of the absolute square of the electric field vector of the radiation by writing

$$I(\boldsymbol{r},t) = \tilde{\boldsymbol{E}}^{*}(\boldsymbol{r},t)\tilde{\boldsymbol{E}}(\boldsymbol{r},t) = |E_0\tilde{e}\exp[2\pi i(\tilde{\boldsymbol{n}}\cdot\boldsymbol{r}/\lambda - \nu t)]|^2 \tag{3.14}$$

where $\tilde{\boldsymbol{n}}$ is a complex vector parallel to the propagation direction of the radiation. If we again take direction of propagation to be the laboratory Z-axis, we can write the intensity of radiation emerging from an isotropic medium of thickness z with complex refraction $\tilde{n}$ by:

$$I(z,\lambda,t) = |E_0\tilde{e}\exp[2\pi i(\tilde{n}z/\lambda - \nu t)]|^2 \tag{3.15}$$

We can write the complex refractive index in terms of its real and imaginary parts as:

$$\tilde{n} = n + in' \tag{3.16}$$

Note that alternative conventions for defining the complex index of refraction often used are $\tilde{n} = n' + in''$ and $\tilde{n} = n + ik$. A complex index of refraction is necessary whenever the radiation has absorption losses in the medium at wavelength, λ. Substituting this expression into Equation (3.15) yields:

$$I(z,\lambda,t) = |E_0\tilde{e}\exp[2\pi i(nz/\lambda - \nu t) - 2\pi n'z/\lambda]|^2 \tag{3.17}$$

where we have separated the imaginary and real parts of the exponent for reasons to be made clear below.

3.2.4.2 Absorption Observables

If we carry out the absolute square of Equation (3.17) the temporal dependence of the radiation, and its dependence on the real part of the refractive index, is lost. All that remains is the dependence of the intensity, for a given wavelength λ and distance z traversed through the medium, on the imaginary part of the refractive index, n'.

$$I(z,\lambda) = E_0^{\,2}\exp(-4\pi n'z/\lambda) \tag{3.18}$$

If we now convert the dependence on wavelength to wavenumber frequency, given as before by $\bar{\nu} = 1/\lambda$, and indicate explicitly those quantities that depend on the wavenumber of the radiation, we have

$$I(\bar{\nu}, z) = I_0(\bar{\nu})e^{-4\pi\bar{\nu}n'(\bar{\nu})z} = I_0(\bar{\nu})10^{-\varepsilon(\bar{\nu})Cz} = I_0(\bar{\nu})e^{-2.303\varepsilon(\bar{\nu})Cz} \tag{3.19}$$

Also included in this equation are relationships from Chapter 2 in Equation (2.54) defining the molar absorption coefficient, or molar absorptivity $\varepsilon(\bar{\nu})$, where C is the concentration of the sample in moles per liter, or its molar density in moles per $1000\,\mathrm{cm}^3$, and z is taken as the pathlength of the sample. Equating the exponents of the terms with base e we find the connection between the material absorbance (loss) property, $n'(\bar{\nu})$, and the molecular absorbance property, $\varepsilon(\bar{\nu})$, to be:

$$4\pi\bar{\nu}n'(\bar{\nu})z = 2.303\varepsilon(\bar{\nu})Cz \tag{3.20}$$

The pathlength of the sample, z, cancels from this equation, and solving for $n'(\bar{\nu})$ gives:

$$n'(\bar{\nu}) = \left(\frac{2.303C}{4\pi\bar{\nu}}\right)\varepsilon(\bar{\nu}) \tag{3.21}$$

3.2.4.3 Circular Dichroism and Ellipticity Observables

For a chiral sample, there is a difference in the complex index of refraction for RCP and LCP radiation. Thus we can write

$$\Delta\tilde{n} = \tilde{n}_L - \tilde{n}_R = (n_L - n_R) + i(n'_L - n'_R) = \Delta n + i\Delta n' \tag{3.22}$$

where we have momentarily suppressed the spectral frequency dependence of these quantities. Also, as will become clear below, Δn is related to optical rotation and $\Delta n'$ to circular dichroism. The relationship between material and molecular absorption properties given in Equation (3.21) can now be given for circular dichroism by an equation in terms of the difference in molar absorptivity, $\Delta\varepsilon(\bar{\nu})$, by:

$$\Delta n'(\bar{\nu}) = \left(\frac{2.303C}{4\pi\bar{\nu}}\right)\Delta\varepsilon(\bar{\nu}) \tag{3.23}$$

The ellipticity angle in radians induced on an incident plane polarized wave of radiation by an absorbing chiral sample is given in radians by the simple relationship (Salvadori and Ciardelli, 1973)

$$\psi_r(\bar{\nu}) = \frac{\pi l}{\lambda}\left[n'_L(\bar{\nu}) - n'_R(\bar{\nu})\right] = \frac{\pi l}{\lambda}\Delta n'(\bar{\nu}) \tag{3.24}$$

where the pathlength, l, and the wavelength, λ, are both given in centimeters. The ellipticity in degrees is easily related to circular dichroism in decadic absorbance units, $\Delta A(\bar{\nu})$, by:

$$\psi_d(\bar{\nu}) = \frac{180l}{\lambda}\Delta n'(\bar{\nu}) = \frac{180l}{\lambda}\left(\frac{2.303C}{4\pi\bar{\nu}}\right)\Delta\varepsilon(\bar{\nu}) = 33\Delta\varepsilon(\bar{\nu})Cl = 33\Delta A(\bar{\nu}) \tag{3.25}$$

Historically, the pathlength and density dependence of $\psi_d(\bar{\nu})$ are removed by defining the specific ellipticity for a neat liquid,

$$[\psi(\bar{\nu})] = \psi_d(\bar{\nu})/\rho l' \tag{3.26}$$

where ρ is the density in $\mathrm{g\,cm^{-3}}$ and l' is the pathlength in decimeters (dm). For solutions, the corresponding definition is:

$$[\psi(\bar{\nu})] = \psi_d(\bar{\nu})/cl' \tag{3.27}$$

where c is the concentration in $\mathrm{g\,cm^{-3}}$ or $\mathrm{g\,mL^{-1}}$. Ellipticity is also reported as molar ellipticity in units of degrees $\mathrm{mol^{-1}\,L\,cm^{-1}}$, given by:

$$[\Theta(\bar{\nu})] = \frac{[\psi(\bar{\nu})]M}{100} = \frac{100}{Cl}\psi_d(\bar{\nu}) \tag{3.28}$$

where M is the molecular weight, or molar mass, of the sample in $\mathrm{g\,mol^{-1}}$. The division by 100 in the definition of molar ellipticity is only a matter of convention and convenience. For the expression following the second equality a factor of 10^4 is required for the conversion of the units of the product of molar concentration and pathlength from $(c/M)l'$, in units of $\mathrm{mol\,cm^{-3}}$ and dm, to Cl in units of $\mathrm{mol\,L^{-1}}$ and cm. Combining Equations (3.28) and (3.25) gives the frequently cited relationship between molar ellipticity and circular dichroism expressed as molar absorptivity,

$$[\Theta(\bar{\nu})] = \frac{100}{Cl}[33\Delta A(\bar{\nu})] = 3300\Delta\varepsilon(\bar{\nu}) \tag{3.29}$$

3.2.4.4 Optical Rotation Angle and Optical Rotatory Dispersion Observables

Optical rotation is a complementary optical activity observable to circular dichroism, which arises from the difference in the real part of the index of refraction for left versus right circularly polarized radiation. For a sample of pathlength l, the observed optical rotation in radians, α_r, can be defined by analogy to Equation (3.24) as:

$$\alpha_r(\bar{\nu}) = \frac{\pi l}{\lambda}[n_L(\bar{\nu}) - n_R(\bar{\nu})] = \frac{\pi l}{\lambda}\Delta n(\bar{\nu}) \tag{3.30}$$

The specific rotations can be defined in degrees in the same way as specific ellipticities in Equations (3.26) and (3.27) by:

$$[\alpha(\bar{\nu})]_\lambda^T = \alpha_d(\bar{\nu})/\rho l' = \alpha_d(\bar{\nu})/cl' \tag{3.31}$$

where it is customary to specify the temperature T in degrees centigrade and wavelength λ in nanometers. Finally, the molar rotation is defined in terms of specific rotation, as before, by:

$$[\Phi(\bar{\nu})]_\lambda^T = \frac{[\alpha(\bar{\nu})]_\lambda^T M}{100} \tag{3.32}$$

where the units of $[\Phi(\bar{\nu})]_\lambda^T$ are the same as molar ellipticity, $[\Theta(\bar{\nu})]$, namely degrees $\mathrm{mol^{-1}\,L\,cm^{-1}}$.

3.3 Classical Forms of Optical Activity

The classical forms of natural optical activity are optical rotation and circular dichroism in the visible and ultraviolet regions of the spectrum. As was noted above, optical rotation (OR) measurements were the earliest measurements of optical activity followed by circular dichroism (CD). The visible and ultraviolet spectral regions are associated with electronic transitions, so these original forms of optical activity are also referred to as electronic optical activity (EOA). With the advent and development of optical activity in the vibrational region of the spectrum, the term vibrational optical activity (VOA) has come into widespread use. Thus, to distinguish CD of electronic origin from CD of vibrational origin the term electronic circular dichroism (ECD), in analogy to VCD, has been gaining wider use. Similarly, the spectrum of OR, termed optical rotatory dispersion (ORD), needs to be distinguished for transitions of electronic versus vibrational origin. This has been a mute point until recently; however, the first reports of optical rotation in vibrational transitions have been published (Lombardi and Nafie, 2009; Rhee *et al.*, 2009), thus giving rise to the terms vibrational optical rotatory dispersion (VORD), or equivalently vibrational circular birefringence (VCB), to distinguish them from electronic optical rotatory dispersion (EORD) or electronic circular birefringence (ECB).

3.3.1 Optical Rotation and Optical Rotatory Dispersion

OR can refer to a measurement of optical rotation for a particular sample as defined above in Equation (3.30). More commonly, it is the specific optical rotation in degrees, given by Equation (3.31), at say the sodium D line (546 nm) using a pathlength of 1 dm, corrected to unit density ($g\,cm^{-3}$) or concentration ($g\,mL^{-1}$), at a specified temperature. For spectral measurements related to the molecular property, such as an optical rotatory dispersion ORD spectrum, the intensity units are usually the molar (specific) rotation given by Equation (3.32).

Because OR is such a simple measurement, its discovery historically preceded that of circular dichroism measurements (Lowry, 1935). The reasons are twofold. The measurement of OR does not require the direct use of circular polarization states. These states of light, for any form of radiation, are more challenging to prepare than the corresponding linear polarization states. Secondly, OR can be measured in the visible region of the spectrum for colorless samples far from their lowest energy ultraviolet electronic absorption bands. This situation is commonly encountered for the organic compounds where optical activity was first observed.

Today OR remains as the simplest measure of optical activity and has become the standard way to distinguish samples of enantiomers from one another experimentally. Unfortunately, however, OR, being a dispersion property of matter, is more difficult to calculate and express theoretically than the corresponding absorption property, CD. The OR at a particular spectral location, is the net effect of all transitions in a molecule, although, as will be shown below, nearby transitions in a spectral sense are more influential than more distant transitions. As a consequence, the measurement of ORD is far less common today than the measurement of CD spectra.

3.3.2 Circular Dichroism

CD can be measured either as the difference in absorbance or absorption coefficient (molar absorptivity) or as an ellipticity measurement. Electronic CD instruments (but not VCD instruments!) usually present ECD spectra in units of ellipticity, but this is actually the result of a conversion to units of traditional usage rather than an actual measurement of the degree of ellipticity imposed by a chiral sample on an incident linearly polarized light beam. The use of ellipticity for the measurement of ECD hides the fact that ECD and the corresponding parent ultraviolet (UV) or visible absorbance spectrum are the corresponding whole and circular differential forms of the same molecular property as there is no logical way to express absorption in ellipticity units.

The routine measurement of CD spectra did not occur until after instrumentation for the measurement of absorption in the UV region became routinely available. Even so, the underlying relationship between CD and ORD spectra was not appreciated until the early 1960s (Moscowitz, 1962), and the first commercial CD instruments became available in the mid-1960s. Once the relationship between CD and ORD had been understood, it was realized that CD spectra were simpler to interpret theoretically and that in principle no new information was available from the entire ORD spectrum that was not equivalently available from the corresponding entire CD spectrum. As a consequence, the measurement of ORD gradually fell into disfavor, although more recently a slight resurgence in the wavelength dependence of OR measurement has emerged for reasons related to verifying the accuracy of quantum mechanical computations of OR, which can now be carried out as a function of the spectral frequency.

3.3.3 Kramers–Kronig Transform Between CD and ORD

The relationship between a CD spectrum and its corresponding ORD spectrum parallels the relationship between the absorption spectrum of a material and its index of refraction. In particular, there is a relationship known as the Kramers–Kronig transform between the real and imaginary parts of any response function for the interaction of radiation with matter (Barron, 2004). In the case of the complex index of refraction, defined above in Equation (3.16), and written as a function of spectral frequency:

$$\tilde{n}(\bar{\nu}) = n(\bar{\nu}) + in'(\bar{\nu}) \tag{3.33}$$

The Kramers–Kronig transform relationship is:

$$n(\bar{\nu}) = 1 + \frac{2}{\pi}\mathscr{P}\int_0^\infty \frac{\bar{\xi}n'(\bar{\xi})}{\bar{\xi}^2 - \bar{\nu}^2}\mathrm{d}\bar{\xi} \tag{3.34}$$

$$n'(\bar{\nu}) = -\frac{2\bar{\nu}}{\pi}\mathscr{P}\int_0^\infty \frac{n(\bar{\xi})}{\bar{\xi}^2 - \bar{\nu}^2}\mathrm{d}\bar{\xi} \tag{3.35}$$

where the integration variable $\bar{\xi}$ has the same units as wavenumbers $\bar{\nu}$, or inverse wavelength, $1/\lambda$, typically in units of cm^{-1}, although the use of any frequency variable leaves these equations in the same form. The additional term of 1 in Equation (3.34) enters because the index of refraction of the vacuum is taken as unity. Here $\mathscr{P}$ designates the Cauchy principal value of the integral, which is evaluated by contour integration in the complex plane where the frequencies are allowed to become complex variables. It can be seen from these relationships that knowledge of the entire spectrum of either $n(\bar{\nu})$ or $n'(\bar{\nu})$ permits evaluation of the other spectrum at any specified frequency $\bar{\nu}$ in its spectrum. The Kramers–Kronig transforms linking $n(\bar{\nu})$ and $n'(\bar{\nu})$ are sometimes referred to as dispersion relationships, because of the connection between the dispersion quantity, in this case the spectrum of the index of refraction, and its corresponding absorption quantity, the absorption spectrum. The relationship is also a result of the principle of causality expressed by means of response theory in which the response of the medium cannot precede its radiation stimulus. The general form of the absorption spectrum is a maximum at the resonance frequency, whereas the index of refraction exhibits anomalous dispersion resulting in negative index values, relative to its nearby off-resonance value, on the high-frequency side of resonance and more positive index values on the low-frequency side. Particular examples of this are provided later in this section.

The Kramers–Kronig transform relationship can easily be extended to CD and ORD spectra as, by Equation (3.22), these are simply the circular polarization differences of the real and imaginary parts of the complex indices of refraction, $n(\bar{\nu})$ and $n'(\bar{\nu})$, featured in the transform relationships above. So we can write:

$$\Phi(\bar{\nu}) = \frac{2}{\pi}\mathscr{P}\int_0^{\infty}\frac{\bar{\xi}\Theta(\bar{\xi})}{\bar{\xi}^2 - \bar{\nu}^2}\mathrm{d}\bar{\xi} \tag{3.36}$$

$$\Theta(\bar{\nu}) = -\frac{2\bar{\nu}}{\pi}\mathscr{P}\int_0^{\infty}\frac{\Phi(\bar{\xi})}{\bar{\xi}^2 - \bar{\nu}^2}\mathrm{d}\bar{\xi} \tag{3.37}$$

where Equation (3.36) expresses the ORD spectrum at any point in terms of the integral over all frequencies of the CD spectrum, whereas Equation (3.37) does just the opposite, the CD spectrum at any point is given in terms of the entire ORD spectrum. We have chosen to express the Kramers–Kronig transform in terms of the molar rotations and ellipticities, but any pair of equivalent ORD and CD observables, such as $\alpha_r(\bar{\nu})$ and $\psi_r(\bar{\nu})$, $[\alpha](\bar{\nu})$ and $[\psi](\bar{\nu})$, or $\Delta n(\bar{\nu})$ and $\Delta n'(\bar{\nu})$, are related by this pair of integral equations.

3.3.4 Lorentzian Dispersion and Absorption Relationships

The Kramers–Kronig transform introduced in the previous section holds for any form of physical quantities related by dispersion relationships, such the refractive index and absorption spectra, independent of the form of their spectral lineshapes. In this section, we present a simple yet powerful relationship between Kramers–Kronig transform pairs under the assumption that the lifetimes of the excited states involved are governed by a simple exponential decay constant, γ_a. Under this assumption, the time dependence of a quantum state a is described by the wavefunction

$$\Psi_a(\mathbf{r},t) = \psi_a(\mathbf{r})\mathrm{e}^{-i\omega_a t}\mathrm{e}^{-\gamma_a t} = \psi_a(\mathbf{r})\mathrm{e}^{-i(\omega_a - i\gamma_a)t} \tag{3.38}$$

In the first relationship, the state is seen to decay with time and in the second relationship we show that this decay can be introduced by adding an imaginary decay term into the frequency of the state a, as $\omega_a - i\gamma_a$. The spectral lineshape of this state is obtained from its Fourier transform

$$\Psi_a(\mathbf{r},\omega) = \frac{1}{2\pi}\int_0^{\infty}\Psi_a(\mathbf{r},t)\mathrm{e}^{i\omega t}\mathrm{d}t \tag{3.39}$$

and by direct integration is just a complex Lorentzian lineshape given by:

$$\Psi_a(\mathbf{r},\omega) = \psi_a(\mathbf{r})\frac{1}{\pi}\left[\frac{1}{(\omega_a - \omega) - i\gamma_a}\right] \tag{3.40}$$

The complex lineshape factor can be expressed in terms of real and imaginary parts as:

$$\tilde{f}_a(\omega) = \frac{1}{\pi}\left[\frac{1}{(\omega_a - \omega) - i\gamma_a}\right] = f_a(\omega) + if'_a(\omega) \tag{3.41}$$

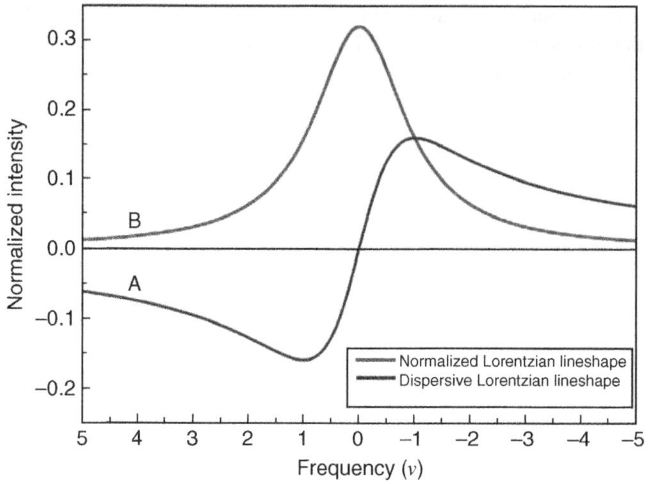

Figure 3.1 *The real (A) and imaginary (B) parts, Equations (3.42) and (3.43), respectively, of the complex normalized Lorentzian lineshape, Equation (3.41), are plotted in frequency units of half-width at half-maximum of the absorption lineshape where the real and imaginary parts cross each other. The band center frequency is chosen to be zero. Reproduced with permission from John Wiley & Sons (Lombardi and Nafie, 2009)*

$$f_a(\omega) = \frac{1}{\pi}\left[\frac{\omega_a - \omega}{(\omega_a - \omega)^2 + \gamma_a^2}\right] \quad (3.42)$$

$$f_a'(\omega) = \frac{1}{\pi}\left[\frac{\gamma_a}{(\omega_a - \omega)^2 + \gamma_a^2}\right] \quad (3.43)$$

These two Lorentzian lineshape functions are shown in Figure 3.1. They are Kramers–Kronig transforms of one another and describe the lineshapes of the refractive index and absorption under the assumption of the exponential time decay of the state a with decay constant, γ_a. The dispersion lineshape $f_a(\omega)$ is anti-symmetric about the resonance frequency, ω_a. It is negative on its high frequency side, zero at the resonance frequency, and positive on its low frequency side. The absorption lineshape was first introduced in Equation (2.56) in the context of the lineshape of the molecular absorption coefficient and its relationship to the dipole strength of a transition. This function is symmetric about the resonance frequency characteristic of spectral absorption bands.

3.3.5 Dipole and Rotational Strengths

Using these lineshapes we can obtain expressions for the Lorentzian lineshapes of dispersion and absorption bands. In particular, we can obtain expressions for the spectra of the real part of the refractive index and absorption by combining Equations (2.56), (3.23), (3.42), and (3.43) to write

$$n(\bar{\nu}) = 1 + \left(\frac{2.303C}{4\pi\bar{\nu}}\right)\frac{8\pi^3 N\bar{\nu}}{3000hc\ln(10)}\sum_a D_a\left[\frac{(\bar{\nu}_a - \bar{\nu})/\pi_a}{(\bar{\nu}_a - \bar{\nu})^2 + \gamma_a^2}\right] \quad (3.44)$$

$$n'(\bar{\nu}) = \left(\frac{2.303C}{4\pi\bar{\nu}}\right)\frac{8\pi^3 N\bar{\nu}}{3000hc\ln(10)}\sum_a D_a\left[\frac{\gamma_a/\pi}{(\bar{\nu}_a-\bar{\nu})^2+\gamma_a^2}\right] \tag{3.45}$$

Here we have used wavenumbers instead of frequency in radians to express the spectral frequency dependence, and D_a is the position or velocity forms of the dipole strength defined previously in Equations (2.17) and (2.21), respectively. The corresponding expressions for CD and ORD are given by:

$$\Delta n(\bar{\nu}) = \left(\frac{2.303C}{4\pi\bar{\nu}}\right)\frac{32\pi^3 N\bar{\nu}}{3000hc\ln(10)}\sum_a R_a\left[\frac{(\bar{\nu}_a-\bar{\nu})/\pi}{(\bar{\nu}_a-\bar{\nu})^2+\gamma_a^2}\right] \tag{3.46}$$

$$\Delta n'(\bar{\nu}) = \left(\frac{2.303C}{4\pi\bar{\nu}}\right)\frac{32\pi^3 N\bar{\nu}}{3000hc\ln(10)}\sum_a R_a\left[\frac{\gamma_a/\pi}{(\bar{\nu}_a-\bar{\nu})^2+\gamma_a^2}\right] \tag{3.47}$$

where R_a is the rotational strength, the optical activity analogue of the dipole strength, D_a, and forms the theoretical basis for CD or ORD intensities. For a general transition from a ground state, 0, to an excited state, a, the dipole and rotational strengths are defined as:

$$D_a = \langle\psi_0|\boldsymbol{\mu}|\psi_a\rangle^* \cdot \langle\psi_a|\boldsymbol{\mu}|\psi_0\rangle = |\langle\psi_a|\boldsymbol{\mu}|\psi_0\rangle|^2 \tag{3.48}$$

$$R_a = \mathrm{Im}[\langle\psi_0|\boldsymbol{\mu}|\psi_a\rangle \cdot \langle\psi_a|\boldsymbol{m}|\psi_0\rangle] \tag{3.49}$$

where $\boldsymbol{m}$ is the magnetic dipole moment operator of the molecule given by:

$$\boldsymbol{m} = \sum_j \frac{e_j}{2m_j c}\boldsymbol{r}_j \times \boldsymbol{p}_j \tag{3.50}$$

The expression given here for the rotational strength, also called the Rosenfeld equation, applies only for a molecule in an isotropic medium or the expression of a single molecule averaged over all orientations. If a single crystal or spatially oriented non-isotropic sample is considered, a more general expression for the rotational strength (and the dipole strength) must be used that includes, in general, the electric quadrupole moment. Notice that the rotational strength is a vector dot (scalar) product of two different transition moment vectors and hence can be either a positive quantity for positive projection of one vector on the other (same direction overlap), or negative for a negative projection (opposite direction overlap), or zero for orthogonal vectors. Only chiral molecules support non-zero values of the rotational strength. For non-chiral molecules with at least an improper axis of rotation, the rotational strength must vanish. Either the electric dipole or magnetic dipole transition moment vanishes for a given transition, or if both are non-zero, they must be orthogonal with zero overlap. We will consider further properties of the rotation strength in the next chapter when we describe the theory of vibrational circular dichroism.

We conclude this section by using these relationships to derive the expression connecting circular dichroism in terms of differential molar absorptivity and the rotational strength. In particular we have:

$$\begin{aligned}\Delta\varepsilon(\bar{\nu}) &= \frac{32\pi^3 N\bar{\nu}}{3000hc\ln(10)}\sum_a R_a\left[\frac{\gamma_a/\pi}{(\bar{\nu}_a-\bar{\nu})^2+\gamma_a^2}\right]\\ &= \frac{\bar{\nu}}{2.236\times10^{-39}}\sum_a R_a f'_a(\bar{\nu})\end{aligned} \tag{3.51}$$

This equation is directly analogous to Equation (2.56), which connects the spectrum $\varepsilon(\bar{\nu})$ of the molar absorption coefficient to the sum of the dipole strengths of a molecule with Lorentzian lineshapes for each transition, reproduced here for comparison,

$$\varepsilon(\bar{\nu}) = \frac{8\pi^3 N\bar{\nu}}{3000hc\ln(10)}\sum_a D_a f'_a(\bar{\nu}) = \frac{\bar{\nu}}{9.184\times 10^{-39}}\sum_a D_a f'_a(\bar{\nu}) \tag{3.52}$$

These equations become of central importance when comparing the results of measurements and calculations of molecular spectral properties. Finally, one of the most important relationships in optical activity is that equating the dimensionless ratio of circular dichroism, $\Delta\varepsilon_a(\bar{\nu})$, to its parent absorption coefficient, $\varepsilon_a(\bar{\nu})$, for a given transition to the corresponding ratio of the rotational and dipole strengths. From the last two equations we can write:

$$g_a = \frac{\Delta\varepsilon_a(\bar{\nu})}{\varepsilon_a(\bar{\nu})} = \frac{4R_a(\bar{\nu})}{D_a(\bar{\nu})} \tag{3.53}$$

The lineshape factors and units are virtually the same for the spectral quantities in the numerator and denominator. The quantity g_a for a spectral band associated with a transition to state a is called the anisotropy ratio and is a measure of the intrinsic chiral strength of a transition. As we will see in Chapter 6, it is also a measure of the expected signal-to-noise ratio in the measurement of a CD spectrum.

3.3.6 Magnetic Optical Activity

For completeness, we describe here, briefly, the phenomenon of magnetic optical activity (MOA), a property of all molecules, in contrast to natural optical activity that arises only from chiral molecules (Buckingham and Stephens, 1966). Magnetic optical activity, like natural optical activity, is manifested in two forms, magnetic optical rotation and magnetic circular dichroism (MCD) (Stephens, 1974). In all cases of MOA, a magnetic field is applied along the direction of propagation of the beam of radiation, either parallel or anti-parallel, that passes through a sample of molecules. The magnetic field induces a response in the molecule, such as splitting of degenerate electronic or vibrational states or couplings of states, which in turn induces a differential response to left and right circularly polarized states of the beam of radiation. Magnetic optical rotation, also known as the Faraday Effect, causes linearly polarized light to rotate to the left or right depending on the region of the spectrum and the direction of the magnetic field. If the field is reversed, the sense of rotation of the linearly polarized light beam reverses. Similarly, for MCD, the signs of the MCD bands reverse upon changing the direction of the magnetic field from parallel to anti-parallel to the radiation beam. For a sample that is chiral, one observes a linear superposition of CD and MCD spectra as $\Delta\varepsilon_{CD}(\bar{\nu}) + \Delta\varepsilon_{MCD}(\bar{\nu})$. The two spectra are easily separated, either by reversing the direction of the magnetic field one obtains $\Delta\varepsilon_{CD}(\bar{\nu}) - \Delta\varepsilon_{MCD}(\bar{\nu})$, or instead by changing the sample from one enantiomer to the other one obtains $-\Delta\varepsilon_{CD}(\bar{\nu}) + \Delta\varepsilon_{MCD}(\bar{\nu})$, thereby reversing the sign of the CD while keeping the MCD spectrum the same sign.

3.4 Newer Forms of Optical Activity

The classical forms of natural optical activity, as we have seen in the previous section, are ECD and EORD in the ultraviolet and visible regions of the spectrum involving electronic transitions. Over the past 40 years, new types of measurements of optical activity outside the classical forms have occurred.

In some cases, overlap exists between the different forms of optical activity, and in others it possible to observe in a single spectrum more than one form of optical activity. In this section, we briefly describe some of these new forms of optical activity before we turn more exclusively to considerations of vibrational optical activity in the following chapters.

3.4.1 Infrared Optical Activity, VCD, and IR-ECD

An obvious extension of ECD and EORD is to measure these spectra toward longer wavelengths of the infrared region. Although a few near-infrared or infrared ORD measurements have been reported, most of the extensions of optical activity into the infrared region were carried with CD measurements. The first such measurements of infrared optical activity were near-infrared and infrared CD measurements of a rare earth metal complex that possessed lower lying electronic transitions between 5000 and 2000 cm^{-1} (Chabay *et al.*, 1972). Next came VCD measurements of transition metal crystals, in both the near-infrared regions for overtones and combination bands (Hsu and Holzwarth, 1973), and then into the infrared region with fundamentals of vibrational transitions as described earlier.

3.4.1.1 VCD–ECD Overlap

Infrared CD measurements have been carried out for some transition metal complexes that possess low-lying infrared transitions (He *et al.*, 2001). These compounds exhibit both infrared ECD (IR-ECD) and VCD. Furthermore, the VCD is seen to be enhanced from borrowing magnetic-dipole transition moment strength from the d–d electronic transitions of the transition metal through a vibronic coupling mechanism. Thus, it is possible to see an *overlap* between ECD and VCD in appropriate molecules in the near-infrared or infrared spectral regions when both types of transitions are present.

Recently, two reports of infrared vibrational optical rotatory dispersion (VORD), also called vibrational circular birefringence (VCB), have appeared. In the first of these, both VCD and VORD are measured simultaneously as in-phase and out-of-phase components of a femtosecond laser-pulse generated spectrum in the CH stretching region, thereby opening a new approach to the VCD and VORD measurement with femtosecond time resolution (Rhee *et al.*, 2009). In the second report, a standard Fourier transform VCD spectrometer was converted to allow measurement of VCB by insertion of a polarizer oriented at 45° after the sample and detecting as a pseudo linear dichroism spectrum at twice the polarization modulator frequency (Lombardi and Nafie, 2009). This last report also included the first quantum mechanical calculation of VCB obtained by merely changing the lineshape function from Lorentzian absorption, Equation (3.43), to Lorentzian dispersion, Equation (3.42). Agreements between calculated and measured VCD and VCB spectra were excellent and equally close, as will be discussed further in later chapters.

3.4.2 Vacuum Ultraviolet and Synchrotron Circular Dichroism

About the same time that ultraviolet–visible ECD was being extended into the near-infrared and infrared regions, efforts were underway to extend ECD measurements to shorter wavelengths (lower that about 180 nm) into the so-called vacuum ultraviolet (VUV) region (Snyder and Johnson, 1978). Here electronic transitions can be accessed that are higher in energy than the π-electron transitions ($n-\pi^*$ and $\pi-\pi^*$) typical of normal UV spectroscopy where a group with π-electrons, called a chromophore, must be present to see absorption bands or ECD spectra. VUV ECD can access excited electronic states where the presence of a standard UV chromophore is not necessary in the molecule. These efforts were supplemented by access to high energy UV and vacuum UV radiation at synchrotron facilities where CD and MCD experiments have been set up and measurements carried out and analyzed (Snyder and Rowe, 1980).

3.4.3 Rayleigh and Raman Optical Activity, RayOA and ROA

Rayleigh optical activity (RayOA) and Raman optical activity (ROA) are distinctly new forms of optical activity involving molecular scattering where there are no classical optical activity analogues. However, it has been shown theoretically that there are connections between OR and RayOA in the forward direction. As pointed out in Chapter 1, there are four forms of circular polarization (ICP, SCP, DCP_I, and DCP_{II}) ROA and RayOA. In addition, four forms of linear polarization (ILP, SLP, DLP_I, and DLP_{II}) ROA and RayOA have been defined and predicted theoretically but have not yet been discovered (Hecht and Nafie, 1990). As electronic Raman scattering exists, particularly for molecules with low-lying electronic states, both electronic and vibrational ROA exist. The former has been observed but only for magnetic-field induced ROA, that is, magnetic ROA, a form of magnetic optical activity. To date, natural electronic resonance ROA has been reported for the first time recently (Merten *et al.*, 2010).

3.4.3.1 ROA Overlaps

From these considerations, it is clear that both electronic and vibrational ROA exist and have appeared in the same spectrum. Furthermore, as Rayleigh scattering overlaps with low-frequency Raman scattering, the potential also exists for overlap of the observation of RayOA in the depolarized Rayleigh wings with low-frequency vibrational ROA spectra. The best form of RayOA for seeing RayOA intensity away from the central Rayleigh line is in-phase dual circular polarization (DCP_I) RayOA as predicted recently in a computational study (Zuber *et al.*, 2008).

3.4.4 Magnetic Vibrational Optical Activity

Although MVOA is not a chiroptical spectroscopy, it is closely related instrumentally and theoretically. In Section 3.3.6, we discussed briefly the concepts of magnetic optical activity in general and MCD in particular. Here we describe the extension of MOA to vibrational transitions. Magnetic ROA has been described theoretically (Barron and Buckingham, 1972) and observed experimentally in vibrational transitions as magnetic resonance ROA (Barron, 1975; Barron, 1977; Barron *et al.*, 1982a) and in split degenerate electronic transitions of transition metal complexes as magnetic electronic ROA (Barron and Meehan, 1979; Barron *et al.*, 1982b). As with natural ROA, MROA is a new form of magnetic optical activity distinct from magnetic optical rotation and MCD.

In an analogous way, MCD has been extended to infrared vibrational transitions as MVCD. It was first reported experimentally as A-term MVCD in the splitting of degenerate vibrational modes of simple threefold symmetric methyl halides (Keiderling, 1981; Devine and Keiderling, 1983), benzene and 1,3,5-trisubstituted derivatives, (Devine and Keiderling, 1984), and metal hexacarbonyls (Devine and Keiderling, 1985). The theory of MVCD was subsequently developed theoretically in terms of the vibronic coupling theory of molecules with degenerate states (Pawlikowski and Keiderling, 1984; Pawlikowski and Pilch, 1992).

More recently, MVCD has been observed in chiral molecules where it appears together with the natural VCD spectrum (Ma, 2007). In this case MVCD, along with infrared electronic MCD (MECD), was observed with a 1.4 Tesla magnetic field in the mid-IR regions using an FT-VCD spectrometer for transition metal complexes with low-lying excited electronic states that enhanced the intensity of both the VCD, as previously reported (He *et al.*, 2001), and the MVCD spectra (Ma, 2007). The MVCD spectra were easily separated from the natural VCD spectra by reversing the direction of the magnetic field with respect to the direction of propagation of the IR beam, as described above for MOA.

The theory of MVCD for chiral molecules lacking degenerate states involves only the B-term mechanism of MVCD, not previously developed in the literature. An elemental theoretical treatment of B-term MVCD, as well as the first B-term MVCD spectra, is presented in Appendix D.

3.4.5 Fluorescence Optical Activity, FDCD and CPL

There are two forms of fluorescence optical activity (FOA). Both are forms of electronic OA as fluorescence is very weak in the near-IR and IR spectral regions. One form of FOA is simply a way of detecting CD intensity in absorption bands that lead to fluorescence by measuring the total fluorescence for absorption by the sample of alternately left and right circularly polarized excitation radiation. This is called fluorescence detected circular dichroism (FDCD) and is simply the CD associated with a fluorescence excitation spectrum (Mueller *et al.*, 2005).

The other form of FOA is termed circularly polarized luminescence (CPL) or circularly polarized emission (CPE) largely for historical reasons (Gussakovsky, 2010). It could also be called circularly polarized fluorescence (CPF). This form of OA was first detected in luminescent transition metal and lanthanide metal complexes. CPF is measured as the difference in the fluorescence emission spectrum for left versus right circularly polarized fluorescence radiation for unpolarized excitation radiation. FDCD is a measure of a ground electronic state property, namely the CD starting from the ground state of the molecule, whereas CPF is a measure an excited electronic state property, namely its preference for left or right circularly polarized fluorescence from that excited state.

3.4.5.1 FOA and ROA Overlap

As fluorescence and Raman scattering can be observed in the same spectrum, it is natural to expect that is it possible to observe FOA and ROA simultaneously. If one measures ICP-ROA the incident (excitation) beam is modulated between the RCP and LCP states and the total ROA intensity is measured. Accompanying such an ROA measurement, as a broad underlying FOA spectrum, would be the FDCD spectrum. Similarly, for an SCP-ROA measurement, the underlying FOA background would be the CPL spectrum because for SCP-ROA one uses unpolarized laser excitation followed by the difference in circular polarization of the Raman scattered and fluorescence radiation. To date, there are no fluorescence analogues of DCP-ROA measurements, but as fluorescence is simply two one-photon transitions, unlike ROA, the fluorescence FOA spectra underlying DCP-ROA spectra would be the sum of the FDCD and CPF spectra for DCP_I-ROA and their difference for DCP_{II}-ROA.

An odd, perhaps unfortunate quirk of FOA and ROA is that they are defined oppositely with respect to RCP and LCP radiation, after some early debate on the subject of which sign convention should be used (Barron and Vrbancich, 1983; Nafie, 1983). All forms of CD and ORD are defined as the intensity of LCP minus that of RCP, and hence positive OA represents a preference for LCP over RCP, whereas ROA is the opposite. Thus, FOA would appear with the opposite sign (positive FOA in the negative ROA intensity direction) in a spectrum that exhibited both FOA and ROA in the same spectrum.

This odd juxtaposition of sign conventions has also arisen in the theory of resonance ROA from a single resonant electronic state (Nafie, 1996). The theory predicts that the ratio of ROA and Raman intensity, for the entire ROA spectrum, is the same to within a factor of two of the anisotropy ratio (CD to absorbance) of the resonant electronic state, but with the *opposite* sign! In this limit the most primitive origin of ROA intensity can be seen as having its origin in the CD of a single electronic state, but also revealed is the difference in sign convention of ROA versus all other forms of molecular optical activity.

3.4.6 Other Forms of Optical Activity

A number of rather exotic forms of OA have been described theoretically, or attempted experimentally, or both. Here we describe a few of those of current or potential interest.

3.4.6.1 X-Ray Circular Dichroism

CD in the far-ultraviolet and X-ray region have been carried out recently. These measurements yield CD associated with excitation of inner shell electrons. These transitions are high localized and report only weakly on the overall chirality of the molecule. Nevertheless they are interesting and provide a new frontier of investigation for CD spectroscopy (Kuberski *et al.*, 2008).

3.4.6.2 Neutron Optical Activity

The theory of neutron optical activity (NOA) was presented over three decades ago (Cox and Richardson, 1977) but to date no measurements of NOA have been reported. The theory is based on the fact that neutrons can possess spin angular momentum relative to their direction of propagation. Experimentally it is possible to modulate the spin polarization of neutrons between opposite angular momentum states, which is all that was assumed in the published theory of NOA. As inelastic neutron scattering has recently become an emerging form of vibrational spectroscopy, free of selection rules and having a heavy bias to the motion of hydrogen nuclei, the potential exists for NOA to complement very nicely all other forms of OA, and in particular the two principal forms of VOA.

3.4.6.3 Far-Infrared and Rotational CD

Circular dichroism in pure rotational transitions (RCD) has been explored theoretically for gas phase samples in the microwave region (Salzman and Polavarapu, 1991). Similarly, CD in the terahertz (far-infrared) region has been explored but not yet reported experimentally. A major difficulty in extending CD measurements to longer wavelengths and lower photon energies, which was overcome in the development of VCD starting from ECD, is that the ratio of CD to parent absorbance, the anisotropy ratio, scales as the ratio of the dimensions of the transition (one to several bond lengths) to the wavelength of the radiation. This is the reason that VCD intensity ratios are in the range of 10^{-3} to 10^{-6} whereas ECD anisotropy ratios are in the range 10^{-2} to 10^{-4}. For terahertz and microwave CD spectra the anisotropy ratio is smaller than that of VCD by another factor of 10 to 100, and given that this ratio is a measure of relative signal-to-noise ratio of measured spectra, prospects for observing CD in this region, at least from individual molecules, are not high. The basic problem in going to longer wavelength radiation is that a molecule has an increasingly difficult time sensing whether the polarization state of the radiation is circulating clockwise or counter clockwise over the dimensions of the molecule or the transition within the molecule.

3.4.6.4 NMR Chiral Discrimination

It is well known the NMR spectroscopy, as currently practiced, has no discrimination for enantiomeric pairs of molecules. In other words, NMR is blind to chirality. Of course diastereomeric interactions between pairs of chiral molecules, such as engendered by NMR shift reagents, do reveal shifts in NMR lines that can be associated with different contributions from pairs of enantiomers, but these are subtle and difficult to predict theoretically. A number of attempts, both theoretical and experimental, have been made to observe optical activity in NMR, such as by exciting the sample with high intensity circularly polarized radiation (Ikalainen *et al.*, 2008). The latest effort, which shows some promise for experimental realization, is the application of an orthogonal electric field while conducting an NMR measurement (Buckingham and Fischer, 2007). Perhaps one day, a form of NMR optical activity will be reported.

References

Arago, D.F. (1811) *Mem. de L'Inst.*, **12**, part 1, 93.

Barron, L.D. (1994) CP violation and molecular physics. *Chem. Phys. Lett.*, **221**, 311–316.

Barron, L.D. (2004). *Molecular Light Scattering and Optical Activity*, 2nd edn, Cambridge University Press, Cambridge.

Barron, L.D., and Buckingham, A.D. (1972) Rayleigh and Raman scattering by molecules in magnetic fields. *Mol. Phys.*, **23**, 145–150.
Barron, L.D. (1975) Magnetic vibrational optical activity in the resonance Raman spectrum of ferrocytochrome c. *Nature*, **275**, 372–374.
Barron, L.D. (1977) Magnetic resonance Raman optical activity of ferrocytochrome c at different laser wavelengths. *Mol. Phys.*, **46**(3), 579–581.
Barron, L.D., and Meehan, C. (1979) Magnetic optical activity in the resonance Raman spectra of metal halides: Optical detection of ground-state Zeeman splitting. *Chem. Phys. Lett.*, **66**(3), 444–448.
Barron, L.D., Meehan, C., and Vrbancich, J. (1982a) Magnetic resonance-Raman optical activity of ferrocytochrome c. *J. Raman Spectrosc.*, **12**(3), 251–261.
Barron, L.D., Vrbancich, J., and Watts, R.S. (1982b) Magnetic resonance-Raman optical activity of $IrBr_6^{2-}$. *Chem. Phys. Lett.*, **89**(1), 71–74.
Barron, L.D., and Vrbancich, J. (1983) On the sign convention for Raman optical activity. *Chem. Phys. Lett.*, **102**, 285–286.
Buckingham, A.D., and Stephens, P.J. (1966) Magnetic optical activity. *Annu. Rev. Phys. Chem.*, **17**, 399–432.
Buckingham, A.D., and Fischer, P. (2007) Direct chiral discrimination in NMR spectroscopy. *Chem. Phys.*, **324**, 111–116.
Cahn, R.S. (1964) *J. Chem. Educ.* **41**, 116–125.
Cahn, R.S., Ingold, C., and Prelog, V. (1966) Specification of molecular chirality. *Agnew. Chem., Int. Ed. Engl.*, **5**, 385–415.
Chabay, I., Hsu, E.C., and Holzwarth, G. (1972) Infrared circular dichroism measurement between 2000 and 5000 cm^{-1}: Pr^{+3} tartrate complexes. *Chem. Phys. Lett.*, **15**, 211–214.
Cox, J.N., and Richardson, F., S. (1977) Differential scattering of spin $+/-$ (1/2) neutrons by chiral systems. *I. Theory. J. Chem. Phys.*, **67**, 5702–5713.
Devine, T.R., and Keiderling, T.A. (1983) Magnetic vibrational circular dichroism of methyl halides in solution. *J. Chem. Phys.*, **79**, 5796–5801.
Devine, T.R., and Keiderling, T.A. (1984) Magnetic vibrational circular dichroism of benzene and 1,3, 5-trisubstituted derivatives. *J. Phys. Chem.*, **88**, 390–394.
Devine, T.R., and Keiderling, T.A. (1985) Magnetic vibrational circular dichroism of metal hexacarbonyls. *J. Chem. Phys.*, **83**, 3749–3754.
Fresnel, A. (1824) *Bull. Soc. Philomath.*, 147.
Gussakovsky, E. (2010) Circularly polarized luminescence (CPL) of proteins and protein complexes. *Rev. Fluoresc.*, **5**, 425–459.
He, Y., Cao, X., Nafie, L.A., and Freedman, T.B. (2001) *Ab initio* VCD calculation of a transition-metal containing molecule and a new intensity enhancement mechanism for VCD. *J. Am. Chem. Soc.*, **123**, 11320–11321.
Hecht, L., and Nafie, L.A. (1990) Linear polarization Raman optical activity: a new form of natural optical activity. *Chem. Phys. Lett.*, **174**, 575–582.
Hsu, E.C., and Holzwarth, G. (1973) Vibrational circular dichroism observed in crystalline α-nickel sulfate hexahydrate and α-zinc selenate hexahydrate between 1900 and 5000 cm^{-1}. *J. Chem. Phys.*, **59**, 4678–4685.
Ikalainen, S., Lantto, P., Manninen, P., and Vaara, J. (2008) Laser-induced nuclear magnetic resonance splitting in hydrocarbons *J. Chem. Phys.*, **129**, 124102.
Keiderling, T.A. (1981) Observation of magnetic vibrational circular dichroism. *J. Chem. Phys.*, **75**, 3639–3641.
Kelvin, L. (1904) *The Baltimore Lectures*. C.J. Clay and Sons, London.
Kuberski, B., Pecul, M., and Szumna, A. (2008) A chiral "frozen" hydrogen bonding in C4-symmetric inherently chiral resorcin[4]arenes: NMR, X-ray, circular dichroism, and theoretical study. *Euro. J. Org. Chem.* 3069–3078.
Le Bel, J.A. (1874) *Bull. Soc. Chim.*, **22**, 337.
Lombardi, R.A., and Nafie, L.A. (2009) Observation and calculation of a new form of vibrational optical activity: Vibrational circular birefringence. *Chirality*, **21**, E277–E286.
Lowry, T.M. (1935) *Optical Rotatory Power*. Longmans, Green and Co., London.
Ma, S. (2007) Mid-infrared and near-infrared vibrational circular dichroism: New methodologies for biological and pharmaceutical applications. Ph.D. thesis, Syracuse University.
Mason, S.F. (1973). The development of theories of optical activity and their applications. In: *Optical Rotatory Dispersion and Circular Dichroism* (eds F. Ciardelli, and P. Salvadori), Heydon & Son, London, pp. 27–40.

Mauskopf, S. (2006) A history of chirality. In: *Chiral Analysis* (eds K.W. Busch, and M.A. Busch), Elsevier, Amsterdam, pp. 3–24.

Merten, C., Li, H., Lu, X. *et al.* (2010) Observation of resonant electronic and non-resonance-enhanced vibrational natural Raman optical activity. *J. Raman Spectrosc.*, **41**, 1273–1275.

Mislow, K. (1966) *Introduction to Stereochemistry*, W.A. Benjamin, Inc., New York.

Moscowitz, A. (1962) *Adv. Chem. Phys.*, **4**, 67.

Mueller, G., Mueller, F.C., Maupin, C.L., and Riehl, J.P. (2005) The measurement of the fluorescence detected circular dichroism (FDCD) from a chiral Eu(III) system. *Chem. Commun.* 3615–3617.

Nafie, L.A. (1983) An alternative view on the sign convention of Raman optical activity. *Chem. Phys. Lett.*, **102**, 287–288.

Nafie, L.A. (1996) Theory of resonance Raman optical activity: The single-electronic state limit. *Chem. Phys.*, **205**, 309–322.

Nafie, L.A., and Freedman, T.B. (2000). Vibrational optical activity theory. In: *Circular Dichroism*, 2nd edn (eds N. Berova, K. Nakanishi, and R. W. Woody), John Wiley & Sons, Inc., New York, pp. 97–131.

Pasteur, L. (1848) *Ann. Chim.*, **24**, 457.

Pawlikowski, M., and Keiderling, T.A. (1984) Vibronic coupling effects in magnetic vibrational circular dichroism. A model formalism for doubly degenerate states. *J. Chem. Phys.*, **81**, 4765–4773.

Pawlikowski, M., and Pilch, M. (1992) Vibronic theory of magnetic vibrational circular dichroism in systems with four-fold symmetry: Theoretical analysis for copper tetraphenylporphyrin. *J. Chem. Phys.*, **96**, 4982–4990.

Quack, M., and Stohner, J. (2005) Parity violation in chiral molecules. *Chimia*, **59**, 530–538.

Rhee, H., Young-Gun, J., Lee, J.-S. *et al.* (2009) Femtosecond characterization of vibrational optical activity of chiral molecules. *Nature*, **458**, 310–313.

Salvadori, P., and Ciardelli, F. (1973) An introduction to chiroptical techniques: Basic principles, definitions and applications. In: *Optical Rotatory Dispersion and Circular Dichroism* (eds F. Ciardelli, and P. Salvadori), Heyden & Sons, Ltd., London, pp. 3–24.

Salzman, W.R., and Polavarapu, P.L. (1991) Calculated rotational strengths and dissymmetry factors for rotational transitions of the chiral deuterated oxiranes, methy- and dimethyl-oxirane, and methylthirrane. *Chem. Phys. Lett.*, **79**, 12.

Snyder, P.A., and Johnson, W.C., Jr. (1978) Circular dichroism of l-borneol. *J. Am. Chem. Soc.*, **100**, 2939–2944.

Snyder, P.A., and Rowe, E.M. (1980) The first use of synchrotron radiation for vacuum ultraviolet circular dichroism measurements. *Nucl. Instrum. Methods*, **172**, 345–349.

Stephens, P.J. (1974) Magnetic circular dichroism. *Annu. Rev. Phys. Chem.*, **25**, 201.

van't Hoff, J.H. (1874) *Chemische Constitutie van Organische Verbindingen*, Utrecht.

Zuber, G., Wipf, P., and Beratan, D.N. (2008) Exploring the optical activity tensor by anisotropic Rayleigh optical activity scattering. *ChemPhysChem.*, **9**, 265–271.

4

Theory of Vibrational Circular Dichroism

In this chapter we combine the theoretical developments of the last two chapters and present the theory of vibrational circular dichroism at a level sufficient to understand ongoing applications of VCD. These include the determination of absolute configuration of small- to medium-sized chiral molecules and the solution conformations of all classes of chiral molecules, including molecules of biological significance such as peptides, proteins, nucleic acids, and carbohydrates.

As explained first in Chapter 3, circular dichroism in general, and VCD in particular, arises from rotational strength, $R = \mathbf{Im}\, \boldsymbol{\mu} \cdot \boldsymbol{m}$, the imaginary part of the vector dot product, of the electric dipole transition moment, $\boldsymbol{\mu}$, with the magnetic dipole transition moment, $\boldsymbol{m}$. As illustrated in Figure 4.1, the electric dipole transition moment in a molecule arises from and is parallel to the sum of linear oscillations of charge while the magnetic dipole transition moment arises from the sum of circulations of charge in a plane and each magnetic moment contribution is perpendicular to its associated plane of charge circulation. If the sense of circular of charge reverses, clockwise to counterclockwise from some perspective for example, then the direction of the associated magnetic dipole moment reverses. This effect is illustrated in Figure 4.1 where positive CD arises from positive motion upward by positive charge along a right-handed helical path. The net upward direction gives rise to an upward electric dipole moment and an upward magnetic dipole moment. For upward motion of a positive charge along a left-handed helical path gives rise again to an upward electric dipole moment but a *downward* magnetic dipole moment and hence negative CD. For oscillatory motion, when the positive charge moves downward, all moments reverse direction and the vector dot product of the two moments remains the same. By comparison vibrational absorption (VA), arises from the dipole strength $D = |\boldsymbol{\mu}|^2$, the absolute square of the electric dipole transition moment that is always positive for all motions of charge.

The theory of VCD at the simplest level can be understood in terms of transition moment derivatives, in a manner parallel to IR vibrational intensities described in Chapter 2. This is illustrated by the following expressions for the position form of the dipole strength, $D^a_{r,gv',gv'}$, adapted from Equations (2.59) and (2.69b), and the rotational strength $R^a_{r,gv',gv'}$; these expressions are proportional to

Vibrational Optical Activity: Principles and Applications, First Edition. Laurence A. Nafie.

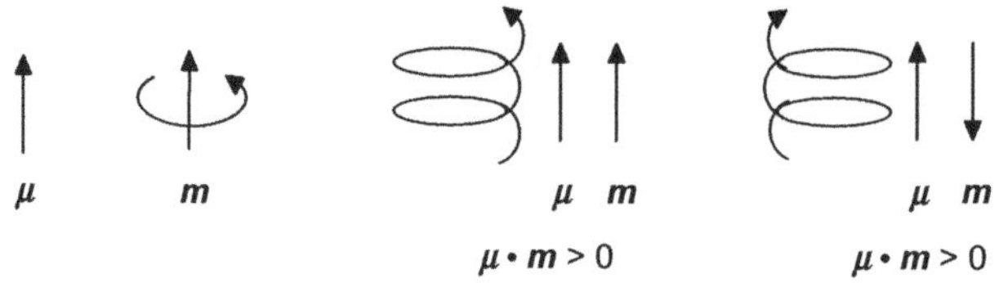

*Figure 4.1 The upward direction of an electric dipole moment, **μ**, arising from the upward motion of positive charge and the upward direction of a magnetic dipole moment, **m**, arising from a counterclockwise circulation of positive charge when viewed from above. Combining these motions for a positive charge moving upward along a right-handed helical path gives rise to parallel electric and magnetic dipole moments and positive CD, whereas the corresponding motion along a left-handed helical path gives rises to anti-parallel electric and magnetic moments and hence negative CD*

VA and VCD intensities, respectively.

$$D^a_{r,gv',gv} = \left| \left(\frac{\partial \langle \boldsymbol{\mu} \rangle}{\partial Q_a} \right)_{P_a=0} \left\langle \phi^a_{gv'} | Q_a | \phi^a_{gv} \right\rangle \right|^2 \tag{4.1}$$

$$R^a_{r,gv',gv'} = \mathrm{Im}\left[\left(\frac{\partial \langle \boldsymbol{\mu} \rangle}{\partial Q_a} \right)_{Q_a=0} \cdot \left(\frac{\partial \langle \boldsymbol{m} \rangle}{\partial P_a} \right)_{P_a=0} \left\langle \phi^a_{gv} | Q_a | \phi^a_{gv'} \right\rangle \left\langle \phi^a_{gv'} | P_a | \phi^a_{gv} \right\rangle \right] \tag{4.2}$$

Analogous expression using the velocity form of the electric dipole transitions moments from Equations (2.60) and (2.71b) are:

$$D^a_{v,gv',gv} = \omega_a^{-2} \left| \left(\frac{\partial \langle \dot{\boldsymbol{\mu}} \rangle}{\partial P_a} \right)_{P_a=0} \left\langle \phi^a_{gv'} | P_a | \phi^a_{gv} \right\rangle \right|^2 \tag{4.3}$$

$$R^a_{v,gv',gv'} = \omega_a^{-1} \mathrm{Re}\left[\left(\frac{\partial \langle \dot{\boldsymbol{\mu}} \rangle}{\partial P_a} \right)_{P_a=0} \cdot \left(\frac{\partial \langle \boldsymbol{m} \rangle}{\partial P_a} \right)_{P_a=0} \left\langle \phi^a_{gv} | P_a | \phi^a_{gv'} \right\rangle \left\langle \phi^a_{gv'} | P_a | \phi^a_{gv} \right\rangle \right] \tag{4.4}$$

From these expressions, it is clear that the magnetic dipole transition moment is a dynamic, not a static, property of the molecule that must couple with nuclear momenta, or velocities, where for normal modes $P_a = \dot{Q}_a$ as normal coordinates are square-root-mass weighted. Dynamic coupling is not available within the Born–Oppenheimer approximation and hence formalism must be used that transcends that level of standard quantum chemistry. As explained in Chapter 2, the velocity form of the dipole moment also requires formalism beyond the BO approximation.

Before embarking on the formal theory of VCD, we mention that a number of models of VCD have been developed that circumvent problems inherent in the formal theory of VCD, which involves problems with the BO approximation just mentioned. A brief description of the simplest of these models can be found in Appendix A and a more thorough description of all of the models of VCD and their relationship to the general theory of VCD has been published a number of years ago (Freedman and Nafie, 1994).

4.1 General Theory of VCD

Previously, expressions have been derived for the absorption intensity of a molecular transition in terms of the dipole strength between electronic states e and g in Equations (2.17) and (2.21).

Subsequently the definition was extended to vibrational transitions from level gv to level gv' of the ath vibrational mode of the ground electronic state, $D^a_{gv,gv'}$, in Equations (2.59) and (2.60), and then for fundamental transitions $g0$ to $g1$ as $D^a_{g0,g1}$, in Equation (2.92). In addition, circular dichroism intensity was defined in terms of the rotational strength, R_a, of the excited state a in Equations (3.48) to (3.50) where comparison with the corresponding expression for the dipole strength, D_a, is also given. In this section, these definitions are extended to include the rotational strength of vibrational transitions within a given electronic state, typically the ground electronic state.

4.1.1 Definitions of VCD Intensity and Rotational Strength

The dipole strength in the position-dipole-moment formulation for a vibrational transition between states $g0$ and $g1$ of the ath normal mode in the ground electronic state is:

$$D^a_{r,g1,g0} = \left|\left\langle \Psi^a_{g1} \middle| \mu_\beta \middle| \Psi^a_{g0} \right\rangle\right|^2 \tag{4.5}$$

where Ψ^a_{g0} and Ψ^a_{g1} are the vibronic wavefunctions of states $g0$ and $g1$, respectively, and μ_β is the βth Cartesian component of the electric dipole moment operator given by:

$$\mu_\beta = \mu^E_\beta + \mu^N_\beta = -\sum_j e r_{j\beta} + \sum_J Z_J e R_{J\beta} \tag{4.6}$$

This equation shows the separate contributions of the electrons and the nuclei to the dipole moment operator as defined previously in Equation (2.63). The corresponding definition of the dipole strength in the velocity dipole formulation is given by:

$$D^a_{v,g1,g0} = \omega_a^{-2}\left|\left\langle \Psi^a_{g1} \middle| \dot{\mu}_\beta \middle| \Psi^a_{g0} \right\rangle\right|^2 \tag{4.7}$$

where the velocity dipole moment operator is given by:

$$\dot{\mu}_\beta = \dot{\mu}^E_\beta + \dot{\mu}^N_\beta = -\sum_j e\dot{r}_{j\beta} + \sum_J Z_J e\dot{R}_{J\beta} = -\sum_j \frac{e}{m} p_{j\beta} + \sum_J \frac{Z_J e}{M_J} P_{J\beta} \tag{4.8}$$

Here the velocity operator is of the form momentum divided by corresponding mass as in

$$\dot{r}_{j\beta} = -(i\hbar/m)\partial/\partial r_{j\beta} = p_{j\beta}/m \tag{4.9a}$$

$$\dot{R}_{J\beta} = -(i\hbar/M_J)\partial/\partial R_{J\beta} = P_{J\beta}/M \tag{4.9b}$$

and the second relationship in Equation (4.8) uses instead the corresponding momentum operators, $p_{j\beta}$ and $P_{J\beta}$. One reason to present the expressions for the velocity form of the dipole moment and dipole strength is to illustrate the close connection between velocity dipole formulation of VA intensities and the expressions needed for the rotational strength and VCD intensities.

The rotational strength for the vibrational transition from $g0$ to $g1$ for the ath normal mode is given by:

$$R^a_{r,g1,g0} = \mathrm{Im}\left[\left\langle \Psi^a_{g0} \middle| \mu_\beta \middle| \Psi^a_{g1} \right\rangle \cdot \left\langle \Psi^a_{g1} \middle| m_\beta \middle| \Psi^a_{g0} \right\rangle\right] \tag{4.10}$$

where Im stands for the imaginary part insuring that the rotational strength is a real quantity required for an observable by quantum mechanics. This is necessary in this particular case because the magnetic dipole moment operator is pure imaginary. In Equation (4.10), the operator m_β is the βth Cartesian component of the magnetic dipole operator given by:

$$m_\beta = m_\beta^E + m_\beta^N = -\sum j \frac{e}{2mc} \varepsilon_{\beta\gamma\delta} r_{j\gamma} p_{j\delta} + \sum_J \frac{Z_J e}{2M_J c} \varepsilon_{\beta\gamma\delta} R_{J\gamma} P_{J\delta} \tag{4.11}$$

This operator is defined in terms of the vector cross product of the position and momentum vectors, which can be seen from the following relationship:

$$\varepsilon_{\beta\gamma\delta} r_{j\gamma} p_{j\delta} = (\boldsymbol{r} \times \boldsymbol{p})_{j\beta} \tag{4.12}$$

where the Einstein convention of repeated Greek subscripts indicates summation over Cartesian directions x, y, and z, and the symbol $\varepsilon_{\beta\gamma\delta}$ is the alternating tensor equal to $+1$ for an even permutation of the order xyz and -1 for an odd permutation. For example, the xth component of the electron contribution to the magnetic dipole operator in Equation (4.11) is given by:

$$m_x^E = -\sum_j \frac{e}{2mc} (r_{jy} p_{jz} - r_{jz} p_{jy}) \tag{4.13}$$

The rotational strength given in Equation (4.10) is the scalar product of two different vectors, the transition moment matrix elements for the electric dipole moment, $\langle \Psi_{g0}^a | \mu_\beta | \Psi_{g1}^a \rangle$, and the magnetic dipole moment, $\langle \Psi_{g1}^a | m_\beta | \Psi_{g0}^a \rangle$. The scalar product is positive if these two vectors have a projection angle between them of less than 90° and negative if the angle is more than 90° up to a maximum of 180°. This corresponds to VCD bands that in some cases are positive and in others negative. By contrast, the dipole strength is the absolute square of the electric dipole transition moment and is always positive, as are vibrational absorption intensities.

The definition of the rotational strength given in Equation (4.9) is based on the position formulation of the electric dipole transition moment. Alternatively, the rotational strength can be defined using the velocity formulation of the rotational strength given by:

$$R_{v,g1,g0}^a = \omega_a^{-1} \mathrm{Re}\left[\left\langle \Psi_{g0}^a \middle| \dot{\mu}_\beta \middle| \Psi_{g1}^a \right\rangle \cdot \left\langle \Psi_{g1}^a \middle| m_\beta \middle| \Psi_{g0}^a \right\rangle\right] \tag{4.14}$$

The real part of this expression is needed because, as seen from Equation (4.9), there are two imaginary operators in this equation, the dipole velocity moment operator and the magnetic dipole moment operator.

4.1.2 Complete Adiabatic Correction to the Born–Oppenheimer Approximation

For completeness, here we briefly review from Chapter 2 the theoretical basis of the nuclear velocity dependence of the electronic wavefunction. The Born–Oppenheimer (BO) approximation is obtained by a separation of the electronic and nuclear motions in a molecule. The BO Hamiltonian operator is given by

$$\mathcal{H}_{BO} = H_E^A(\boldsymbol{R}) + (T_N)_N = H_E^A(\boldsymbol{R}) - \sum_J \frac{\hbar^2}{M_J}\left(\frac{\partial^2}{\partial R_{J,\alpha}^2}\right)_{nucl} \tag{4.15}$$

The total Hamiltonian operator of the molecule is separated into an electronic part and a nuclear part consisting of the nuclear kinetic energy operator that acts only the nuclear wavefunction, as indicated by the subscript *nucl*. Restricting the operation of the nuclear wavefunction is the key step in the BO approximation, because otherwise is it could also operate on the electronic wavefunction, as it does when one goes beyond the BO approximation. The BO adiabatic wavefunction yields the adiabatic energy levels of the molecule E_{ev}^A by factorization of the wavefunction into a product of an electronic wavefunction, with parametric (but not operator) dependence on the nuclear positions, and a nuclear wavefunction,

$$\mathcal{H}_{BO}\Psi_{ev}^A(\boldsymbol{r},\boldsymbol{R}) = E_{ev}^A\Psi_{ev}^A(\boldsymbol{r},\boldsymbol{R}) = E_{ev}^A\psi_e^A(\boldsymbol{r},\boldsymbol{R})\phi_{ev}(\boldsymbol{R}) \tag{4.16}$$

The adiabatic electronic wavefunction, $\psi_e^A(\boldsymbol{r},\boldsymbol{R})$, is solved by using the adiabatic electronic Hamiltonian, $H_E^A(\boldsymbol{R})$, which has parametric dependence on the nuclear positions. This yields a Schrödinger equation with an energy $E_e^A(\boldsymbol{R})$ having a parametric dependence on the nuclear positions,

$$H_E^A(\boldsymbol{R})\psi_e^A(\boldsymbol{r},\boldsymbol{R}) = E_e^A(\boldsymbol{R})\psi_e^A(\boldsymbol{r},\boldsymbol{R}) \tag{4.17}$$

This equation is solved for particular choices of nuclear positions as parameters, where the nuclear positions can literally be represented by numbers with appropriate units. The energy $E_e^A(\boldsymbol{R})$ becomes a potential energy surface defined by its second derivatives with respect to nuclear positions, and this is used for the solution of the nuclear wavefunction, which after integration over the nuclear coordinates yields the vibronic energy levels, E_{ev}^A, of the molecular wavefunction in Equation (4.16):

$$\left[(T_N)_N + E_e^A(\boldsymbol{R})\right]\phi_{ev}(\boldsymbol{R}) = E_{ev}^A\phi_{ev}(\boldsymbol{R}) \tag{4.18}$$

$$E_{ev}^A = \langle\phi_{ev}(\boldsymbol{R})|\left[(T_N)_N + E_e^A(\boldsymbol{R})\right]|\phi_{ev}(\boldsymbol{R})\rangle \tag{4.19}$$

We note that the nuclear positions, $\boldsymbol{R}$, play the role of *classical* variables in the solution of the electronic wave equation in Equation (4.17) and then *quantum mechanical* position operators when the nuclear wave equation and vibronic energy levels are obtained in Equations (4.18) and (4.19).

To proceed beyond the BO approximation, we now allow a term in which one of the derivatives in the nuclear kinetic energy operator acts on the electronic wavefunction while the other remains to operate on the nuclear wavefunction. The non-BO Hamilton is given by:

$$\mathcal{H}_{NBO} = \mathcal{H}_{BO} + (T_N)_{EN} = \mathcal{H}_{BO} - \sum_J \frac{\hbar^2}{M_J}\left(\frac{\partial}{\partial R_{J,\alpha}}\right)_{elec}\left(\frac{\partial}{\partial R_{J,\alpha}}\right)_{nucl} \tag{4.20}$$

The solution to this Hamiltonian is in general not a wavefunction factorable into a product of separate electronic and nuclear wavefunctions, and the new perturbation term serves to couple nuclear velocities to the velocities of the electrons, clearly beyond the BO approximation.

As was first demonstrated in 1983, it is possible to retain the factorable, adiabatic nature of the molecular wavefunction by back-converting the derivative of the nuclear kinetic energy that operates on the nuclear wavefunction from a quantum mechanical operator to a classical nuclear velocity coordinate as a new parametric variable of the electronic wavefunction (Nafie, 1983). This is accomplished explicitly here using the relationship $(\partial/\partial R_{J\beta})_{nucl} = iM_J\dot{R}_{J\beta}/\hbar$ presented in Equation (2.45) seen also in Equation (4.9b) above to give:

$$\mathcal{H}_{CA} = \mathcal{H}_{BO} + \left[T_N(\dot{\boldsymbol{R}})\right]_E = H_E^A(\boldsymbol{R}) + (T_N)_N - i\hbar\sum_J \left(\frac{\partial}{\partial R_{J,\alpha}}\right)_{elec} \dot{R}_{J,\alpha} \tag{4.21}$$

We call this new Hamiltonian operator, $\mathcal{H}_{CA}$, the complete adiabatic (CA) Hamiltonian.

The first and third terms in the second part of this equation constitute the CA *electronic* Hamiltonian operator

$$H_E^{CA}(\boldsymbol{R},\dot{\boldsymbol{R}}) = H_E^A(\boldsymbol{R}) - i\hbar\frac{\partial}{\partial \boldsymbol{R}}\cdot\dot{\boldsymbol{R}} \tag{4.22}$$

The solution of the Schrödinger equation with this operator is the CA electronic wavefunction, $\tilde{\psi}_e^{CA}(\boldsymbol{r},\boldsymbol{R},\dot{\boldsymbol{R}})$, which carries parametric dependence on both the nuclear positions and velocities.

$$\left(H_E^A(\boldsymbol{R}) - i\hbar\frac{\partial}{\partial \boldsymbol{R}}\cdot\dot{\boldsymbol{R}}\right)\tilde{\psi}_e^{CA}(\boldsymbol{r},\boldsymbol{R},\dot{\boldsymbol{R}}) = E_e^{CA}(\boldsymbol{R},\dot{\boldsymbol{R}})\tilde{\psi}_e^{CA}(\boldsymbol{r},\boldsymbol{R},\dot{\boldsymbol{R}}) \tag{4.23}$$

It has been demonstrated previously that the imaginary perturbation term in Equations (4.22) and (4.23) adds a description of electron *current* density to the electronic wavefunction. Further, it has been shown that the velocity perturbation does not make a first-order contribution to the electronic energy potential, that is, $E_e^{CA}(\boldsymbol{R},\dot{\boldsymbol{R}}) = E_e^A(\boldsymbol{R})$ through first order in $\dot{\boldsymbol{R}}$, as the electrostatic potential is not affected by instantaneous values of any nuclear velocities and hence does not depend on the velocities, $\dot{\boldsymbol{R}}$. As a further consequence, the vibrational wavefunctions are in no way affected by the change in the electronic wavefunction from adiabatic within the BO approximation to complete adiabatic beyond the BO approximation. In essence, the CA approximation merely adds a dynamic equivalent to the nuclear dependence of the motion of the nuclei, but does not affect the equilibrium conformational structure, the electronic or vibrational energy levels, or any other property of the molecule depending on its energy. Any molecular property that is affected by nuclear velocities is calculated through the vibrational wavefunction after the nuclear velocities are converted back into quantum mechanical operators and integrated over all nuclear coordinate space. This is directly analogous to what was done for nuclear positions where initially the electronic dependence on $\boldsymbol{R}$ is parametric, but the final molecular property integrated over nuclear coordinates requires that all nuclear positions are $\boldsymbol{R}$ integrated over all possible values as quantum mechanical variables (operators) of the nuclear wavefunction matrix element as in Equation (4.19). We next consider further formalities of the CA wavefunction.

4.1.3 Derivation of the Complete Adiabatic Wavefunction

The full CA wavefunction including the vibrational part is written as presented previously in Chapter 2 by Equations (2.47) and (2.48) reproduced here for convenience:

$$\tilde{\Psi}_{ev}^{CA}(\boldsymbol{r},\boldsymbol{R},\dot{\boldsymbol{R}}) = \tilde{\psi}_e^{CA}(\boldsymbol{r},\boldsymbol{R},\dot{\boldsymbol{R}})\phi_{ev}(\boldsymbol{R}) \tag{4.24}$$

where the electronic part of the CA wavefunction can be written in terms of real and imaginary parts as:

$$\tilde{\psi}_e^{CA}(\boldsymbol{r},\boldsymbol{R},\dot{\boldsymbol{R}}) = \psi_e^A(\boldsymbol{r},\boldsymbol{R}) + i\psi_e^{CA}(\boldsymbol{r},\boldsymbol{R},\dot{\boldsymbol{R}})$$
$$= \psi_e^0(\boldsymbol{r}) + \sum_J \left(\frac{\partial\psi_e^A(\boldsymbol{r})}{\partial \boldsymbol{R}_J}\right)_{R=0} \boldsymbol{R}_J + i\sum_J \left(\frac{\partial\psi_e^{CA}(\boldsymbol{r})}{\partial \dot{\boldsymbol{R}}_J}\right)_{\substack{R=0\\ \dot{R}=0}} \dot{\boldsymbol{R}}_J + \ldots \quad (4.25)$$

We now derive a more compact expression for the CA wavefunction presented in Equation (2.86) by means of vibronic coupling theory showing a key approximation invoked along the way. The real adiabatic part of the wavefunction is derived starting from the Herzberg–Teller expansion given by:

$$\psi_e^A(\boldsymbol{r},\boldsymbol{R}) = \psi_e^0(\boldsymbol{r}) - \sum_J \sum_{s\neq e} \frac{\langle\psi_s^0|(\partial H_E^A/\partial R_{J,\alpha})_0|\psi_e^0\rangle}{E_s^0 - E_e^0} \psi_s^0(\boldsymbol{r}) R_{J,\alpha}$$
$$= \psi_e^0(\boldsymbol{r}) + \sum_{s\neq e} \sum_J \langle\psi_s^0|(\partial\psi_e^A/\partial R_{J,\alpha})_0\rangle R_{J,\alpha}\psi_s^0(\boldsymbol{r}) \quad (4.26)$$

Here we have simplified the Hertzberg–Teller expansion by recognizing that the perturbation term in the matrix element can be obtained from the first derivative of the $\psi_e^A(\boldsymbol{r},\boldsymbol{R})$ with respect to the nuclear coordinates. If desired, the sum over s in the second part of this equation may be carried out to closure to yield a Taylor series expansion through first order of $\psi_e^A(\boldsymbol{r},\boldsymbol{R})$ about the equilibrium nuclear position.

$$\psi_e^A(\boldsymbol{r},\boldsymbol{R}) = \psi_e^0(\boldsymbol{r}) + \sum_J (\partial\psi_e^A(\boldsymbol{r})/\partial R_{J,\alpha})_0 R_{J,\alpha} \quad (4.27)$$

Turning our attention to the imaginary CA term, we first write the non-BO wavefunction in terms of first-order perturbation theory using Equation (4.20) as the source of the perturbation,

$$\Psi_{ev}^{NBO}(\boldsymbol{r},\boldsymbol{R}) = \psi_e^A(\boldsymbol{r},\boldsymbol{R})\phi_{ev}(\boldsymbol{R})$$
$$+ \sum_{su\neq ev} \sum_J \left(\frac{\hbar^2}{M_J}\right) \frac{\langle\psi_s^A|\partial/\partial R_{J,\alpha}|\psi_e^A\rangle\langle\phi_{su}|\partial/\partial R_{J,\alpha}|\phi_{ev}\rangle}{E_{su} - E_{ev}} \psi_s^A(\boldsymbol{r},\boldsymbol{R})\phi_{su}(\boldsymbol{R}) \quad (4.28)$$

The unperturbed wavefunctions are taken to be the usual BO adiabatic product wavefunctions with the two perturbations from Equation (4.20) operating separately on the electronic and nuclear wavefunctions. This perturbation expansion is often called the Born–Oppenheimer non-adiabatic expansion. The wavefunction, $\Psi_{ev}^{NBO}(\boldsymbol{r},\boldsymbol{R})$, is clearly a non-separable, non-adiabatic wavefunction. The perturbation operator in the vibrational matrix element is next back-converted from a quantum mechanical operator into a classical variable. This gives a pre-CA (pCA) complex wavefunction that is still not separable as the sum is over all vibronic sublevels of the excited states, and the electronic and vibrational parts of the wavefunction are inextricably intermingled,

$$\Psi_{ev}^{pCA}(\boldsymbol{r},\boldsymbol{R},\dot{\boldsymbol{R}}) = \psi_e^A(\boldsymbol{r},\boldsymbol{R})\phi_{ev}(\boldsymbol{R})$$
$$+ i\hbar \sum_{su\neq ev} \sum_J \frac{\langle\psi_s^A|\partial/\partial R_{J,\alpha}|\psi_e^A\rangle\langle\phi_{su}|\phi_{ev}\rangle\dot{R}_{J,\alpha}}{E_{su} - E_{ev}} \psi_s(\boldsymbol{r},\boldsymbol{R})\phi_{su}(\boldsymbol{R}) \quad (4.29)$$

We now introduce the approximation that vibronic detail in the energy denominator is not important relative to the energy spacing between the state e and all other states s, where e will be assigned to the ground electronic state. This allows the substitution of the pure electronic energy denominator, $E_s^0 - E_e^0$, for the vibronic energy denominator, $E_{su} - E_{ev}$. This is an excellent approximation provided there are no low-lying electronic states in the molecule. The development of the theory of VCD in the presence of low-lying electronic states has been published (Nafie, 2004) and is summarized in Appendix C. With the approximation of neglecting excited state vibronic detail, we can sum over the excited vibrational states su to closure, that is, $\sum_u |\phi_{su}\rangle\langle\phi_{su}| = 1$. This yields a complex *factorable*, and hence *adiabatic*, wavefunction, which, because of its dependence on nuclear velocities, is beyond the BO approximation.

$$\tilde{\Psi}_{ev}^{CA}(\boldsymbol{r},\boldsymbol{R},\dot{\boldsymbol{R}}) = \left[\psi_e^A(\boldsymbol{r},\boldsymbol{R}) + i\hbar\sum_{s\neq e}\sum_J \frac{\langle\psi_s^0|(\partial\psi_e^A/\partial R_{J,\alpha})_0\rangle}{E_s^0 - E_e^0}\psi_s^0(\boldsymbol{r})\dot{R}_{J,\alpha}\right]\phi_{ev}(\boldsymbol{R}) \tag{4.30}$$

If we now substitute Equation (4.26) into this equation and combine the two perturbations terms we have:

$$\tilde{\Psi}_{ev}^{CA}(\boldsymbol{r},\boldsymbol{R},\dot{\boldsymbol{R}}) = \left[\psi_e^0(\boldsymbol{r}) + \sum_{s\neq e}\sum_J \left\langle\psi_s^0\left|\left(\frac{\partial\psi_e^A}{\partial R_{J,\alpha}}\right)_0\right.\right\rangle\psi_s^0(\boldsymbol{r})\left(R_{J,\alpha} + \frac{i\dot{R}_{J,\alpha}}{\omega_{se}^0}\right)\right]\phi_{ev}(\boldsymbol{R}) \tag{4.31}$$

where we have written $\omega_{se}^0 = (E_s^0 - E_e^0)/\hbar$. This is the final compact expression for the CA wavefunction that will be used in the following sections for the derivation of all vibrational intensities.

4.1.4 Vibronic Coupling Theory of VCD and IR Intensity

In this section we develop the basic vibronic coupling expressions for VCD intensity that forms the basis for all of the various ways of formulating VCD intensities. In doing so we will follow the development of IR intensity theory in Chapter 2 by presenting the rotational strength and dipole strengths in parallel so that the departures needed for VCD theory are clearly seen relative to the standard theory of IR vibrational absorption.

The transition moments needed for the definitions of the dipole and rotational strengths defined above in Equations (4.5), (4.7), (4.10) and (4.14) are given below for the position and velocity forms of the electric dipole transition moment and the magnetic dipole transition moment between vibrational states $g0$ and $g1$ of the ground electronic state,

$$\left\langle\Psi_{g0}^a\left|\mu_\beta\right|\Psi_{g1}^a\right\rangle = \sum_J P_{r,\alpha\beta}^J S_{J\alpha,a}\left\langle\phi_{g1}^a\left|Q_a\right|\phi_{g0}^a\right\rangle \tag{4.32}$$

$$\left\langle\Psi_{g0}^a\left|\dot{\mu}_\beta\right|\Psi_{g1}^a\right\rangle = \sum_J P_{v,\alpha\beta}^J S_{J\alpha,a}\left\langle\phi_{g1}^a\left|P_a\right|\phi_{g0}^a\right\rangle \tag{4.33}$$

$$\left\langle\Psi_{g0}^a\left|m_\beta\right|\Psi_{g1}^a\right\rangle = \sum_J M_{\alpha\beta}^J S_{J\alpha,a}\left\langle\phi_{g1}^a\left|P_a\right|\phi_{g0}^a\right\rangle \tag{4.34}$$

where the S-vector converting Cartesian into normal coordinates is given by:

$$S_{J\alpha,a} = \left(\frac{\partial R_{J\alpha}}{\partial Q_a}\right)_{Q=0} = \left(\frac{\partial \dot{R}_{J\alpha}}{\partial \dot{Q}_a}\right)_{\dot{Q}=0} = \left(\frac{\partial \dot{R}_{J\alpha}}{\partial P_a}\right)_{P=0} \tag{4.35}$$

The atomic polar tensors (APTs), $P^J_{r,\alpha\beta}$ and $P^J_{v,\alpha\beta}$, were introduced in Chapter 2. Here we introduced a new tensor $M^J_{\alpha\beta}$ called the atomic axial tensor (AAT), which contains information about the generation of magnetic dipole moments in response to nuclear motion (actually velocities, not displacements). As with the transition moments as a whole, the APTs and AATs can be written as the sum of an electronic contribution and nuclear contribution, given by the following expressions:

$$P^J_{r,\alpha\beta} = E^J_{v,\alpha\beta} + N^J_{\alpha\beta} \quad \text{or} \quad P^J_{v,\alpha\beta} = E^J_{v,\alpha\beta} + N^J_{\alpha\beta} \tag{4.36}$$

$$M^J_{\alpha\beta} = I^J_{\alpha\beta} + J^J_{\alpha\beta} \tag{4.37}$$

The electronic contributions to the transition moments are the key focus of interest. They contain information about the motion of electronic charge in response to nuclear motion as changes in charge density with nuclear displacement or current density with nuclear velocities. The nuclear contributions are trivial by comparison and require only information on the charge of the nucleus. For example, the electronic and nuclear position electric-dipole transitions moments are:

$$\left\langle \Psi^a_{g0} \middle| \mu^E_\beta \middle| \Psi^a_{g1} \right\rangle = \sum_J E^J_{r,\alpha\beta} S_{J\alpha,a} \left\langle \phi^a_{g1} \middle| Q_a \middle| \phi^a_{g0} \right\rangle \tag{4.38}$$

$$\left\langle \Psi^a_{g0} \middle| \mu^N_\beta \middle| \Psi^a_{g1} \right\rangle = \sum_J N^J_{\alpha\beta} S_{J\alpha,a} \left\langle \phi^a_{g1} \middle| Q_a \middle| \phi^a_{g0} \right\rangle \tag{4.39}$$

$$N^J_{\alpha\beta} = \left(\frac{\partial \mu^N_\beta}{\partial R_{J\alpha}} \right)_{R=0} = Z_J e \delta_{\alpha\beta} \tag{4.40}$$

The nuclear APT, $N^J_{\alpha\beta}$, is simply the charge $Z_J e$ of the nucleus at the location of the nucleus whereas the electronic APT, $E^J_{r,\alpha\beta}$, is much more complex. The corresponding electronic and nuclear transition moments for the magnetic dipole moment are

$$\left\langle \Psi^a_{g0} \middle| m^E_\beta \middle| \Psi^a_{g1} \right\rangle = \sum_J I^J_{\alpha\beta} S_{J\alpha,a} \left\langle \phi^a_{g1} \middle| P_a \middle| \phi^a_{g0} \right\rangle \tag{4.41}$$

$$\left\langle \Psi^a_{g0} \middle| m^N_\beta \middle| \Psi^a_{g1} \right\rangle = \sum_J J^J_{\alpha\beta} S_{J\alpha,a} \left\langle \phi^a_{g1} \middle| P_a \middle| \phi^a_{g0} \right\rangle \tag{4.42}$$

$$J^J_{\alpha\beta} = \left(\frac{\partial m^N_\beta}{\partial \dot{R}_{J\alpha}} \right)_{\dot{R}=0} = \frac{Z_J e}{2c} \varepsilon_{\alpha\beta\gamma} R^0_{J\gamma} \tag{4.43}$$

The nuclear AAT, $J^J_{\alpha\beta}$, describes the angular motion of a nucleus with charge $Z_J e$ on moment arm $R^0_{J\gamma}$ from the origin, whereas the electronic AAT, $I^J_{\alpha\beta}$, describes the magnetic moment moments set up throughout the molecule by virtue of the αth Cartesian component of the nuclear velocity vector, $\dot{R}_{J\alpha}$.

We now focus on the electronic contributions to the electric and magnetic dipole transition moments in terms of vibronic coupling expressions, and present the final expressions that form the basis of all formulations of VCD intensities. Substituting the expression for the CA wavefunction in Equation (4.31) into the electric dipole transition moment in Equation (4.38), we have:

$$\left\langle \Psi^a_{g1} \middle| \mu^E_\beta \middle| \Psi^a_{g0} \right\rangle = 2 \sum_{e \neq g} \sum_J \left\langle \psi^0_g \middle| \mu^E_\beta \middle| \psi^0_e \right\rangle \left\langle \psi^0_e \middle| (\partial \psi^A_g / \partial R_{J,\alpha})_0 \right\rangle S_{J\alpha,a} \left\langle \phi^a_{g1} \middle| Q_a \middle| \phi^a_{g0} \right\rangle \tag{4.44}$$

and again from Equation (4.38) the electronic contribution to the APT is:

$$E^J_{r,\alpha\beta} = 2\sum_{e\neq g}\left\langle \psi^0_g \middle| \mu^E_\beta \middle| \psi^0_e \right\rangle \left\langle \psi^0_e \middle| (\partial \psi^A_g / \partial R_{J,\alpha})_0 \right\rangle \tag{4.45}$$

The corresponding equations for electronic part of magnetic dipole transition moment and AAT are given by:

$$\left\langle \Psi^a_{g1} \middle| m^E_\beta \middle| \Psi^a_{g0} \right\rangle = 2i\hbar \sum_{e\neq g}\sum_{J} \frac{\left\langle \psi^0_g \middle| m^E_\beta \middle| \psi^0_e \right\rangle \left\langle \psi^0_e \middle| (\partial \psi^A_g / \partial R_{J,\alpha})_0 \right\rangle S_{J\alpha,a} \left\langle \phi^a_{g1} \middle| P_a \middle| \phi^a_{g0} \right\rangle}{E^0_e - E^0_g} \tag{4.46}$$

$$I^J_{\alpha\beta} = 2i\hbar \sum_{e\neq g} \frac{\left\langle \psi^0_g \middle| m^E_\beta \middle| \psi^0_e \right\rangle \left\langle \psi^0_e \middle| (\partial \psi^A_g / \partial R_{J,\alpha})_0 \right\rangle}{E^0_e - E^0_g} \tag{4.47}$$

Finally, for completeness, we provide the corresponding velocity form of the electronic APT, which can be seen to be closely related to $I^J_{\alpha\beta}$:

$$\left\langle \Psi^a_{g1} \middle| \dot{\mu}^E_\beta \middle| \Psi^a_{g0} \right\rangle = 2i\hbar \sum_{e\neq g}\sum_{A} \frac{\left\langle \psi^0_g \middle| \dot{\mu}^E_\beta \middle| \psi^0_e \right\rangle \left\langle \psi^0_e \middle| (\partial \psi^A_g / \partial R_{J,\alpha})_0 \right\rangle S_{J\alpha,a} \left\langle \phi^a_{g1} \middle| P_a \middle| \phi^a_{g0} \right\rangle}{E^0_e - E^0_g} \tag{4.48}$$

$$E^J_{v,\alpha\beta} = 2i\hbar \sum_{e\neq g} \frac{\left\langle \psi^0_g \middle| \dot{\mu}^E_\beta \middle| \psi^0_e \right\rangle \left\langle \psi^0_e \middle| (\partial \psi^A_g / \partial R_{A,\alpha})_0 \right\rangle}{E^0_e - E^0_g} \tag{4.49}$$

Formally, the position and velocity forms of the APT are equal. Equality of the electronic APTs in Equations (4.45) and (4.49) requires that

$$\left\langle \psi^0_e \middle| \dot{\mu}^E_\beta \middle| \psi^0_g \right\rangle = i\omega^0_{eg} \left\langle \psi^0_e \middle| \mu^E_\beta \middle| \psi^0_g \right\rangle \tag{4.50}$$

This is the standard hypervirial relationship, which holds for exact wavefunctions, between the matrix element of an operator and that of its time derivative first presented in Equation (2.90), and the corresponding relationship for the vibrational transition matrix elements given previously in Equation (2.41) is:

$$\left\langle \phi^a_{g1} \middle| P_a \middle| \phi^a_{g0} \right\rangle = \left\langle \phi^a_{g1} \middle| \dot{Q}_a \middle| \phi^a_{g0} \right\rangle = i\omega_a \left\langle \phi^a_{g1} \middle| Q_a \middle| \phi^a_{g0} \right\rangle \tag{4.51}$$

The Cartesian components of the transition-moment derivatives, given in Equations (4.1) to (4.4), can be expressed in terms of vibronic coupling theory by the following relationships:

$$\begin{aligned}
\left(\frac{\partial\langle\mu_\beta\rangle}{\partial Q_a}\right)_{Q=0} &= \sum_J \left(\frac{\partial\langle\mu_\beta\rangle}{\partial R_{J,\alpha}}\right)_{R=0} S_{J\alpha,a} = \sum_J P^J_{r,\alpha\beta} S_{J\alpha,a} = \sum_J \left(E^J_{r,\alpha\beta} + N^J_{\alpha\beta}\right) S_{J\alpha,a} \\
&= \sum_J \left(2\sum_{e\neq g}\left\langle \psi^0_g \middle| \mu^E_\beta \middle| \psi^0_e \right\rangle \left\langle \psi^0_e \middle| (\partial \psi^A_g / \partial R_{J,\alpha})_0 \right\rangle + Z_J e\right) S_{J\alpha,a} \\
&= 2\left\langle \psi^0_g \middle| \mu^E_\beta \middle| (\partial \psi^A_g / \partial Q_a)_0 \right\rangle + \left(\partial \mu^N_\beta / \partial Q_a\right)_0
\end{aligned} \tag{4.52}$$

$$\left(\frac{\partial\langle\dot{\mu}_\beta\rangle}{\partial P_a}\right)_{P_a=0} = \sum_J \left(\frac{\partial\langle\dot{\mu}_\beta\rangle}{\partial \dot{R}_{J,\alpha}}\right)_{\dot{R}=0} S_{J\alpha,a} = \sum_J P^J_{v,\alpha\beta} S_{J\alpha,a} = \sum_J \left(E^J_{v,\alpha\beta} + N^J_{\alpha\beta}\right) S_{J\alpha,a}$$
$$= \sum_J \left(2i \sum_{e\neq g} \left\langle \psi_g^0 \middle| \dot{\mu}_\beta^E \middle| \psi_e^0 \right\rangle \left\langle \psi_e^0 \middle| (\partial\psi_g^A/\partial R_{J,\alpha})_0 \right\rangle \left(\omega_{eg}^0\right)^{-1} + Z_J e\right) S_{J\alpha,a} \tag{4.53}$$
$$= 2\left\langle \psi_g^0 \middle| \dot{\mu}_\beta^E \middle| (\partial\tilde{\psi}_g^{CA}/\partial P_a)_0 \right\rangle + \left(\partial\dot{\mu}_\beta^N/\partial P_a\right)_0$$

$$\left(\frac{\partial\langle m_\beta\rangle}{\partial P_a}\right)_{P_a=0} = \sum_J \left(\frac{\partial\langle m_\beta\rangle}{\partial \dot{R}_{J,\alpha}}\right)_{\dot{R}=0} S_{J\alpha,a} = \sum_J M^J_{\alpha\beta} S_{J\alpha,a} = \sum_J \left(I^J_{\alpha\beta} + J^J_{\alpha\beta}\right) S_{J\alpha,a}$$
$$= \sum_J \left(2i \sum_{e\neq g} \left\langle \psi_g^0 \middle| m_\beta^E \middle| \psi_e^0 \right\rangle \left\langle \psi_e^0 \middle| (\partial\psi_g^A/\partial R_{J,\alpha})_0 \right\rangle \left(\omega_{eg}^0\right)^{-1} + \frac{Z_J e}{2c} \varepsilon_{\alpha\beta\gamma} R^0_{J\gamma}\right) S_{J\alpha,a}$$
$$= 2\left\langle \psi_g^0 \middle| m_\beta^E \middle| (\partial\tilde{\psi}_g^{CA}/\partial P_a)_0 \right\rangle + \left(\partial m_\beta^N/\partial P_a\right)_0 \tag{4.54}$$

The last expressions in Equations (4.53) and (4.54) involve three steps: (i) converting the nuclear momentum derivative into a nuclear velocity derivative, (ii) summing over all electronic states e and converting this sum to a nuclear momentum derivative, and (iii) summing over all nuclei J to convert, using $S_{J\alpha,a}$, Cartesian derivatives into normal coordinate derivatives.

4.1.5 Origin Dependence of the Rotational Strength

The presentation of the theory of VCD has included at each step the velocity form of the electric dipole transition moment. The principal reason for doing this is to emphasize two central ideas. One is that IR vibrational absorption intensities can also be envisioned as arising from electronic current density in molecules driven by, and in phase with, the velocities of the nuclei. Another reason is that the velocity formulation of the rotational strength is the most natural formulation of rotational strength and is independent of the choice of origin of the coordinates. By contrast, the position formulation is not the more natural form and as a consequence is origin dependent. The origin dependence of the position formulation can be circumvented by using a distributed origin gauge. This can be derived directly or by using gauge-invariant atomic orbital (GIAO) basis sets for the calculation of VCD intensities. Below, we first demonstrate this origin dependence and then show that by using the velocity form of the electric dipole transition moment in the calculation of the rotational strength, the origin dependence vanishes. Secondly, we demonstrate that with the use of a distributed origin gauge, rather than a common origin gauged, this problem can be circumvented.

4.1.5.1 General Description of Origin Dependence

The basic idea with origin dependence of the rotational strength is fairly simple. The magnetic dipole transition moment operator can be written in general form as:

$$\boldsymbol{m}(\boldsymbol{0}) = \frac{e}{2mc}(\boldsymbol{r} \times \boldsymbol{p}) = \frac{1}{2c}(\boldsymbol{r} \times \dot{\boldsymbol{\mu}}) \tag{4.55}$$

where the position vector, $\boldsymbol{r}$, is measured relative to the origin at $\boldsymbol{R} = \mathbf{0}$. If the origin is translated from $\mathbf{0}$ to $\mathbf{0}' = \mathbf{0} + \boldsymbol{T}$, the position vector changes from $\boldsymbol{r} - \mathbf{0} = \boldsymbol{r}$ to $\boldsymbol{r} - \mathbf{0}' = \boldsymbol{r} - \boldsymbol{T}$ and the magnetic dipole moment changes from

$$\boldsymbol{m}(\mathbf{0}') = \boldsymbol{m}(\mathbf{0}) - \boldsymbol{T} \times \dot{\boldsymbol{\mu}}/2c \tag{4.56}$$

The position form of the rotational strength then changes in the following way when the origin of coordinates is changed.

$$R^a_{r,g1,g0}(\mathbf{0}) = \mathrm{Im}\left[\langle \Psi_{g0}|\boldsymbol{\mu}|\Psi_{g1}\rangle \cdot \langle \Psi_{g1}|\boldsymbol{m}|\Psi_{g0}\rangle\right] \tag{4.57}$$

$$R^a_{r,g1,g0}(\mathbf{0}') = \mathrm{Im}\left[\langle \Psi_{g0}|\boldsymbol{\mu}|\Psi_{g1}\rangle \cdot \langle \Psi_{g1}|\boldsymbol{m}|\Psi_{g0}\rangle - \boldsymbol{T} \times \langle \Psi_{g0}|\boldsymbol{\mu}|\Psi_{g1}\rangle \cdot \langle \Psi_{g1}|\dot{\boldsymbol{\mu}}|\Psi_{g0}\rangle/2c\right] \tag{4.58}$$

Hence, the position form of the rotational strength is intrinsically origin dependent unless the position transition moment, $\langle \Psi_{g1}|\boldsymbol{\mu}|\Psi_{g0}\rangle$, is exactly parallel (or anti-parallel) to the velocity transition moment, $\langle \Psi_{g1}|\dot{\boldsymbol{\mu}}|\Psi_{g0}\rangle$, since *a vector triple product,* $\boldsymbol{a} \times \boldsymbol{b} \cdot \boldsymbol{c}$, *vanishes if any two vectors are parallel.* For exact wavefunctions, or inexact wavefunctions defined using a complete (infinite) set of basis functions, these two transition moments in fact are parallel, as expressed by the hypervirial relationship in Equation (4.50), but all wavefunctions of molecules with more than one electron are approximate and defined with finite basis sets, and hence, for all practical cases, the position form of the rotational strength possesses some degree of origin dependence. As a result, additional steps, to be described below, are required to eliminate this dependence.

On the other hand, there is no origin dependence for the velocity formulation of the rotational strength that is used regardless of the quality of the wavefunction. Using the velocity form of the electric dipole transition moment, we have:

$$R^a_{v,g1,g0}(\mathbf{0}) = \omega_a^{-1}\mathrm{Im}\left[\langle \Psi_{g0}|\dot{\boldsymbol{\mu}}|\Psi_{g1}\rangle \cdot \langle \Psi_{g1}|\boldsymbol{m}|\Psi_{g0}\rangle\right] \tag{4.59}$$

$$R^a_{v,g1,g0}(\mathbf{0}') = \omega_a^{-1}\mathrm{Im}\left[\langle \Psi_{g0}|\dot{\boldsymbol{\mu}}|\Psi_{g1}\rangle \cdot \langle \Psi_{g1}|\boldsymbol{m}|\Psi_{g0}\rangle - \boldsymbol{T} \times \langle \Psi_{g0}|\dot{\boldsymbol{\mu}}|\Psi_{g1}\rangle \cdot \langle \Psi_{g1}|\dot{\boldsymbol{\mu}}|\Psi_{g0}\rangle/2c\right] \tag{4.60}$$

Here the origin-dependent term always vanishes because it contains two identical velocity dipole transition moments. Origin dependence can only be completely eliminated from rotational strength calculations by using the velocity form of the rotational strength. The origin dependence can effectively be suppressed, however, by using the position form of the rotational strength with what is called the distributed origin gauge.

4.1.5.2 Distributed Origin Gauge and Effective Origin Independence

To this point we have considered the expressions for the rotational strength and the AAT in terms of a single origin, called the common origin (CO) gauge, where the word gauge means choice or measure of the origin. In this section, we describe an alternative formalism called the distributed origin (DO) gauge that was developed to address the origin dependence of the position form of the rotational strength. A starting point for considerations is the expression for the AAT in the CA approximation in the CO gauge using $\mathbf{0}$ as the common molecular origin.

$$M^J_{\alpha\beta}(\mathbf{0}) = \left[\partial\left\langle \tilde{\psi}_g^{CA}|m_\beta(\mathbf{0})|\tilde{\psi}_g^{CA}\right\rangle \Big/ \partial\dot{R}_{J,\alpha}\right]_0 = 2\left\langle \psi_g^0|m_\beta(\mathbf{0})|(\partial\tilde{\psi}_g^{CA} \Big/ \partial\dot{R}_{J,\alpha})_0\right\rangle \tag{4.61}$$

Moving the CO to $\mathbf{0}'$ we have, using Equation (4.56) and Cartesian tensor notation,

$$\begin{aligned} M^J_{\alpha\beta}(\mathbf{0}') &= 2\left\langle \psi^0_g \middle| m_\beta(\mathbf{0}') \middle| (\partial\tilde{\psi}^{CA}_g/\partial\dot{R}_{J,\alpha})_0 \right\rangle \\ &= M^J_{\alpha\beta}(\mathbf{0}) - (\varepsilon_{\beta\gamma\delta}T_\gamma/c)\left\langle \psi^0_g \middle| \dot{\mu}_\delta \middle| (\partial\tilde{\psi}^{CA}_g/\partial\dot{R}_{J,\alpha})_0 \right\rangle \end{aligned} \tag{4.62}$$

where we have used $m_\beta = \varepsilon_{\beta\gamma\delta} r_\gamma \dot{\mu}_\delta/c$. In the DO gauge, a different origin is chosen for each for AAT associated with nucleus J. In particular, $\boldsymbol{R}_J$ is chosen as the origin for $M^J_{\alpha\beta}$ so that

$$\left(M^J_{\alpha\beta}(\boldsymbol{R}_J)\right)^{DO} = \left(M^J_{\alpha\beta}(\mathbf{0})\right)^{CO} - (\varepsilon_{\beta\gamma\delta}R_{J,\gamma}/c)\left\langle \psi^0_g \middle| \dot{\mu}_\delta \middle| (\partial\tilde{\psi}^{CA}_g/\partial\dot{R}_{J,\alpha})_0 \right\rangle \tag{4.63}$$

Rearranging, the expression for the AAT in the CO gauge in terms of the DO AAT is:

$$\left(M^J_{\alpha\beta}(\mathbf{0})\right)^{CO} = \left(M^J_{\alpha\beta}(\boldsymbol{R}_J)\right)^{DO} + (\varepsilon_{\beta\gamma\delta}R_{J,\gamma}/c)\left\langle \psi^0_g \middle| \dot{\mu}_\delta \middle| (\partial\tilde{\psi}^{CA}_g/\partial\dot{R}_{J,\alpha})_0 \right\rangle \tag{4.64}$$

In more compact notation we can re-write this equation as:

$$M^J_{\alpha\beta} = \left(M^J_{\alpha\beta}\right)^J + \varepsilon_{\beta\gamma\delta}R_{J,\gamma}P^J_{v,\alpha\delta}/2c \tag{4.65}$$

A final step is to substitute the position form of the APT for the velocity form,

$$M^J_{\alpha\beta} \cong \left(M^J_{\alpha\beta}\right)^J + \varepsilon_{\beta\gamma\delta}R_{J,\gamma}P^J_{r,\alpha\delta}/2c \tag{4.66}$$

This is only exact for exact wavefunctions or wavefunction expressed in terms of complete, rather than finite, basis sets. Nevertheless, it does produce an origin-independent expression for the rotational strength as we now show.

In terms of APTs and AATs, the rotational strength in the position formulation is:

$$R^a_{r,g1,g0} = \sum_{J,J'} \left[P^J_{r,\alpha\beta}S_{J\alpha,a}\right]\left[M^{J'}_{\alpha'\beta}S_{J'\alpha',a}\right] \tag{4.67}$$

where summations over all repeated subscripts, α, α', and β are implied. Substitution of Equation (4.66) into (4.67) yields:

$$R^a_{r,g1,g0} = \sum_{J,J'} \left\{ \left[P^J_{r,\alpha\beta}S_{J\alpha,a}\right]\left[\left(M^{J'}_{\alpha'\beta}\right)^{J'} S_{J'\alpha',a}\right] + \left(\varepsilon_{\beta\gamma\delta}R_{J',\gamma}P^J_{r,\alpha\beta}P^{J'}_{r,\alpha'\delta}\right)S_{J\alpha,a}S_{J'\alpha',a}/2c \right\} \tag{4.68}$$

The second term in this equation vanishes because it represents a vector triple product with two identical vectors, the APT from the electric dipole transition moment, $P^J_{r,\alpha\beta}$, and the same APT from the origin dependence of the DO gauge, $P^{J'}_{r,\alpha'\delta}$. This means that the rotational strength is now independent of any choice of common origin of the molecule as the nuclear position component in the second term would change from $R_{J',\gamma}$ to $R_{J',\gamma} + T_\gamma$ and the second term still vanishes. However, if the location of the distributed origins were to change, the rotational strength due to the first term in the rotational strength would change. There is no compelling reason to change the choice of the location of the distributed origins because it is a natural choice for building wavefunctions in terms of atomic orbital basis sets, as described later in this chapter. The important point to keep in mind is that the rotational strength in Equation (4.68) is only independent of the choice of the common origin and not of the selected set of distributed origins. So in practice, origin independence has been achieved by using the DO gauge, but on a fundamental level, origin dependence remains in the choice of distributed origins. It is also important to bear in mind the approximation associated with Equation (4.66), which makes the DO gauge molecular origin independent as shown in Equation (4.68).

4.2 Formulations of VCD Theory

In the previous section, we developed a general theory of VCD intensity by defining the rotational strength for a vibrational transition and then developing the expressions for the rotational strength at the lowest level of correction to the Born–Oppenheimer approximation. This necessitated the introduction of a perturbation expansion of the electronic wavefunction involving, in principle, an infinite summation over all excited states of a molecule. Because it is impractical to carry out this infinite summation, a variety of approaches have been published for formulating the general theory of VCD in a form amenable to practical calculations.

4.2.1 Average Excited-State Energy Approximation

The first and simplest of these approximations was published as an additional section to the publication of the first complete vibronic coupling theory of VCD (Nafie and Freedman, 1983). This method involves using a single average energy, E^0_{Ave}, for all the excited electronic states *e* of the molecule, and the denominator in the expression for the electronic contribution to the AAT given in Equation (4.47) becomes $E^0_{Ave} - E^0_g$. By making this approximation the summation over all excited states may be carried out to closure, and the equation for the AAT becomes:

$$I^J_{\alpha\beta} = 2i\hbar \frac{\left\langle \psi^0_g \middle| m^E_\beta \middle| (\partial\psi_g / \partial R_{J,\alpha})_0 \right\rangle}{E^0_{Ave} - E^0_g} \tag{4.69}$$

This approximation is known to be crude and has never been implemented, in part because a superior method using magnetic field perturbation theory was published, along with the first *ab initio* calculations of VCD by Stephens and co-workers (Stephens, 1985; Lowe *et al.*, 1986).

4.2.2 Magnetic Field Perturbation Theory

The method of magnetic field perturbation (MFP) also circumvents the need to evaluate explicitly the summation of all excited electronic states of the molecule. The theoretical basis of the MFP method employs first-order perturbation theory of a non-degenerate electronic wavefunction by a magnetic field, $\boldsymbol{H}$, in which case the complex ground state electronic wavefunction can be written as:

$$\tilde{\psi}^{HA}_g(\boldsymbol{r},\boldsymbol{R},\boldsymbol{H}) = \psi^A_g(\boldsymbol{r},\boldsymbol{R}) + i\psi^H_g(\boldsymbol{r},\boldsymbol{R}_0,\boldsymbol{H}) \tag{4.70}$$

The perturbation of the molecule by a magnetic field adds a new imaginary term to the wavefunction that is associated with the generation of electron *current* density in the same way that perturbation of the molecule by nuclear velocities adds a new imaginary term to the wavefunction, as seen in Equation (2.48) in Chapter 2 and above. This new term can be expressed through first-order in magnetic field dependence by quantum mechanical perturbation theory, which can be obtained from Equation (4.24) by using the electron perturbation operator $-\boldsymbol{m}^E \cdot \boldsymbol{H}$ instead of $(\partial H_E^A/\partial R_{J,\alpha})_0 R_{J,\alpha}$,

$$\tilde{\psi}_g^{HA}(\boldsymbol{r},\boldsymbol{R},\boldsymbol{H}) = \psi_g^A(\boldsymbol{r},\boldsymbol{R}) + \sum_{e\neq g} \frac{\langle \psi_e^0 | \boldsymbol{m}^E | \psi_g^0 \rangle_{H=0}}{E_e^0 - E_g^0} \cdot \boldsymbol{H}\psi_e^0(\boldsymbol{r},0) \tag{4.71}$$

Here the magnetic dipole moment is pure imaginary. If we develop the first-order dependence of the adiabatic wavefunction $\psi_g^A(\boldsymbol{r},\boldsymbol{R})$ on nuclear position, we can write the complex magnetic field perturbed wavefunction to first order in its dependence on nuclear position and magnetic field as:

$$\tilde{\psi}_g^{HA}(\boldsymbol{r},\boldsymbol{R},\boldsymbol{H}) = \psi_g^0 + \sum_{e\neq g}\sum_J \left\langle \psi_e^0 \left| \left(\frac{\partial \psi_g^A}{\partial R_{J,\alpha}}\right)_0 \right.\right\rangle \psi_e^0 R_{J,\alpha} + \sum_{e\neq g} \frac{\langle \psi_e^0 | \boldsymbol{m}^E | \psi_g^0 \rangle_{H=0}}{E_e^0 - E_g^0} \cdot \boldsymbol{H}\psi_e^0 \tag{4.72}$$

Taking the derivative of $\tilde{\psi}_g^A(\boldsymbol{r},\boldsymbol{R},\boldsymbol{H})$ with respect to the perturbing magnetic field, we have:

$$\left(\frac{\partial \tilde{\psi}_g^{HA}}{\partial H_\beta}\right)_{H=0} = i\left(\frac{\partial \psi_g^H}{\partial H_\beta}\right)_{H=0} = \sum_{e\neq g} \frac{\langle \psi_e^0 | m_\beta^E | \psi_g^0 \rangle}{E_e^0 - E_g^0} \psi_e^0 \tag{4.73}$$

The electronic contribution to the AAT in Equation (4.47) can be re-written in a form similar to that of Equation (4.71) as:

$$\begin{aligned} I_{\alpha\beta}^J &= 2i\hbar \sum_{e\neq g} \frac{\langle \psi_g^0 | m_\beta^E | \psi_e^0 \rangle \langle \psi_e^0 | (\partial \psi_g^A/\partial R_{J,\alpha})_0 \rangle}{E_e^0 - E_g^0} \\ &= -2i\hbar \left\langle \left(\frac{\partial \psi_g^A}{\partial R_{J,\alpha}}\right)_{R=0} \left| \sum_{e\neq g} \frac{\langle \psi_e^0 | m_\beta^E | \psi_g^0 \rangle}{E_e^0 - E_g^0} \right| \psi_e^0 \right\rangle \end{aligned} \tag{4.74}$$

where we have used that $\langle \psi_g^0 | m_\beta^E | \psi_e^0 \rangle = -\langle \psi_e^0 | m_\beta^E | \psi_g^0 \rangle$. Using Equation (4.73) in this equation allows us to write

$$I_{\alpha\beta}^J = -2i\hbar \left\langle \left(\frac{\partial \psi_g^A}{\partial R_{J,\alpha}}\right) \middle| \left(\frac{\partial \tilde{\psi}_g^{HA}}{\partial H_\beta}\right) \right\rangle_{R,H=0} = -2i\hbar \left\langle \left(\frac{\partial \tilde{\psi}_g^{HA}}{\partial R_{J,\alpha}}\right) \middle| \left(\frac{\partial \tilde{\psi}_g^{HA}}{\partial H_\beta}\right) \right\rangle_{R,H=0} \tag{4.75}$$

The definitions for the MFP AAT used here differs by a factor of $-2i\hbar$ from those originally defined by Stephens. In the Stephens definition, the imaginary overlap integral is defined directly as the AAT as:

$$\left(I_{\alpha\beta}^J\right)_{Stephens} = \left\langle \left(\frac{\partial \tilde{\psi}_g^{HA}}{\partial R_{J,\alpha}}\right) \middle| \left(\frac{\partial \tilde{\psi}_g^{HA}}{\partial H_\beta}\right) \right\rangle_{R,H=0} = \frac{i}{2\hbar} I_{\alpha\beta}^J \tag{4.76}$$

The corresponding nuclear contribution, defined this chapter in Equation (4.3), is defined by Stephens as:

$$\left(J^{J}_{\alpha\beta}\right)_{Stephens} = \frac{i}{2\hbar} J^{J}_{\alpha\beta} = \frac{iZ_J e}{4\hbar c} \varepsilon_{\alpha\beta\gamma} R^{0}_{J\gamma} \tag{4.77}$$

is also an imaginary quantity. The Stephens' definition of the AAT is a particular approach to the calculation of VCD that avoids the explicit sum over excited states. The formalism associated with the perturbation of the wavefunction by the magnetic field is a computational convenience without a physical or heuristic relationship to the actual mechanism by which AATs and VCD intensities arise. We prefer definitions of both position and velocity forms of real, rather than imaginary, APTs and the AAT using derivatives of nuclear position and velocity coordinates, as given in Equations (4.52)–(4.54), not in terms of a static perturbing magnetic field, which is not present during VCD measurements. In particular, there exist very real correlations between nuclear positions and electron densities, as described by the adiabatic Born–Oppenheimer approximation, and between nuclear velocities and electron current density, as described by the complete adiabatic non-Born–Oppenheimer approximation. We will defer further description of vibrational current density to later in this chapter.

4.2.3 Sum-Over-States Vibronic Coupling Theory

A few years after the implementation of VCD calculations using MFP formalism, the theory of VCD was implemented in its most direct formulation, namely vibronic coupling theory (VCT) using a summation-over-states (SOS) (Dutler and Rauk, 1989). The infinite summation over all excited electronic states was carried out using a set of finite basis functions and a well-defined finite set of single electronic excitations between Hartree–Fock occupied and unoccupied molecular orbitals. Although the sum-over-states is evaluated explicitly using Equation (4.47), and the sum is avoided in the MFP formalism, both the SOS and MFP formalism formally yield the same result if the same finite basis set, electronic excitations, and origin-dependence formalism are used in the first case for the explicit sum-over-excited-states and in the second case to evaluate the magnetic-field perturbed electronic wavefunction by coupled perturbed Hartree–Fock theory.

4.2.4 Nuclear Velocity Perturbation Theory

A second perturbation formalism has been proposed but not yet implemented. Instead of perturbing the molecule with a magnetic field and creating electron currents through a new imaginary term for the electronic wavefunction, the molecule is perturbed by the velocities of the nuclei, which through the CA electronic wavefunction adds a new imaginary term to the electronic wavefunction, as determined from Equation (4.31).

$$\tilde{\psi}^{CA}_{g}(\mathbf{r}, \mathbf{R}, \dot{\mathbf{R}}) = \psi^{0}_{g}(\mathbf{r}) + \sum_{e \neq g} \sum_{J} \left\langle \psi^{0}_{e} \middle| \left(\frac{\partial \psi^{A}_{g}}{\partial R_{J,\alpha}} \right)_{0} \right\rangle \psi^{0}_{e}(\mathbf{r}) \left(R_{J,\alpha} + \frac{i\dot{R}_{J,\alpha}}{\omega^{0}_{eg}} \right) \tag{4.78}$$

From this equation, it can be seen that the difference in the derivatives of the CA electronic wavefunction with respect to nuclear position, $R_{J,\alpha}$, and velocity, $\dot{R}_{J,\alpha}$, namely a factor of i/ω^{0}_{eg} for the latter, are very closely related. In particular, for the nuclear position derivative we can write

$$\left(\frac{\partial \tilde{\psi}^{CA}_{g}}{\partial R_{J,\alpha}} \right)_{\mathbf{R}=0} = \left(\frac{\partial \psi^{A}_{g}}{\partial R_{J,\alpha}} \right)_{\mathbf{R}=0} = \sum_{e \neq g} \left\langle \psi^{0}_{e} \middle| \left(\frac{\partial \psi^{A}_{g}}{\partial R_{J,\alpha}} \right)_{0} \right\rangle \psi^{0}_{e}(\mathbf{r}) \tag{4.79}$$

If we sum over all excited states e to closure we have simply an identity relationship. Similarly for the nuclear velocity derivative we have:

$$\left(\frac{\partial\tilde{\psi}_g^{CA}}{\partial\dot{R}_{J,\alpha}}\right)_{\dot{R}=0} = \sum_{e\neq g}\left\langle\psi_e^0\left|\left(\frac{\partial\psi_g^A}{\partial R_{J,\alpha}}\right)_0\right.\right\rangle\psi_e^0(\boldsymbol{r})\left(i/\omega_{eg}^0\right) \tag{4.80}$$

Here we cannot sum over the excited states e to closure because of the frequency denominator $\omega_{eg}^0 = \left(E_e^0 - E_g^0\right)/\hbar$, which depends on the energy of the individual excited states e. Comparing Equations (4.79) and (4.80), we can write more generally that

$$(\partial\tilde{\psi}_g^{CA}/\partial\dot{R}_{J\alpha})_0 = \left(i/\omega_{eg}^0\right)\left(\partial\psi_g^A/\partial R_{J\alpha}\right)_0 \tag{4.81}$$

provided that both sides of this equation include a summation over all excited electronic states. As a result, we can write the electronic AAT adapted from Equation (4.47) as:

$$I_{\alpha\beta}^J = 2i\sum_{e\neq g}\left\langle\psi_g^0\left|m_\beta^E\right|\psi_e^0\right\rangle\left\langle\psi_e^0\left|\left(\partial\psi_g^A/\partial R_{J\alpha}\right)_0\right.\right\rangle\left(\omega_{eg}^0\right)^{-1} \tag{4.82}$$

We next write this expression with explicit dependence on the nuclear velocity derivative,

$$I_{\alpha\beta}^J = 2\sum_{e\neq g}\left\langle\psi_g^0\left|m_\beta^E\right|\psi_e^0\right\rangle\left\langle\psi_e^0\left|(\partial\tilde{\psi}_g^{CA}/\partial\dot{R}_{J\alpha})_0\right.\right\rangle \tag{4.83}$$

and then in a more compact notation by summing over all electronic excited states to closure and thus avoiding reference to the summation over excited states as:

$$I_{\alpha\beta}^J = 2\left\langle\psi_g^0\left|m_\beta^E\right|\left(\partial\tilde{\psi}_g^{CA}/\partial\dot{R}_{J\alpha}\right)\right\rangle_{R,\dot{R}=0} \tag{4.84}$$

This equation for the AAT in terms of velocity perturbed electronic wavefunction is comparable in simplicity to the AAT expressed in terms of a magnetic field perturbed electronic wavefunction given in Equation (4.75),

$$I_{\alpha\beta}^J = -2i\hbar\left\langle\left(\frac{\partial\psi_g^A}{\partial R_{J,\alpha}}\right)\middle|\left(\frac{\partial\tilde{\psi}_g^{AH}}{\partial H_\beta}\right)\right\rangle_{R,H=0} \tag{4.85}$$

4.2.5 Energy Second-Derivative Theory

The AAT formalisms of MFP and NVP can be unified by using a mixed second-derivative formalism involving the dependence of the electronic energy of the molecule on both the magnetic field and nuclear velocities. If the adiabatic BO electronic Hamiltonian is perturbed by the nuclear velocities, as in Equation (4.22), and by a magnetic field, one can write

$$\tilde{H}_E^{HCA}(\boldsymbol{R},\dot{\boldsymbol{R}},\boldsymbol{H}) = H_E^A(\boldsymbol{R}) - i\hbar\left(\frac{\partial}{\partial\boldsymbol{R}}\right)_E\cdot\dot{\boldsymbol{R}} - \boldsymbol{m}^E\cdot\boldsymbol{H} \tag{4.86}$$

We can then write the energy of the electronic wavefunction associated with this Hamiltonian having dependence on the magnetic field and nuclear velocities in the following way:

$$E_g(\boldsymbol{R},\dot{\boldsymbol{R}},\boldsymbol{H}) = \left\langle \tilde{\psi}_e^{HCA}(\boldsymbol{r},\boldsymbol{R},\dot{\boldsymbol{R}},\boldsymbol{H}) \middle| \tilde{H}_E^{HCA}(\boldsymbol{R},\dot{\boldsymbol{R}},\boldsymbol{H}) \middle| \tilde{\psi}_e^{HCA}(\boldsymbol{r},\boldsymbol{R},\dot{\boldsymbol{R}},\boldsymbol{H}) \right\rangle \tag{4.87}$$

Next we seek derivatives of this energy with respect to the perturbing magnetic field and nuclear velocities. As the energy is a real quantity, we must restrict our results to expressions with one derivative of the imaginary perturbed Hamiltonian followed by one imaginary derivative of the perturbed electronic wavefunction. We start by taking the derivative of the Hamilton with respect to magnetic field giving

$$\left(\frac{\partial E_g(\boldsymbol{R},\dot{\boldsymbol{R}},\boldsymbol{H})}{\partial H_\beta}\right)_{H=0} = -\left\langle \tilde{\psi}_e^{CA}(\boldsymbol{r},\boldsymbol{R},\dot{\boldsymbol{R}}) \middle| m_\beta^E \middle| \tilde{\psi}_e^{CA}(\boldsymbol{r},\boldsymbol{R},\dot{\boldsymbol{R}},\boldsymbol{H}) \right\rangle_{H=0} \tag{4.88}$$

Setting the magnetic field equal to zero reduces the electronic wavefunction to the CA approximation. Now the derivative of this expression is taken with respect to a component of the nuclear velocity. This yields:

$$I_{\alpha\beta}^J = -\left(\frac{\partial^2 E_g(\boldsymbol{R},\dot{\boldsymbol{R}},\boldsymbol{H})}{\partial \dot{R}_{J,\alpha}\partial H_\beta}\right)_{H,\dot{R},R=0} = 2\left\langle \psi_g^0 \middle| m_\beta^E \middle| (\partial\tilde{\psi}_g^{CA}/\partial\dot{R}_{J\alpha}) \right\rangle_{\dot{R},R=0} \tag{4.89}$$

This mixed second derivative of the energy is the negative of our definition of the electronic AAT in the CA-NVP formulation in Equation (4.84).

If we reverse the order of taking the derivatives, we can write the first derivative of the electronic energy with respect to nuclear velocity by taking that derivative of the perturbation operator in Equation (4.86). Thus, from Equation (4.87) we have:

$$\left(\frac{\partial E_g(\boldsymbol{R},\dot{\boldsymbol{R}},\boldsymbol{H})}{\partial \dot{R}_{J,\alpha}}\right)_{\dot{R}=0} = -i\hbar\left\langle \tilde{\psi}_e^{HA}(\boldsymbol{r},\boldsymbol{R},\dot{\boldsymbol{R}},\boldsymbol{H}) \middle| \frac{\partial}{\partial R_{J,\alpha}} \middle| \tilde{\psi}_e^{HA}(\boldsymbol{r},\boldsymbol{R},\dot{\boldsymbol{R}},\boldsymbol{H}) \right\rangle_{\dot{R}=0} \tag{4.90}$$

If we now take the derivative of the each of the wavefunctions in this expression with respect to magnetic field we find that the negative of this second-order mixed derivative is exactly our definition of the AAT in the MFP formulation, given in Equation (4.75),

$$I_{\alpha\beta}^J = -\left(\frac{\partial^2 E_g(\boldsymbol{R},\dot{\boldsymbol{R}},\boldsymbol{H})}{\partial H_\beta \partial \dot{R}_{J,\alpha}}\right)_{H,\dot{R},R=0} = 2i\hbar\left\langle \frac{\partial\tilde{\psi}_e^{HA}}{\partial H_\beta} \middle| \frac{\partial\psi_e^A}{\partial R_{J,\alpha}} \right\rangle_{H,R=0} = -2i\hbar\left\langle \frac{\partial\psi_e^A}{\partial R_{J,\alpha}} \middle| \frac{\partial\tilde{\psi}_e^{HA}}{\partial H_\beta} \right\rangle_{H,R=0} \tag{4.91}$$

We can therefore conclude that the NVP and MFP formulations of the AAT are simply different orders of differentiation of the mixed second-order derivative of the electronic energy with respect to nuclear velocity and magnetic field. Using more abbreviated notation we have the MFP electronic AAT as:

$$I_{\alpha\beta}^J = -\left(\frac{\partial^2 E_g}{\partial H_\beta \partial \dot{R}_{J,\alpha}}\right)_{0,0} = -2i\hbar\left\langle \frac{\partial\psi_e^A}{\partial R_{J,\alpha}} \middle| \frac{\partial\tilde{\psi}_e^{HA}}{\partial H_\beta} \right\rangle_{0,0} \tag{4.92}$$

and the NVP electronic AAT

$$I^J_{\alpha\beta} = -\left(\frac{\partial^2 E_g}{\partial \dot{R}_{J,\alpha}\partial H_\beta}\right)_{0,0} = 2\left\langle \psi^0_g \middle| m^E_\beta \middle| \frac{\partial \tilde{\psi}^{CA}_g}{\partial \dot{R}_{J\alpha}} \right\rangle_{0,0} \tag{4.93}$$

where both formulations avoid explicit consideration of the summation of excited electronic states. These formulae reveal a deep unity between the two principal formulations of the AAT and VCD intensities.

Similar expressions also hold for the electronic contribution to the velocity form of the APT, $E^J_{v,\alpha\beta}$, for the vector field perturbation energy $-\dot{\boldsymbol{\mu}}\cdot \boldsymbol{A}/c$ namely,

$$E^J_{v,\alpha\beta} = -c\left(\frac{\partial^2 E_g}{\partial A_\beta \partial \dot{R}_{J,\alpha}}\right)_{0,0} = 2i\hbar c\left\langle \frac{\partial \tilde{\psi}^{FA}_e}{\partial A_\beta} \middle| \frac{\partial \psi^A_e}{\partial R_{J,\alpha}} \right\rangle_{0,0} \tag{4.94}$$

and the NVP electronic AAT

$$E^J_{v,\alpha\beta} = -c\left(\frac{\partial^2 E_g}{\partial \dot{R}_{J,\alpha}\partial A_\beta}\right)_{0,0} = 2\left\langle \psi^0_g \middle| \dot{\mu}^E_\beta \middle| \frac{\partial \tilde{\psi}^{CA}_g}{\partial \dot{R}_{J\alpha}} \right\rangle_{0,0} \tag{4.95}$$

where A_β is the βth Cartesian component of the vector potential associated with interaction of light with matter. Finally, for the perturbation of a molecule by an electric field $\boldsymbol{F}$ according to the interaction energy $-\boldsymbol{\mu}\cdot\boldsymbol{F}$ and by nuclear position according to $\left(\frac{\partial H^A_E}{\partial R_{J,\alpha}}\right)_{R=0} R_{J,\alpha}$, the corresponding energy expressions for the position form of electronic part of the APT are given by:

$$E^J_{r,\alpha\beta} = -\left(\frac{\partial^2 E_g}{\partial F_\beta \partial R_{J,\alpha}}\right)_{0,0} = 2\left\langle \frac{\partial \psi^{FA}_e}{\partial F_\beta} \middle| \frac{\partial H^A_E}{\partial R_{J,\alpha}} \middle| \psi^0_g \right\rangle_{0,0} \tag{4.96}$$

$$E^J_{r,\alpha\beta} = -\left(\frac{\partial^2 E_g}{\partial R_{J,\alpha}\partial F_\beta}\right)_{0,0} = 2\left\langle \psi^0_g \middle| \mu^E_\beta \middle| \frac{\partial \psi^A_g}{\partial R_{J\alpha}} \right\rangle_{0,0} \tag{4.97}$$

In principle, either of these last two equations can be used for the calculation of IR vibrational absorption intensities. The first is the electric field derivative of the energy gradient of the molecule and is a common computational route to VA intensities, whereas the second equation conforms more closely to the more common view of the origin of VA intensity, namely the nuclear position dependence of the electric dipole moment of the molecule.

4.2.6 Other Formulations of VCD Theory

There have been several additional formulations of the electronic AAT. These have involved concepts such as nuclear velocity-dependent electronic property surfaces (Buckingham *et al.*, 1987), non-local susceptibilities (Nafie and Freedman, 1987) and shielding tensors (Hunt and Harris, 1991). All of these take advantage of the relationships between nuclear motion, as opposed to static position, and corresponding induced fields and magnetic dipole moments in the molecule. Each offers a somewhat different view of the physical origin of the electronic contribution to the vibrational magnetic dipole moment in a molecule, which is zero in the BO approximation and that has its origins in the inter-relationship between electric and magnetic phenomena as described by Maxwell's equations. In this chapter, we will focus on the simplest of such descriptions, namely the derivative of the magnetic moment of the molecule with respect to nuclear velocity without generalizing the formalism to include property surfaces, shielding tensors or non-local susceptibilities, all of which have an origin in the sum-over-states (SOS) and nuclear velocity perturbation (NVP) expressions for the perturbed magnetic dipole moment of the molecule. We also include the magnetic field perturbation (MVP) theory for its dominating role to date in the numerical calculation of the AAT and VCD intensity.

4.3 Atomic Orbital Level Formulations of VCD Intensity

A more complete understanding of the theory of VCD, and also vibrational absorption, intensities requires continuing the development of the formalism to the level of atomic orbital basis sets. Here, we encounter concepts associated with changes in population densities at the atomic orbital level as well as gauge origin invariance. The focus will continue to be on the electronic parts of the APT and AAT tensors since the nuclear parts are trivial and the same for all formulations of VA and VCD intensities.

4.3.1 Atomic Orbital Basis Descriptions of Transition Moments

We begin with the description of the adiabatic (superscript A) ground-state (subscript g) electronic wavefunction down to the level of atomic orbitals (AOs). The wavefunction is first written, in its simplest form, as an anti-symmetrized product of j molecular orbitals (MOs), $\chi^A_{g,j}$, represented by a single determinant

$$\psi^A_g(\boldsymbol{r},\boldsymbol{R}) \cong \overline{\prod_j} \chi^A_{g,j}(\boldsymbol{r},\boldsymbol{R}) \tag{4.98}$$

The bar over the product symbol means the wavefunction is anti-symmetric with respect to the interchange of any two electrons. For a closed-shell molecule there are two spin-paired electrons for each MO in the product of MOs. Each MO can in turn be written as a linear combination of atomic orbitals (LCAOs), φ^B_μ, summed over all atoms B and over all AOs μ on B,

$$\chi^A_{g,j}(\boldsymbol{r},\boldsymbol{R}) = \sum_B \sum_\mu^B c^A_{\mu j}(\boldsymbol{R})\varphi^B_\mu(\boldsymbol{r}-\boldsymbol{R}_B) \tag{4.99}$$

The AO $\varphi^B_\mu(\boldsymbol{r}-\boldsymbol{R}_B)$ is centered at the location $\boldsymbol{R}_B$ of atom B and its weight in the LCAO expansion is represented by an adiabatic AO coefficient, $c^A_{\mu j}(\boldsymbol{R})$, which changes its value as the position of atom B in the molecule changes, for example during vibrational motion. The AOs can be further expressed as a linear combination of basis functions that differ according to which basis set is chosen for a calculation, but we will not consider this level of functional detail until Chapter 9.

4.3.1.1 *Position Form of the Electronic APT*

The expression for the electronic contribution to the position-dipole APT can be obtained by substitution of Equation (4.99) into the definition of the APT from Equation (2.74) as:

$$\begin{aligned} E^J_{r,\alpha\beta} = \left(\frac{\partial}{\partial R_{J,\alpha}}\left\langle\psi^A_g\middle|\mu^E_{r,\beta}\middle|\psi^A_g\right\rangle\right)_{R=0} &= -2e\left(\frac{\partial}{\partial R_{J,\alpha}}\sum_j\left\langle\chi^A_{g,j}\middle|r_{j,\beta}\middle|\chi^A_{g,j}\right\rangle\right)_{R=0} \\ &= -2e\left(\frac{\partial}{\partial R_{J,\alpha}}\sum_j\sum_B\sum_\mu^B\sum_D\sum_\nu^D c^A_{\mu j}c^A_{\nu j}\left\langle\varphi^B_\mu\middle|r_\beta\middle|\varphi^D_\nu\right\rangle\right)_{R=0} \end{aligned} \tag{4.100}$$

When the anti-symmetrized product of MOs is substituted for the wavefunction, each orbital j takes its turn with the corresponding electron j in definition of the electric dipole moment operator, $\mu^E_{r,\beta} = -e\sum_j r_{j\beta}$ from Equation (4.6), and all other orbitals, without an operator for its electrons, integrate to unity. As a result, for the APT in Equation (4.100), the sum over electron positions, j, in the operator, $\mu^E_{r,\beta}$, and the two products of MOs, one for each wavefunction, ψ^A_g, reduce to a single sum over each orbital/electron j. An initial factor of two appears in these expressions from the two spin-paired electrons present in each occupied MO. In the final step of Equation (4.100) LCAO expansions for each MO are inserted where we no longer retain the subscript j on the electron coordinate as only one orbital at a time is ever considered. This final expression has two sets of equivalent summations over all AOs, one labeled $B\mu$ and the other labeled $D\nu$. If we carry out the derivative of the AOs and their expansion coefficients in Equation (4.100), we obtain after combining equivalent terms from derivatives of the B and D summations,

$$E^J_{r,\alpha\beta} = -4e\sum_j\sum_B^B\sum_\mu\sum_D^D\sum_\nu\left[c_{\mu j,0}\left(\frac{\partial c^A_{\nu j}}{\partial R_{J,\alpha}}\right)_0\left\langle\varphi^B_{\mu,0}\middle|r_\beta\middle|\varphi^D_{\nu,0}\right\rangle + c_{\mu j,0}c_{\nu j,0}\left\langle\varphi^B_{\mu,0}\middle|r_\beta\middle|\left(\frac{\partial\varphi^D_\nu}{\partial R_{J,\alpha}}\right)_0\right\rangle\right] \tag{4.101}$$

A final expression is obtained by recognizing from the form of the AO $\varphi^D_\nu(\boldsymbol{r}-\boldsymbol{R}_D)$ that the derivative of this orbital is only non-zero if operated on by a nuclear derivative of the same nucleus. In addition, the nuclear position derivative and the electron position derivative of an AO are equal and opposite in sign, as an infinitesimal displacement of an AO on nucleus J in the positive direction, while keeping the electron position fixed, is the same as displacing the electron of the AO in the opposite direction while keeping the AO and nucleus J fixed. Thus we can write:

$$\sum_D\left(\frac{\partial\varphi^D_\nu}{\partial R_{J,\alpha}}\right)_0 = \sum_D\left(\frac{\partial\varphi^D_\nu}{\partial R_{J,\alpha}}\right)_0\delta_{DJ} = \left(\frac{\partial\varphi^J_\nu}{\partial R_{J,\alpha}}\right)_0 = -\left(\frac{\partial\varphi^J_{\nu,0}}{\partial r_\alpha}\right) \tag{4.102}$$

Substituting Equation (4.102) into Equation (4.101) yields:

$$E^J_{r,\alpha\beta} = -4e\sum_j\sum_B^B\sum_\mu\left\{\sum_D^D\sum_\nu\left[c_{\mu j,0}\left(\frac{\partial c^A_{\nu j}}{\partial R_{J,\alpha}}\right)_0\left\langle\varphi^B_{\mu,0}\middle|r_\beta\middle|\varphi^D_{\nu,0}\right\rangle - \sum_\nu^J c_{\mu j,0}c_{\nu j,0}\left\langle\varphi^B_{\mu,0}\middle|r_\beta\frac{\partial}{\partial r_\alpha}\middle|\varphi^J_{\nu,0}\right\rangle\right]\right\} \tag{4.103}$$

It is convenient to define adiabatic AO populations, $P^A_{\mu\nu}$, as a summation j over the doubly occupied LCAO coefficients by:

$$P^A_{\mu\nu}(\boldsymbol{R}) = 2\sum_j c^A_{\mu j}(\boldsymbol{R})c^A_{\nu j}(\boldsymbol{R}) \tag{4.104}$$

$$\begin{aligned}\left(\frac{\partial P^A_{\mu\nu}}{\partial R_{J,\alpha}}\right)_{R=0} &= 2\sum_j\left[c_{\mu j,0}\left(\frac{\partial c^A_{\nu j}}{\partial R_{J,\alpha}}\right)_{R=0} + c_{\nu j,0}\left(\frac{\partial c^A_{\mu j}}{\partial R_{J,\alpha}}\right)_{R=0}\right]\\ &= 4\sum_j c_{\mu j,0}\left(\frac{\partial c^A_{\nu j}}{\partial R_{J,\alpha}}\right)_0\end{aligned} \tag{4.105}$$

where the second equality can be written, as in Equation (4.101) for example, due to the equivalence of the summations over μ and ν. The final expression for the APT now becomes

$$E^J_{r,\alpha\beta} = -e\sum_B^B\sum_\mu\left\{\sum_D^D\sum_\nu\left[\left(\frac{\partial P^A_{\mu\nu}}{\partial R_{J,\alpha}}\right)_0\left\langle\varphi^B_{\mu,0}\middle|r_\beta\middle|\varphi^D_{\nu,0}\right\rangle - 2\sum_\nu^J P_{\mu\nu,0}\left\langle\varphi^B_{\mu,0}\middle|r_\beta\frac{\partial}{\partial r_\alpha}\middle|\varphi^J_{\nu,0}\right\rangle\right]\right\} \tag{4.106}$$

From this equation it is clear that the electronic APT has two principal terms. The first involves the change in the population of $\mu\nu$-pairs of AOs times the position matrix element of those AO pairs. The second term involves the equilibrium population of the $\mu\nu$-pair of AOs times a position matrix element where one of the AOs must be located at J and that AO is displaced along the nuclear-coordinate displacement direction. More simply, the first term is a population derivative term while the second is a matrix-element derivative term for displacement of nucleus J in the αth Cartesian direction.

4.3.1.2 *Velocity Form of the Electronic APT*

In order to write a corresponding expression for the velocity form of the APT, $E^J_{v,\alpha\beta}$, one needs the effect of perturbation of the MO by either a vector potential A_β or the nuclear velocities, $\dot{R}_{J,\alpha}$, to provide a source of electron current density in the molecule. For perturbation by nuclear velocities, the complex molecular orbitals, in the CA approximation, take the form

$$\tilde{\psi}^{CA}_g(\boldsymbol{r},\boldsymbol{R},\boldsymbol{R}) \cong \prod_j \tilde{\chi}^{CA}_{g,j}(\boldsymbol{r},\boldsymbol{R},\boldsymbol{R})$$

$$\tilde{\chi}^{CA}_{g,j}(\boldsymbol{r},\boldsymbol{R},\dot{\boldsymbol{R}}) = \sum_B^B\sum_\mu \tilde{c}^{CA}_{\mu j}(\boldsymbol{R},\dot{\boldsymbol{R}})\varphi^B_\mu(\boldsymbol{r}-\boldsymbol{R}_B) \tag{4.107}$$

where the nuclear velocity dependence is carried by the imaginary part of the complex CA AO coefficient, and the AOs remain the same without any nuclear velocity dependence. Shortly, we will show how this velocity dependence can be introduced by means of a complex velocity gauge exponential factor. Substitution of the expressions for the CA MOs in Equation (4.107) into the definition of the velocity form of the electronic APT, $E^J_{v,\alpha\beta}$, from Equation (2.79) we obtain

$$\begin{aligned} E^J_{v,\alpha\beta} &= 2\left\langle\psi^0_g\middle|\dot{\mu}^E_\beta\middle|(\partial\tilde{\psi}^{CA}_g/\partial\dot{R}_{J\alpha})_0\right\rangle \\ &= -4e\sum_j\sum_B^B\sum_\mu\left\{\sum_D^D\sum_\nu\left[c_{\mu j,0}\left(\frac{\partial\tilde{c}^{CA}_{\nu j}}{\partial\dot{R}_{J,\alpha}}\right)_{\dot{R}=0}\left\langle\varphi^B_{\mu,0}\middle|\frac{\hbar}{im}\frac{\partial}{\partial r_\beta}\middle|\varphi^D_{\nu,0}\right\rangle\right]\right\} \\ &= -e\sum_B^B\sum_\mu\left\{\sum_D^D\sum_\nu\left[\left(\frac{\partial\tilde{P}^{CA}_{\mu\nu}}{\partial\dot{R}_{J,\alpha}}\right)_0\left\langle\varphi^B_{\mu,0}\middle|\frac{\hbar}{im}\frac{\partial}{\partial r_\beta}\middle|\varphi^D_{\nu,0}\right\rangle\right]\right\} \end{aligned} \tag{4.108}$$

Here the complex population, $\tilde{P}^{CA}_{\mu\nu}$, is given in terms of its real (A) and imaginary parts (CA) as:

$$\tilde{P}^{CA}_{\mu\nu}(\boldsymbol{R},\dot{\boldsymbol{R}}) = P^A_{\mu\nu}(\boldsymbol{R}) + iP^{CA}_{\mu\nu}(\boldsymbol{R}_0,\dot{\boldsymbol{R}}) = 2\sum_j \tilde{c}^{CA^*}_{\mu j}(\boldsymbol{R},\dot{\boldsymbol{R}})\tilde{c}^{CA}_{\nu j}(\boldsymbol{R},\dot{\boldsymbol{R}}) \tag{4.109}$$

$$\tilde{c}^{CA}_{\mu j}(\boldsymbol{R},\dot{\boldsymbol{R}}) = c^A_{\mu j}(\boldsymbol{R}) + ic^{CA}_{\mu j}(\boldsymbol{R}_0,\dot{\boldsymbol{R}}) \tag{4.110}$$

where the real part depends on nuclear positions and the imaginary part on nuclear velocities. From these definitions, we have the following first-derivative relationships between AO coefficients and their AO pair-wise populations,

$$\tilde{P}_{\mu\nu}(\boldsymbol{R},\dot{\boldsymbol{R}}) = P_{\mu\nu,0} + \sum_J \left(\frac{\partial P^A_{\mu\nu}}{\partial R_{J,\alpha}}\right)_{0,0} \cdot R_{J,\alpha} + i\sum_J \left(\frac{\partial P^{CA}_{\mu\nu}}{\partial \dot{R}_{J,\alpha}}\right)_{0,0} \cdot \dot{R}_{J,\alpha} \tag{4.111}$$

$$\begin{aligned}\left(\frac{\partial P^A_{\mu\nu}}{\partial R_{J,\alpha}}\right) &= 2\sum_j \left[c_{\mu j,0}\left(\frac{\partial c^A_{\nu j}}{\partial R_{J,\alpha}}\right) + \left(\frac{\partial c^A_{\mu j}}{\partial R_{J,\alpha}}\right)c_{\nu j,0}\right] \\ &= 4\sum_j c_{\mu j,0}\left(\frac{\partial c^A_{\nu j}}{\partial R_{J,\alpha}}\right) = 4\sum_j c_{\nu j,0}\left(\frac{\partial c^A_{\mu j}}{\partial R_{J,\alpha}}\right)\end{aligned} \tag{4.112}$$

$$\begin{aligned}\left(\frac{\partial P^{CA}_{\mu\nu}}{\partial \dot{R}_{J,\alpha}}\right) &= 2\sum_j \left[c_{\mu j,0}\left(\frac{\partial c^{CA}_{\nu j}}{\partial \dot{R}_{J,\alpha}}\right) - \left(\frac{\partial c^{CA}_{\mu j}}{\partial \dot{R}_{J,\alpha}}\right)c_{\nu j,0}\right] \\ &= 4\sum_j c_{\mu j,0}\left(\frac{\partial c^{CA}_{\nu j}}{\partial \dot{R}_{J,\alpha}}\right) = -4\sum_j \left(\frac{\partial c^{CA}_{\mu j}}{\partial \dot{R}_{J,\alpha}}\right)c_{\nu j,0}\end{aligned} \tag{4.113}$$

for equivalent LCAO summations over $B\mu$ and $D\nu$. Note that $\left(\partial P^A_{\mu\nu}/\partial R_{J,\alpha}\right)$ is symmetric in $\mu\nu$ interchange whereas $\left(\partial P^{CA}_{\mu\nu}/\partial \dot{R}_{J,\alpha}\right)$ is anti-symmetric. Finally, substituting into Eq. (4.108) we write an expression for $E^J_{v,\alpha\beta}$ which is clearly real and is most like the population derivative term for the final expression for the electronic position APT, $E^J_{r,\alpha\beta}$, in Equation (4.106),

$$E^J_{v,\alpha\beta} = -e\sum_B\sum_\mu^B\sum_D\sum_\nu^D \left[\left(\frac{\partial P^{CA}_{\mu\nu}}{\partial \dot{R}_{A\alpha}}\right)_0 \left\langle \varphi^B_{\mu,0}\middle|\frac{\hbar}{m}\frac{\partial}{\partial r_\beta}\middle|\varphi^D_{\nu,0}\right\rangle\right] \tag{4.114}$$

Because in this equation there is no matrix element derivative term (the second term) in Equation (4.106), the full weight of describing the velocity dependence is carried by the population derivative term, which is inherently not as accurate as including such a term. This problem is eliminated when velocity gauge atomic orbitals are used, as demonstrated in the following section.

We next turn to the vector field perturbation formalism. Here again the MO carries the A-dependence by means of a new imaginary contribution to the AO population coefficient, $\tilde{c}^{FA}_{\mu j} = c^A_{\mu j} + ic^{FA}_{\mu j}$,

$$\tilde{\chi}^{FA}_{g,j}(\boldsymbol{r},\boldsymbol{R},\boldsymbol{A}) = \sum_B\sum_\mu^B \tilde{c}^{FA}_{\mu j}(\boldsymbol{R},\boldsymbol{A})\varphi^B_\mu(\boldsymbol{r}-\boldsymbol{R}_B) \tag{4.115}$$

Following analogous notation and definitions, we can write the expression for $E^J_{v,\alpha\beta}$ from Equation (4.94) with vector field perturbation as:

$$\begin{aligned}E^J_{v,\alpha\beta} &= -2i\hbar c\left\langle\frac{\partial\psi^A_e}{\partial R_{J,\alpha}}\middle|\frac{\partial\tilde{\psi}^{FA}_e}{\partial A_\beta}\right\rangle_{0,0} \\ &= -4i\hbar c\sum_j\sum_B\sum_\mu^B\sum_D\sum_\nu^D\left\{\left[\left(\frac{\partial c^A_{\mu j}}{\partial R_{J,\alpha}}\right)_{R=0}\left\langle\varphi^B_{\mu,0}\middle|\varphi^D_{\nu,0}\right\rangle + c_{\mu j,0}\left\langle\frac{\partial\varphi^B_\mu}{\partial R_{J,\alpha}}\middle|\varphi^D_{\nu,0}\right\rangle\right]\left(\frac{\partial\tilde{c}^{FA}_{\nu j}}{\partial A_\beta}\right)_{A=0}\right\}\end{aligned} \tag{4.116}$$

4.3.1.3 Electronic AAT

By even closer analogy we can write the corresponding expressions for velocity and magnetic field perturbed AATs using MOs in the absence of velocity gauge factors. The entire dependence of the nuclear velocity or magnetic field is carried by the imaginary part of the complex AO coefficients. The expression for the electronic part of the AAT for the velocity perturbation is given by:

$$I^J_{\alpha\beta} = -4e\sum_j \sum_B^B \sum_\mu \sum_D^D \sum_\nu \left[c_{\mu j,0} \left(\frac{\partial c^{CA}_{\nu j}}{\partial \dot{R}_{J,\alpha}} \right)_{\dot{R}=0} \left\langle \varphi^B_{\mu,0} \middle| \varepsilon_{\beta\gamma\delta} \frac{\hbar}{2mc} r_\gamma \frac{\partial}{\partial r_\delta} \middle| \varphi^D_{\nu,0} \right\rangle \right] \tag{4.117}$$

and for the magnetic field perturbation, the AAT is:

$$I^J_{\alpha\beta} = -4i\hbar\sum_j \sum_B^B \sum_\mu \sum_D^D \sum_\nu \left\{ \left[\left(\frac{\partial c^A_{\mu j}}{\partial R_{J,\alpha}} \right)_{R=0} \left\langle \varphi^B_{\mu,0} \middle| \varphi^D_{\nu,0} \right\rangle + c_{\mu j,0} \left\langle \frac{\partial \varphi^B_{\mu}}{\partial R_{J,\alpha}} \middle| \varphi^D_{\nu,0} \right\rangle \right] \left(\frac{\partial \tilde{c}^{HA}_{\nu j}}{\partial H_\beta} \right)_{H=0} \right\} \tag{4.118}$$

At this point it is interesting to compare the vibronic coupling formalism for the AAT involving a sum over all defined excited electronic states with the two perturbed wavefunction formalisms just presented. If the expressions for the AAT in Equations (4.117) and (4.118) are compared with the vibronic coupling expression in Equation (4.47) using the Herzberg–Teller expansion for the adiabatic wavefunction $\psi^A_e(\boldsymbol{r}, \boldsymbol{R})$ in Equation (4.1.24) one can write

$$I^J_{\alpha\beta} = -2i\hbar \sum_{e\neq g} \frac{\left\langle \psi^0_g \middle| m^E_\beta \middle| \psi^0_e \right\rangle \left\langle \psi^0_e \middle| (\partial H^A_E/\partial R_{J,\alpha})_0 \middle| \psi^0_g \right\rangle}{(E^0_e - E^0_g)^2} \tag{4.119}$$

Therefore, the magnetic part of the AAT can either be calculated using only imaginary AO coefficient derivatives, as in Equations (4.117) and (4.118), or by means of a sum over all excited states with no coefficient derivatives, as in Equation (4.119). As we shall see below, when gauge velocity factors are included in the AOs, a balance is obtained between these extremes where AO coefficient derivatives are used but are not responsible for carrying the entire perturbation dependence of the AAT.

4.3.2 Velocity Dependent Atomic Orbitals

A particle moving with constant velocity or momentum is described in quantum mechanics by the wavefunction for a plane wave,

$$\psi(\boldsymbol{r}) = A\exp(i\boldsymbol{k}\cdot\boldsymbol{r}) \tag{4.120}$$

where the wavevector $\boldsymbol{k}$ is defined in terms of the wavelength, λ, momentum, $\boldsymbol{p}$, or velocity, $\mathbf{v}$, of the particle of mass m as:

$$\boldsymbol{k} = 2\pi\boldsymbol{u}/\lambda = \boldsymbol{p}/\hbar = m\mathbf{v}/\hbar \tag{4.121}$$

and where $\boldsymbol{u}$ is the unit vector in the direction of motion of the particle. For an atom or molecule moving at constant velocity $\mathbf{v}$, the wavefunction for the electrons, $\psi_e(\boldsymbol{r}, \boldsymbol{R})$, can be written as:

$$\psi_e(\boldsymbol{r}, \boldsymbol{R}, \mathbf{v}) = \psi_e(\boldsymbol{r}, \boldsymbol{R})\exp(im\mathbf{v}\cdot\boldsymbol{r}/\hbar) \tag{4.122}$$

where the complex exponential function serves as a velocity gauge function that describes the motion of the electrons in the uniformly moving atom or molecule.

4.3.2.1 Field Adiabatic Velocity Gauge

We can extend the idea of uniform motion of a molecule to a molecule in a uniform vector potential field. If an electromagnetic vector potential, $\boldsymbol{A}$, perturbs a molecule, the presence of the field acts as though the molecule as a whole is undergoing uniform motion at a velocity

$$\mathbf{v} = -\frac{eA}{mc} \tag{4.123}$$

Inserting this velocity into Equation (4.122) yields

$$\tilde{\psi}_g^{FA}(\boldsymbol{r}, \boldsymbol{R}, \boldsymbol{A}) = \psi_g^A(\boldsymbol{r}, \boldsymbol{R})\exp(-ie\boldsymbol{A}\cdot\boldsymbol{r}/\hbar c) \tag{4.124}$$

Extending this to MO and AO levels gives

$$\tilde{\chi}_{g,j}^{FA}(\boldsymbol{r}, \boldsymbol{R}, \boldsymbol{A}) = \sum_B \sum_{\mu}^{B} c_{\mu j}^{A}(\boldsymbol{R})\varphi_{\mu}^{B}(\boldsymbol{r} - \boldsymbol{R}_B)\exp(-ie\boldsymbol{A}\cdot\boldsymbol{r}/\hbar c) \tag{4.125}$$

In the following section, we will use this wavefunction to find a connection between the position and velocity forms of the electronic APT and hence IR absorption intensities. Introduction of the velocity gauge factor simplifies the AO description as coefficients perturbed by the vector field, used in Equation (4.115), are not needed because only the molecule as a whole is affected by the perturbation.

4.3.2.2 Complete Adiabatic Nuclear Velocity Gauge

The idea of gauge velocity can be extended to the velocity the individual nuclei, $\dot{\boldsymbol{R}}_J$, can impart on an AO associated with it as:

$$\varphi_{\mu}^{J}(\boldsymbol{r}, \boldsymbol{R}, \dot{\boldsymbol{R}}) = \varphi_{\mu}^{J}(\boldsymbol{r}, \boldsymbol{R})\exp(im\dot{\boldsymbol{R}}_J\cdot\boldsymbol{r}/\hbar) \tag{4.126}$$

Here the gauge relationship must be applied at the level of AOs when describing the motion of electron densities on individual nuclear centers. Using nuclear velocity-dependent AOs, it is now possible to write the velocity dependence of the CA electron wavefunction of a molecule in its ground electronic state as:

$$\begin{aligned}
\tilde{\psi}_g^{CA}(\boldsymbol{r}, \boldsymbol{R}, \dot{\boldsymbol{R}}) &\cong \prod_j \tilde{\chi}_{j,g}^{CA}(\boldsymbol{r}, \boldsymbol{R}, \dot{\boldsymbol{R}}) \\
\tilde{\chi}_{j,g}^{CA}(\boldsymbol{r}, \boldsymbol{R}, \dot{\boldsymbol{R}}) &= \sum_B \sum_{\mu}^{B} \tilde{c}_{\mu j}^{CA}(\boldsymbol{R}, \dot{\boldsymbol{R}})\tilde{\varphi}_{\mu}^{B}(\boldsymbol{r}, \boldsymbol{R}, \dot{\boldsymbol{R}}) \\
&= \sum_B \sum_{\mu}^{B} \tilde{c}_{\mu j}^{CA}(\boldsymbol{R}, \dot{\boldsymbol{R}})\tilde{\varphi}_{\mu}^{B}(\boldsymbol{r} - \boldsymbol{R}_B)\exp(im\dot{\boldsymbol{R}}_B\cdot\boldsymbol{r}/\hbar)
\end{aligned} \tag{4.127}$$

This wavefunction is called the nuclear velocity gauge complete adiabatic (NVG-CA) electronic wavefunction. Each AO in the molecular orbital expansion is centered at its atomic nucleus and thereby carries its nuclear-position dependence, $\boldsymbol{R}_B$, and is multiplied by a nuclear-velocity gauge function that moves the AO with a plane-wave function at the nuclear velocity, $\dot{\boldsymbol{R}}_B$. The complex expansion coefficients depend on both the nuclear positions and velocities and are required because the nuclear velocity perturbation differs from nucleus to nucleus and is not uniform across the entire molecule.

4.3.3 Field Adiabatic Velocity Gauge Transition Moments

In this section we consider the expression for the electronic contribution to the velocity form of the APT given in Equation (4.116) when the field adiabatic velocity gauge wavefunction in Equation (4.125) is used instead of Equation (4.115) as before. Using that the derivative of the field adiabatic MO with respect to A_β is:

$$\frac{\partial \tilde{\chi}_{g,j}^{FA}}{\partial A_\beta} = (-ier_\beta/\hbar c)\sum_{D,\nu} c_{\nu j}^A \varphi_\nu^D \exp(-ie\boldsymbol{A}\cdot\boldsymbol{r}/\hbar c) \tag{4.128}$$

we can write that

$$\begin{aligned} E_{v,\alpha\beta}^J &= -2i\hbar c\left\langle \frac{\partial \psi_e^A}{\partial R_{J,\alpha}} \middle| \frac{\partial \tilde{\psi}_e^{FA}}{\partial A_\beta} \right\rangle_{0,0} \\ &= -4i\hbar c \sum_j \sum_B^B \sum_\mu \sum_D^D \sum_\nu \left\{ \left[\left(\frac{\partial c_{\mu j}^A}{\partial R_{J,\alpha}} \right) c_{\nu j}^A \left\langle \varphi_\mu^B \middle| (-ier_\beta/\hbar c)\varphi_\nu^D \exp(-ie\boldsymbol{A}\cdot\boldsymbol{r}/\hbar c) \right\rangle \right.\right. \\ &\quad \left.\left. + c_{\nu j}^A c_{\mu j}^A \left\langle \frac{\partial \varphi_\mu^B}{\partial R_{J,\alpha}} \middle| (-ier_\beta/\hbar c)\varphi_\nu^D \exp(-ie\boldsymbol{A}\cdot\boldsymbol{r}/\hbar c)/\hbar c \right\rangle \right] \right\}_{R=0,A=0} \end{aligned} \tag{4.129}$$

Setting the values of the nuclear positions and vector potential equal to zero gives

$$E_{v,\alpha\beta}^J = -4e \sum_j \sum_B^B \sum_\mu \sum_D^D \sum_\nu \left\{ \left[\left(\frac{\partial c_{\mu j}^A}{\partial R_{J,\alpha}} \right) c_{\nu j}^A \left\langle \varphi_{\mu,0}^B \middle| r_\beta \middle| \varphi_{\nu,0}^D \right\rangle + c_{\mu j,0} c_{\nu j,0} \left\langle \left(\frac{\partial \varphi_\mu^B}{\partial R_{J,\alpha}} \right)_0 \middle| r_\beta \middle| \varphi_{\nu,0}^D \right\rangle \right] \right. \tag{4.130}$$

This equation is identical to the position form of the electronic APT given in Equation (4.101) bearing in mind the equivalence of the $B\mu$- and $D\nu$-summations. Thus, we conclude that the gauge transformation of the vector potential exactly converts the *velocity* form of the electronic APT into its *position* form. As we shall see in the next section, the situation is not so simple for the magnetic field perturbation.

4.3.4 Gauge Invariant Atomic Orbitals and AATs

We now show that the introduction of gauge velocity dependence into the AOs provides not only benefits of computational accuracy but also confers origin independence to the calculation of the rotational strength as described in Section 4.1.5. First, however, we consider whether a uniform velocity gauge associated with a magnetic field can produce a position formulation expression for $I_{\alpha\beta}^J$ that is equivalent to the velocity-dependent form used above for $E_{v,\alpha\beta}^J$. The vector field associated with a static magnetic field is:

$$\boldsymbol{A} = \frac{1}{2}\boldsymbol{H}\times\boldsymbol{r} \tag{4.131}$$

and the field adiabatic velocity gauge electronic wavefunction from Equation (4.124) is:

$$\tilde{\psi}_g^{HA}(\boldsymbol{r}, \boldsymbol{R}, \boldsymbol{H}) = \psi_g^A(\boldsymbol{r}, \boldsymbol{R})\exp(-ie\boldsymbol{H} \times \boldsymbol{r} \cdot \boldsymbol{r}/2\hbar c) \tag{4.132}$$

The velocity gauge factor contains a vector triple product with two identical vectors and hence is zero for all values of $\boldsymbol{r}$. This reduces the exponential to unity and means that there is no position formulation equivalent of the electronic AAT.

If instead, we use a velocity gauge factor specific to each nucleus by first writing the position vector as the sum of vectors to the nuclear position and from there to the electron position, $\boldsymbol{r} = \boldsymbol{R}_J + \boldsymbol{r}_J$. The magnetic field adiabatic MO is written in terms of velocity gauge AOs as:

$$\tilde{\chi}_{g,j}^{HA}(\boldsymbol{r}, \boldsymbol{R}, \boldsymbol{H}) = \sum_B \sum_\mu^B \tilde{c}_{\mu j}^{J,HA}(\boldsymbol{R}, \boldsymbol{H})\varphi_\mu^B(\boldsymbol{r} - \boldsymbol{R}_B)\exp(-ie\boldsymbol{H} \times \boldsymbol{R}_B \cdot \boldsymbol{r}/2\hbar c) \tag{4.133}$$

The AOs in this equation are in fact gauge-invariant AOs (GIAOs) as this form of the velocity gauge leads to an electronic AAT in the same form as the DO gauge AO presented above, which was shown to yield an origin-independent rotational strength. The AO complex coefficients, $\tilde{c}_{\mu j}^{J,HA}$, show that the origin of the magnetic field perturbation is at each atomic nucleus. Using the MO comprised of GIAOs, the electronic AAT is given by:

$$\begin{aligned}
I_{\alpha\beta}^J &= -2i\hbar\left\langle \frac{\partial \psi_e^A}{\partial R_{J,\alpha}} \middle| \frac{\partial \tilde{\psi}_e^{HA}}{\partial H_\beta} \right\rangle_{0,0} \\
&= -4i\hbar \sum_j \sum_B \sum_\mu^B \sum_D \sum_\nu^D \Bigg\{ \Bigg[\left(\frac{\partial c_{\mu j}^A}{\partial R_{J,\alpha}}\right) \left\langle \varphi_\mu^B \middle| \varphi_\nu^D \exp(-i\phi_{J,H}) \right\rangle \\
&\quad + c_{\mu j} \left\langle \frac{\partial \varphi_\mu^B}{\partial R_{J,\alpha}} \middle| \varphi_\nu^D \exp(-i\phi_{J,H}) \right\rangle \Bigg] \left(\frac{\partial \tilde{c}_{\nu j}^{J,HA}}{\partial H_\beta}\right) \\
&\quad + \Bigg[\left(\frac{\partial c_{\mu j}^A}{\partial R_{J,\alpha}}\right) c_{\nu j}^A \left\langle \varphi_\mu^B \middle| \left(\frac{-ie\varepsilon_{\beta\gamma\delta} R_{J,\gamma} r_\delta}{2\hbar c}\right) \varphi_\nu^D \exp(-i\phi_{J,H}) \right\rangle \\
&\quad + c_{\nu j}^A c_{\mu j}^A \left\langle \frac{\partial \varphi_\mu^B}{\partial R_{J,\alpha}} \middle| \left(\frac{-ie\varepsilon_{\beta\gamma\delta} R_{J,\gamma} r_\delta}{2\hbar c}\right) \varphi_\nu^D \exp(-i\phi_{J,H}) \right\rangle \Bigg] \Bigg\}_{R=0,H=0}
\end{aligned} \tag{4.134}$$

where we have abbreviated the gauge exponent to be the phase angle $\phi_{J,H} = e\boldsymbol{H} \times \boldsymbol{R}_J \cdot \boldsymbol{r}/2\hbar c$. Consolidating terms, we have

$$\begin{aligned}
= -4i\hbar \sum_j \sum_B \sum_\mu^B \sum_D \sum_\nu^D \Bigg\{ \Bigg[\left(\frac{\partial c_{\mu j}^A}{\partial R_{J,\alpha}}\right)_{R=0} \left\langle \varphi_{\mu,0}^B \middle| \varphi_{\nu,0}^D \right\rangle + c_{\mu j,0} \left\langle \left(\frac{\partial \varphi_\mu^B}{\partial R_{J,\alpha}}\right) \middle| \varphi_\nu^D \right\rangle_{R=0} \Bigg] \left(\frac{\partial \tilde{c}_{\nu j}^{J,HA}}{\partial H_\beta}\right)_{H=0} \\
+ \left(\frac{-ie}{2\hbar c}\right) \varepsilon_{\beta\gamma\delta} R_{J,\gamma} \Bigg[c_{\nu j}^A \left(\frac{\partial c_{\mu j}^A}{\partial R_{J,\alpha}}\right)_{R=0} \left\langle \varphi_{\mu,0}^B | r_\delta | \varphi_{\nu,0}^D \right\rangle + c_{\nu j,0}^A c_{\mu j,0}^A \left\langle \frac{\partial \varphi_\mu^B}{\partial R_{J,\alpha}} | r_\delta | \varphi_\nu^D \right\rangle_{R=0} \Bigg] \Bigg\}
\end{aligned} \tag{4.135}$$

This expression finally can be written as:

$$I^J_{\alpha\beta} = -2i\hbar\left\langle\left(\frac{\partial\psi^A_e}{\partial R_{J,\alpha}}\right)_0\middle|\left(\frac{\partial\tilde{\psi}^{HA}_e}{\partial H_\beta}\right)_0\right\rangle_J + \varepsilon_{\beta\gamma\delta}R_{J,\gamma,0}E^J_{r,\alpha\delta}/2c \tag{4.136}$$

where the subscript J on the overlap matrix element refers to the origin at nucleus J. This equation is the same as Equation (4.66) for the DO gauge expressions for the electronic AAT obtained by an entirely different derivational route.

4.3.5 Complete Adiabatic Nuclear Velocity Gauge Transition Moments

We conclude Section 4.3 with a description of electronic velocity APTs and AATs. Including a velocity gauge factor brings the expression for the velocity APT, $E^J_{r,\alpha\beta}$, into a remarkably close form to that of the position APT, $E^J_{r,\alpha\beta}$. In fact, the forms of these two expressions are so close that all computational elements may be exchangeable between the two expressions, leading to an underlying computational unity to both forms of electronic APTs and opening a new computational route to electronic AATs. This conjecture, however, remains to be verified by computation.

4.3.5.1 Velocity APT with Nuclear Velocity Gauge Atomic Orbitals

If we use the CA-NVG electronic wavefunction given in Equation (4.127) with the nuclear velocity gauge exponential factor, in addition to nuclear-velocity-dependent coefficients, we have the CA MO given by:

$$\tilde{\chi}^{CA}_{\mu j}(\boldsymbol{r},\boldsymbol{R},\dot{\boldsymbol{R}}) = \sum_B\sum_\mu^B \tilde{c}^{CA}_{\mu j}(\boldsymbol{R},\dot{\boldsymbol{R}})\varphi^B_\mu(\boldsymbol{r}-\boldsymbol{R}_B)\exp(im\dot{\boldsymbol{R}}_B\cdot\boldsymbol{r}/\hbar) \tag{4.137}$$

The expression for the electronic APT in the velocity formulation from Equation (4.108) becomes

$$\begin{aligned}E^J_{v,\alpha\beta} &= 2\left\langle\psi^0_g\middle|\dot{\mu}^E_\beta\middle|(\partial\tilde{\psi}^{CA}_g/\partial\dot{R}_{J\alpha})_0\right\rangle\\ &= -4e\sum_j\sum_B\sum_\mu^B\left\{\sum_D\sum_\nu^D\left[c_{\mu j,0}\left(\frac{\partial\tilde{c}^{CA}_{\nu j}}{\partial\dot{R}_{J,\alpha}}\right)_{\dot{R}=0}\left\langle\varphi^B_{\mu,0}\middle|\frac{\hbar}{im}\frac{\partial}{\partial r_\beta}\middle|\varphi^D_{\nu,0}\right\rangle\right.\right.\\ &\quad\left.\left.+\sum_\nu^J c_{\mu j,0}c_{\nu j,0}\left\langle\varphi^B_{\mu,0}\middle|\frac{\hbar}{im}\frac{\partial}{\partial r_\beta}\middle|\frac{im}{\hbar}r_\alpha\varphi^J_{\nu,0}\right\rangle\right]\right\}\end{aligned} \tag{4.138}$$

In this expression, the first term is identical in form to that in Equation (4.108) but now the coefficient derivatives do not have to carry the entire description of the nuclear velocity derivative because we can also take the derivative of the AO with respect to nuclear velocity. This is the second term in Equation (4.138) where we have left the results of taking the derivative of $\tilde{\varphi}^D_\nu$ with respect to $\dot{R}_{J,\alpha}$ in explicit form and have evaluated the expression at the equilibrium nuclear position and zero nuclear velocities as:

$$\left\{\sum_D\frac{\partial[\varphi^D_\nu(\boldsymbol{r}-\boldsymbol{R}_D)\exp(im\dot{\boldsymbol{R}}_D\cdot\boldsymbol{r}/\hbar)]}{\partial\dot{R}_{J,\alpha}}\right\}_{R=0,\dot{R}=0} = (imr_\alpha/\hbar)\varphi^J_{\nu,0} \tag{4.139}$$

Equation (4.138) can be simplified slightly be canceling the factor $im/\hbar$ with its inverse from the second term and rearranging the second integral using the identity

$$\left\langle \varphi^B_{\mu,0} \left| \frac{\partial}{\partial r_\beta} \right| r_\alpha \varphi^J_{\nu,0} \right\rangle = -\left\langle r_\alpha \varphi^J_{\nu,0} \left| \frac{\partial}{\partial r_\beta} \right| \varphi^B_{\mu,0} \right\rangle \tag{4.140}$$

as the matrix element of the operator $\partial/\partial r_\beta$ is odd upon interchange of wavefunctions. Finally, we come to the expression for $E^J_{v,\alpha\beta}$ after making one minor change in the first term of Equation (4.138) by using the pure imaginary part CA coefficient $ic^{CA}_{\nu j}$ instead of the more general complex quantity $\tilde{c}^{CA}_{\nu j}$ as defined in Equation (4.110),

$$E^J_{v,\alpha\beta} = -4e\sum_j^{}\sum_B^{}\sum_\mu^{B}\left\{\sum_D^{}\sum_\nu^{D}\left[c_{\mu j,0}\left(\frac{\partial c^{CA}_{\nu j}}{\partial \dot{R}_{J,\alpha}}\right)_{\dot{R}=0}\left\langle \varphi^B_{\mu,0}\left|\frac{\hbar}{m}\frac{\partial}{\partial r_\beta}\right|\varphi^D_{\nu,0}\right\rangle\right.\right.$$
$$\left.\left.-\sum_\nu^{J} c_{\mu j,0}c_{\nu j,0}\left\langle \varphi^J_{\nu,0}\left| r_\alpha \frac{\partial}{\partial r_\beta}\right|\varphi^B_{\mu,0}\right\rangle\right]\right\} \tag{4.141}$$

The corresponding expression for the position form of the electronic APT from Equation (4.103) reproduced here for convenience of comparison is:

$$E^J_{r,\alpha\beta} = -4e\sum_j^{}\sum_B^{}\sum_\mu^{B}\left\{\sum_D^{}\sum_\nu^{D}\left[c_{\mu j,0}\left(\frac{\partial c^A_{\nu j}}{\partial R_{J,\alpha}}\right)_0\left\langle \varphi^B_{\mu,0}|r_\beta|\varphi^D_{\nu,0}\right\rangle - \sum_\nu^{J} c_{\mu j,0}c_{\nu j,0}\left\langle \varphi^B_{\mu,0}\left| r_\beta \frac{\partial}{\partial r_\alpha}\right|\varphi^J_{\nu,0}\right\rangle\right]\right\} \tag{4.103}$$

These two equations for the electronic APT are remarkably similar. They represent two equivalent ways of arriving at the same quantity. The first terms have similar but closely related AO coefficient derivatives multiplied by integrals between pairs of AOs involving the velocity operator in one case and the position operator in the other. The second terms differ only in the exchange of the positions of the AOs and the interchange of Cartesian subscripts. In essence, the second terms are just different summation relationships of the same set of integrals with respect to the total APT expression. Until Equation (4.141) can be evaluated computationally we hypothesize that the two major groupings of terms in each equation may be equal to one another, that the coefficient derivatives $\partial c^A_{\nu j}/\partial R_{J,\alpha}$ and $\partial c^A_{\nu j}/\partial \dot{R}_{J,\alpha}$ are closely related one another. In other words, the coefficients, and hence the infinitesimal change in the values of AOs in the two expressions are identical for infinitesimal changes in nuclear displacement and nuclear velocity. This is equivalent to saying that in the CA approximation the changes in electron density resulting from relatively slow nuclear velocities exactly follow instantaneous static changes in the nuclear position. The CA approximation does not allow any departures or slippage of electron densities at any point in time due to nuclear velocities. This is the same as taking the limit of arbitrarily slow but non-zero nuclear velocities. Thus we postulate that the dependence of the AO coefficients on nuclear positions and nuclear velocities is closely related, as constrained by the conservation of charge and current density in a closed molecular system of charge density and current density.

The corresponding expression for the CA-NVG expression for the electronic AAT can be written by inspection of Equation (4.141) and (4.117) we have that

$$I^J_{\alpha\beta} = -4e\sum_j\sum_B^B\sum_\mu^D\left\{\sum_D\sum_\nu\left[c_{\mu j,0}\left(\frac{\partial c^{CA}_{\nu j}}{\partial \dot{R}_{J,\alpha}}\right)_{\dot{R}=0}\left\langle\varphi^B_{\mu,0}\middle|\varepsilon_{\beta\gamma\delta}\frac{\hbar}{m}r_\gamma\frac{\partial}{\partial r_\delta}\middle|\varphi^D_{\nu,0}\right\rangle\right.\right.$$
$$\left.\left.-\sum_\nu^J c_{\mu j,0}c_{\nu j,0}\left\langle\varphi^J_{\nu,0}\middle|r_\alpha\varepsilon_{\beta\gamma\delta}r_\gamma\frac{\partial}{\partial r_\delta}\middle|\varphi^B_{\mu,0}\right\rangle\right]\right\} \tag{4.142}$$

This expression has not yet been used to calculate $I^J_{\alpha\beta}$ but could so easily be, given the hypothesized equality

$$\left(\frac{\partial c^{CA}_{\nu j}}{\partial \dot{R}_{J,\alpha}}\right)_{\dot{R}=0} \Longleftrightarrow \left(\frac{\partial c^{A}_{\nu j}}{\partial R_{J,\alpha}}\right)_{R=0} \tag{4.143}$$

4.4 Transition Current Density and VCD Intensities

In Chapter 2 we derived expressions for the transition current density (TCD) associated with any pair of quantum states, and the formalism was applied to transitions between pure electronic states. We now extend that formalism to vibrational transitions with the idea of visualizing the motion of charge during the transition. Generally, vibrational transitions are depicted for a given normal mode in terms of nuclear displacement vectors of a given magnitude and direction on each nuclear center in the molecule. In Figure 4.2 we illustrate the nuclear displacement vectors for the methine CH stretching mode of the molecule (*S*)-methyl-d_3-lactate *Cd*$_3$ (Freedman *et al.*, 2000). The diagram on the left is a simple structural diagram while on the right is a view of the geometry optimized structure of the molecule taken from a Gaussian DFT calculation. In both diagrams the only significant displacement vectors are displacements of the methine hydrogen in one direction and the asymmetric carbon atom in the opposite direction.

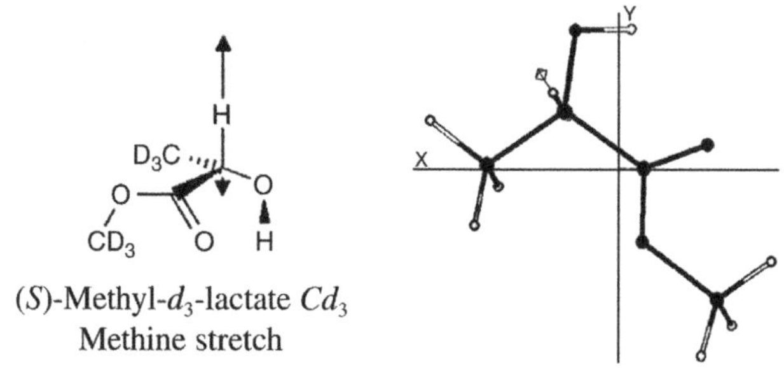

Figure 4.2 Two views of the nuclear displacement vectors for the methine C–H stretching mode of the molecule (S)-methyl-d_3-lactate Cd$_3$ (Freedman et al., 2000). Reproduced with permission from the American Chemical Society (Freedman and Nafie, 2000)

These nuclear displacement vectors are what we have defined as S-vectors which were expressed in terms of Cartesian components in Equation (4.35) and here expressed as full Cartesian vectors,

$$\boldsymbol{S}_{J,a} = \left(\frac{\partial \boldsymbol{R}_J}{\partial Q_a}\right)_{Q=0} = \left(\frac{\partial \dot{\boldsymbol{R}}_J}{\partial \dot{Q}_a}\right)_{\dot{Q}=0} = \left(\frac{\partial \dot{\boldsymbol{R}}_J}{\partial P_a}\right)_{P=0} \tag{4.144}$$

Each vector $\boldsymbol{S}_{J,a}$ describes the displacement of nucleus J in normal mode a. Because the BO approximation provides only changes in electron density with changes in electron position, the BO level of theory does not provide any *vector* information about the electronic motion that parallels that of the nuclear displacement vectors. On the other hand, if these vectors are viewed, not as nuclear displacement vectors, but equivalently as nuclear *velocity* vectors, then we will show below that the CA approximation provides beautiful maps of electron TCD, $\boldsymbol{J}^a_{ev',ev}(\boldsymbol{r})$, which accompany nuclear velocity, or displacement, vectors for vibrational transitions between vibrational levels v and v' of electronic state e. The electron TCD is nothing more than a map of electron density in vector motion, but it reveals in some cases surprising information about where electrons are going in molecules during vibrational motion. We show below that there is close connection between the electronic contribution to the velocity form of the electric dipole transition moment and $\boldsymbol{J}^a_{g1,g0}(\boldsymbol{r})$, namely the vibrational analogue of Equation (2.22)

$$\left\langle \Psi^a_{g1} \middle| \dot{\boldsymbol{\mu}}^E \middle| \Psi^a_{g0} \right\rangle = -ie \int \boldsymbol{J}^a_{g0,g1}(\boldsymbol{r}) \mathrm{d}\boldsymbol{r} \tag{4.145}$$

Note that the convention for the order of subscripts for states for $g0$ to $g1$ is reversed from that of matrix elements compared with that for current density. From this equation it is clear that vibrational electron TCD is the integrand with respect to the space of the molecule of the velocity form of the electric dipole transition moment. $\boldsymbol{J}^a_{g0,g1}(\boldsymbol{r})$ is the desired vector field of the electronic motion that is the equivalent of the nuclear velocity vectors, $\boldsymbol{S}_{J,a}$ for normal mode a.

4.4.1 Relationship Between Vibrational TCD and VA Intensity

The electron probability density and the electron current density for electronic state g, to first order in vibrational position and velocity, are given from Equations (2.49) and (2.50) as:

$$\begin{aligned} \rho^A_g(\boldsymbol{r}, \boldsymbol{R}) = \psi^A_g(\boldsymbol{r})\psi^A_g(\boldsymbol{r}) &\cong \rho^0_g(\boldsymbol{r}) + \sum_J \left(\frac{\partial \rho^A_g(\boldsymbol{r})}{\partial \boldsymbol{R}_J}\right)_0 \cdot \boldsymbol{R}_J \\ &= \psi^0_g(\boldsymbol{r})^2 + 2\sum_J \psi^0_g(\boldsymbol{r}) \left(\frac{\partial \psi^A_g(\boldsymbol{r}, \boldsymbol{R})}{\partial \boldsymbol{R}_J}\right)_0 \cdot \boldsymbol{R}_J \end{aligned} \tag{4.146}$$

$$\begin{aligned} j^{CA}_g(\boldsymbol{r}, \boldsymbol{R}, \dot{\boldsymbol{R}}) &= \frac{\hbar}{2mi}\left[\tilde{\psi}^{CA*}_g(\boldsymbol{r})\nabla\tilde{\psi}^{CA}_g(\boldsymbol{r}) - \tilde{\psi}^{CA}_g(\boldsymbol{r})\nabla\tilde{\psi}^{CA*}_g(\boldsymbol{r})\right] \cong \sum_J \left(\frac{\partial j^{CA}_g(\boldsymbol{r})}{\partial \dot{\boldsymbol{R}}_J}\right)_{0,0} \cdot \dot{\boldsymbol{R}}_J \\ &= -\frac{\hbar}{m}\sum_J \left[\left(\frac{\partial \psi^{CA}_g(\boldsymbol{r})}{\partial \dot{\boldsymbol{R}}_J}\right)_{0,0} \nabla\psi^0_g(\boldsymbol{r}) - \psi^0_g(\boldsymbol{r})\nabla\left(\frac{\partial \psi^{CA}_g(\boldsymbol{r})}{\partial \dot{\boldsymbol{R}}_J}\right)_{0,0}\right] \cdot \dot{\boldsymbol{R}}_J \end{aligned} \tag{4.147}$$

where we have used Equation (2.28) for the definition of the imaginary part of $\tilde{\psi}_e^{CA}$, namely ψ_e^{CA}. The probability density is a scalar field describing the electron probability density for any point in the molecule for any particular position of the nuclei. The current density, by contrast, is a vector field describing the magnitude and direction of electron current density at any point in the space of the molecule arising from non-zero nuclear velocities. The quantities obey the following conservation relationship from Equation (2.52), which connects changes in electron probability density at any point in space with the gradient of the change in current density with the corresponding change in nuclear velocity at the same point in space.

$$-\nabla \cdot \left(\frac{\partial j_g^{CA}(\boldsymbol{r})}{\partial \dot{\boldsymbol{R}}_J}\right)_{\substack{\dot{R}=0\\R=0}} = \left(\frac{\partial \rho_g^A(\boldsymbol{r})}{\partial \boldsymbol{R}_J}\right)_{R=0} \tag{4.148}$$

One can imagine a small volume element in the space of the molecule. Any changes in the electron charge density in the volume element due to a displacement of nucleus J from its equilibrium position, must be compensated by electron current density flowing in or out through the surface surrounding the volume element due to the velocity (the change in the velocity from zero) of nucleus J. The validity of Equation (4.148) is demonstrated in Appendix B, where it is shown that this equation can be reduced to a sum of the transition continuity equation for each of the excited states of the molecule.

To find the TCD of the electrons associated with a particular vibrational transition, such as between $g0$ and $g1$ for normal a, we must specify the transition between vibrational states with the appropriate pair of vibrational wavefunctions and integrate the current density over the normal coordinates. An additional factor of i is needed because TCD is always a real quantity whereas the velocity form of the electric dipole transition moment is pure imaginary. This is accomplished by the definition

$$\begin{aligned}
\boldsymbol{J}^a_{g0,g1}(\boldsymbol{r}) &\equiv i\left\langle \phi^a_{g1}(Q_a)\middle| j_g^{CA}(\boldsymbol{r}, \boldsymbol{R}, \dot{\boldsymbol{R}})\middle|\phi^a_{g0}(Q_a)\right\rangle \\
&= i\sum_J \left(\frac{\partial j_g^{CA}(\boldsymbol{r})}{\partial \dot{\boldsymbol{R}}_J}\right)_{0,0} \cdot \left(\frac{\partial \dot{\boldsymbol{R}}_J}{\partial \dot{Q}_a}\right)_0 \left\langle \phi^a_{g1}(Q_a)\middle|\dot{Q}_a\middle|\phi^a_{g0}(Q_a)\right\rangle
\end{aligned} \tag{4.149}$$

Substituting for all three of the quantities in the summation over J gives:

$$\boldsymbol{J}^a_{g0,g1}(\boldsymbol{r}) = \frac{h}{m}\sum_J \left[\psi_g^0(\boldsymbol{r})\nabla\left(\frac{\partial \psi_g^{CA}(\boldsymbol{r})}{\partial \dot{\boldsymbol{R}}_J}\right)_{0,0} - \left(\frac{\partial \psi_g^{CA}(\boldsymbol{r})}{\partial \dot{\boldsymbol{R}}_J}\right)_{0,0} \nabla\psi_g^0(\boldsymbol{r})\right] \cdot \boldsymbol{S}_{J,a}\left(\frac{\hbar\omega_a}{2}\right)^{1/2} \tag{4.150}$$

If we now substitute the derivative of the CA electronic wavefunction with respect to nuclear velocity from the definition of the CA electronic wavefunction in Equations (2.87) or (4.31)

$$\left(\frac{\partial \psi_g^{CA}(\boldsymbol{r})}{\partial \dot{\boldsymbol{R}}_J}\right)_0 = \sum_{e\neq g} \boldsymbol{C}^{J,0}_{eg}\psi_e^0(\boldsymbol{r})/\omega^0_{eg} \tag{4.151}$$

The vibrational transition current density in Equation (4.150) becomes:

$$
\begin{aligned}
\boldsymbol{J}^{a}_{g0,g1}(\boldsymbol{r}) &= \sum_{J}\sum_{e\neq g}\frac{C^{J,0}_{eg}}{\omega^{0}_{eg}}\left[\frac{\hbar}{2m}\left(\psi^{0}_{g}(\boldsymbol{r})\nabla\psi^{0}_{e}(\boldsymbol{r})-\psi^{0}_{e}(\boldsymbol{r})\nabla\psi^{0}_{g}(\boldsymbol{r})\right)\right]\cdot\boldsymbol{S}_{J,a}(2\hbar\omega_{a})^{1/2} \\
&= (2\hbar\omega_{a})^{1/2}\sum_{J}\sum_{e\neq g}\frac{C^{J,0}_{eg}\cdot\boldsymbol{S}_{J,a}}{\omega^{0}_{eg}}\boldsymbol{J}^{0}_{ge}(\boldsymbol{r}) \qquad (4.152)
\end{aligned}
$$

where the quantity in square brackets in the first equation above is the electron transition current density between pure electronic states e and g defined in Equation (2.19),

$$\boldsymbol{J}^{0}_{ge}(\boldsymbol{r}) = \frac{\hbar}{2m}\left[\psi^{0}_{g}(\boldsymbol{r})\nabla\psi^{0}_{e}(\boldsymbol{r})-\psi^{0}_{e}(\boldsymbol{r})\nabla\psi^{0}_{g}(\boldsymbol{r})\right] \qquad (4.153)$$

If the vibronic coupling matrix element $C^{0}_{eg,a}$ is defined in normal coordinates by carrying out the summation over J as:

$$C^{0}_{eg,a} = \sum_{J}\boldsymbol{C}^{J,0}_{eg}\cdot\boldsymbol{S}_{J,a} = \sum_{J}\left\langle\psi^{0}_{e}\middle|(\partial\psi^{A}_{g}/\partial\boldsymbol{R}_{J})_{0}\right\rangle\cdot\boldsymbol{S}_{J,a} = \left\langle\psi^{0}_{e}\right\rangle(\partial\psi^{A}_{g}/\partial Q_{a})_{0}\big\rangle \qquad (4.154)$$

Equation (4.152) for the electron TCD can now be written as:

$$\boldsymbol{J}^{a}_{g0,g1}(\boldsymbol{r}) = (2\hbar\omega_{a})^{1/2}\sum_{e\neq g}\frac{C^{0}_{eg,a}}{\omega^{0}_{eg}}\boldsymbol{J}^{0}_{ge}(\boldsymbol{r}) \qquad (4.155)$$

In Figure 4.3 we show a map of $\boldsymbol{J}^{a}_{g0,g1}(\boldsymbol{r})$ with its corresponding nuclear displacement or nuclear velocity vectors for the same methine CH stretching mode of the molecule (*S*)-methyl-d_3-lactate *Cd*$_3$ as shown in Figure 4.2 (Freedman *et al.*, 2000). Here, the field of small arrows corresponds to vibrational TCD vectors. The length and direction of the TCD vectors shows the strength and direction of the TCD field at the location of the TCD vector. What is surprising is the degree of electron motion

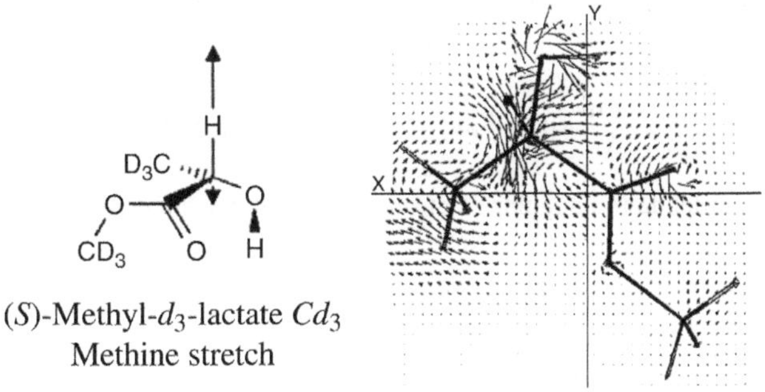

Figure 4.3 The nuclear velocity vectors for the methine C–H stretching mode of the molecule (S)-methyl-d$_3$-lactate Cd$_3$ are illustrated on the left as in Figure 4.2 and the nuclear velocity vectors and the corresponding map of the vibrational electron TCD vector field is illustrated on the right (Freedman et al., 2000). Reproduced with permission from the American Chemical Society (Freedman and Nafie, 2000)

away from the methine C–H bond. In particular there is strong circulation about the oxygen of the hydroxyl (OH) group corresponding to a strong vibrational magnetic dipole moment.

In order to derive Equation (4.145) from Equation (4.155), we write from Equation (4.48)

$$\left\langle \Psi^a_{g1} \middle| \dot{\boldsymbol{\mu}}^E \middle| \Psi^a_{g0} \right\rangle = -2i \sum_{e \neq g} \frac{\left\langle \psi^0_e \middle| \dot{\mu}^E_\beta \middle| \psi^0_g \right\rangle C^0_{eg,a} \left\langle \phi^a_{g1} \middle| \dot{Q}_a \middle| \phi^a_{g0} \right\rangle}{\omega^0_{eg}} \tag{4.156}$$

In writing this equation we note that the velocity dipole transition moment is odd with respect to interchange of wavefunctions. Using Equation (2.22) for the electron TCD for pure electronic transitions

$$\left\langle \Psi^a_{g1} \middle| \dot{\boldsymbol{\mu}}^E \middle| \Psi^a_{g0} \right\rangle = \sum_{e \neq g} \frac{\left(-ie \int_{-\infty}^{\infty} \boldsymbol{J}^0_{ge}(\boldsymbol{r}) \mathrm{d}\boldsymbol{r} \right) C^0_{eg,a} (2\hbar\omega_a)^{1/2}}{\omega^0_{eg}} = -ie \int \boldsymbol{J}^a_{g0,g1}(\boldsymbol{r}) \mathrm{d}\boldsymbol{r} \tag{4.157}$$

and from which Equation (4.145) follows by comparison with Equation (4.155). The relationships above define $\boldsymbol{J}^a_{g1,g0}(\boldsymbol{r})$ and establish the relationship of vibration TCD to the velocity formulation of VA absorption intensities. Finally, to emphasize that $\boldsymbol{J}^a_{g0,g1}(\boldsymbol{r})$ is closely related to the integrand of the velocity dipole vibrational transition moment, we introduce the following notation, which will be used more extensively in the next section:

$$\dot{\boldsymbol{\mu}}^{E,a}_{g1,g0}(\boldsymbol{r}) = -\dot{\boldsymbol{\mu}}^{E,a}_{g0,g1}(\boldsymbol{r}) = -ie\boldsymbol{J}^a_{g0,g1}(\boldsymbol{r}) \tag{4.158}$$

Note again that the convention for the order of states is reversed from matrix elements compared with that for current density.

4.4.2 Relationship Between Vibrational TCD and VCD Intensity

Given the relationship in Equation (4.55) between the velocity form of the electric dipole moment operator and the magnetic dipole moment operator, it is straightforward to write the magnetic dipole transition moment in terms of the vibrational current density as an extension of Equation (4.157),

$$\left\langle \Psi^a_{g1} \middle| \boldsymbol{m}^E \middle| \Psi^a_{g0} \right\rangle = -\frac{ie}{2c} \int \boldsymbol{r} \times \boldsymbol{J}^a_{g0,g1}(\boldsymbol{r}) \mathrm{d}\boldsymbol{r} \tag{4.159}$$

from whence we can define the angular electron vibrational current density as:

$$\boldsymbol{m}^{E,a}_{g1,g0}(\boldsymbol{r}) = -\frac{ie}{2c} \boldsymbol{r} \times \boldsymbol{J}^a_{g0,g1}(\boldsymbol{r}) \tag{4.160}$$

This quantity is obviously origin dependent, but this dependence can be partially suppressed, as before, by distributing the origin among the nuclei of the molecule as:

$$\boldsymbol{m}^{E,a}_{g1,g0}(\boldsymbol{r}) = -\frac{ie}{2c} \sum_J (\boldsymbol{R}_{J,0} + \boldsymbol{r}_J) \times \boldsymbol{J}^a_{g0,g1}(\boldsymbol{r}) \tag{4.161}$$

The term that depends on the position of the Jth nucleus has the same moment arm as the nuclear contribution to the magnetic dipole transition density given by:

$$m^{N,a}_{g1,g0}(r) = \frac{ie}{2c}\sum_J R_{J,0} \times Z_J S_{J,a}(2\hbar\omega_a)^{1/2}\delta(r - R_{J,0}) \tag{4.162}$$

The charge density of the nucleus is located only at the nuclear position and hence its spatial dependence in the molecule is expressed simply as a delta function. If the electronic and nuclear contributions are combined we have

$$\begin{aligned} m^a_{g1,g0}(r) &= m^{E,a}_{g1,g0}(r) + m^{N,a}_{g1,g0}(r) \\ &= -\frac{ie}{2c}\sum_J \left\{ R_{J,0} \times \left[J^a_{g0,g1}(r) - Z_J S_{J,a}(2\hbar\omega_a)^{1/2}\delta(r - R_{J,0}) \right] + r_J \times J^a_{g0,g1}(r) \right\} \end{aligned} \tag{4.163}$$

This is the total (electron plus nuclear) TCD for the transition $g0$ to $g1$ in the ath normal mode. We can use this density together with the corresponding total velocity dipole TCD to express the velocity form of the rotational strength from Equation (4.14) as:

$$\begin{aligned} R^a_{v,g1,g0} &= \omega_a^{-1}\mathrm{Re}\left[\dot{\mu}^a_{g0,g1} \cdot m^a_{g1,g0}\right] \\ &= \omega_a^{-1}\mathrm{Re}\left[\int \dot{\mu}^a_{g0,g1}(r)dr \cdot m^a_{g1,g0}\right] = \omega_a^{-1}\mathrm{Re}\left[\dot{\mu}^a_{g0,g1} \cdot \int m^a_{g1,g0}(r)dr\right] \end{aligned} \tag{4.164}$$

If we remove the integrations over the space of the molecule from both sides, we can identify a quantity called the rotational strength density,

$$R^a_{v,g1,g0}(r) = \omega_a^{-1}\mathrm{Re}\left[\dot{\mu}^a_{g0,g1}(r) \cdot m^a_{g1,g0}\right] = \omega_a^{-1}\mathrm{Re}\left[\dot{\mu}^a_{g0,g1} \cdot m^a_{g1,g0}(r)\right] \tag{4.165}$$

This is a scalar field that is either a positive or a negative number at each point in the space of the molecule to express the contribution of that spatial position to the total rotational strength for that vibrational transition. A corresponding expression can be written for the velocity form of the dipole strength, which again has electronic and nuclear contributions for the constituent vector TCDs,

$$D^a_{v,g1,g0}(r) = \omega_a^{-2}\mathrm{Re}\left[\dot{\mu}^a_{g0,g1}(r) \cdot \dot{\mu}^a_{g1,g0}\right] = \omega_a^{-2}\mathrm{Re}\left[\dot{\mu}^a_{g0,g1} \cdot \dot{\mu}^a_{g1,g0}(r)\right] \tag{4.166}$$

The velocity dipole strength density is a scalar field that is positive at every point in the molecule representing the contribution of that location to the total VA absorption intensity for the vibrational transition considered. The corresponding position dipole strength density can be defined as the integrand of the integral over the space of the molecule that yields the position dipole strength. However, as was discussed in Chapter 2, such a quantity is different for each choice of origin. The only origin-free quantities are the velocity dipole TCD and velocity dipole strength. They, and only they, show the true location of charge motion leading to vibrational intensities in the molecule. The rotational strength density is origin independent but the magnetic dipole TCD, or angular TCD, is origin dependent. The dipole and rotational strength densities have not yet been calculated but would be of considerable interest for the insight they would provide regarding the spatial origin of VA and VCD intensities.

References

Buckingham, A.D., Fowler, P.W., and Galwas, P.A. (1987) Velocity-dependent property surfaces and the theory of vibrational circular dichroism. *Chem. Phys.*, **112**, 1–14.

Dutler, R., and Rauk, A. (1989) Calculated infrared absorption and vibrational circular dichroism intensities of oxirane and its deuterated analogues. *J. Am. Chem. Soc.*, **111**, 6957–6966.

Freedman, T.B., and Nafie, L.A. (1994). Theoretical formalism and models for vibrational circular dichroism intensity. In: *Modern Nonlinear Optics, Part 3* (eds M. Evans, and S. Kielich), John Wiley & Sons, Inc., New York, pp. 207–263.

Freedman, T.B., Lee, E., and Nafie, L.A. (2000) Vibrational current density in (*S*)-methyl lactate: Visualizing the origin of the methine-stretching vibrational circular dichroism intensity. *J. Phys. Chem. A*, **104**, 3944–3951.

Hunt, K.L.C., and Harris, R.A. (1991) Vibrational circular dichroism and electric-field shielding tensors: a new physical interpretation based on nonlocal susceptibility densities. *J. Chem. Phys.*, **94**, 6995–7002.

Lowe, M.A., Segal, G.A., and Stephens, P.J. (1986) The theory of vibrational circular dichroism: *trans*-1, 2-dideuteriocyclopropane. *J. Am. Chem. Soc.*, **108**, 248–256.

Nafie, L.A. (1983) Adiabatic behavior beyond the Born–Oppenheimer approximation. Complete adiabatic wavefunctions and vibrationally induced electronic current density. *J. Chem. Phys.*, **79**, 4950–4957.

Nafie, L.A. (2004) Theory of vibrational circular dichroism and infrared absorption: extension to molecules with low-lying excited electronic states. *J. Phys. Chem. A*, **108**, 7222–7231.

Nafie, L.A., and Freedman, T.B. (1983) Vibronic coupling theory of infrared vibrational intensities. *J. Chem. Phys.*, **78**, 7108–7116.

Nafie, L.A., and Freedman, T.B. (1987) Vibrational circular dichroism theory: Formulation defining magnetic atomic polar tensors and vibrational nuclear magnetic shielding tensors. *Chem. Phys. Lett.*, **134**, 225–232.

Stephens, P.J. (1985) Theory of vibrational circular dichroism. *J. Phys. Chem.*, **89**, 748–752.

5

Theory of Raman Optical Activity

In this chapter, the theoretical developments of Raman scattering in Chapter 2 will be combined with the concepts of optical activity in Chapter 3 to present the theory of Raman optical activity (ROA) at various levels of approximation. The goal is to provide a complete account of the basic theory with derivations of expressions needed to understand the theoretical basis of existing experimental applications in addition to reports on new theoretical extensions of ROA. As with the previous chapter, we begin with a relatively simple overview of the theory. This will be followed by the next simplest description of ROA, the far-from resonance (FFR) theory. Next, the general unrestricted (GU) theory of ROA and Raman will be presented that embraces all levels of the theory of ROA. Finally, important limits of the GU theory will be derived, including the FFR theory, the near-resonance (NR) theory, the single electronic state (SES) strong resonance theory, and the two electronic state (TES) resonance theory.

5.1 Comparison of ROA to VCD Theory

The theory of vibrational ROA and Raman scattering is different in many ways from the theory of infrared vibrational circular dichroism (VCD) and vibrational absorption (VA), and yet these two forms of vibrational optical activity (VOA) possess a high level similarity. To make clear these similarities and differences we extend the introductory treatment of VCD and ROA given in Chapter 1 with the theoretical definitions that we have developed thus far for vibrational spectroscopy, chirality, and VCD in Chapters 2, 3, and 4, respectively.

There are many possible forms of ROA. The most accessible at this time is scattered circular polarization (SCP) ROA, as this is provided in the first and to date only commercially available ROA instrument. From Chapter 1, the general definition of SCP-ROA for a vibrational transition from $g0$ to $g1$ of the ground electronic state of vibrational normal mode a is given by:

$$\text{SCP-ROA:}\quad (\Delta I^{\alpha})^{a}_{g1,g0} = \left[(I^{\alpha}_{R})^{a}_{g1,g0} - (I^{\alpha}_{L})^{a}_{g1,g0} \right] \tag{5.1}$$

Vibrational Optical Activity: Principles and Applications, First Edition. Laurence A. Nafie.

This form of ROA consists of fixed, non-elliptically (including non-circularly) polarized, or unpolarized, incident radiation and is followed by the measurement of the difference in the intensity for right minus left circularly polarized (RCP minus LCP) Raman scattered radiation. The available commercial SCP-ROA instrument uses unpolarized incident radiation (superscript U) and back-scattering (180°) scattering geometry. The corresponding theoretical expression is given by:

$$[I_R^U(180°) - I_L^U(180°)]^a_{g1,g0} = \frac{8K}{c}\left[12\left(\beta(G')^2\right)^a_{g1,g0} + 4\left(\beta(A)^2\right)^a_{g1,g0}\right] \tag{5.2}$$

where K is a constant and $\beta(G')^2$ and $\beta(A)^2$ are ROA invariants that will be defined in the following section. Briefly, they are magnetic dipole and electric quadrupole anisotropic ROA invariants in the far-from resonance (FFR) approximation that are the analogues of the anisotropic Raman invariant $\beta(\alpha)^2$ introduced in Equation (2.135). The equations corresponding to (5.1) and (5.2) for SCP-Raman are given by:

$$\text{SCP-Raman:} \quad (I^\alpha)^a_{g1,g0} = \left[(I_R^\alpha)^a_{g1,g0} + (I_L^\alpha)^a_{g1,g0}\right] \tag{5.3}$$

$$[I_R^U(180°) + I_R^U(180°)]^a_{g1,g0} = 4K\left[45(\alpha^2)^a_{g1,g0} + 7\left(\beta(\alpha)^2\right)^a_{g1,g0}\right] \tag{5.4}$$

The theory of VCD at this same level of theory is comparatively simple. There is only one invariant for VCD, the rotational strength, and one for the corresponding VA intensity, the dipole strength. The experimental measurement geometry does not need to be specified for the absorption of the IR beam transmitted through the sample. The definition for VCD is given by the following expressions:

$$\text{VCD-rotational strength:} \quad (\Delta A)^a_{g1,g0} = \left[(A_L)^a_{g1,g0} - (A_R)^a_{g1,g0}\right] \tag{5.5}$$

$$R^a_{r,g1,g0} = K'\left[(\varepsilon_L)^a_{g0,g1} - (\varepsilon_R)^a_{g0,g1}\right] = \mathrm{Im}\left[(\boldsymbol{\mu})^a_{g0,g1} \cdot (\boldsymbol{m})^a_{g1,g0}\right] \tag{5.6}$$

where K' is a constant. The measured definitions are taken from Chapter 1, and the second part of Equation (5.6) was first defined in Equation (4.2). The subscript r on the symbol for the rotational strength refers to the position form of the electric dipole moment operator, $\boldsymbol{\mu}$. The corresponding expressions for the VA intensity are given by:

$$\text{VA-dipole strength:} \quad (A)^a_{g1,g0} = \frac{1}{2}\left[(A_L)^a_{g1,g0} + (A_R)^a_{g1,g0}\right] \tag{5.7}$$

$$D^a_{r,g1,g0} = K'\frac{1}{2}\left[(\varepsilon_L)^a_{g0,g1} + (\varepsilon_R)^a_{g0,g1}\right] = \left|(\boldsymbol{\mu})^a_{g1,g0}\right|^2 \tag{5.8}$$

It can be seen again that ROA follows the right minus left convention whereas VCD adopts the left minus right convention, common to all other forms of optical activity. Also, the Raman intensity is defined as the sum of the RCP and LCP intensities whereas vibrational absorption is defined as the average of LCP and RCP intensities.

In the following section, we describe a broader introduction to the theory of ROA but still restricted to the FFR approximation. Before doing so, we mention that after the FFR theory of ROA was published (Barron and Buckingham, 1971), a number of models of ROA intensity were described as conceptual aides to understanding the origin of ROA intensity. Some of these models also found use before full quantum chemistry calculations of ROA appeared in order to carry out simple calculations of ROA. As mentioned briefly in Appendix A, these models included the two-group model (Barron and Buckingham, 1975), the methyl torsion model (Barron and Buckingham, 1979), the perturbed generate oscillator model (Barron and Buckingham, 1975), the atom-dipole interaction model (Prasad and Nafie, 1979), and the bond polarizability model (Barron *et al.*, 1986; Escribano *et al.*, 1987). We will not describe these models here as they are no longer in active use and have been described in detail previously (Barron and Buckingham, 1975; Barron, 2004).

5.2 Far-From Resonance Theory (FFR) of ROA

Before describing the theory or ROA in its full generality, we begin by considering the fundamentals of what is sometimes called the original theory of ROA. Recall from Chapter 1 that incident circular polarization (ICP-ROA) was the only form of ROA measured from its discovery in 1973 (Barron *et al.*, 1973) until the first SCP-ROA measurements in 1988 (Spencer *et al.*, 1988), which in turn was followed by DCP-ROA theory in the following year (Nafie and Freedman, 1989) and the first DCP-ROA measurements a few years later (Che *et al.*, 1991). During the early period of ICP-ROA, with one exception (Hug, 1982), all measurements were made with right-angle scattering geometry. The original theory is constructed in the far-from resonance (FFR) approximation of Raman scattering as described in Section 2.4.7 of Chapter 2.

5.2.1 Right-Angle ROA Scattering

The original theory of ROA was developed for right-angle ICP-ROA in which the laser radiation is switched between RCP and LCP states. The existence of ROA on the scattered radiation was also described in terms of the excess of RCP or LCP in the scattered light for each vibrational band, called the degree of circular polarization, but no method for carrying out such a measurement was offered until SCP-ROA was first measured by the intensity difference of RCP and LCP scattered Raman radiation. There are two principal forms of ICP-ROA for right-angle scattering, polarized and depolarized, corresponding to whether a polarization analyzer placed in the scattered beam is perpendicular or parallel, respectively, to the scattering plane. Because depolarized ICP-ROA has much lower Raman intensity in the polarized bands where artifacts are most difficult to control, this form of ROA was used for virtually all ROA measurements until the late 1980s. In terms of the ROA invariants mentioned above, the *depolarized* form of ROA is given by:

$$\text{ICP}_D\text{-ROA}\,(90^\circ):\quad I_Z^R(90^\circ) - I_Z^L(90^\circ) = \frac{4K}{c}\left[6\beta(G')^2 - 2\beta(A)^2\right] \tag{5.9}$$

$$\text{ICP}_D\text{-Raman}\,(90^\circ):\quad I_Z^R(90^\circ) + I_Z^L(90^\circ) = 2K\left[6\beta(\alpha)^2\right] \tag{5.10}$$

Equation (5.10) has been presented previously in Equation (2.116) for depolarized Raman right-angle Raman scattering using unpolarized incident radiation. These equations are the same because $I_Z^R(90^\circ) + I_Z^L(90^\circ) = I_Z^X(90^\circ) + I_Z^Y(90^\circ) = I_Z^U(90^\circ)$. For simplicity, we have not included the labels for the initial and final states nor the normal mode excited in the Raman transition. These can easily be added for any particular case as needed. A factor $1/c$ appears in Equation (5.9) that does not appear in Equation (5.10), because traditionally SI units are used to describe the theory of ROA, and the expression for the magnetic dipole moment in ROA (see below) differs by a factor of c from the corresponding expressions in CGS units used for the theory of VCD. In order to conform to existing convention, we use SI units for ROA theory and CGS units for VCD theory. The corresponding expressions for the more difficult *polarized* right-angle ICP intensities are given by:

$$\text{ICP}_P\text{-ROA } (90^\circ): \quad I_X^R(90^\circ) - I_X^L(90^\circ) = \frac{4K}{c}\left[45\alpha G' + 7\beta(G')^2 + \beta(A)^2\right] \tag{5.11}$$

$$\text{ICP}_P\text{-Raman } (90^\circ): \quad I_X^R(90^\circ) + I_X^L(90^\circ) = 2K\left[45\alpha^2 + 7\beta(\tilde{\alpha})^2\right] \tag{5.12}$$

The expression for the Raman intensity is the same as that given for unpolarized right-angle scattering in Equation (2.137). It can also be seen that this Raman intensity is exactly half that given for unpolarized *backscattering* SCP-Raman intensity above in Equation (5.4).

The equations for the two Raman invariants and three ROA invariants in the FFR approximation are given by:

$$\alpha^2 = \frac{1}{9}\alpha_{\alpha\alpha}\alpha_{\beta\beta} \qquad \beta(\alpha)^2 = \frac{1}{2}\left(3\alpha_{\alpha\beta}\alpha_{\alpha\beta} - \alpha_{\alpha\alpha}\alpha_{\beta\beta}\right) \tag{5.13}$$

for Raman scattering, and by:

$$\alpha G' = \frac{1}{9}\alpha_{\alpha\alpha}G'_{\beta\beta} \tag{5.14}$$

$$\beta(G')^2 = \frac{1}{2}\left(3\alpha_{\alpha\beta}G'_{\alpha\beta} - \alpha_{\alpha\alpha}G'_{\beta\beta}\right) \tag{5.15}$$

$$\beta(A)^2 = \frac{1}{2}\omega_0\alpha_{\alpha\beta}\varepsilon_{\alpha\gamma\delta}A_{\gamma,\delta\beta} \tag{5.16}$$

for ROA. Here we have used Cartesian tensor notation and the Einstein summation convention first defined in Equation (4.12) where repeated Greek subscripts in a symbol or product of symbols implies summation over Cartesian axes x, y, and z, and $\varepsilon_{\alpha\gamma\delta}$ is the alternating tensor that equals $+1$ for even permutations of x, y, and z and -1 for their odd permutation. Thus, for example, $\alpha G'$ consists of a sum of nine terms written explicitly from Equation (5.14) as $(\alpha_{xx}G'_{xx} + \alpha_{xx}G'_{yy} + \ldots + \alpha_{zz}G'_{zz})/9$. The Raman invariants in Equation (5.13) are the same as given in Equations (2.134) and (2.135). The three types of scattering tensors appearing in these equations are the polarizability

$$\alpha_{\alpha\beta} = \frac{2}{\hbar}\sum_{e\neq g}\frac{\omega_{eg}^0}{\left(\omega_{eg}^0\right)^2 - \omega_0^2}\mathrm{Re}\left[\langle g|\hat{\mu}_\alpha|e\rangle\langle g|\hat{\mu}_\beta|g\rangle\right] \tag{5.17}$$

for Raman scattering as given in Equation (2.133), and

$$G'_{\alpha\beta} = -\frac{2}{\hbar}\sum_{e\neq g}\frac{\omega_0}{\left(\omega_{eg}^0\right)^2 - \omega_0^2}\mathrm{Im}\left[\langle g|\hat{\mu}_\alpha|e\rangle\langle e|\hat{m}_\beta|g\rangle\right] \tag{5.18}$$

$$A_{\alpha,\beta\gamma} = \frac{2}{\hbar}\sum_{e\neq g}\frac{\omega_{eg}^0}{\left(\omega_{eg}^0\right)^2 - \omega_0^2}\mathrm{Re}\left[\langle g|\hat{\mu}_\alpha|e\rangle\langle e|\hat{\Theta}_{\beta\gamma}|g\rangle\right] \tag{5.19}$$

$G'_{\alpha\beta}$ and $A_{\alpha,\beta\gamma}$ are referred to as the magnetic dipole optical activity tensor and the electric quadrupole optical activity tensor, respectively. Using these invariants, we can write intensity expressions for ROA and Raman that cover all possible polarization modulations and scattering geometries in the FFR approximation. The relevant electric dipole, magnetic dipole, and electric quadrupole operators in SI units are given by:

$$\hat{\mu}_\alpha = -e\sum_j r_{j\alpha} \tag{5.20}$$

$$\hat{m}_\alpha = -\frac{e}{2m}\sum_j \varepsilon_{\alpha\beta\gamma} r_{j\beta} p_{j\gamma} \tag{5.21}$$

$$\hat{\Theta}_{\alpha\beta} = -\frac{e}{2}\sum_j (3r_{j\alpha}r_{j\beta} - r_{j\alpha}r_{j\alpha}) \tag{5.22}$$

where for Raman and ROA scattering summations are needed only for the electrons of the molecule. Unlike, IR and VCD, the nuclei are too small to have measurable contributions to the polarizability of the molecule. As mentioned above, the SI definition of the magnetic dipole moment differs by a factor of $1/c$ compared with the CGS definition of this operator used for VCD.

All the tensors given in this section are presented without vibrational state labels. As a result, these expressions apply equally well to Rayleigh optical activity (RayOA) and ROA. To add the vibrational labels for a fundamental transition in the ground electronic state between levels $g0$ and $g1$ for normal mode a we write

$$\begin{aligned}(\alpha_{\alpha\beta})^a_{g1,g0} &= \left\langle \phi^a_{g1}|\alpha_{\alpha\beta}|\phi^a_{g0}\right\rangle = \left\langle \phi^a_{g1}\left|\frac{2}{\hbar}\sum_{e\neq g}\frac{\omega^0_{eg}\mathrm{Re}\left[\left\langle g|\hat{\mu}_\alpha|e\right\rangle\left\langle e|\hat{\mu}_\beta|g\right\rangle\right]}{\left(\omega^0_{eg}\right)^2-\omega_0^2}\right|\phi^a_{g0}\right\rangle \\ &= \frac{2}{\hbar}\sum_{e\neq g}\frac{\omega^0_{eg}\mathrm{Re}\left[\langle g|\hat{\mu}_\alpha|e\rangle_0\left\langle e|\hat{\mu}_\beta|g\right\rangle_0^{Q_a} + \langle g|\hat{\mu}_\alpha|e\rangle_0^{Q_a}\left\langle e|\hat{\mu}_\beta|g\right\rangle_0\right]}{\left(\omega^0_{eg}\right)^2-\omega_0^2}\left\langle \phi^a_{g1}\left|Q_a\right|\phi^a_{g0}\right\rangle\end{aligned} \tag{5.23}$$

In the second line of Equation (5.23), each term contains a first derivative of an electronic matrix element with respect to Q_a using notation introduced previously in Equation (2.107). In this way the vibrational matrix element for first-order nuclear position dependence can be separated from the electronic matrix elements. Similar expressions for the ROA tensors $\left(G'_{\alpha\beta}\right)^a_{g1,g0}$ and $\left(A_{\alpha,\beta\gamma}\right)^a_{g1,g0}$ can easily be written.

5.2.2 Backscattering ROA

The advantages of backscattering in ROA were first recognized by Hug in 1982 and then implemented some years later. As a result of these advantages, measurements of ROA in right-angle scattering are now rarely undertaken, and then only to address questions of theoretical interest. The first backscattering ROA experiment measurements with a modern CCD camera (Barron *et al.*, 1989) used ICP with no polarization discrimination in the scattered beam:

$$\mathrm{ICP}_U\ (180°):\quad I^R_U(180°) - I^L_U(180°) = \frac{8K}{c}\left[12\beta(G')^2 + 4\beta(A)^2\right] \tag{5.24}$$

$$I^R_U(180°) + I^L_U(180°) = 4K\left[45\alpha^2 + 7\beta(\alpha)^2\right] \tag{5.25}$$

If one compares these expressions with those for unpolarized backscattering SCP-ROA in Equations (5.2) and (5.4), one can see that, in the FFR approximation, these two experiments give exactly the same results. In addition, if one compares ICP or SCP backscattering with ICP or SCP right-angle scattering in Equations (5.9)–(5.12), one sees two significant differences. Firstly, the backscattering Raman spectra are the same as *polarized* right angle scattering, but *twice* as intense, and secondly, the backscattering ROA invariants are the same relative magnitudes but *four* times as intense as *depolarized* right-angle scattering, *and* the two invariant contributions *add* in backscattering rather than *subtract* in right-angle scattering. As a result, there is a huge advantage to measuring ROA in backscattering. Compared with right-angle scattering ICP and SCP intensities, unpolarized backscattering ICP and SCP are an odd mixture of polarized (Raman) and depolarized (ROA) scattering. When in-phase DCP (DCP_I) backscattering is considered, this odd mixture of polarized Raman and depolarized ROA disappears in favor of pure depolarized scattering for both Raman and ROA, as we consider next.

Even though unpolarized ICP- and SCP-ROA are identical in the FFR approximation, not only for backscattering, but for all scattering geometries, there is a form of ROA that differs from these two in its Raman scattering, namely DCP_I-ROA,

$$\text{DCP}_\text{I}\,(180°):\quad I_R^R(180°) - I_L^L(180°) = \frac{8K}{c}\left[12\beta(G')^2 + 4\beta(A)^2\right] \tag{5.26}$$

$$I_R^R(180°) + I_R^L(180°) = 4K\left[6\beta(\alpha)^2\right] \tag{5.27}$$

DCP_I-ROA is measured by adding synchronous circular polarization detection in the scattered light to an ICP-ROA measurement, or adding synchronous circular polarization modulation of the incident laser radiation instead of unpolarized incident radiation, to an SCP-ROA measurement. Comparing the last two sets of intensity expressions, one can see that DCP_I-ROA intensities are identical with those of unpolarized ICP- or SCP-ROA, but the corresponding Raman intensity is much less. The reason is that, compared with unpolarized SCP-ROA, for example, only part of the scattered radiation is measured in DCP_I-ROA. The part not measured is the other form of DCP-ROA, namely out-of-phase dual circular polarization (DCP_{II})-ROA,

$$\text{DCP}_\text{II}\,(180°):\quad I_L^R(180°) - I_R^L(180°) = 0 \tag{5.28}$$

$$I_L^R(180°) + I_R^L(180°) = 4K\left[45\alpha^2 + \beta(\alpha)^2\right] \tag{5.29}$$

For this form of Raman scattering, the ROA vanishes in the FFR approximation and the parent Raman intensity is highly polarized, even more polarized than traditional polarized Raman scattering. If one simply adds the ROA and Raman intensities for both forms of DCP scattering, one obtains the corresponding unpolarized ICP- or SCP-ROA intensities given in Equations (5.24) and (5.25). Further, one can see that, unlike unpolarized backscattering ICP or SCP, DCP_I-Raman and -ROA intensities are both pure depolarized. On the other hand, as mentioned above, DCP_{II}-Raman intensities are highly polarized.

5.2.3 Forward and Magic Angle Scattering ROA

There are two other forms of ROA that have been measured and provide unique spectral information. The first is ROA in forward scattering. This has been performed using ICP scattering and yields the following intensities:

$$\text{ICP}_U\ (0°):\quad I_U^R(0°) - I_U^L(0°) = \frac{8K}{c}\left[90\alpha G' + 2\beta(G')^2 - 2\beta(A)^2\right] \tag{5.30}$$

$$I_U^R(0°) + I_U^L(0°) = 4K\left[45\alpha^2 + 7\beta(\alpha)^2\right] \tag{5.31}$$

Here the Raman scattering is the same as unpolarized backscattering ICP, intense and polarized compared with right-angle scattering, and the ROA is weaker with a subtractive combination of ROA invariants that gives relatively more weight to the electric quadrupole ROA invariant.

If instead one considers right-angle ICP-ROA using a magic angle of 54.7° from the vertical for the analyzing polarizer (the same angle as the magic spinning angle used for the measurement of NMR of solids), one obtains a sum of ⅔ polarized and ⅓ depolarized ROA intensity. This yields a cancellation of the electric quadrupole ROA invariant, and the ICP scattering intensities is this case are

$$\text{ICP}_*\ (90°):\quad I_*^R(90°) - I_*^L(90°) = \frac{8K}{c}\left[\frac{45}{3}\alpha G' + \frac{10}{3}\beta(G')^2\right] \tag{5.32}$$

$$I_*^R(90°) + I_*^L(90°) = 4K\left[\frac{45}{3}\alpha^2 + \frac{10}{3}\beta(\alpha)^2\right] \tag{5.33}$$

Thus for right-angle ROA scattering with the analyzer at the magic angle, the only contributions to ROA intensity are from the two magnetic dipole ROA invariants. This is analogous to VCD where for isotropic solutions the electric quadrupole transition moments make no contribution to the observed intensity.

5.3 General Unrestricted (GU) Theory of ROA

The general theory of ROA, comparable in scope to the general theory of Raman scattering presented in Chapter 2, is fairly complex. There it was shown that the general theory of Raman scattering has three invariants, α^2, $\beta_S(\tilde{\alpha})^2$, and $\beta_A(\tilde{\alpha})^2$, instead of the two found in the simpler FFR theory, which has only α^2 and $\beta(\tilde{\alpha})^2$. However, the general unrestricted (GU) theory of ROA involves a total of *ten* ROA invariants instead of the *three* that appear in the FFR theory highlighted in the previous section. One needs the GU theory of ROA to describe all possible ROA experiments when it is no longer sufficient to assume that the incident laser radiation is very far-from resonance with the lowest allowed electronic state of the molecule. Most applications of ROA have not advanced to this state of sophistication, but the GU theory is presented here as it is the origin of all levels of approximation of linear ROA theory.

5.3.1 ROA Tensors

The theory of Raman scattering presented in Chapter 2 was formulated at the general unrestricted level, and subsequently these equations were reduced to the theory of Raman in the FFR approximation used above to introduce the theory of ROA. We return to consider Equation (2.101) for Raman scattering intensity as a function of the complex polarization state vectors for the incident and scattered photons, $\tilde{e}^i_\alpha$ and $\tilde{e}^d_\alpha$, respectively, but here the more general complex scattering tensor, $\tilde{a}_{\alpha\beta}$, will be used instead of the complex polarizability, $\tilde{\alpha}_{\alpha\beta}$, as used previously. The GU formalism for ROA begins with the following equation (Hecht and Nafie, 1991; Nafie and Che, 1994):

$$I(\tilde{e}^d, \tilde{e}^i) = 90K\left\langle \left|\tilde{e}^{d*}_\alpha \tilde{a}_{\alpha\beta}\tilde{e}^i_\beta\right|^2 \right\rangle \tag{5.34}$$

The leading term in the expansion of $\tilde{a}_{\alpha\beta}$ is $\tilde{\alpha}_{\alpha\beta}$, and the next four terms are the complex optical activity tensors as given by:

$$\tilde{a}_{\alpha\beta} = \tilde{\alpha}_{\alpha\beta} + \frac{1}{c}\left[\varepsilon_{\gamma\delta\beta}n^i_\delta \tilde{G}_{\alpha\gamma} + \varepsilon_{\gamma\delta\alpha}n^d_\delta \tilde{\mathscr{G}}_{\gamma\beta} + \frac{i}{3}\left(\omega_0 n^i_\gamma \tilde{A}_{\alpha,\gamma\beta} - \omega_R n^d_\gamma \tilde{\mathscr{A}}_{\beta,\gamma\alpha}\right)\right] \tag{5.35}$$

and where

$$K = \frac{1}{90}\left(\frac{\omega_R^2 \mu_0 \tilde{E}^{(0)}}{4\pi R}\right)^2 \tag{5.36}$$

As explained previously, K is a constant that depends on the intensity of the incident laser radiation. Here, ω_R is the angular frequency of the Raman (or Rayleigh) scattered light, μ_0 is the magnetic permeability of free space, $\tilde{E}^{(0)}$ is the effective electric field strength of the incident laser radiation of frequency ω_0, and R is the distance from the scattering source to the detector. The angular brackets in Equation (5.34) designate an average over all angles of orientation of the molecule relative to the laboratory frame of reference in the case of non-oriented sample molecules. Again, repeated Greek subscripts in a term imply summation over the Cartesian directions x, y, and z.

Equation (5.34) has nine terms, after executing the Einstein summation convention, within the vertical brackets, and these brackets designate the absolute value of the complex quantities within the brackets. The tilde above a quantity, such as a polarization vector or a scattering tensor, indicates that this quantity is in general complex. The star superscript for the polarization vector of the scattered light designates the complex conjugate. The general scattering tensor $\tilde{a}_{\alpha\beta}$ is given through the lowest-order in the breakdown of the dipole approximation leading to magnetic dipole and electric quadrupole contributions. The leading term is the polarizability tensor, first given in Equation (2.103), and is responsible for ordinary Raman (and Rayleigh) scattering. The tensors in square brackets are the optical activity tensors, or in the case of ROA, the ROA tensors. The first two are magnetic dipole–electric dipole ROA tensors and the second two are electric quadrupole–electric dipole ROA tensors. The vectors n^i_α and n^d_α are the propagation vectors for the incident and scattered light, respectively. Three additional terms (not shown) may be added to Equation (5.35) to account for electric field induced birefringence and the effects of finite cone of collection of the Raman scattering and ROA.

The Raman polarizability tensor is given by:

$$\tilde{\alpha}_{\alpha\beta}(\omega_0) = \frac{1}{\hbar}\sum_{j\neq m,n}\left[\frac{\langle\tilde{m}|\hat{\mu}_\alpha|\tilde{j}\rangle\langle\tilde{j}|\hat{\mu}_\beta|\tilde{n}\rangle}{\omega_{jn} - \omega_0 - i\Gamma_j} + \frac{\langle\tilde{m}|\hat{\mu}_\beta|\tilde{j}\rangle\langle\tilde{j}|\hat{\mu}_\alpha|\tilde{n}\rangle}{\omega_{jm} + \omega_0 + i\Gamma_j}\right] \tag{5.37}$$

where, as before, the summation is over all excited electronic state wavefunctions, $\tilde{j}$, except for the initial and final states, $\tilde{n}$ and $\tilde{m}$, respectively. The wavefunctions of all states are taken to be complex in the general case as denoted by a tilda above the state label. The states $\tilde{n}$ and $\tilde{m}$ differ by a vibrational quantum of energy, and the terms ω_{jm} and ω_{jn} are the angular frequency differences between the states $\tilde{j}$ and $\tilde{n}$ or $\tilde{m}$. The terms $i\Gamma_j$ are imaginary terms proportional to the spectral width of the excited state $\tilde{j}$, and hence inversely proportional to the lifetime of that state. The opposite-sign convention has been followed for the $i\Gamma_j$ terms, determined recently to be preferred based on a number of physical and phenomenological arguments (Long, 2002). The first term in Equation (5.37) is called the resonance term as the difference between the jn-transition frequency and the laser frequency vanishes at the resonance condition, and the second term is the non-resonance term. The electric-dipole moment operator $\hat{\mu}_\alpha$ has been defined previously in Equation (5.36). Finally, we note that the polarizability, $\tilde{\alpha}_{\alpha\beta}(\omega_0)$, is a function of the frequency of the incident laser radiation, ω_0. This dependence is particularly important as resonance is approached; however, for simplicity this additional labeling will be suppressed in development of expressions beyond initial definitions unless explicitly needed.

The matrix elements in Equation (5.37) involving the operator $\hat{\mu}_\beta$ represents the interaction of the molecule with the incident radiation, while the matrix elements with the operator $\hat{\mu}_\alpha$ represents the interaction of the molecule with the scattered radiation. This can be seen most easily by inspection of Equation (5.34). The matrix element products in each term in Equation (5.37) can be read from right to left in a time-ordered sense, and hence the resonance term describes the molecule interacting first with a laser photon and subsequently creating a scattered photon, whereas the non-resonance terms reverse the time order of the those two events. In Chapter 2 we previously illustrated the time order of these events graphically with time-ordered diagrams in Figure 2.1.

The four ROA tensors differ from the Raman polarizability tensor by substitution of a higher-order operator for an electric-dipole operator in Equation (5.37). The two operators needed for ROA are the magnetic-dipole moment operator and the electric-quadruple moment operator $\hat{m}_\alpha$ and $\hat{\theta}_{\alpha\beta}$ defined above in Equations (5.21) and (5.22). The resulting ROA tensors, including dependence on the incident laser radiation are given by:

$$\tilde{G}_{\alpha\beta}(\omega_0) = \frac{1}{\hbar}\sum_{j\neq m,n}\left[\frac{\langle\tilde{m}|\hat{\mu}_\alpha|\tilde{j}\rangle\langle\tilde{j}|\hat{m}_\beta|\tilde{n}\rangle}{\omega_{jn}-\omega_0-i\Gamma_j}+\frac{\langle\tilde{m}|\hat{m}_\beta|\tilde{j}\rangle\langle\tilde{j}|\hat{\mu}_\alpha|\tilde{n}\rangle}{\omega_{jn}+\omega_R+i\Gamma_j}\right] \tag{5.38}$$

$$\tilde{\mathscr{G}}_{\alpha\beta}(\omega_0) = \frac{1}{\hbar}\sum_{j\neq m,n}\left[\frac{\langle\tilde{m}|\hat{m}_\alpha|\tilde{j}\rangle\langle\tilde{j}|\hat{\mu}_\beta|\tilde{n}\rangle}{\omega_{jn}-\omega_0-i\Gamma_j}+\frac{\langle\tilde{m}|\hat{\mu}_\beta|\tilde{j}\rangle\langle\tilde{j}|\hat{m}_\alpha|\tilde{n}\rangle}{\omega_{jn}+\omega_R+i\Gamma_j}\right] \tag{5.39}$$

$$\tilde{A}_{\alpha,\beta\gamma}(\omega_0) = \frac{1}{\hbar}\sum_{j\neq m,n}\left[\frac{\langle\tilde{m}|\hat{\mu}_\alpha|\tilde{j}\rangle\langle\tilde{j}|\hat{\Theta}_{\beta\gamma}|\tilde{n}\rangle}{\omega_{jn}-\omega_0-i\Gamma_j}+\frac{\langle\tilde{m}|\hat{\Theta}_{\beta\gamma}|\tilde{j}\rangle\langle\tilde{j}|\hat{\mu}_\alpha|\tilde{n}\rangle}{\omega_{jn}+\omega_R+i\Gamma_j}\right] \tag{5.40}$$

$$\tilde{\mathscr{A}}_{\alpha,\beta\gamma}(\omega_0) = \frac{1}{\hbar}\sum_{j\neq m,n}\left[\frac{\langle\tilde{m}|\hat{\Theta}_{\beta\gamma}|\tilde{j}\rangle\langle\tilde{j}|\hat{\mu}_\alpha|\tilde{n}\rangle}{\omega_{jn}-\omega_0-i\Gamma_j}+\frac{\langle\tilde{m}|\hat{\mu}_\alpha|\tilde{j}\rangle\langle\tilde{j}|\hat{\Theta}_{\beta\gamma}|\tilde{n}\rangle}{\omega_{jn}+\omega_R+i\Gamma_j}\right] \tag{5.41}$$

Here we have written $\omega_{jm}+\omega_0$ as $\omega_{jn}+\omega_R$ to emphasize the frequency of the scattered light. The expressions given above define the tensors used for the description of all forms of polarized Raman scattering through first-order in the magnetic-dipole and electric-quadrupole interaction of light with matter. It can be seen that in the GU theory there are two magnetic-dipole optical activity tensors, $\tilde{G}_{\alpha\beta}$ and $\tilde{\mathscr{G}}_{\alpha\beta}$, and two electric-quadrupole optical activity tensors $\tilde{A}_{\alpha,\beta\gamma}$ and $\tilde{\mathscr{A}}_{\alpha,\beta\gamma}$. In the Roman font tensors, the magnetic-dipole or electric-quadrupole operators are associated with the incident radiation while for the script tensors, these operators are associated with the scattered radiation. These four complex tensors are sufficient to describe all forms of ROA to first order beyond the dipole approximation. In Figure 5.1 we present, in analogy to

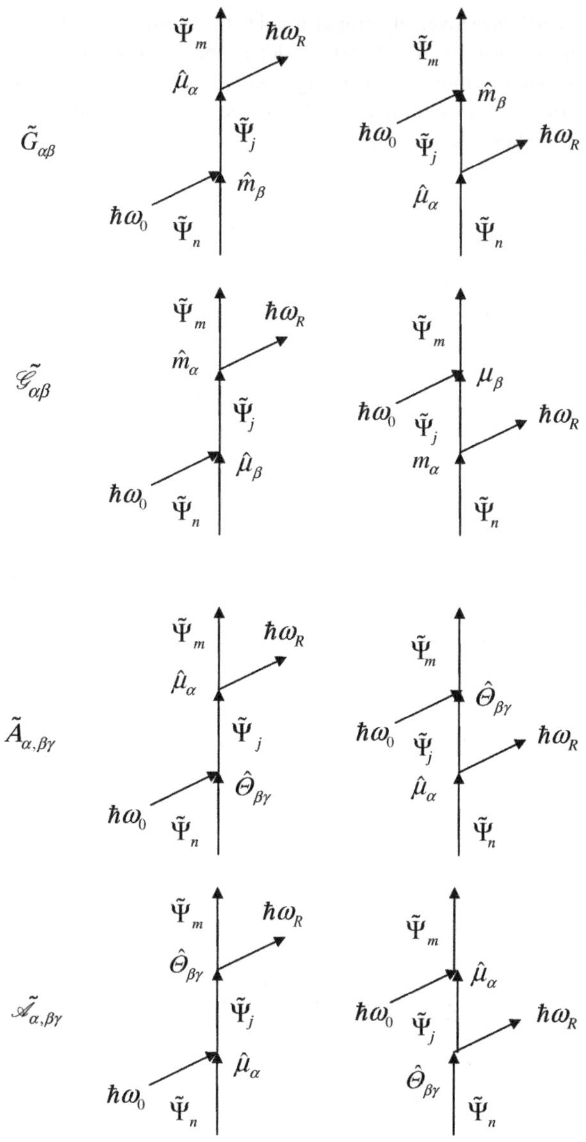

Figure 5.1 *Time-ordered diagrams of the four ROA tensors representing the resonance term (left) and non-resonance term (right) of Equations (5.38)–(5.41). The time axis is vertical and the space axis is horizontal. Molecular state vectors are represented by vertical arrows that move in time but not in space. Diagonal arrows represent photons traveling both in time and in space. The interaction vertices are labeled by operators with Greek subscripts that represent the polarization direction of the photon interaction that changes the state vectors. In the resonance term, the molecule first destroys the incident photon, $\hbar\omega_0$, and then creates the scattered photon, $\hbar\omega_R$, whereas in the non-resonance term the time order of the photon interactions is reversed*

Figure 2.1, time-ordered diagrams to illustrate graphically the four ROA tensors given in Equations (5.38)–(5.41).

5.3.2 Forms of ROA

If the polarization states of the incident and scattered radiation are specified, one can construct theoretical expressions for the various forms of ROA intensities that can be measured. Considering first circular polarization, ROA intensities can be obtained from pairs of intensity expressions that differ only in the change in the polarization state of the incident beam, the scattered beam, or both beams between right and left circular states.

The fundamental forms of CP-ROA observables are classified by polarization states of the incident and scattering radiation and by the scattering angle, ξ. There are four different forms of CP ROA given by:

$$\text{ICP-ROA:} \quad \Delta I_\alpha(\xi) = I_\alpha^R(\xi) - I_\alpha^L(\xi) \tag{5.42}$$

$$\text{SCP-ROA:} \quad \Delta I^\alpha(\xi) = I_R^\alpha(\xi) - I_L^\alpha(\xi) \tag{5.43}$$

$$\text{DCP}_\text{I}\text{-ROA:} \quad \Delta I_\text{I}(\xi) = I_R^R(\xi) - I_L^L(\xi) \tag{5.44}$$

$$\text{DCP}_\text{II}\text{-ROA:} \quad \Delta I_\text{II}(\xi) = I_L^R(\xi) - I_R^L(\xi) \tag{5.45}$$

In the case of ICP- and SCP-ROA, the polarization state α is any fixed linear value or the unpolarized state. The standard choices are unpolarized, linearly polarized parallel to the scattering plane (depolarized), or linearly polarized perpendicular to the scattering plane (polarized). The common scattering angles are 90° (right-angle scattering) 180° (backscattering), and 0° (forward scattering).

There are also four forms of linear polarization (LP) ROA:

$$\text{ILP-ROA:} \quad \Delta I_\alpha(\xi) = I_\alpha^+(\xi) - I_\alpha^-(\xi) \tag{5.46}$$

$$\text{SLP-ROA:} \quad \Delta I^\alpha(\xi) = I_+^\alpha(\xi) - I_-^\alpha(\xi) \tag{5.47}$$

$$\text{DLP}_\text{I}\text{-ROA:} \quad \Delta I_\text{I}(\xi) = I_+^+(\xi) - I_+^+(\xi) \tag{5.48}$$

$$\text{DLP}_\text{II}\text{-ROA:} \quad \Delta I_\text{II}(\xi) = I_-^+(\xi) - I_+^-(\xi) \tag{5.49}$$

Instead of being defined between left and right circular polarization states, LP-ROA is defined as the difference between different orientations of linear polarization states, the optimum of which differ by ±45°. In the definitions above + refers to a linear polarization state that is +45° relative to the orientation of some fixed linear polarization state α. Similarly, − refers to a linear polarization at −45° from the fixed linear polarization state α. The difference between the CP and LP forms of ROA is similar to the difference between the CD and optical rotation (OR) forms of single photon optical activity. For OR, if light with a vertical linear polarization state (v) enters the sample and the difference in transmitted intensity $I_+^\text{v}(\xi) - I_-^\text{v}(\xi)$ were measured, for small rotations, the difference would be directly related to the sign and magnitude of the OR.

5.3.3 CP-ROA Invariants

The GU theory of ROA embraces all possible polarization experiments, scattering geometries, and degrees of resonance Raman intensity enhancement. ROA and Raman intensity are proportional to the square of a quantity involving a general tensor, $\tilde{a}_{\alpha\beta}$, and two complex polarization vectors, as expressed in Equation (5.34). For Raman scattering, only the square of the polarizability is needed, whereas ROA intensity arises from the products of the polarizability and one the ROA tensors. The ROA tensors are approximately three orders of magnitude smaller than the polarizability, and hence an ROA spectrum is approximately three orders of magnitude smaller than its parent Raman spectrum. The square of any two ROA tensors would be six orders of magnitude smaller than the parent Raman spectrum and is too small to measure. As noted above, the Greek subscripts of the tensors refer to the molecular axis system. However, for both Raman and ROA, linear combinations of products of tensors can be found that do not vary with the choice of the orientation of the molecular coordinate frame relative to the laboratory coordinate frame. Such combinations are called invariants. As shown in Chapter 2, all Raman intensities from samples of randomly oriented molecules can be expressed in terms of only three invariants, called the isotropic Raman invariant and the symmetric and anti-symmetric anisotropic Raman invariants given by:

$$\alpha^2 = \frac{1}{9}\left(\tilde{\alpha}_{\alpha\alpha}\right)^S\left(\tilde{\alpha}_{\beta\beta}\right)^{S*} \tag{5.50}$$

$$\beta_S(\tilde{\alpha})^2 = \frac{3}{2}\left(\tilde{\alpha}_{\alpha\beta}\right)^S\left(\tilde{\alpha}_{\alpha\beta}\right)^{S*} - \frac{1}{2}\left(\tilde{\alpha}_{\alpha\alpha}\right)^S\left(\tilde{\alpha}_{\beta\beta}\right)^{S*} \tag{5.51}$$

$$\beta_A(\tilde{\alpha})^2 = \frac{3}{2}\left(\tilde{\alpha}_{\alpha\beta}\right)^A\left(\tilde{\alpha}_{\alpha\beta}\right)^{A*} \tag{5.52}$$

where the symmetric and anti-symmetric forms of the tensors here and below are given by:

$$\left(T_{\alpha\beta}\right)^S = \frac{1}{2}\left[\left(T_{\alpha\beta}\right) + \left(T_{\beta\alpha}\right)\right] \tag{5.53}$$

$$\left(T_{\alpha\beta}\right)^A = \frac{1}{2}\left[\left(T_{\alpha\beta}\right) - \left(T_{\beta\alpha}\right)\right] \tag{5.54}$$

The three Raman invariants are pure real quantities as each term in Equations (5.50)–(5.52) is a complex tensor multiplied by its own complex conjugate.

For CP-ROA there are ten invariants, five associated with the Roman-font tensors, $\left(\alpha G, \beta_S(\tilde{G})^2, \beta_A(\tilde{G})^2, \beta_S(\tilde{A})^2 \text{ and } \beta_A(\tilde{A})^2\right)$ and five with the script-font tensors, $\left(\alpha\mathscr{G}, \beta_S(\tilde{\mathscr{G}})^2, \beta_A(\tilde{\mathscr{G}})^2, \beta_S(\tilde{\mathscr{A}})^2 \text{and } \beta_A(\tilde{\mathscr{A}})^2\right)$. The Roman ROA tensors are given by:

$$\alpha G = \frac{1}{9}\mathrm{Im}\left[\left(\tilde{\alpha}_{\alpha\alpha}\right)^S\left(\tilde{G}_{\beta\beta}\right)^{S*}\right] \tag{5.55}$$

$$\beta_S(\tilde{G})^2 = \frac{1}{2}\mathrm{Im}\left[3\left(\tilde{\alpha}_{\alpha\beta}\right)^S\left(\tilde{G}_{\alpha\beta}\right)^{S*} - \left(\tilde{\alpha}_{\alpha\alpha}\right)^S\left(\tilde{G}_{\beta\beta}\right)^{S*}\right] \tag{5.56}$$

$$\beta_A(\tilde{G})^2 = \frac{1}{2}\mathrm{Im}\left[3\left(\tilde{\alpha}_{\alpha\beta}\right)^A\left(\tilde{G}_{\alpha\beta}\right)^{A*}\right] \tag{5.57}$$

$$\beta_S(\tilde{A})^2 = \frac{1}{2}\omega_0\mathrm{Im}\left\{i\left(\tilde{\alpha}_{\alpha\beta}\right)^S\left[\varepsilon_{\alpha\gamma\delta}\left(\tilde{A}_{\gamma,\delta\beta}\right)\right]^{S*}\right\} \tag{5.58}$$

$$\beta_A(\tilde{A})^2 = \frac{1}{2}\omega_0 \mathrm{Im}\left\{ i(\tilde{\alpha}_{\alpha\beta})^A \left\{ \left[\varepsilon_{\alpha\gamma\delta}(\tilde{A}_{\gamma,\delta\beta})\right]^{A*} + \left[\varepsilon_{\alpha\beta\gamma}(\tilde{A}_{\delta,\gamma\delta})\right]^{A*} \right\} \right\} \tag{5.59}$$

The corresponding five script tensor invariants are the same in form as those for the Roman tensors, and ω_R replaces ω_0 in the expressions for the electric quadrupole optical activity invariants. For completeness the script invariants are:

$$\alpha\mathscr{G} = \frac{1}{9}\mathrm{Im}\left[(\tilde{\alpha}_{\alpha\alpha})^S(\tilde{\mathscr{G}}_{\beta\beta})^{S*}\right] \tag{5.60}$$

$$\beta_S(\tilde{\mathscr{G}})^2 = \frac{1}{2}\mathrm{Im}\left[3(\tilde{\alpha}_{\alpha\beta})^S(\tilde{\mathscr{G}}_{\alpha\beta})^{S*} - (\tilde{\alpha}_{\alpha\alpha})^S(\tilde{\mathscr{G}}_{\beta\beta})^{S*}\right] \tag{5.61}$$

$$\beta_A(\tilde{\mathscr{G}})^2 = \frac{1}{2}\mathrm{Im}\left[3(\tilde{\alpha}_{\alpha\beta})^A(\tilde{\mathscr{G}}_{\alpha\beta})^{A*}\right] \tag{5.62}$$

$$\beta_S(\tilde{\mathscr{A}})^2 = \frac{1}{2}\omega_R \mathrm{Im}\left\{ i(\tilde{\alpha}_{\alpha\beta})^S \left[\varepsilon_{\alpha\gamma\delta}(\tilde{\mathscr{A}}_{\gamma,\delta\beta})\right]^{S*} \right\} \tag{5.63}$$

$$\beta_A(\tilde{A})^2 = \frac{1}{2}\omega_R \mathrm{Im}\left\{ i(\tilde{\alpha}_{\alpha\beta})^A \left\{ \left[\varepsilon_{\alpha\gamma\delta}(\tilde{\mathscr{A}}_{\gamma,\delta\beta})\right]^{A*} + \left[\varepsilon_{\alpha\beta\gamma}(\tilde{\mathscr{A}}_{\delta,\gamma\delta})\right]^{A*} \right\} \right\} \tag{5.64}$$

All of the different CP-ROA experiments can be expressed in terms of these ten invariants. The ROA intensity for any experiment is expressed as a linear combination of some or all of the ten ROA invariants. Although sets of experiments can be devised to isolate all three ordinary Raman invariants, only six distinct combinations of ROA invariants have so far been isolated theoretically.

5.3.4 CP-ROA Observables and Invariant Combinations

In order to determine the ROA and Raman intensity expressions for any possible experiment, the following information is needed. The angle ξ of the scattered beam relative to the incident beam and the polarization states of either the incident or scattered beam if they are not modulated between RCP and LCP states are required. Any fixed polarization state must be either unpolarized or pure linearly polarized as any circular polarization content of the fixed polarization state results in unequal *Raman* intensities for the two CP modulated states, as described in Chapter 2, because the resulting reversal ratio or degree of circularity would overwhelm an attempted measurement of the ROA. If the fixed polarization is not unpolarized, it must be linearly polarized and the angle of that polarization state relative to the vertical direction must be specified. In Figure 5.2, the angle θ specifies the linear polarization state of the incident beam while the angle ϕ specifies the polarization angle of the scattered beam. Expressed as a function of these experimental angles, ICP- and SCP-ROA, and Raman intensities are given by:

$$\text{ICP:}\quad I_\phi^R(\xi) - I_\phi^L(\xi) = \frac{4K}{c}\left(I_1 + \mathscr{I}_2\cos\xi - I_3\sin^2\xi\,\sin^2\phi\right) \tag{5.65}$$

$$I_\phi^R(\xi) + I_\phi^L(\xi) = 2K\left(R_1 - R_3\sin^2\xi\,\sin^2\phi\right) \tag{5.66}$$

$$\text{SCP:}\quad I_R^\theta(\xi) - I_L^\theta(\xi) = \frac{4K}{c}\left(\mathscr{I}_1 + I_2\cos\xi - \mathscr{I}_3\sin^2\xi\,\sin^2\theta\right) \tag{5.67}$$

$$I_R^\theta(\xi) + I_L^\theta(\xi) = 2K\left(R_1 - R_3\sin^2\xi\,\sin^2\theta\right) \tag{5.68}$$

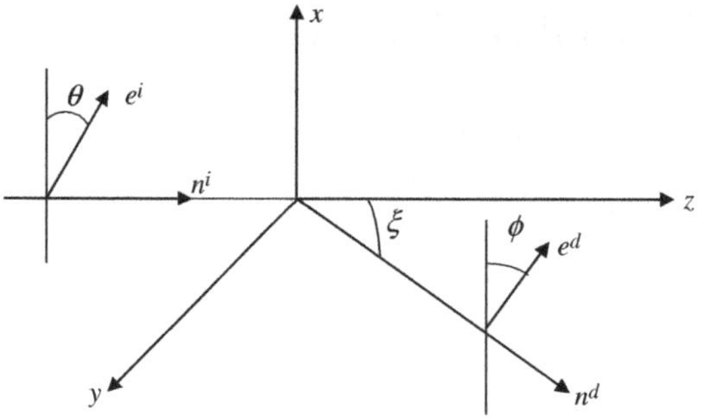

Figure 5.2 *Scattering diagram showing coordinate axes, incident and scattered propagation vectors, incident and scattered linear polarization vectors, the angle of scattering in the yz-plane from the forward direction, and the angles of any linear polarization states with respect to the x-direction*

To obtain the expressions of *unpolarized* ICP or SCP intensities, the sum of vertical (0°) and horizontal (90°) linear polarization states yields the desired result assuming that the analyzing polarizer is removed for ICP intensities and an unpolarized source of radiation is available that is converted by a polarizer for polarized SCP experiments. For DCP-ROA and Raman intensities, there is no dependence on the angle of any linear polarization states, as there are none. These intensities are given by:

$$\mathrm{DCP_I}: \quad I_R^R(\xi) - I_L^L(\xi) = \frac{4K}{c}\left[I_1 + \mathscr{I}_1 + (I_2 + \mathscr{I}_2)\cos\xi - \frac{1}{2}(I_3 + \mathscr{I}_3)\sin^2\xi\right] \tag{5.69}$$

$$I_R^R(\xi) + I_L^L(\xi) = 2K\left(R_1 + R_2\cos\xi - \frac{1}{2}R_3\sin^2\xi\right) \tag{5.70}$$

$$\mathrm{DCP_{II}}: \quad I_L^R(\xi) - I_R^L(\xi) = \frac{4K}{c}\left[I_1 - \mathscr{I}_1 - (I_2 - \mathscr{I}_2)\cos\xi - \frac{1}{2}(I_3 - \mathscr{I}_3)\sin^2\xi\right] \tag{5.71}$$

$$I_L^R(\xi) + I_R^L(\xi) = 2K\left(R_1 - R_2\cos\xi - \frac{1}{2}R_3\sin^2\xi\right) \tag{5.72}$$

The symbols I, $\mathscr{I}$ and R in these equations represent combinations of Roman and script ROA and Raman invariants, respectively. The combinations of Roman ROA invariants are:

$$I_1 = 45\alpha G + 7\beta_S(\tilde{G})^2 + 5\beta_A(\tilde{G})^2 + \beta_S(\tilde{A})^2 - \beta_A(\tilde{A})^2 \tag{5.73}$$

$$I_2 = 45\alpha G - 5\beta_S(\tilde{G})^2 + 5\beta_A(\tilde{G})^2 - 3\beta_S(\tilde{A})^2 - \beta_A(\tilde{A})^2 \tag{5.74}$$

$$I_3 = 45\alpha G + \beta_S(\tilde{G})^2 - 5\beta_A(\tilde{G})^2 + 3\beta_S(\tilde{A})^2 - 3\beta_A(\tilde{A})^2 \tag{5.75}$$

and the combinations of script ROA invariants are:

$$\mathcal{I}_1 = -45\alpha\mathcal{G} - 7\beta_S(\tilde{\mathcal{G}})^2 - 5\beta_A(\tilde{\mathcal{G}})^2 + \beta_S(\tilde{\mathcal{A}})^2 + \beta_A(\tilde{\mathcal{A}})^2 \tag{5.76}$$

$$\mathcal{I}_2 = -45\alpha\mathcal{G} + 5\beta_S(\tilde{\mathcal{G}})^2 - 5\beta_A(\tilde{\mathcal{G}})^2 - 3\beta_S(\tilde{\mathcal{A}})^2 + \beta_A(\tilde{\mathcal{A}})^2 \tag{5.77}$$

$$\mathcal{I}_3 = -45\alpha\mathcal{G} - \beta_S(\tilde{\mathcal{G}})^2 + 5\beta_A(\tilde{\mathcal{G}})^2 + 3\beta_S(\tilde{\mathcal{A}})^2 + 3\beta_A(\tilde{\mathcal{A}})^2 \tag{5.78}$$

Finally, the combinations of Raman invariants are:

$$R_1 = 45\alpha^2 + 7\beta_S(\tilde{\alpha})^2 + 5\beta_A(\tilde{\alpha})^2 \tag{5.79}$$

$$R_2 = 45\alpha^2 - 5\beta_S(\tilde{\alpha})^2 + 5\beta_A(\tilde{\alpha})^2 \tag{5.80}$$

$$R_3 = 45\alpha^2 + \beta_S(\tilde{\alpha})^2 - 5\beta_A(\tilde{\alpha})^2 \tag{5.81}$$

5.3.5 Backscattering CP-ROA Observables

From an experimental point of view, the most important ROA experiments are unpolarized backscattering ICP, SCP, and DCP_I. The expressions for ROA and Raman intensities for these experiments are:

$$\mathrm{ICP}_U(180^\circ):\quad I_U^R(180^\circ) - I_U^L(180^\circ) = \frac{8K}{c}\Big[45\alpha G + 7\beta_S(\tilde{G})^2 + 5\beta_A(\tilde{G})^2 + \beta_S(\tilde{A})^2 - \beta_A(\tilde{A})^2 + 45\alpha\mathcal{G} - 5\beta_S(\tilde{\mathcal{G}})^2 + 5\beta_A(\tilde{\mathcal{G}})^2 + 3\beta_S(\tilde{\mathcal{A}})^2 - \beta_A(\tilde{\mathcal{A}})^2\Big] \tag{5.82}$$

$$I_U^R(180^\circ) + I_U^L(180^\circ) = 4K\left[45\alpha^2 + 7\beta_S(\tilde{\alpha})^2 + 5\beta_A(\tilde{\alpha})^2\right] \tag{5.83}$$

$$\mathrm{SCP}_U(180^\circ):\quad I_L^U(180^\circ) - I_R^U(180^\circ) = \frac{8K}{c}\Big[-45\alpha G + 5\beta_S(\tilde{G})^2 - 5\beta_A(\tilde{G})^2 + 3\beta_S(\tilde{A})^2 + \beta_A(\tilde{A})^2 - 45\alpha\mathcal{G} - 7\beta_S(\tilde{\mathcal{G}})^2 - 5\beta_A(\tilde{\mathcal{G}})^2 + \beta_S(\tilde{\mathcal{A}})^2 + \beta_A(\tilde{\mathcal{A}})^2\Big] \tag{5.84}$$

$$I_R^U(180^\circ) + I_L^U(180^\circ) = 4K\left[45\alpha^2 + 7\beta_S(\tilde{\alpha})^2 + 5\beta_A(\tilde{\alpha})^2\right] \tag{5.85}$$

$$\mathrm{DCP_I}(180^\circ):\quad I_R^R(180^\circ) - I_L^L(180^\circ) = \frac{8K}{c}\left[6\beta_S(\tilde{G})^2 + 2\beta_S(\tilde{A})^2 - 6\beta_S(\tilde{\mathcal{G}})^2 + 2\beta_S(\tilde{\mathcal{A}})^2\right] \tag{5.86}$$

$$I_R^R(180^\circ) + I_L^L(180^\circ) = 4K\left[6\beta_S(\tilde{\alpha})^2\right] \tag{5.87}$$

Both ICP(180°) and SCP(180°) Raman and ROA experiments use all three Raman invariants and all ten CP-ROA invariants. Although ICP and SCP Raman intensities are identical, the corresponding ROA are different from one another. By contrast, additional circular polarization discrimination gives $\mathrm{DCP_I}(180°)$ Raman and ROA a remarkable level of simplicity, only four CP-ROA invariants and only one Raman invariant. Only the symmetric anisotropic invariants contribute to all $\mathrm{DCP_I}(180°)$ intensities. $\mathrm{DCP_{II}}(180°)$ intensities can be obtained by subtracting intensities for $\mathrm{DCP_I}(180°)$ from those for ICP_U, namely:

$$\mathrm{DCP_{II}}(180°):\quad I_L^R(180°) - I_R^L(180°) = \frac{8K}{c}\Big[45\alpha G + \beta_S(\tilde{G})^2 + 5\beta_A(\tilde{G})^2 - \beta_S(\tilde{A})^2 - \beta_A(\tilde{A})^2 + 45\alpha\mathscr{G} + \beta_S(\tilde{\mathscr{G}})^2 + 5\beta_A(\tilde{\mathscr{G}})^2 + \beta_S(\tilde{\mathscr{A}})^2 - \beta_A(\tilde{\mathscr{A}})^2\Big] \tag{5.88}$$

$$I_L^R(180°) + I_R^L(180°) = 4K\Big[45\alpha^2 + \beta_S(\tilde{\alpha})^2 + 5\beta_A(\tilde{\alpha})^2\Big] \tag{5.89}$$

$\mathrm{DCP_{II}}(180°)$ ROA intensities can also be obtained by subtracting $\mathrm{SCP}_U(180°)$ intensities from $\mathrm{DCP_I}(180°)$intensities. Similarly, $\mathrm{DCP_I}(180°)$ can be obtained from one-half the sum of $\mathrm{ICP}_U(180°)$ plus $\mathrm{SCP}_U(180°)$, while $\mathrm{DCP_{II}}(180°)$ is one-half the correspond difference $\mathrm{ICP}_U(180°)$ minus $\mathrm{SCP}_U(180°)$.

A list of common CP-ROA and Raman intensities is provided in Table 5.1, which gives the numerical factors for each of the possible ten CP-ROA invariants and three Raman invariants for each form of ROA and the major experimental geometries. The four examples of backscattering ROA given above are listed in this table along with many others that are not listed. The entire table can be generated from the four sets of CP-ROA and Raman intensities given in Section 5.3.4 as a function of scattering angle and linear polarization states. Also listed are the combinations of invariants from Equations (5.73)–(5.81) used for each invariant entry and also the ROA FFR invariants.

5.3.6 LP-ROA Invariants

Instead of measuring Raman intensity differences between orthogonal states of CP radiation, one can also measure intensity differences between orthogonal LP polarization states (Hecht and Nafie, 1990; Nafie and Che, 1994). In this case, the five Roman LP-ROA invariants are given by real parts, instead of the imaginary parts, of the tensor combinations in Equations (5.55)–(5.59), namely:

$$(\alpha G)' = \frac{1}{9}\mathrm{Re}\Big[(\tilde{\alpha}_{\alpha\alpha})^S(\tilde{G}_{\beta\beta})^{S*}\Big] \tag{5.90}$$

$$\beta_S'(\tilde{G})^2 = \frac{1}{2}\mathrm{Re}\Big[3(\tilde{\alpha}_{\alpha\beta})^S(\tilde{G}_{\alpha\beta})^{S*} - (\tilde{\alpha}_{\alpha\alpha})^S(\tilde{G}_{\beta\beta})^{S*}\Big] \tag{5.91}$$

$$\beta_A'(\tilde{G})^2 = \frac{1}{2}\mathrm{Re}\Big[3(\tilde{\alpha}_{\alpha\beta})^A(\tilde{G}_{\alpha\beta})^{A*}\Big] \tag{5.92}$$

$$\beta_S'(\tilde{A})^2 = \frac{1}{2}\omega_0\mathrm{Re}\Big\{i(\tilde{\alpha}_{\alpha\beta})^S\big[\varepsilon_{\alpha\gamma\delta}(\tilde{A}_{\gamma,\delta\beta})\big]^{S*}\Big\} \tag{5.93}$$

$$\beta_A'(\tilde{A})^2 = \frac{1}{2}\omega_0\mathrm{Re}\Big\{i(\tilde{\alpha}_{\alpha\beta})^A\Big\{\big[\varepsilon_{\alpha\gamma\delta}(\tilde{A}_{\gamma,\delta\beta})\big]^{A*} + \big[\varepsilon_{\alpha\beta\gamma}(\tilde{A}_{\delta,\gamma\delta})\big]^{A*}\Big\}\Big\} \tag{5.94}$$

Table 5.1 *Invariants for selected forms of Raman and ROA*

		Raman				ROA-GU											ROA-FFR		
		α^2 2K	$\beta_S(\alpha)^2$ 2K	$\beta_A(\alpha)^2$ 2K		αG 4K/c	$\beta_S(G)^2$ 4K/c	$\beta_A(G)^2$ 4K/c	$\beta_S(A)^2$ 4K/c	$\beta_A(A)^2$ 4K/c		αG 4K/c	$\beta_S(G)^2$ 4K/c	$\beta_A(G)^2$ 4K/c	$\beta_S(A)^2$ 4K/c	$\beta_A(A)^2$ 4K/c	αG 4K/c	$\beta(G')^2$ 4K/c	$\beta(A)^2$ 4K/c
ICP																			
0°-U	$2R_1$	90	14	10	$2I_1$	90	14	10	2	-2	$2\mathcal{I}_2$	-90	10	-10	-6	2	180	4	-4
90°-P	R_1	45	7	5	I_1	45	7	5	1	-1	NA	0	0	0	0	0	45	7	1
90°-DP	R_1-R_3	0	6	10	I_1-I_3	0	6	10	-2	2	NA	0	0	0	0	0	0	6	-2
90°-Magic	$R_1-R_3/3$	90/3	20/3	20/3	$I_1-I_3/3$	90/3	20/3	20/3	0	0	NA	0	0	0	0	0	90/3	20/3	0
180°-U	$2R_1$	90	14	10	$2I_1$	90	14	10	2	-2	$-2\mathcal{I}_2$	90	-10	10	6	-2	0	24	8
SCP																			
0°-U	$2R_1$	90	14	10	$2I_2$	90	-10	10	-6	-2	$2\mathcal{I}_1$	-90	-14	-10	2	2	180	4	-4
90°-P	R_1	45	7	5	NA	0	0	0	0	0	$\mathcal{I}_1$	-45	-7	-5	1	1	45	7	1
90°-DP	R_1-R_3	0	6	10	NA	0	0	0	0	0	$\mathcal{I}_1-\mathcal{I}_3$	0	-6	-10	-2	-2	0	6	-2
90°-Magic	$R_1-R_3/3$	90/3	20/3	20/3	NA	0	0	0	0	0	$\mathcal{I}_1-\mathcal{I}_3/3$	$-90/3$	$-20/3$	$-20/3$	0	0	90/3	20/3	0
180°-U	$2R_1$	90	14	10	$-2I_2$	-90	10	-10	6	2	$2\mathcal{I}_1$	-90	-14	-10	2	2	0	24	8
DCP$_I$																			
0°	R_1+R_2	90	2	10	I_1+I_2	90	2	10	-2	-2	$\mathcal{I}_1+\mathcal{I}_2$	-90	-2	-10	-2	2	180	4	-4
90°	$R_1-R_3/2$	45/2	13/2	15/2	$I_1-I_3/2$	45/2	13/2	15/2	$-1/2$	1/2	$\mathcal{I}_1-\mathcal{I}_3/2$	$-45/2$	$-13/2$	$-15/2$	$-1/2$	$-1/2$	90	13	-1
180°	R_1-R_2	0	12	0	I_1-I_2	0	12	0	4	0	$\mathcal{I}_1-\mathcal{I}_2$	0	-12	0	4	0	0	24	8
DCP$_{II}$																			
0°	R_1-R_2	0	12	0	I_1-I_2	0	12	0	4	0	$-\mathcal{I}_1+\mathcal{I}_2$	0	12	0	-4	0	0	0	0
90°	$R_1-R_3/2$	45/2	13/2	15/2	$I_1-I_3/2$	45/2	13/2	15/2	$-1/2$	1/2	$-\mathcal{I}_1+\mathcal{I}_3/2$	45/2	13/2	15/2	1/2	1/2	0	0	0
180°	R_1+R_2	90	2	10	I_1+I_2	90	2	10	-2	-2	$-\mathcal{I}_1-\mathcal{I}_2$	90	2	10	2	-2	0	0	0

where the LP-ROA invariants are distinguished from the CP-ROA invariants by a prime symbol. The five script LP-ROA invariants follow in the same way from Equations (5.90)–(5.94) with ω_R replacing ω_0 in the last two equations.

5.3.7 LP-ROA Observables and Invariant Combinations

The theory of LP-ROA is much simpler than that for CP-ROA. There is only one invariant combination for the Roman invariants and one for the script invariants. These are:

$$I_4 = 45(\alpha G)' + \beta_S'(\tilde{G})^2 - 5\beta_A'(\tilde{G})^2 - \beta_S'(\tilde{A})^2 + \beta_A'(\tilde{A})^2 \qquad (5.95)$$

$$\mathcal{I}_4 = 45(\alpha\mathcal{G})' + \beta_S'(\tilde{\mathcal{G}})^2 - 5\beta_A'(\tilde{\mathcal{G}})^2 + \beta_S'(\tilde{\mathcal{A}})^2 + \beta_A'(\tilde{\mathcal{A}})^2 \qquad (5.96)$$

From these two combinations, all forms of LP-ROA can be written as:

$$\text{ILP:} \quad I_\phi^+(\xi) - I_\phi^-(\xi) = \frac{4K}{c}\left\{I_4 - \mathcal{I}_4\cos\xi - \left[I_4\left(1+\cos^2\xi\right) - 2\mathcal{I}_4\cos\xi\right]\sin^2\phi\right\} \qquad (5.97)$$

$$I_\phi^+(\xi) + I_\phi^-(\xi) = 2K\left(R_1 - R_3\sin^2\xi\sin^2\phi\right) \qquad (5.98)$$

$$\text{SLP:} \quad I_+^\theta(\xi) - I_-^\theta(\xi) = \frac{4K}{c}\left\{\mathcal{I}_4 - I_4\cos\xi - \left[\mathcal{I}_4\left(1+\cos^2\xi\right) - 2I_4\cos\xi\right]\sin^2\theta\right\} \qquad (5.99)$$

$$I_+^\theta(\xi) + I_+^\theta(\xi) = 2K\left(R_1 - R_3\sin^2\xi\sin^2\theta\right) \qquad (5.100)$$

$$\text{DLP}_\text{I}: \quad I_+^+(\xi) - I_-^-(\xi) = \frac{2K}{c}[I_4 + \mathcal{I}_4]\sin^2\xi \qquad (5.101)$$

$$I_+^+(\xi) + I_-^-(\xi) = K\left[2R_1 + R_3\left(2\cos\xi - \sin^2\xi\right)\right] \qquad (5.102)$$

$$\text{DLP}_\text{II}: \quad I_-^+(\xi) - I_+^-(\xi) = \frac{2K}{c}[I_4 - \mathcal{I}_4]\sin^2\xi \qquad (5.103)$$

$$I_-^+(\xi) + I_+^-(\xi) = K\left[2R_1 - R_3\left(2\cos\xi + \sin^2\xi\right)\right] \qquad (5.104)$$

Using these equations, LP-ROA and Raman intensities for any desired experimental setup can be specified. ILP and SLP-ROA intensities differ by the exchange of Roman and script LP-ROA invariants, and DLP_I and DLP_II differ only in the relative sign of these two invariants. We note that ILP and SLP Raman intensities are the same as ICP and SCP Raman intensities and both forms of DLP-ROA have maximum values for right-angle scattering and vanish in back and forward scattering. LP-ROA is predicted to be zero in the FFR approximation and only observable sufficiently close to resonance that the damping term $i\Gamma_{ev}$ becomes important.

5.4 Vibronic Theories of ROA

The transition from the GU theory of ROA to the FFR theory passes through the near resonance (NR) theory and various related levels of approximations. All departures from the GU theory require a vibronic theory of ROA that starts with the adiabatic approximation to separate the molecular wavefunction into a product of electronic and vibrational parts. We first consider the GU theory of ROA expressed in terms of vibronic wavefunctions for a Raman Stokes transition between the zeroth and first vibrational levels, $g0$ and $g1$, of normal mode a of the ground electronic state.

5.4.1 General Unrestricted Vibronic ROA Theory

General vibronic expressions for the polarizability and optical activity tensors at the GU level of theory can be written by using the following notation for the states $|\tilde{n}\rangle$, $|\tilde{m}\rangle$, and $|\tilde{j}\rangle$ that appear in Equation (5.37) for the GU polarizability and Equations (5.38)–(5.41) for the optical tensors,

$$|\tilde{n}\rangle = |\tilde{\Psi}_n(\mathbf{r},\mathbf{R},\dot{\mathbf{R}})\rangle = \left|\tilde{\Psi}_{g0}^{CA}(\mathbf{r},\mathbf{R},\dot{\mathbf{R}})\right\rangle = \left|\tilde{\psi}_g^{CA}(\mathbf{r},\mathbf{R},\dot{\mathbf{R}})\phi_{g0}(\mathbf{R})\right\rangle = |\tilde{g}\rangle|\phi_{g0}\rangle \tag{5.105}$$

$$|\tilde{m}\rangle = |\tilde{\Psi}_m(\mathbf{r},\mathbf{R},\dot{\mathbf{R}})\rangle = \left|\tilde{\Psi}_{g1}^{CA}(\mathbf{r},\mathbf{R},\dot{\mathbf{R}})\right\rangle = \left|\tilde{\psi}_g^{CA}(\mathbf{r},\mathbf{R},\dot{\mathbf{R}})\phi_{g1}(\mathbf{R})\right\rangle = |\tilde{g}\rangle|\phi_{g1}\rangle \tag{5.106}$$

$$|\tilde{j}\rangle = |\tilde{\Psi}_j(\mathbf{r},\mathbf{R},\dot{\mathbf{R}})\rangle = \left|\tilde{\Psi}_{ev}^{CA}(\mathbf{r},\mathbf{R},\dot{\mathbf{R}})\right\rangle = \left|\tilde{\psi}_e^{CA}(\mathbf{r},\mathbf{R},\dot{\mathbf{R}})\phi_{ev}(\mathbf{R})\right\rangle = |\tilde{e}\rangle|\phi_{ev}\rangle \tag{5.107}$$

This set of notation takes into account the possibility of complex CA wavefunctions and nuclear dependence of these wavefunctions on nuclear positions and nuclear momenta. Using this notation in Equation (5.37) yields the polarizability tensor given by:

$$\left(\tilde{\alpha}_{\alpha\beta}\right)^a_{g1,g0} = \frac{1}{\hbar}\sum_{ev}\left[\frac{\left\langle\phi^a_{g1}\right|\langle\tilde{g}|\hat{\mu}_\alpha|\tilde{e}\rangle|\phi_{ev}\rangle\langle\phi_{ev}|\langle\tilde{e}|\hat{\mu}_\beta|\tilde{g}\rangle\left|\phi^a_{g0}\right\rangle}{\omega_{ev,g0}-\omega_0-i\Gamma_{ev}} + \frac{\left\langle\phi^a_{g1}\right|\langle\tilde{g}|\hat{\mu}_\beta|\tilde{e}\rangle|\phi_{ev}\rangle\langle\phi_{ev}|\langle\tilde{e}|\hat{\mu}_\alpha|\tilde{g}\rangle\left|\phi^a_{g0}\right\rangle}{\omega_{ev,g0}+\omega_R+i\Gamma_{ev}}\right] \tag{5.108}$$

Here we have again written the non-resonant frequency denominator as $\omega_{ev,g0}+\omega_R+i\Gamma_{ev}$ instead of its exact equivalent $\omega_{ev,g1}+\omega_0+i\Gamma_{ev}$ in Equation (5.37) to emphasize the different resonant roles of the frequencies of the incident and scattered radiation, ω_0 and ω_R, respectively, in subsequent approximations of this equation. Aside from invoking the CA approximation, this expression for the Raman polarizability tensor has the full generality and flexibility to describe any form of spontaneous Raman scattering at the dipole level of approximation. The corresponding expressions for the vibronic Raman optical activity tensors are:

$$\left(\tilde{G}_{\alpha\beta}\right)^a_{g1,g0} = \frac{1}{\hbar}\sum_{ev}\left[\frac{\left\langle\phi^a_{g1}\right|\langle\tilde{g}|\hat{\mu}_\alpha|\tilde{e}\rangle|\phi_{ev}\rangle\langle\phi_{ev}|\langle\tilde{e}|\hat{m}_\beta|\tilde{g}\rangle\left|\phi^a_{g0}\right\rangle}{\omega_{ev,g0}-\omega_0-i\Gamma_{ev}} + \frac{\left\langle\phi^a_{g1}\right|\langle\tilde{g}|\hat{m}_\beta|\tilde{e}\rangle|\phi_{ev}\rangle\langle\phi_{ev}|\langle\tilde{e}|\hat{\mu}_\alpha|\tilde{g}\rangle\left|\phi^a_{g0}\right\rangle}{\omega_{ev,g0}+\omega_R+i\Gamma_{ev}}\right] \tag{5.109}$$

$$\left(\tilde{\mathscr{G}}_{\alpha\beta}\right)^a_{g1,g0} = \frac{1}{\hbar}\sum_{ev}\left[\frac{\left\langle\phi^a_{g1}\right|\langle\tilde{g}|\hat{m}_\alpha|\tilde{e}\rangle|\phi_{ev}\rangle\langle\phi_{ev}|\langle\tilde{e}|\hat{\mu}_\beta|\tilde{g}\rangle\left|\phi^a_{g0}\right\rangle}{\omega_{ev,g0}-\omega_0-i\Gamma_{ev}} + \frac{\left\langle\phi^a_{g1}\right|\langle\tilde{g}|\hat{\mu}_\beta|\tilde{e}\rangle|\phi_{ev}\rangle\langle\phi_{ev}|\langle\tilde{e}|\hat{m}_\alpha|\tilde{g}\rangle\left|\phi^a_{g0}\right\rangle}{\omega_{ev,g0}+\omega_R+i\Gamma_{ev}}\right] \tag{5.110}$$

$$(\tilde{A}_{\alpha,\beta\gamma})^{a}_{g1,g0} = \frac{1}{\hbar}\sum_{ev}\left[\frac{\left\langle \phi^{a}_{g1}\middle|\langle\tilde{g}|\hat{\mu}_{\alpha}|\tilde{e}\rangle|\phi_{ev}\rangle\langle\phi_{ev}|\langle\tilde{e}|\hat{\Theta}_{\beta\gamma}|\tilde{g}\rangle\middle|\phi^{a}_{g0}\right\rangle}{\omega_{ev,g0}-\omega_{0}-i\Gamma_{ev}} + \frac{\left\langle \phi^{a}_{g1}\middle|\langle\tilde{g}|\hat{\Theta}_{\beta\gamma}|\tilde{e}\rangle|\phi_{ev}\rangle\langle\phi_{ev}|\langle\tilde{e}|\hat{\mu}_{\alpha}|\tilde{g}\rangle\middle|\phi^{a}_{g0}\right\rangle}{\omega_{ev,g0}+\omega_{R}+i\Gamma_{ev}}\right] \tag{5.111}$$

$$(\tilde{\mathscr{A}}_{\alpha,\beta\gamma})^{a}_{g1,g0} = \frac{1}{\hbar}\sum_{ev}\left[\frac{\left\langle \phi^{a}_{g1}\middle|\langle\tilde{g}|\hat{\Theta}_{\beta\gamma}|\tilde{e}\rangle|\phi_{ev}\rangle\langle\phi_{ev}|\langle\tilde{e}|\hat{\mu}_{\alpha}|\tilde{g}\rangle\middle|\phi^{a}_{g0}\right\rangle}{\omega_{ev,g0}-\omega_{0}-i\Gamma_{ev}} + \frac{\left\langle \phi^{a}_{g1}\middle|\langle\tilde{g}|\hat{\mu}_{\alpha}|\tilde{e}\rangle|\phi_{ev}\rangle\langle\phi_{ev}|\langle\tilde{e}|\hat{\Theta}_{\beta\gamma}|\tilde{g}\rangle\middle|\phi^{a}_{g0}\right\rangle}{\omega_{ev,g0}+\omega_{R}+i\Gamma_{ev}}\right] \tag{5.112}$$

5.4.2 Vibronic Levels of Approximation

There are several levels of approximation between the vibronic GU theory in Equations (5.108)–through (5.112) and the FFR theory introduced earlier in Equations (5.17)–(5.19) and (5.23). These levels involve the treatment or inclusion of the following: (i) vibronic detail in the frequency denominators, (ii) the summation over vibrational sublevels of the excited states, (iii) the imaginary vibronic bandwidth terms in the frequency denominators, (iv) the nuclear position dependence of the electronic wavefunction, and (v) the nuclear velocity dependence of the electronic wavefunction through the use of the CA approximation, first introduced in Chapter 2. These levels of approximation apply primarily to the case of non-resonance with any particular electronic state and culminate in the FFR theory. The case of strong resonance with a single electronic state (SES), with and without vibronic coupling to a nearby electronic state, introduced in Chapter 2, is obtained from the GU theory along different lines and will be treated separately below.

In this section on vibronic levels of approximation, the focus will be on approximations to the full GU vibronic theory. Specifically, the Raman polarizability tensor will be used to define the various levels of approximation between the GU theory in Equation (5.108) and corresponding polarizability tensor in the FFR theory in Equation (5.23). Along this path, a new intermediate level of vibronic theory, called the near-resonance (NR) theory, is introduced. Briefly, the NR theory retains the generality of the GU theory while possessing the simplicity of excited state detail of the FFR theory. The four ROA tensors, given at the GU level in Equations (5.109)–(5.112), can be written by direct analogy to the corresponding Raman polarizability tensor for each level of the theory.

5.4.3 Near Resonance Vibronic Raman Theory

Returning to Equation (5.108) we consider the situation where the incident laser radiation is close enough to the lowest electronic state of the molecule to distinguish the resonance responses of the molecule to the frequencies of the incident *versus* the scattered radiation. Thus the separate frequencies of the incident and scattered radiation, ω_0 and ω_R, present in Equation (5.108) are retained. We seek the simplest way to carry out a summation over the vibrational sublevels of the excited electronic states without losing a description of the difference in frequencies between the incident and scattered radiation. For each excited electronic state of the molecule there is a set of vibrational normal modes based on the potential energy surface of that excited state. These modes in general are linear combinations of the normal vibrational modes of the ground electronic state. Nevertheless, they bear some resemblance in terms of what nuclear motions are at higher frequencies versus lower frequencies. Each set of vibrational wavefunctions for the various electronic states of the molecules is a complete set of wavefunctions spanning the same Hilbert space of nuclear displacement coordinates. Without losing all the vibronic detail of the molecule, one could substitute each set of *excited* state vibrational wavefunctions with the set of *ground* state vibrational wavefunctions (Nafie, 2008). This leads to the following simplification of Equation (5.108),

$$(\tilde{\alpha}_{\alpha\beta})^{a}_{g1,g0} = \frac{1}{\hbar}\sum_{ev}\left[\frac{\left\langle \phi^{a}_{g1}\middle|\langle\tilde{g}|\hat{\mu}_{\alpha}|\tilde{e}\rangle\middle|\phi^{a}_{gv}\right\rangle\left\langle \phi^{a}_{gv}\middle|\langle\tilde{e}|\hat{\mu}_{\beta}|\tilde{g}\rangle\middle|\phi^{a}_{g0}\right\rangle}{\omega_{ev,g0}-\omega_{0}-i\Gamma_{ev}} + \frac{\left\langle \phi^{a}_{g1}\middle|\langle\tilde{g}|\hat{\mu}_{\beta}|\tilde{e}\rangle\middle|\phi^{a}_{gv}\right\rangle\left\langle \phi^{a}_{gv}\middle|\langle\tilde{e}|\hat{\mu}_{\alpha}|\tilde{g}\rangle\middle|\phi^{a}_{g0}\right\rangle}{\omega_{ev,g0}+\omega_{R}+i\Gamma_{ev}}\right] \tag{5.113}$$

We next seek expressions for the nuclear coordinate dependence of the electronic matrix elements in order to determine the allowed intermediate vibronic states. The ground-state CA electronic wavefunctions are written in normal coordinates, as given previously in Cartesian coordinates in Equation (2.86), by:

$$\begin{aligned}|\tilde{g}\rangle = \psi^{CA}_{g}(Q_a, P_a) &= \psi_{g,0} + \left(\frac{\partial\tilde{\psi}_g}{\partial Q_a}\right)_{0,0} Q_a + \left(\frac{\partial\tilde{\psi}_g}{\partial P_a}\right)_{0,0} P_a + \ldots \\ &= \psi_{g,0} + \sum_{s} C^{a}_{sg,0}\psi_{s,0}(Q_a + iP_a/\omega^{0}_{sg}) + \ldots\end{aligned} \tag{5.114}$$

$$\begin{aligned}\langle\tilde{g}| = \psi^{CA*}_{g}(Q_a, P_a) &= \psi_{g,0} + \left(\frac{\partial\tilde{\psi}^{*}_g}{\partial Q_a}\right)_{0,0} Q_a + \left(\frac{\partial\tilde{\psi}^{*}_g}{\partial P_a}\right)_{0,0} P_a + \ldots \\ &= \psi_{g,0} + \sum_{s} C^{a}_{sg,0}\psi_{s,0}(Q_a - iP_a/\omega^{0}_{sg}) + \ldots\end{aligned} \tag{5.115}$$

where the excited state vibronic coupling integral is given by:

$$C^{a}_{sg,0} = \left\langle \psi_{s,0}|(\partial\psi_g/\partial Q_a)_0\right\rangle \tag{5.116}$$

Analogous expressions can be written for $|\tilde{e}\rangle$ and $\langle\tilde{e}|$. Using these expressions, the first order nuclear coordinate dependence of the electronic matrix elements in Equation (5.113) can be written using notation established above and in Equation (2.107) as:

$$\begin{aligned}\left(\frac{\partial}{\partial Q_a}\langle\tilde{e}|\hat{\mu}_{\alpha}|\tilde{g}\rangle\right)_0 &= \left\langle \psi_{e,0}|\hat{\mu}_{\alpha}|\left(\frac{\partial\psi_g}{\partial Q_a}\right)_0\right\rangle + \left\langle\left(\frac{\partial\psi_e}{\partial Q_a}\right)_0|\hat{\mu}_{\alpha}|\psi_{g,0}\right\rangle \\ &= \langle\tilde{e}|\hat{\mu}_{\alpha}|\tilde{g}\rangle^{Q_a} = \sum_{s}\left[\langle e|\hat{\mu}_{\alpha}|s\rangle_0 C^{a}_{sg,0} + \langle s|\hat{\mu}_{\alpha}|g\rangle_0 C^{a}_{se,0}\right]\end{aligned} \tag{5.117}$$

$$\begin{aligned}\left(\frac{\partial}{\partial P_a}\langle\tilde{e}|\hat{\mu}_{\alpha}|\tilde{g}\rangle\right)_0 &= \left\langle \psi_{e,0}|\hat{\mu}_{\alpha}|\left(\frac{\partial\tilde{\psi}_g}{\partial P_a}\right)_0\right\rangle + \left\langle\left(\frac{\partial\tilde{\psi}_e}{\partial P_a}\right)_0|\hat{\mu}_{\alpha}|\psi_{g,0}\right\rangle \\ &= \langle\tilde{e}|\hat{\mu}_{\alpha}|\tilde{g}\rangle^{P_a} = \sum_{s} i\left[\langle e|\hat{\mu}_{\alpha}|s\rangle_0 C^{a}_{sg,0}/\omega^{0}_{sg} - \langle s|\hat{\mu}_{\alpha}|g\rangle_0 C^{a}_{se,0}/\omega^{0}_{se}\right]\end{aligned} \tag{5.118}$$

Substituting these expressions for the CA electronic wavefunctions into Equation (5.113) and separating out the vibrational integrals yields

$$
\begin{aligned}
(\tilde{\alpha}_{\alpha\beta})^a_{g1,g0} = \frac{1}{\hbar}\sum_{e,s}\Bigg[&\frac{\langle g|\hat{\mu}_\alpha|e\rangle_0\langle e|\hat{\mu}_\beta|s\rangle_0 C^a_{sg,0}}{\omega_{e1,g0}-\omega_0-i\Gamma_e} \\
&+\frac{\langle g|\hat{\mu}_\beta|e\rangle_0\langle e|\hat{\mu}_\alpha|s\rangle_0 C^a_{sg,0}}{\omega_{e1,g0}+\omega_R+i\Gamma_e}\Bigg]\left\langle\phi^a_{g1}\middle|\phi^a_{g1}\right\rangle\left\langle\phi^a_{g1}\middle|Q_a+\frac{iP_a}{\omega^0_{sg}}\middle|\phi^a_{g0}\right\rangle \\
&+\left[\frac{\langle g|\hat{\mu}_\alpha|e\rangle_0 C^a_{se,0}\langle s|\hat{\mu}_\beta|g\rangle_0}{\omega_{e1,g0}-\omega_0-i\Gamma_e}+\frac{\langle g|\hat{\mu}_\beta|e\rangle_0 C^a_{se,0}\langle s|\hat{\mu}_\alpha|g\rangle_0}{\omega_{e1,g0}+\omega_R+i\Gamma_e}\right]\left\langle\phi^a_{g1}\middle|\phi^a_{g1}\right\rangle\left\langle\phi^a_{g1}\middle|Q_a-\frac{iP_a}{\omega^0_{se}}\middle|\phi^a_{g0}\right\rangle \\
&+\left[\frac{\langle g|\hat{\mu}_\alpha|s\rangle_0 C^a_{se,0}\langle e|\hat{\mu}_\beta|g\rangle_0}{\omega_{e0,g0}-\omega_0-i\Gamma_e}+\frac{\langle g|\hat{\mu}_\beta|s\rangle_0 C^a_{se,0}\langle e|\hat{\mu}_\alpha|g\rangle_0}{\omega_{e0,g0}+\omega_R+i\Gamma_e}\right]\left\langle\phi^a_{g1}\middle|Q_a+\frac{iP_a}{\omega^0_{se}}\middle|\phi^a_{g0}\right\rangle\left\langle\phi^a_{g0}\middle|\phi^a_{g0}\right\rangle \\
&+\left[\frac{C^a_{sg,0}\langle s|\hat{\mu}_\alpha|e\rangle_0\langle e|\hat{\mu}_\beta|g\rangle_0}{\omega_{e0,g0}-\omega_0-i\Gamma_e}+\frac{C^a_{sg,0}\langle s|\hat{\mu}_\beta|e\rangle_0\langle e|\hat{\mu}_\alpha|g\rangle_0}{\omega_{e0,g0}+\omega_R+i\Gamma_e}\right]\left\langle\phi^a_{g1}\middle|Q_a-\frac{iP_a}{\omega^0_{sg}}\middle|\phi^a_{g0}\right\rangle\left\langle\phi^a_{g0}\middle|\phi^a_{g0}\right\rangle
\end{aligned}
\tag{5.119}
$$

Each of the four major groupings of terms represents the substitution of a particular electronic state in the resonance and non-resonance terms in Equation (5.108). The first two such groupings of terms represent substitutions into the first matrix elements from the right and is associated with a vibrational transition from the level 0 to 1 for that matrix element followed by no change in vibrational state, 1 to 1, for the second matrix element. The third and fourth groupings of terms in Equation (5.119) correspond to substitution into the second matrix elements in Equation (5.108) and correspond to no change from 0 to 0 in the first vibrational matrix element followed by a change from 0 to 1 in the second matrix element. The energy denominators involving the excited vibronic state $e1$ in the first two groupings of terms can be converted into the vibronic state $e0$ by using $\omega_{e1,g0}-\omega_0=\omega_{e0,g0}-\omega_R=\omega^0_{eg}-\omega_R$ and $\omega_{e1,g0}+\omega_R=\omega_{e0,g0}+\omega_0=\omega^0_{eg}+\omega_0$. Making these substitutions and setting the matrix elements $\langle\phi^a_{g1}|\phi^a_{g1}\rangle$ and $\langle\phi^a_{g0}|\phi^a_{g0}\rangle$ to unity for normalized vibrational wavefunctions yields

$$
\begin{aligned}
(\tilde{\alpha}_{\alpha\beta})^a_{g1,g0} = \frac{1}{\hbar}\sum_{e,s}\Bigg\{&\left[\frac{\langle g|\hat{\mu}_\alpha|e\rangle_0\langle e|\hat{\mu}_\beta|s\rangle_0 C^a_{sg,0}}{\omega^0_{eg}-\omega_R-i\Gamma_e}+\frac{\langle g|\hat{\mu}_\beta|e\rangle_0\langle e|\hat{\mu}_\alpha|s\rangle_0 C^a_{sg,0}}{\omega^0_{eg}+\omega_0+i\Gamma_e}\right]\left\langle\phi^a_{g1}\middle|Q_a+\frac{iP_a}{\omega^0_{sg}}\middle|\phi^a_{g0}\right\rangle \\
&+\left[\frac{\langle g|\hat{\mu}_\alpha|e\rangle_0 C^a_{se,0}\langle s|\hat{\mu}_\beta|g\rangle_0}{\omega^0_{eg}-\omega_R-i\Gamma_e}+\frac{\langle g|\hat{\mu}_\beta|e\rangle_0 C^a_{se,0}\langle s|\hat{\mu}_\alpha|g\rangle_0}{\omega^0_{eg}+\omega_0+i\Gamma_e}\right]\left\langle\phi^a_{g1}\middle|Q_a-\frac{iP_a}{\omega^0_{se}}\middle|\phi^a_{g0}\right\rangle \\
&+\left[\frac{\langle g|\hat{\mu}_\alpha|s\rangle_0 C^a_{se,0}\langle e|\hat{\mu}_\beta|g\rangle_0}{\omega^0_{eg}-\omega_0-i\Gamma_e}+\frac{\langle g|\hat{\mu}_\beta|s\rangle_0 C^a_{se,0}\langle e|\hat{\mu}_\alpha|g\rangle_0}{\omega^0_{eg}+\omega_R+i\Gamma_e}\right]\left\langle\phi^a_{g1}\middle|Q_a+\frac{iP_a}{\omega^0_{se}}\middle|\phi^a_{g0}\right\rangle \\
&+\left[\frac{C^a_{sg,0}\langle s|\hat{\mu}_\alpha|e\rangle_0\langle e|\hat{\mu}_\beta|g\rangle_0}{\omega^0_{eg}-\omega_0-i\Gamma_e}+\frac{C^a_{sg,0}\langle s|\hat{\mu}_\beta|e\rangle_0\langle e|\hat{\mu}_\alpha|g\rangle_0}{\omega^0_{eg}+\omega_R+i\Gamma_e}\right]\left\langle\phi^a_{g1}\middle|Q_a-\frac{iP_a}{\omega^0_{sg}}\middle|\phi^a_{g0}\right\rangle\Bigg\}
\end{aligned}
\tag{5.120}
$$

This equation can be further simplified by using a compressed notation for the electronic and vibrational matrix elements, the relationship between matrix elements of the normal mode momentum and position operators, namely, $(P_a)_{10}^a = i\omega_a(Q_a)_{10}^a$, and a simple algebraic relationship between incident and the Stokes scattered radiation frequencies in terms of a correction term as $\omega_R = \omega_0 - \omega_a = \omega_0(1 - \omega_a/\omega_0)$. This gives

$$
\begin{aligned}
(\tilde{\alpha}_{\alpha\beta})^a_{g1,g0} = \frac{1}{\hbar}\sum_{e,s}\Bigg\{ & \left[\frac{(\hat{\mu}_\alpha)^0_{ge}(\hat{\mu}_\beta)^0_{es}C^a_{sg,0}}{\omega^0_{eg} - \omega_0(1-\omega_a/\omega_0) - i\Gamma_e} + \frac{(\hat{\mu}_\beta)^0_{ge}(\hat{\mu}_\alpha)^0_{es}C^a_{sg,0}}{\omega^0_{eg} + \omega_0 + i\Gamma_e}\right](Q_a)_{10}\left(1 - \frac{\omega_a}{\omega^0_{sg}}\right) \\
& + \left[\frac{(\hat{\mu}_\alpha)^0_{ge}C^a_{se,0}(\hat{\mu}_\beta)^0_{sg}}{\omega^0_{eg} - \omega_0(1-\omega_a/\omega_0) - i\Gamma_e} + \frac{(\hat{\mu}_\beta)^0_{ge}C^a_{se,0}(\hat{\mu}_\alpha)^0_{sg}}{\omega^0_{eg} + \omega_0 + i\Gamma_e}\right](Q_a)_{10}\left(1 + \frac{\omega_a}{\omega^0_{se}}\right) \\
& + \left[\frac{(\hat{\mu}_\alpha)^0_{gs}C^a_{se,0}(\hat{\mu}_\beta)^0_{eg}}{\omega^0_{eg} - \omega_0 - i\Gamma_e} + \frac{(\hat{\mu}_\beta)^0_{gs}C^a_{se,0}(\hat{\mu}_\alpha)^0_{eg}}{\omega^0_{eg} + \omega_0(1-\omega_a/\omega_0) + i\Gamma_e}\right](Q_a)_{10}\left(1 - \frac{\omega_a}{\omega^0_{se}}\right) \\
& + \left[\frac{C^a_{sg,0}(\hat{\mu}_\alpha)^0_{se}(\hat{\mu}_\beta)^0_{eg}}{\omega^0_{eg} - \omega_0 - i\Gamma_e} + \frac{C^a_{sg,0}(\hat{\mu}_\beta)^0_{se}(\hat{\mu}_\alpha)^0_{eg}}{\omega^0_{eg} + \omega_0(1-\omega_a/\omega_0) + i\Gamma_e}\right](Q_a)_{10}\left(1 + \frac{\omega_a}{\omega^0_{sg}}\right)\Bigg\}
\end{aligned}
\tag{5.121}
$$

Equation (5.121) completes the development of the Raman polarizability tensor for the NR theory in its most general form (Nafie, 2008). In this form, the NR theory includes imaginary damping (linewidth or inverse lifetime) terms, $i\Gamma_e$, nuclear momentum correction terms associated with the CA electronic wavefunction, $1 \pm \omega_a/\omega^0_{s,g/e}$, and also the conversion factor between the frequencies of the incident and scattered radiation, $1 - \omega_a/\omega_0$, which is the hallmark of the NR theory. This form of the NR Raman polarizability tensor makes clear the nature of the various first-order corrections included in the full NR theory. In particular, both the NR approximation and the CA nuclear velocity correction terms involve ratios of the vibrational frequency to either the incident radiation frequency (NR) or an electronic transition frequency (CA). As will be shown, this form of the polarizability tensor facilitates the reduction from this expression to various levels of approximation of the NR theory and further to the FFR theory. The corresponding four ROA tensors given in vibronic form in Equation (5.109)–(5.112) can be easily written for Equation (5.121) by making the appropriate substitution of the electron operators.

5.4.4 Levels of the Near Resonance Raman Theory

Equation (5.121) may be reduced to simpler levels of approximation in several steps. The most obvious simplification involves removing the imaginary electronic damping terms, $i\Gamma_e$, as they merely provide breadth to the electronic resonances and are needed only when strong resonance with an individual excited electronic state occurs. Further, the damping term is needed only in the resonance term to avoid the occurrence of a zero resonance denominator. Its appearance in the non-resonance term, the linewidth term is largely superfluous. The damping terms have no effect on the number of Raman or ROA invariants and contain no frequency dependence that might modify the nature or the degree of resonance of either the incident or scattered radiation. They are needed only in the passage from near resonance to strong resonance involving one or a small number of excited electronic states. Therefore, in the NR theory where all electronic states are included in the summation over excited electronic states, the damping terms can easily be deleted.

There are eight terms in Equation (5.121) inside the square brackets. If we omit the damping terms and group pairs of terms with the same energy denominator as terms 3 and 1, 4 and 2, 7 and 5, and 8 and 6, first for the Q_a terms and then for the $\omega_a Q_a/\omega_{sg/se}$terms, we obtain:

$$\begin{aligned}
(\tilde{\alpha}_{\alpha\beta})^a_{g1,g0} = \frac{1}{\hbar}\sum_{e,s}\Bigg\{ & \Bigg[\frac{(\hat{\mu}_\alpha)^0_{ge}\left[(\hat{\mu}_\beta)^0_{sg}C^a_{se,0} + (\hat{\mu}_\beta)^0_{es}C^a_{sg,0}\right]}{\omega^0_{eg} - \omega_R} + \frac{(\hat{\mu}_\beta)^0_{ge}\left[(\hat{\mu}_\alpha)^0_{sg}C^a_{se,0} + (\hat{\mu}_\alpha)^0_{es}C^a_{sg,0}\right]}{\omega^0_{eg} + \omega_0} \\
& + \frac{\left[(\hat{\mu}_\alpha)^0_{se}C^a_{sg,0} + (\hat{\mu}_\alpha)^0_{gs}C^a_{se,0}\right](\hat{\mu}_\beta)^0_{eg}}{\omega^0_{eg} - \omega_0} + \frac{\left[(\hat{\mu}_\beta)^0_{se}C^a_{sg,0} + (\hat{\mu}_\beta)^0_{gs}C^a_{se,0}\right](\hat{\mu}_\alpha)^0_{eg}}{\omega^0_{eg} + \omega_R}\Bigg](Q_a)_{10} \\
& + \Bigg[\frac{(\hat{\mu}_\alpha)^0_{ge}\left[(\hat{\mu}_\beta)^0_{sg}C^a_{se,0}/\omega^0_{se} - (\hat{\mu}_\beta)^0_{es}C^a_{sg,0}/\omega^0_{sg}\right]}{\omega^0_{eg} - \omega_R} \\
& + \frac{(\hat{\mu}_\beta)^0_{ge}\left[(\hat{\mu}_\alpha)^0_{sg}C^a_{se,0}/\omega^0_{se} - (\hat{\mu}_\alpha)^0_{es}C^a_{sg,0}/\omega^0_{sg}\right]}{\omega^0_{eg} + \omega_0} \\
& + \frac{\left[(\hat{\mu}_\alpha)^0_{se}C^a_{sg,0}/\omega^0_{sg} - (\hat{\mu}_\alpha)^0_{gs}C^a_{se,0}/\omega^0_{se}\right](\hat{\mu}_\beta)^0_{eg}}{\omega^0_{eg} - \omega_0} \\
& + \frac{\left[(\hat{\mu}_\beta)^0_{se}C^a_{sg,0}/\omega^0_{sg} - (\hat{\mu}_\beta)^0_{gs}C^a_{se,0}/\omega^0_{se}\right](\hat{\mu}_\alpha)^0_{eg}}{\omega^0_{eg} + \omega_R}\Bigg]\omega_a(Q_a)_{10}\Bigg\}
\end{aligned} \tag{5.122}$$

where the Raman Stokes frequency ω_R is used explicitly rather than $\omega_0(1-\omega_a/\omega_0)$ for simplicity. Inspection of this equation reveals close similarity among the four resonance terms and the same similarity among the four non-resonance terms. In particular, if the vibrational perturbation occurs for the matrix element involving the dipole operator associated with the incident radiation $\hat{\mu}_\beta$-terms in square brackets, the photon frequency in the denominator is ω_R, whereas if the vibrational perturbation is with the matrix element of the scattered radiation $\hat{\mu}_\alpha$-terms in square brackets, then the photon frequency in the denominator is ω_0. If the terms in Equation (5.122) are rearranged such that terms with the *scattered* radiation frequency in the denominator appear first followed by the terms with the *incident* radiation frequency in the denominator for both the $(Q_a)_{10}$ and the $(P_a)_{10} = i\omega(Q_a)_{10}$ set of terms, one obtains

$$\begin{aligned}
(\tilde{\alpha}_{\alpha\beta})^a_{g1,g0} = \frac{1}{\hbar}\sum_{e}\Bigg\{ & \left[\frac{(\hat{\mu}_\alpha)^0_{ge}(\hat{\mu}_\beta)^{Q_a}_{eg}}{\omega^0_{eg} - \omega_R} + \frac{(\hat{\mu}_\beta)^{Q_a}_{ge}(\hat{\mu}_\alpha)^0_{eg}}{\omega^0_{eg} + \omega_R} + \frac{(\hat{\mu}_\alpha)^{Q_a}_{ge}(\hat{\mu}_\beta)^0_{eg}}{\omega^0_{eg} - \omega_0} + \frac{(\hat{\mu}_\beta)^0_{ge}(\hat{\mu}_\alpha)^{Q_a}_{eg}}{\omega^0_{eg} + \omega_0}\right](Q_a)_{10} \\
& + \left[\frac{(\hat{\mu}_\alpha)^0_{ge}(\hat{\mu}_\beta)^{P_a}_{eg}}{\omega^0_{eg} - \omega_R} + \frac{(\hat{\mu}_\beta)^{P_a}_{ge}(\hat{\mu}_\alpha)^0_{eg}}{\omega^0_{eg} + \omega_R} + \frac{(\hat{\mu}_\alpha)^{P_a}_{ge}(\hat{\mu}_\beta)^0_{eg}}{\omega^0_{eg} - \omega_0} + \frac{(\hat{\mu}_\beta)^0_{ge}(\hat{\mu}_\alpha)^{P_a}_{eg}}{\omega^0_{eg} + \omega_0}\right](P_a)_{10}\Bigg\}
\end{aligned} \tag{5.123}$$

where we have used the notation for the derivative of the electronic matrix element with respect to the normal coordinate of mode a from Equations (5.117) and (5.118) as:

$$(\hat{\mu}_\alpha)_{eg}^{Q_a} = \langle e|\hat{\mu}_\alpha|g\rangle^{Q_a} = \sum_s \left[(\hat{\mu}_\alpha)_{es}^0 C_{sg,0}^a + (\hat{\mu}_\alpha)_{sg}^0 C_{se,0}^a\right] = (\hat{\mu}_\alpha)_{ge}^{Q_a} \tag{5.124a}$$

$$(\hat{\mu}_\alpha)_{eg}^{P_a} = \langle e|\hat{\mu}_\alpha|g\rangle^{P_a} = i\sum_s \left[(\hat{\mu}_\alpha)_{es}^0 C_{sg,0}^a/\omega_{sg}^0 - (\hat{\mu}_\alpha)_{sg}^0 C_{se,0}^a/\omega_{se}^0\right] = -(\hat{\mu}_\alpha)_{ge}^{P_a} \tag{5.124b}$$

The key distinguishing feature of the NR approximation is the clear absence of symmetry of the Raman polarizability tensor with respect to interchange of Cartesian subscripts α and β. This symmetry is broken because of the appearance of ω_R instead of ω_0 in Equation (5.123) as would be the case in the corresponding FFR expression. Equation (5.123) can be reduced yet further by considering the following relationship for the Hermitian properties of the electronic matrix element which, for real wavefunctions and operators is:

$$(\hat{\mu}_\alpha)_{ge}^0(\hat{\mu}_\beta)_{eg}^{Q_a} = (\hat{\mu}_\alpha)_{eg}^{0*}(\hat{\mu}_\beta)_{ge}^{Q_a*} = (\hat{\mu}_\beta)_{ge}^{Q_a}(\hat{\mu}_\alpha)_{eg}^0 \tag{5.125}$$

A similar relationship holds for the Hermitian properties of the matrix elements involved with terms associated with the CA correction terms as this electronic contribution is pure imaginary

$$(\hat{\mu}_\alpha)_{ge}^0(\hat{\mu}_\beta)_{eg}^{P_a} = (\hat{\mu}_\alpha)_{eg}^{0*}(\hat{\mu}_\beta)_{ge}^{P_a*} = -(\hat{\mu}_\beta)_{ge}^{P_a}(\hat{\mu}_\alpha)_{eg}^0 \tag{5.126}$$

Using these relationships, Equation (5.123) can be written as:

$$\begin{aligned}(\tilde{\alpha}_{\alpha\beta})_{g1,g0}^a = \frac{2}{\hbar}\sum_e \Bigg\{ & \left[\frac{\omega_{eg}^0 \,\mathrm{Re}\left[(\hat{\mu}_\alpha)_{ge}^0(\hat{\mu}_\beta)_{eg}^{Q_a}\right]}{(\omega_{eg}^0)^2-\omega_R^2} + \frac{\omega_{eg}^0 \,\mathrm{Re}\left[(\hat{\mu}_\alpha)_{ge}^{Q_a}(\hat{\mu}_\beta)_{eg}^0\right]}{(\omega_{eg}^0)^2-\omega_0^2}\right](Q_a)_{10} \\ & + \left[\frac{\omega_R \,\mathrm{Im}\left[(\hat{\mu}_\alpha)_{ge}^0(\hat{\mu}_\beta)_{eg}^{P_a}\right]}{(\omega_{eg}^0)^2-\omega_R^2} + \frac{\omega_0 \,\mathrm{Im}\left[(\hat{\mu}_\alpha)_{ge}^{P_a}(\hat{\mu}_\beta)_{eg}^0\right]}{(\omega_{eg}^0)^2-\omega_0^2}\right]\omega_a(Q_a)_{10}\Bigg\}\end{aligned} \tag{5.127}$$

From this equation, we can reach two lower levels of approximation by either (i) relaxing the NR level part the theory to the FFR approximation while keeping the CA correction terms or (ii) keeping the NR approximation and eliminating the CA correction terms. Of these two, it is most logical to eliminate the CA correction terms and retain the NR level of approximation as the CA approximation corrections are in general smaller than those associated with the NR approximation. In particular, the NR correction term, ω_a/ω_0, is generally larger than the CA correction term $\omega_a/\omega_{s,g/e}^0$ unless there is an excited state e with a lower transition frequency than the photon frequency, in which case a strong resonance theory may be more appropriate, or if a pair of excited states e and s are closer in frequency to each other than the energy of the photon frequency.

Nevertheless, for completeness we consider both limits, first by elimination of the NR terms only and then by elimination of the CA terms only. For the first case, we reduce the polarizability expression in Equation (5.127) to the FFR approximation but keep the CA correction terms. Eliminating the NR correction, the CA-FFR expression for the polarizability is:

$$\left(\tilde{\alpha}_{\alpha\beta}\right)^{a}_{g1,g0} = \frac{2}{\hbar}\sum_{e}\left\{\left[\frac{\omega^{0}_{eg}\,\mathrm{Re}\left[(\hat{\mu}_{\alpha})^{0}_{ge}(\hat{\mu}_{\beta})^{Q_a}_{eg} + (\hat{\mu}_{\alpha})^{Q_a}_{ge}(\hat{\mu}_{\beta})^{0}_{eg}\right]}{\left(\omega^{0}_{eg}\right)^{2} - \omega_0^{2}}\right](Q_a)_{10}\right.$$
$$\left. + \left[\frac{\omega_0\,\mathrm{Im}\left[(\hat{\mu}_{\alpha})^{0}_{ge}(\hat{\mu}_{\beta})^{P_a}_{eg} + (\hat{\mu}_{\alpha})^{P_a}_{ge}(\hat{\mu}_{\beta})^{0}_{eg}\right]}{\left(\omega^{0}_{eg}\right)^{2} - \omega_0^{2}}\right]\omega_a(Q_a)_{10}\right\} \tag{5.128}$$

If the nuclear momentum CA correction terms, proportional to $\omega_a(Q_a)_{10}$, are dropped from this equation, it reduces to the expression for the polarizability in the FFR approximation in Equation (5.23). From the definitions of real and imaginary parts and the relationships in Equation (5.125) and (5.126) one can see that the usual BO term (FFR theory) is symmetric upon interchange of subscripts α and β, whereas the CA correction term is correspondingly anti-symmetric. As a result, the number of Raman invariants remains at three, as in the GU theory. However, even though there are anti-symmetric components in the ROA tensors at this level of theory, without discrimination between the frequencies of the incident and scattered radiation, there is no discrimination between the Roman and script ROA tensors. As a result, there is a reduction in the number of ROA invariants from the ten present in the GU theory to five in the CA-FFR theory. The details are available elsewhere (Nafie, 2008) and will not be presented here because it is unlikely that one would want to use the CA-FFR while ignoring generally more important terms provided by the NR level of theory.

The simplest level of the NR theory is obtained from Equation (5.127) by elimination of the CA correction terms while keeping the NR distinction between ω_R and ω_0. This gives

$$\left(\alpha_{\alpha\beta}\right)^{a}_{g1,g0} = \frac{2}{\hbar}\sum_{e}\omega^{0}_{eg}\,\mathrm{Re}\left[\frac{\left[(\hat{\mu}_{\alpha})^{0}_{ge}(\hat{\mu}_{\beta})^{Q_a}_{eg}\right]}{\left(\omega^{0}_{eg}\right)^{2} - \omega_R^{2}} + \frac{\left[(\hat{\mu}_{\alpha})^{Q_a}_{ge}(\hat{\mu}_{\beta})^{0}_{eg}\right]}{\left(\omega^{0}_{eg}\right)^{2} - \omega_0^{2}}\right](Q_a)_{10} \tag{5.129}$$

This expression is very nearly the FFR approximation. The only difference is that the photon frequency is that of the incident radiation ω_0 when the nuclear coordinate derivative is associated with the matrix element of scattered radiation, $(\hat{\mu}_{\alpha})^{Q_a}_{ge}$, and the scattered radiation ω_R when the nuclear coordinate derivative is associated with the matrix element of incident radiation, $(\hat{\mu}_{\beta})^{Q_a}_{ge}$. Eliminating this distinction by substituting ω_0 for ω_R in this equation again yields the FFR theory for the Raman polarizability tensor given in Equation (5.23).

Several important points are worth noting. Even though the NR polarizability is pure real, it is not symmetric upon interchange of subscripts α and β. As a result, there is a non-zero anti-symmetric Raman invariant and hence the number of Raman invariants remains at three, just as in the GU theory. Another point is that the NR theory converges to the FFR theory as the vibrational frequency approaches zero. Hence the NR theory should be more noticeable and more important for higher frequency vibrations than for lower frequency vibrations. Incidentally, this is also true for the CA terms and hence even Equation (5.127) reduces to the FFR theory for low frequency vibrational modes. A third point, as shown below, all ten ROA invariants present in the GU theory are retained in the NR theory. Hence the NR theory describes a difference between ICP- and SCP-ROA intensities, non-zero $\mathrm{DCP_{II}}$-ROA intensities, and also other ROA intensity differences that are lost in the FFR approximation. Finally, we note that the NR Raman and ROA intensities can be calculated from the average of the two FFR type of calculations, one at the scattered radiation frequency for the first term in Equation (5.129) and one at the incident radiation frequency for the second term in this equation.

5.4.5 Near Resonance Theory of ROA

Starting from the Raman polarizability in the NR approximation in Equation (5.129), the NR theory of the four ROA tensors can be written as:

$$\left(G'_{\alpha\beta}\right)^{a}_{g1,g0} = -\frac{2}{\hbar}\sum_{e}\mathrm{Im}\left[\frac{\omega_R\left[(\hat{\mu}_\alpha)^0_{ge}(\hat{m}_\beta)^{Q_a}_{eg}\right]}{(\omega^0_{eg})^2-\omega_R^2}+\frac{\omega_0\left[(\hat{\mu}_\alpha)^{Q_a}_{ge}(\hat{m}_\beta)^0_{eg}\right]}{(\omega^0_{eg})^2-\omega_0^2}\right](Q_a)_{10} \tag{5.130}$$

$$\left(\mathscr{G}'_{\alpha\beta}\right)^{a}_{g1,g0} = -\frac{2}{\hbar}\sum_{e}\mathrm{Im}\left[\frac{\omega_R\left[(\hat{m}_\alpha)^0_{ge}(\hat{\mu}_\beta)^{Q_a}_{eg}\right]}{(\omega^0_{eg})^2-\omega_R^2}+\frac{\omega_0\left[(\hat{m}_\alpha)^{Q_a}_{ge}(\hat{\mu}_\beta)^0_{eg}\right]}{(\omega^0_{eg})^2-\omega_0^2}\right](Q_a)_{10} \tag{5.131}$$

$$\left(A_{\alpha,\beta\gamma}\right)^{a}_{g1,g0} = \frac{2}{\hbar}\sum_{e}\omega^0_{eg}\mathrm{Re}\left[\frac{\left[(\hat{\mu}_\alpha)^0_{ge}(\hat{\Theta}_{\beta\gamma})^{Q_a}_{eg}\right]}{(\omega^0_{eg})^2-\omega_R^2}+\frac{\left[(\hat{\mu}_\alpha)^{Q_a}_{ge}(\hat{\Theta}_{\beta\gamma})^0_{eg}\right]}{(\omega^0_{eg})^2-\omega_0^2}\right](Q_a)_{10} \tag{5.132}$$

$$\left(\mathscr{A}_{\alpha,\beta\gamma}\right)^{a}_{g1,g0} = \frac{2}{\hbar}\sum_{e}\omega^0_{eg}\mathrm{Re}\left[\frac{\left[(\hat{\Theta}_{\beta\gamma})^0_{ge}(\hat{\mu}_\alpha)^{Q_a}_{eg}\right]}{(\omega^0_{eg})^2-\omega_R^2}+\frac{\left[(\hat{\Theta}_{\beta\gamma})^{Q_a}_{ge}(\hat{\mu}_\alpha)^0_{eg}\right]}{(\omega^0_{eg})^2-\omega_0^2}\right](Q_a)_{10} \tag{5.133}$$

where the unprimed and primed tensors, without the complex tilda over-marks, are defined in terms of the general complex tensor by:

$$\tilde{T} = T - iT' \tag{5.134}$$

Each of these four ROA tensors is distinct and there are no relationships of equivalence between the Roman and script font tensors for the same transition. As a result, the GU level of tensor diversity is maintained in the NR theory. In particular, all ten CP-ROA invariants are distinct and non-zero. Therefore, the entire theoretical formalism of the GU theory of CP-ROA applies to the NR theory of CP-ROA. As this is true of the simplest level of NR theory, it also applies equally well when CA correction terms are added and also the imaginary damping terms.

5.4.6 Reduction of the Near Resonance Theory to the Far-From Resonance Theory of ROA

It is instructive to follow the reduction of the simple NR theory of ROA to the FFR theory presented in Section 5.2. If ω_R is set equal to ω_0 for the Raman polarizability in Equation (5.129), the two terms present there transform into one another using $(\hat{\mu}_\alpha)^0_{eg} = (\hat{\mu}_\alpha)^0_{ge}$ when the subscripts α and β are interchanged and hence

$$\left(\alpha_{\alpha\beta}\right)^{a}_{g1,g0} = \left(\alpha_{\beta\alpha}\right)^{a}_{g1,g0} \tag{5.135}$$

From this equation, it follows from the definition of the anti-symmetric part of a tensor in Equation (5.54) that

$$\left[\left(\alpha_{\alpha\beta}\right)^{a}_{g1,g0}\right]^{A} = \frac{1}{2}\left[\left(\alpha_{\alpha\beta}\right)^{a}_{g1,g0} - \left(\alpha_{\beta\alpha}\right)^{a}_{g1,g0}\right] = 0 \tag{5.136}$$

As a consequence, the anti-symmetric Raman invariant in Equation (5.52) and the four anti-symmetric ROA invariants in Equations (5.57), (5.59), (5.62) and (5.64) all vanish as each one contains the anti-symmetric polarizability as a multiplicative factor. In particular,

$$\beta_A(\tilde{\alpha})^2, \beta_A(\tilde{G})^2, \beta_A(\tilde{A})^2, \beta_A(\tilde{\mathscr{G}})^2, \beta_A(\tilde{\mathscr{A}})^2 = 0 \tag{5.137}$$

The symmetric combinations of the ROA tensors in the FFR approximation can be deduced from the simple NR-ROA tensors in Equations (5.130)–(5.133). After setting ω_R to ω_0, the following equations can be written as:

$$\begin{aligned}\left[(G'_{\alpha\beta})^a_{g1,g0}\right]^S &= \frac{1}{2}\left[(G'_{\alpha\beta})^a_{g1,g0} + (G'_{\beta\alpha})^a_{g1,g0}\right] \\ &= -\frac{1}{2}\left[(\mathscr{G}'_{\beta\alpha})^a_{g1,g0} + (\mathscr{G}'_{\alpha\beta})^a_{g1,g0}\right] = -\left[(\mathscr{G}'_{\alpha\beta})^a_{g1,g0}\right]^S\end{aligned} \tag{5.138}$$

$$\begin{aligned}\left[\varepsilon_{\alpha\gamma\delta}(A_{\gamma,\delta\beta})^a_{g1,g0}\right]^S &= \frac{1}{2}\left[\varepsilon_{\alpha\gamma\delta}(A_{\gamma,\delta\beta})^a_{g1,g0} + \varepsilon_{\beta\gamma\delta}(A_{\gamma,\delta\alpha})^a_{g1,g0}\right] \\ &= \frac{1}{2}\left[\varepsilon_{\alpha\gamma\delta}(\mathscr{A}_{\gamma,\delta\beta})^a_{g1,g0} + \varepsilon_{\beta\gamma\delta}(\mathscr{A}_{\gamma,\delta\alpha})^a_{g1,g0}\right] = \left[\varepsilon_{\alpha\gamma\delta}(\mathscr{A}_{\gamma,\delta\beta})^a_{g1,g0}\right]^S\end{aligned} \tag{5.139}$$

Here we have used relationships such as $(\hat{m}_\beta)^0_{eg} = -(\hat{m}_\beta)^0_{ge}$ in Equation (5.138) and $(\hat{\Theta}_{\delta\beta})^0_{eg} = (\hat{\Theta}_{\delta\beta})^0_{ge}$ in Equation (5.139) to complete the equality between the symmetric combinations of Roman and script font tensors. These equations then lead to the following reduction in the number of symmetric ROA invariants from six in the GU theory to three in the FFR limit,

$$[\alpha G]^a_{g1,g0} = -[\alpha\mathscr{G}]^a_{g1,g0} = [\alpha G']^a_{g1,g0} \tag{5.140}$$

$$\left[\beta_S(\tilde{G})^2\right]^a_{g1,g0} = -\left[\beta_S(\tilde{\mathscr{G}})^2\right]^a_{g1,g0} = \left[\beta(G')^2\right]^a_{g1,g0} \tag{5.141}$$

$$\left[\beta_S(\tilde{A})^2\right]^a_{g1,g0} = \left[\beta_S(\tilde{\mathscr{A}})^2\right]^a_{g1,g0} = \left[\beta(A)^2\right]^a_{g1,g0} \tag{5.142}$$

The distinction between Roman and script invariants is no longer required, and as only symmetric invariants are involved, the superscript S can be dropped. Accordingly, the symmetric anisotropic invariant of the Raman tensor is simplified to:

$$\left[\beta_S(\alpha)^2\right]^a_{g1,g0} = \left[\beta(\alpha)^2\right]^a_{g1,g0} \tag{5.143}$$

The equations for the two Raman invariants and three ROA invariants in the FFR limit are:

$$[\alpha^2]^a_{g1,g0} = \frac{1}{9}(\alpha_{\alpha\alpha})^a_{g1,g0}(\alpha_{\beta\beta})^a_{g1,g0} \tag{5.144}$$

$$\left[\beta(\alpha)^2\right]^a_{g1,g0} = \frac{1}{2}\left[3\left(\alpha_{\alpha\beta}\right)^a_{g1,g0}\left(\alpha_{\alpha\beta}\right)^a_{g1,g0} - \left(\alpha_{\alpha\alpha}\right)^a_{g1,g0}\left(\alpha_{\beta\beta}\right)^a_{g1,g0}\right] \tag{5.145}$$

$$[\alpha G']^a_{g1,g0} = \frac{1}{9}\left(\alpha_{\alpha\alpha}\right)^a_{g1,g0}\left(G'_{\beta\beta}\right)^a_{g1,g0} \tag{5.146}$$

$$\left[\beta(G')^2\right]^a_{g1,g0} = \frac{1}{2}\left[3\left(\alpha_{\alpha\beta}\right)^a_{g1,g0}\left(G'_{\alpha\beta}\right)^a_{g1,g0} - \left(\alpha_{\alpha\alpha}\right)^a_{g1,g0}\left(G'_{\beta\beta}\right)^a_{g1,g0}\right] \tag{5.147}$$

$$\left[\beta(A)^2\right]^a_{g1,g0} = \frac{1}{2}\omega_0\left(\alpha_{\alpha\beta}\right)^a_{g1,g0}\varepsilon_{\alpha\gamma\delta}\left(A_{\gamma,\delta\beta}\right)^a_{g1,g0} \tag{5.148}$$

Finally, the Raman and ROA tensors in the FFR, first given in simplified form in Equations (5.42)–(5.44), are written here using the notation developed for the NR theory in Equations (5.129)–(5.133) as:

$$\left(\alpha_{\alpha\beta}\right)^a_{g1,g0} = \frac{2}{\hbar}\sum_e \frac{\omega^0_{eg}}{\left(\omega^0_{eg}\right)^2 - \omega_0^2}\mathrm{Re}\left[\left(\hat{\mu}_\alpha\right)_{ge}\left(\hat{\mu}_\beta\right)_{eg}\right]^{Q_a}_0 (Q_a)_{10} \tag{5.149}$$

$$\left(G'_{\alpha\beta}\right)^a_{g1,g0} = -\frac{2}{\hbar}\sum_e \frac{\omega_0}{\left(\omega^0_{eg}\right)^2 - \omega_0^2}\mathrm{Im}\left[\left(\hat{\mu}_\alpha\right)_{ge}\left(\hat{m}_\beta\right)_{eg}\right]^{Q_a}_0 (Q_a)_{10} \tag{5.150}$$

$$\left(A_{\alpha,\beta\gamma}\right)^a_{g1,g0} = \frac{2}{\hbar}\sum_e \frac{\omega^0_{eg}}{\left(\omega^0_{eg}\right)^2 - \omega_0^2}\mathrm{Re}\left[\left(\hat{\mu}_\alpha\right)_{ge}\left(\hat{\Theta}_{\beta\gamma}\right)_{eg}\right]^{Q_a}_0 (Q_a)_{10} \tag{5.151}$$

5.5 Resonance ROA Theory

The theory of resonance ROA has been developed previously only for the case of resonance with a single excited electronic state (SES). This level of theory corresponds to the so-called Albrecht A-term resonance Raman scattering (Champion and Albrecht, 1982). The predictions of this theory have been confirmed both experimentally and more recently computationally. The SES theory of ROA predicts an ROA spectrum of a single sign, positive or negative, with the same relative intensities for all the bands as the parent resonance Raman spectrum. The SES-ROA intensity depends completely on the circular dichroism of the single excited electronic state. In particular, the ratio of the electronic CD to electronic absorbance of the parent resonant electronic state is the same, but opposite in sign, as the ratio of the SES-ROA to SES-Raman intensities for every point throughout the entire spectrum. At this primitive level, ROA in its simplest manifestation is exposed as nothing more than a reflection of the intensity and sign of the CD of the resonant excited electronic state. In this section, the details of the SES-ROA theory will be presented as well as the extension of this theory to two participating excited electronic states, either in the so-called Albrect B-term scattering where a second excited state participates by vibronic coupling or when two electronic states are simultaneously in resonance but are not coupled via vibronic coupling.

5.5.1 Strong Resonance in the Single Electronic State (SES) Limit

As first presented in Chapter 2 starting with Equation (2.151), the strong resonance limit of the Raman polarizability involves only the contribution of a single electronic state, e. For a vibrational transition from $g0$ to $g1$ in the ground electronic state, the Raman polarizability in the strong resonance limit is given by:

$$\left(\tilde{\alpha}_{\alpha\beta}\right)^a_{g1,g0} = \frac{1}{\hbar}\sum_v \frac{\langle\phi^a_{g1}|\langle g|\hat{\mu}_\alpha|e\rangle|\phi_{ev}\rangle\langle\phi_{ev}|\langle e|\hat{\mu}_\beta|g\rangle|\phi^a_{g0}\rangle}{\omega_{ev,g0} - \omega_0 - i\Gamma_{ev}} \qquad (5.152)$$

Here the summation over all excited electronic states is reduced to a single resonant state for which the frequency denominator $\omega_{ev,g0} - \omega_0$ is very small relative to that for all other electronic states. In this limit, the non-resonant term for the resonant state e makes an even smaller contribution than most of the resonant terms, already ignored, of electronic states not in strong resonance, and hence it can easily be dropped. The Raman tensor is still complex and the imaginary vibronic linewidth term Γ_{ev} is critical to the validity of the theory. As shown in Equations (2.152)–(2.155), the resonant Raman polarizability tensor in Equation (5.152) can further be expressed in terms of A-term, B-term, and C-term contributions, where the latter two involve vibronic coupling between the resonant state e and other nearby electronic states in the molecule, and hence are multiple-state resonant expressions. The A-term involves only a single electronic state and hence gives rise to the SES vibronic theory of ROA as described in further detail below. The vibronic theory of Raman and ROA can also be extended to B-term and C-term contributions. Of these, the B-term terms are the most likely to be significant except for molecules with low-lying electronic states where C-term contributions may be important. The extension of the SES theory to B-terms, while still in resonance with a single electronic state, involves contributions from one or more additional excited electronic states, and this extension will be described after the SES theory of ROA is presented.

If the electric-dipole transition moment of the single resonant electronic state is taken to lie in the z-direction, there is only one non-zero element in the Raman polarizability tensor, namely,

$$\left(\tilde{\alpha}_{zz}\right)^a_{g1,g0} = \frac{1}{\hbar}\sum_v \frac{\left\langle\phi^a_{g1}\middle|\langle g|\hat{\mu}_z|e\rangle|\phi_{ev}\rangle\langle\phi_{ev}|\langle e|\hat{\mu}_z|g\rangle\middle|\phi^a_{g0}\right\rangle}{\omega_{ev,g0} - \omega_0 - i\Gamma_{ev}} \qquad (5.153)$$

Using this single zz-tensor element, the Raman invariants in Equations (5.50)–(5.52) can be evaluated to give:

$$\left[\alpha^2\right]^a_{g1,g0} = \frac{1}{9}\left|\left(\tilde{\alpha}_{zz}\right)^a_{g1,g0}\right|^2 \qquad (5.154)$$

$$\left[\beta_S(\tilde{\alpha})^2\right]^a_{g1,g0} = \left|\left(\tilde{\alpha}_{zz}\right)^a_{g1,g0}\right|^2 \qquad (5.155)$$

$$\left[\beta_A(\tilde{\alpha})^2\right]^a_{g1,g0} = 0 \qquad (5.156)$$

As a result, the three Raman invariants of the GU theory reduce effectively to only one Raman invariant in the SES theory. Similarly, with only a single z-polarized excited electronic state and only one non-zero Raman polarizability element, $\tilde{\alpha}_{zz}$, only the zz-component of magnetic-dipole ROA tensor can enter any of the ROA invariant expressions given in Equations (5.55) to (5.59) and Equations (5.60) to (5.64), and this single tensor element is

$$\left(\tilde{G}_{zz}\right)^a_{g1,g0} = \frac{1}{\hbar}\sum_v \frac{\left\langle\phi^a_{g1}\middle|\langle g|\hat{\mu}_z|e\rangle|\phi_{ev}\rangle\langle\phi_{ev}|\langle e|\hat{m}_z|g\rangle\middle|\phi^a_{g0}\right\rangle}{\omega_{ev,g0} - \omega_0 - i\Gamma_{ev}} = -\left(\tilde{\mathscr{G}}_{zz}\right)^a_{g1,g0} \qquad (5.157)$$

The two symmetric magnetic-dipole ROA invariants in Equations (5.55) and (5.56) and Equations (5.60) and (5.61) are then given by:

$$[\alpha G]^a_{g1,g0} = -[\alpha \mathscr{G}]^a_{g1,g0} = \frac{1}{9}\mathrm{Im}\left[(\tilde{\alpha}_{zz})(\tilde{G}_{zz})^*\right]^a_{g1,g0} \tag{5.158}$$

$$\left[\beta_S(\tilde{G})^2\right]^a_{g1,g0} = -\left[\beta_S(\tilde{\mathscr{G}})^2\right]^a_{g1,g0} = \mathrm{Im}\left[(\tilde{\alpha}_{zz})(\tilde{G}_{zz})^*\right]^a_{g1,g0} \tag{5.159}$$

The symmetric electric quadrupole invariant, given in Equation (5.58) and also in Equation (5.63), is constrained by the forms of the Raman polarizability, $\tilde{\alpha}_{\alpha\beta} = \tilde{\alpha}_{zz}$, and the electric-quadrupole ROA tensor, $\tilde{A}_{\gamma,\delta\beta} = \tilde{A}_{z,\delta z}$; however, this invariant also contains the factor $\varepsilon_{\alpha\gamma\delta}$, which is constrained from above to be $\varepsilon_{zz\delta}$, and which must always vanish by the definition of the alternating tensor explained for Equation (5.16) and also Equation (4.12). As a result, the symmetric electric quadrupole invariant vanishes and cannot contribute to SES-ROA intensity,

$$\left[\beta_S(\tilde{A})^2\right]^a_{g1,g0} = \left[\beta_S(\tilde{\mathscr{A}})^2\right]^a_{g1,g0} = 0 \tag{5.160}$$

Finally, the anti-symmetric ROA invariants depend on anti-symmetry in the Raman polarizability tensor, which is clearly not present if $\tilde{\alpha}_{zz}$ is the only non-zero tensor element, and hence all the anti-symmetric ROA invariants are zero:

$$\left[\beta_A(\tilde{G})^2\right]^a_{g1,g0} = \left[\beta_A(\tilde{\mathscr{G}})^2\right]^a_{g1,g0} = \left[\beta_A(\tilde{A})^2\right]^a_{g1,g0} = \left[\beta_A(\tilde{\mathscr{A}})^2\right]^a_{g1,g0} = 0 \tag{5.161}$$

As the only two non-zero ROA invariants, Equations (5.158) and (5.159), are proportional to $\mathrm{Im}\left(\tilde{\alpha}_{zz}\tilde{G}_{zz}^{\ *}\right)$, there is essentially only one unique non-zero invariant for SES-ROA intensity. In addition, it can be shown the Raman invariant is proportional to the square of the electronic dipole strength for the resonant electronic state, and the ROA is proportional to the product of the electronic circular dichroism (CD), or theoretically the rotational strength, and the electronic dipole strength of this resonant state. To see this, the Raman polarizability in Equation (5.153) is first separated into pure electronic and vibronic parts as:

$$(\tilde{\alpha}_{zz})^a_{g1,g0} = \frac{1}{\hbar}\langle g|\hat{\mu}_z|e\rangle\langle e|\hat{\mu}_z|g\rangle \sum_v \frac{\langle\phi^a_{g1}|\phi_{ev}\rangle\langle\phi_{ev}|\phi^a_{g0}\rangle}{\omega_{ev,g0} - \omega_0 - i\Gamma_{ev}} \tag{5.162}$$

Next, the electronic part is just the absolute square of the electronic dipole transition moment of the resonant state, while the incident frequency vibronic part can be separated into real and imaginary parts as:

$$(\tilde{\alpha}_{zz})^a_{g1,g0} = \frac{1}{\hbar}|\langle g|\hat{\mu}|e\rangle|^2 \sum_v \left\langle\phi^a_{g1}|\phi_{ev}\right\rangle\left\langle\phi_{ev}|\phi^a_{g0}\right\rangle \times \left[T(\omega_{ev,g0} - \omega_0) + iS(\omega_{ev,g0} - \omega_0)\right] \tag{5.163}$$

where the lineshape factors are:

$$T(\omega_{ev,g0} - \omega_0) = \frac{(\omega_{ev,g0} - \omega_0)}{(\omega_{ev,g0} - \omega_0)^2 + \Gamma^2_{ev}} \tag{5.164}$$

$$S(\omega_{ev,g0} - \omega_0) = \frac{\Gamma_{ev}}{(\omega_{ev,g0} - \omega_0)^2 + \Gamma_{ev}^2} \tag{5.165}$$

With these expressions we can write

$$\left|(\tilde{\alpha}_{zz})^a_{g1,g0}\right|^2 = (1/\hbar)^2 (D_{eg})^2 U(\omega_0, \omega_a) \tag{5.166}$$

where the dipole strength for the resonant electronic state is:

$$D_{eg} = |\langle g|\hat{\mu}|e\rangle|^2 \tag{5.167}$$

and the excitation profile lineshape factor is:

$$U(\omega_0, \omega_a) = \left|\sum_v \left\langle \phi^a_{g1}|\phi_{ev}\right\rangle \left\langle \phi_{ev}\middle|\phi^a_{g0}\right\rangle [T(\omega_{ev,g0} - \omega_0) + iS(\omega_{ev,g0} - \omega_0)]\right|^2 \tag{5.168}$$

These lineshape factors can be obtained directly from the shape of the electronic band profile as the absorption lineshape for the resonance electronic state is given by:

$$\varepsilon(\omega_0) = (1/\hbar)(D_{eg}) \sum_v \left|\left\langle \phi_{ev}\middle|\phi^a_{g0}\right\rangle\right|^2 S(\omega_0, \omega_a) \tag{5.169}$$

and $S(\omega_0, \omega_a)$ and $T(\omega_0, \omega_a)$ are Kramers–Kronig transforms of one another. The vibronic sum is a complete sum over all non-zero overlap integrals and formally includes multimode terms from all occupied normal modes and their energies even if these modes do not change vibrational levels during the transitions. The corresponding expressions for the ROA tensor is:

$$(\tilde{G}_{zz})^a_{g1,g0} = -\frac{1}{\hbar} \mathrm{Im}\left(\langle g|\hat{\boldsymbol{\mu}}|e\rangle \cdot \langle e|\hat{\boldsymbol{m}}|g\rangle\right) \sum_v \left\langle \phi^a_{g1}|\phi_{ev}\right\rangle \left\langle \phi_{ev}\middle|\phi^a_{g0}\right\rangle \times [T(\omega_{ev,g0} - \omega_0) + iS(\omega_{ev,g0} - \omega_0)] \tag{5.170}$$

while the ROA invariant is given by:

$$\mathrm{Im}\left[(\tilde{\alpha}_{zz})(\tilde{G}_{zz})^*\right]^a_{g1,g0} = -(1/\hbar)^2 (D_{eg} R_{eg}) U(\omega_0, \omega_a) \tag{5.171}$$

where the rotational strength of the resonant excited electronic state is:

$$R_{eg} = \mathrm{Im}(\langle g|\hat{\boldsymbol{\mu}}|e\rangle \cdot \langle e|\hat{\boldsymbol{m}}|g\rangle) \tag{5.172}$$

The minus sign in Equation (5.171) comes from the use of the complex conjugate of the ROA magnetic optical activity tensor as $(\tilde{G}_{zz})^*$. A more fundamental reason for the appearance of the minus sign is the difference in the sign convention for ROA, which is the intensity difference for RCP minus LCP radiation, whereas all other forms of optical activity use the LCP minus RCP sign convention.

The expressions for DCP_I ROA and Raman in the SES limit are then

$$I_R^R(180°) - I_L^L(180°) = \frac{96K}{c}\mathrm{Im}\left[(\tilde{\alpha}_{zz})(\tilde{G}_{zz})^*\right]^a_{g1,g0} \tag{5.173}$$

$$I_R^R(180°) + I_L^L(180°) = 24K\left|(\tilde{\alpha}_{zz})^a_{g1,g0}\right|^2 \tag{5.174}$$

Taking the ratio of these equations and using Equations (5.166) and (5.171) one obtains

$$\frac{I_R^R(180°) - I_L^L(180°)}{I_R^R(180°) + I_L^L(180°)} = -\frac{4}{c}\frac{R_{eg}D_{eg}}{(D_{eg})^2} = -\frac{4}{c}\frac{\mathrm{Im}(\langle g|\hat{\mu}|e\rangle \cdot \langle e|\hat{m}|g\rangle)}{|\langle g|\hat{\mu}|e\rangle|^2} = -g_{eg} \tag{5.175}$$

DCP_I-ROA is the form of ROA that gives the largest ratio, by a factor of at least two, between ROA intensity and the anisotropy ratio. For example, unpolarized backscattering ICP and SCP intensities obey the relationships

$$\frac{I_U^R(180°) - I_U^L(180°)}{I_U^R(180°) + I_U^L(180°)} = \frac{I_R^U(180°) - I_L^U(180°)}{I_R^U(180°) + I_L^U(180°)} = -\frac{g_{eg}}{2} \tag{5.176}$$

From these equations one can see a direct connection between RROA in the SES limit and the electronic CD of the resonant electronic state, and this reveals the source of ROA intensity at its deepest, simplest, and most fundamental level. This simplicity, however, is lost when more than one excited state is responsible for the observation of ROA intensity.

5.5.2 Strong Resonance Involving Two Excited Electronic States

To extend the theory of resonance ROA beyond the SES theory requires the involvement of at least one additional excited electronic state. If only one additional excited state is involved, two different mechanisms, or both, can occur. One is single A-term scattering for both excited states and the other is activation of B-term scattering involving the vibronic coupling of the new state with the original single electronic state. Of these two possibilities, the second will occur first as it does not require close resonant proximity of the second excited electronic state. If the second state is not close enough to have a significant contribution from its own A-term scattering mechanism, it can contribute via the B-term mechanism to the overall resonance of the resonant electronic state.

5.5.2.1 *TES Theory With a Single B-Term Contributing State (TES-B)*

For the B-term case, we consider a single excited state, e, and one nearby excited state, s, which is not close enough to resonance to have its own A-term scattering. This level of theory is called the two-excited-state B-state (TES-B) theory. This extension is the simplest first step to understanding how ROA spectra arise with both positive and negative signed bands. Using the expression for the contribution of B-term scattering developed from Equations (2.151) to (2.159), the TES-B Raman polarizability is given by:

$$(\tilde{\alpha}_{\alpha\beta})^a_{g1,g0} = (A_{\alpha\beta})^a_{g1,g0} + (B_{\alpha\beta})^a_{g1,g0} \tag{5.177}$$

$$(A_{\alpha\beta})^a_{g1,g0} = \frac{1}{\hbar}\langle g|\hat{\mu}_\alpha|e\rangle_0 \left\langle e|\hat{\mu}_\beta|g\right\rangle_0 \sum_v \frac{\langle \phi^a_{g1}|\phi_{ev}\rangle\langle \phi_{ev}|\phi^a_{g0}\rangle}{\omega_{ev,g0} - \omega_0 - i\Gamma_{ev}} \tag{5.178}$$

$$(B_{\alpha\beta})^a_{g1,g0} = \frac{1}{\hbar^2}\left[\langle g|\hat{\mu}_\alpha|s\rangle_0 \frac{h^a_{se,0}}{\omega^0_{es}} \left\langle e|\hat{\mu}_\beta|g\right\rangle_0 \sum_v \frac{\langle \phi^a_{g1}|Q_a|\phi_{ev}\rangle\langle \phi_{ev}|\phi^a_{g0}\rangle}{\omega_{ev,g0} - \omega_0 - i\Gamma_{ev}}\right.$$
$$\left. + \langle g|\hat{\mu}_\alpha|e\rangle_0 \frac{h^a_{es,0}}{\omega^0_{es}} \left\langle s|\hat{\mu}_\beta|g\right\rangle_0 \sum_v \frac{\langle \phi^a_{g1}|\phi_{ev}\rangle\langle \phi_{ev}|Q_a|\phi^a_{g0}\rangle}{\omega_{ev,g0} - \omega_0 - i\Gamma_{ev}}\right] \tag{5.179}$$

where from Equation (2.158) and the definition of Herzberg–Teller coupling

$$h^a_{se,0} = \left\langle \psi^0_s|(\partial H_E/\partial Q_a)_0|\psi^0_e\right\rangle = h^a_{es,0} \tag{5.180}$$

and where the symmetry with respect to state e and s arises from the fact the operator is Hermitian rather than anti-Hermitian. If, as in the case of SES Raman and ROA, it is assumed that the transition moment of the resonant electronic state lies along the z-axis of the molecule, the polarizability expression becomes

$$(\tilde{\alpha}_{\alpha\beta})^a_{g1,g0} = \frac{1}{\hbar^2}\left[\hbar\langle g|\hat{\mu}_z|e\rangle_0 \langle e|\hat{\mu}_z|g\rangle_0 \sum_v \frac{\left\langle \phi^a_{g1}\cdot|\phi_{ev}\right\rangle\cdot\left\langle \phi_{ev}|\phi^a_{g0}\right\rangle}{\omega_{ev,g0} - \omega_0 - i\Gamma_{ev}}\right.$$
$$+ \langle g|\hat{\mu}_\alpha|s\rangle_0 \frac{h^a_{se,0}}{\omega^0_{es}} \langle e|\hat{\mu}_z|g\rangle_0 \sum_v \frac{\left\langle \phi^a_{g1}|Q_a|\phi_{ev}\right\rangle\cdot\left\langle \phi_{ev}|\phi^a_{g0}\right\rangle}{\omega_{ev,g0} - \omega_0 - i\Gamma_{ev}}$$
$$\left. + \langle g|\hat{\mu}_z|e\rangle_0 \frac{h^a_{es,0}}{\omega^0_{es}} \left\langle s|\hat{\mu}_\beta|g\right\rangle_0 \sum_v \frac{\left\langle \phi^a_{g1}|\phi_{ev}\right\rangle\cdot\left\langle \phi_{ev}|Q_a|\phi^a_{g0}\right\rangle}{\omega_{ev,g0} - \omega_0 - i\Gamma_{ev}}\right] \tag{5.181}$$

Here, their direction of the transition moment for state s, namely$\langle g_0|\hat{\mu}_\alpha|s_0\rangle$, is in general not oriented along the z-direction and hence there are non-zero x-, y- and z-components for this transition moment. This removes the constraint on the non-zero components of the polarizability tensor, and hence the subscripts α and β can represent any Cartesian direction. The symmetry of $\tilde{\alpha}_{\alpha\beta}$ with respect to interchange of subscripts α and β is also lost because the summations over vibronic sublevels v for the terms with α and β subscripts are not the same and represent different vibronic lineshapes. The normal modes Q_a, important for B-term scattering in general, are different from those important for A-term scattering because of the presence of the operator Q_a in the B-term vibrational matrix elements. If the molecule has symmetry higher than point group C_1, then the vibration modes for A-term scattering are formally restricted to totally symmetric modes while B-term scattering has predominantly non-totally symmetric modes.

Because all restrictions present in the SES resonant limit are lost when a single additional state enters the formalism, the theory of ROA reverts back to the GU theory of Section 5.3 where there are four distinct ROA tensors and ten ROA invariants, instead of essentially only one ROA tensor and one ROA invariant for the SES theory. The four distinct ROA tensors that accompany the Raman polarizability tensor in Equation (5.181) are

$$(\tilde{G}_{\alpha\beta})^a_{g1,g0} = \frac{1}{\hbar^2}\left[\hbar\langle g|\hat{\mu}_z|e\rangle_0\langle e|\hat{m}_\beta|g\rangle_0\sum_v \frac{\langle\phi^a_{g1}|\phi_{ev}\rangle\langle\phi_{ev}|\phi^a_{g0}\rangle}{\omega_{ev,g0}-\omega_0-i\Gamma_{ev}}\right.$$
$$+\langle g|\hat{\mu}_\alpha|s\rangle_0\frac{h^a_{se,0}}{\omega^0_{es}}\langle e|\hat{m}_\beta|g\rangle_0\sum_v \frac{\langle\phi^a_{g1}|Q_a|\phi_{ev}\rangle\langle\phi_{ev}|\phi^a_{g0}\rangle}{\omega_{ev,g0}-\omega_0-i\Gamma_{ev}}$$
$$\left.+\langle g|\hat{\mu}_z|e\rangle_0\frac{h^a_{es,0}}{\omega^0_{es}}\langle s|\hat{m}_\beta|g\rangle_0\sum_v \frac{\langle\phi^a_{g1}|\phi_{ev}\rangle\langle\phi_{ev}|Q_a|\phi^a_{g0}\rangle}{\omega_{ev,g0}-\omega_0-i\Gamma_{ev}}\right] \tag{5.182}$$

$$(\tilde{\mathscr{G}}_{\alpha\beta})^a_{g1,g0} = \frac{1}{\hbar^2}\left[\hbar\langle g|\hat{m}_\alpha|e\rangle_0\langle e|\hat{\mu}_z|g\rangle_0\sum_v \frac{\langle\phi^a_{g1}|\phi_{ev}\rangle\langle\phi_{ev}|\phi^a_{g0}\rangle}{\omega_{ev,g0}-\omega_0-i\Gamma_{ev}}\right.$$
$$+\langle g|\hat{m}_\alpha|s\rangle_0\frac{h^a_{se,0}}{\omega^0_{es}}\langle e|\hat{\mu}_z|g\rangle_0\sum_v \frac{\langle\phi^a_{g1}|Q_a|\phi_{ev}\rangle\langle\phi_{ev}|\phi^a_{g0}\rangle}{\omega_{ev,g0}-\omega_0-i\Gamma_{ev}}$$
$$\left.+\langle g|\hat{m}_\alpha|e\rangle_0\frac{h^a_{es,0}}{\omega^0_{es}}\langle s|\hat{\mu}_\beta|g\rangle_0\sum_v \frac{\langle\phi^a_{g1}|\phi_{ev}\rangle\langle\phi_{ev}|Q_a|\phi^a_{g0}\rangle}{\omega_{ev,g0}-\omega_0-i\Gamma_{ev}}\right] \tag{5.183}$$

$$(\tilde{A}_{\alpha,\beta\gamma})^a_{g1,g0} = \frac{1}{\hbar^2}\left[\hbar\langle g|\hat{\mu}_z|e\rangle_0\langle e|\hat{\Theta}_{\beta\gamma}|g\rangle_0\sum_v \frac{\langle\phi^a_{g1}|\phi_{ev}\rangle\langle\phi_{ev}|\phi^a_{g0}\rangle}{\omega_{ev,g0}-\omega_0-i\Gamma_{ev}}\right.$$
$$+\langle g|\hat{\mu}_\alpha|s\rangle_0\frac{h^a_{se,0}}{\omega^0_{es}}\langle e|\hat{\Theta}_{\beta\gamma}|g\rangle_0\sum_v \frac{\langle\phi^a_{g1}|Q_a|\phi_{ev}\rangle\langle\phi_{ev}|\phi^a_{g0}\rangle}{\omega_{ev,g0}-\omega_0-i\Gamma_{ev}}$$
$$\left.+\langle g|\hat{\mu}_z|e\rangle_0\frac{h^a_{es,0}}{\omega^0_{es}}\langle s|\hat{\Theta}_{\beta\gamma}|g\rangle_0\sum_v \frac{\langle\phi^a_{g1}|\phi_{ev}\rangle\langle\phi_{ev}|Q_a|\phi^a_{g0}\rangle}{\omega_{ev,g0}-\omega_0-i\Gamma_{ev}}\right] \tag{5.184}$$

$$(\tilde{\mathscr{A}}_{\alpha,\beta\gamma})^a_{g1,g0} = \frac{1}{\hbar^2}\left[\hbar\langle g|\hat{\Theta}_{\beta\gamma}|e\rangle_0\langle e|\hat{\mu}_z|g\rangle_0\sum_v \frac{\langle\phi^a_{g1}|\phi_{ev}\rangle\langle\phi_{ev}|\phi^a_{g0}\rangle}{\omega_{ev,g0}-\omega_0-i\Gamma_{ev}}\right.$$
$$+\langle g|\hat{\Theta}_{\beta\gamma}|s\rangle_0\frac{h^a_{se,0}}{\omega^0_{es}}\langle e|\hat{\mu}_z|g\rangle_0\sum_v \frac{\langle\phi^a_{g1}|Q_a|\phi_{ev}\rangle\langle\phi_{ev}|\phi^a_{g0}\rangle}{\omega_{ev,g0}-\omega_0-i\Gamma_{ev}}$$
$$\left.+\langle g|\hat{\Theta}_{\beta\gamma}|e\rangle_0\frac{h^a_{es,0}}{\omega^0_{es}}\langle s|\hat{\mu}_\alpha|g\rangle_0\sum_v \frac{\langle\phi^a_{g1}|\phi_{ev}\rangle\langle\phi_{ev}|Q_a|\phi^a_{g0}\rangle}{\omega_{ev,g0}-\omega_0-i\Gamma_{ev}}\right] \tag{5.185}$$

Note that in the TES-B theory there is no constraint on the direction of either the magnetic dipole or electric quadrupole transition moments, even for the resonant excited state e, as in $\langle e|\hat{m}_\beta|g\rangle$ and $\langle e|\Theta_{\beta\gamma}|g\rangle$. The only constraint is for the matrix elements of the electric dipole transition moment involving the resonant electronic state e, $\langle e|\hat{\mu}_z|g\rangle = \langle g|\hat{\mu}_z|e\rangle$. The TES-B theory is easily extended to multiple B-state contributions, but with only a single electronic state in strong resonance with the incident laser photon energy.

5.5.2.2 *TES Theory with two A-Term Contributing States (TES-A)*

For the case of two resonant electronic states independently contributing significant levels of A-term resonance Raman scattering intensity, one has the Raman polarizability equal to the sum of two A-term SES expressions. In addition, one must consider the B-term contributions from each of these states as the two states must be close enough in energy to have their own A-term contributions for a particular choice of incident laser photon energy. As only two states are in strong resonance, this level of theory is referred to as the two electronic state (TES-A) theory. In the TES theory, the Raman polarizability is written as the sum of the two electronic state resonances, for states e_1 and e_2, and each with their own A- and B-term contributions,

$$
\begin{aligned}
\left(\tilde{\alpha}_{\alpha\beta}\right)^a_{g1,g0} = \frac{1}{\hbar^2}\Bigg[& \hbar\langle g|\hat{\mu}_z|e_1\rangle_0\langle e_1|\hat{\mu}_z|g\rangle_0\sum_v\frac{\left\langle\phi^a_{g1}|\phi_{e_1v}\right\rangle\left\langle\phi_{e_1v}|\phi^a_{g0}\right\rangle}{\omega_{e_1v,g0}-\omega_0-i\Gamma_{e_1v}} \\
& +\langle g|\hat{\mu}_\alpha|e_2\rangle_0\frac{h^a_{e_2e_1,0}}{\omega^0_{e_1e_2}}\langle e_1|\hat{\mu}_z|g\rangle_0\sum_v\frac{\left\langle\phi^a_{g1}|Q_a|\phi_{e_1v}\right\rangle\left\langle\phi_{e_1v}|\phi^a_{g0}\right\rangle}{\omega_{e_1v,g0}-\omega_0-i\Gamma_{e_1v}} \\
& +\langle g|\hat{\mu}_z|e_1\rangle_0\frac{h^a_{e_1e_2,0}}{\omega^0_{e_1e_2}}\left\langle e_2|\hat{\mu}_\beta|g\right\rangle_0\sum_v\frac{\left\langle\phi^a_{g1}|\phi_{e_1v}\right\rangle\left\langle\phi_{e_1v}|Q_a|\phi^a_{g0}\right\rangle}{\omega_{e_1v,g0}-\omega_0-i\Gamma_{ev}} \\
& +\hbar\langle g|\hat{\mu}_\alpha|e_2\rangle_0\left\langle e_2|\hat{\mu}_\beta|g\right\rangle_0\sum_v\frac{\left\langle\phi^a_{g1}|\phi_{e_2v}\right\rangle\left\langle\phi_{e_2v}|\phi^a_{g0}\right\rangle}{\omega_{e_2v,g0}-\omega_0-i\Gamma_{e_2v}} \\
& +\langle g|\hat{\mu}_z|e_1\rangle_0\frac{h^a_{e_1e_2,0}}{\omega^0_{e_2e_1}}\left\langle e_2|\hat{\mu}_\beta|g\right\rangle_0\sum_v\frac{\left\langle\phi^a_{g1}|Q_a|\phi_{e_2v}\right\rangle\left\langle\phi_{e_2v}|\phi^a_{g0}\right\rangle}{\omega_{e_2v,g0}-\omega_0-i\Gamma_{e_1v}} \\
& +\langle g|\hat{\mu}_\alpha|e_2\rangle_0\frac{h^a_{e_2e_1,0}}{\omega^0_{e_2e_1}}\langle e_1|\hat{\mu}_z|g\rangle_0\sum_v\frac{\left\langle\phi^a_{g1}|\phi_{e_2v}\right\rangle\left\langle\phi_{e_2v}|Q_a|\phi^a_{g0}\right\rangle}{\omega_{e_2v,g0}-\omega_0-i\Gamma_{ev}}\Bigg]
\end{aligned}
\tag{5.186}
$$

While the direction of the electronic transition moment for one state can be chosen to be the z-axis, the transition moment of the second resonant state is in general in any arbitrary direction relative to the z-axis. Because there are only two electronic states, their B-terms are in some sense closely related. In particular they have opposite signs for the inter-state energy denominator, $\omega^0_{e_2e_1} = -\omega^0_{e_1e_2}$ and as Herzberg–Teller matrix elements are symmetric, $h^a_{e_1e_2,0} = h^a_{e_2e_1,0}$, the corresponding B-terms have a canceling character to them. Thus if two electronic states both have strong resonance, and the are no additional nearby non-resonant but coupled electronic states, the resonance Raman scattering may be dominated by the two A-terms, and their respective B-terms may interfere and not contribute as strongly. It is straightforward to write the corresponding four ROA tensors corresponding to TES-A polarizability given here in Equation (5.186) as was carried out explicitly for the TES-B theory in

Equations (5.182)–(5.185). Further development of these theoretical expressions, and their computation implementation, can be carried out when experimental RROA spectra become available that show significant departure from the SES theory of RROA.

References

Barron, L.D. (2004) *Molecular Light Scattering and Optical Activity*, 2nd edn, Cambridge University Press, Cambridge.

Barron, L.D., and Buckingham, A.D. (1971) Rayleigh and Raman scattering from optically active molecules. *Mol. Phys.*, **20**, 1111–1119.

Barron, L.D., Bogaard, M.P., and Buckingham, A.D. (1973) Raman scattering of circularly polarized light by optically active molecules. *J. Am. Chem. Soc.*, **95**, 603–605.

Barron, L.D., and Buckingham, A.D. (1975) Rayleigh and Raman optical activity. *Ann. Rev. Phys. Chem.*, **26**, 381–396.

Barron, L.D., and Buckingham, A.D. (1979) The inertial contribution to vibrational optical activity in methyl torsion modes. *J. Am. Chem. Soc.*, **101**(8), 1979–1987.

Barron, L.D., Escribano, J.R., and Torrance, J.F. (1986) Polarized Raman optical activity and the bond polarizability model. *Mol. Phys.*, **57**(3), 653–660.

Barron, L.D., Hecht, L., Hug, W., and MacIntosh, M.J. (1989) Backscattered Raman optical activity with CCD detector. *J. Am. Chem. Soc.*, **111**, 8731–8732.

Champion, P.M., and Albrecht, A.C. (1982) Resonance Raman scattering: the multimode problem and transform methods. *Annu. Rev. Phys. Chem.*, **33**, 353–376.

Che, D., Hecht, L., and Nafie, L.A. (1991) Dual and incident circular polarization Raman optical activity backscattering of (–)-*trans*-pinane. *Chem. Phys. Lett.*, **180**, 182–190.

Escribano, J.R., Freedman, T.B., and Nafie, L.A. (1987) Bond polarizability and vibronic coupling theory of Raman optical activity: The electric-dipole magnetic dipole optical activity tensor. *J. Chem. Phys.*, **87**, 3366–3374.

Hecht, L., and Nafie, L.A. (1990) Linear polarization Raman optical activity: A new form of natural optical activity. *Chem. Phys. Lett.*, **174**, 575–582.

Hecht, L., and Nafie, L.A. (1991) Theory of natural Raman optical activity I. Complete circular polarization formalism. *Mol. Phys.*, **72**, 441–469.

Hug, W. (1982). Instrumental and theoretical advances in Raman optical activity. In: *Raman Spectroscopy* (ed. J. Lascombe), John Wiley & Sons, Ltd, Chichester, pp. 3–12.

Long, D.A. (2002) *The Raman Effect: A Unified Treatment of the Theory of Raman Scattering by Molecules*. John Wiley & Sons, Ltd., Chichester.

Nafie, L.A. (2008) Theory of Raman scattering and Raman optical activity. Near resonance theory and levels of approximation. *Theor. Chem. Acc.*, **119**, 39–55.

Nafie, L.A., and Freedman, T.B. (1989) Dual circular polarization Raman optical activity. *Chem. Phys. Lett.*, **154**, 260–266.

Nafie, L.A., and Che, D. (1994) Theory and measurement of Raman optical activity. In: *Modern Nonlinear Optics, Part 3* (eds M. Evans,and S. Kielich), John Wiley & Sons, Inc., New York, pp. 105–149.

Prasad, P.L., and Nafie, L.A. (1979) Atom dipole interaction model for Raman optical activity. Reformulation and its comparison to the general two group model. *J. Chem. Phys.*, **70**, 5582–5588.

Spencer, K.M., Freedman, T.B., and Nafie, L.A. (1988) Scattered circular polarization Raman optical activity. *Chem. Phys. Lett.*, **149**, 367–374.

6

Instrumentation for Vibrational Circular Dichroism

The chapters on VCD and ROA instrumentation represent a shift in focus from the theory of VOA to its measurement. Since the early days of the discovery of VCD and ROA, instrumentation for their measurement has developed dramatically in sophistication and efficiency. In the mid-1970s, when ROA and VCD were first measured, instrumentation was restricted to grating scanning spectrometers and spectra were obtained from strip-chart recorders with ink on paper. Spectra had to be redrawn by graphic artists on translucent paper that could be photographed against a white background and submitted as glossy black on white pictures for journal publication. Today, VCD and ROA are usually measured across a wide range of vibrational frequencies simultaneously, either as Fourier transform or multi-channel detector spectra, and recorded as digital files in a computer that controls the measurement process. Subsequently the raw data files are formatted for presentation and publication. Although VCD and ROA had origins in similar instrumentation, aside from the laser excitation needed for Raman scattering, their subsequent evolution followed different pathways and timelines of progress. The description of VCD instrumentation begins at the simplest level, which corresponds closely to the instrumentation used for the discovery of VCD.

6.1 Polarization Modulation Circular Dichroism

Circular dichroism and circular birefringence are relatively small optical phenomena compared with their parent spectroscopic quantities: absorption and the index of refraction. In general, the measurement of circular dichroism cannot be accomplished by separately measuring the intensity of a beam using first left (LCP) and then right circularly polarized (RCP) radiation. Instead, the beam is modulated rapidly between LCP and RCP states using a polarization modulator. In principle, the modulator can either be a square-wave or a sine-wave device. The sine-wave devices in the form of photoelastic modulators (PEMs) have become standard for both electronic CD (ECD) measurements in the ultraviolet (UV)–visible and near-infrared (near-IR) regions using dispersive instruments and also VCD in the IR and near-IR regions using Fourier transform (FT) instruments,

Vibrational Optical Activity: Principles and Applications, First Edition. Laurence A. Nafie.

with some overlap between ECD and VCD in the near-IR and IR regions, sometimes in the same spectral measurement.

In the sections that follow, the focus will be on the analysis of the *optical* pathway with less emphasis on the *electronic* pathway and how the intensities are actually measured. This is because the electronic level of consideration is often specific to the instrument being used, and changes in signal analysis are continuously taking place as the technology of the electronic pathway processing improves over time. On the other hand, the optical analysis is fundamental to the phenomena being measured and not as subject to change.

6.1.1 Instrumental Measurement of Circular Dichroism

The measurement of ECD or VCD can be described in general terms without regard to whether the underlying instrument is a dispersive scanning instrument or an FT-instrument. The block diagram in Figure 6.1 illustrates the optical layout of a CD instrument, in this case, an infrared CD instrument.

The linear polarizer (P) is set at an angle of 45° from the stress axes of the photoelastic modulator (PEM). Usually the polarizer is in the vertical or horizontal orientation and the stress axes of the PEM are at a 45° angle with respect to these orientations. The PEM oscillates sinusoidally with a phase angle α_M oscillating at a frequency ω_M in the tens of kilohertz (kHz) range. The sample (S) is placed immediately after the PEM and the beam is then focused onto a detector (D). For CD measurements the PEM oscillates between primarily right (RCP) and left circularly polarized (LCP) states at the PEM frequency. The intensity of the IR beam at the detector is separated electronically by frequency range, high and low, into two pathways. The high-frequency AC-path carries the VCD spectrum and passes first through a lock-in amplifier (LIA) referenced to the PEM frequency, ω_M. As the light chopper (LC) interrupts the light beam at the frequency, ω_C, the output of the PEM LIA can be passed through a second LIA tuned to the light chopper frequency. This step eliminates stray high-frequency electronic signals that may be present near the instrument. The output of the AC-path is the AC-transmission spectrum, $I_{AC}(\bar{\nu})$. The low-frequency DC-path passes through a separate LIA tuned to the chopper frequency, the output of which is just the ordinary transmission spectrum of the sample, $I_{DC}(\bar{\nu})$. These

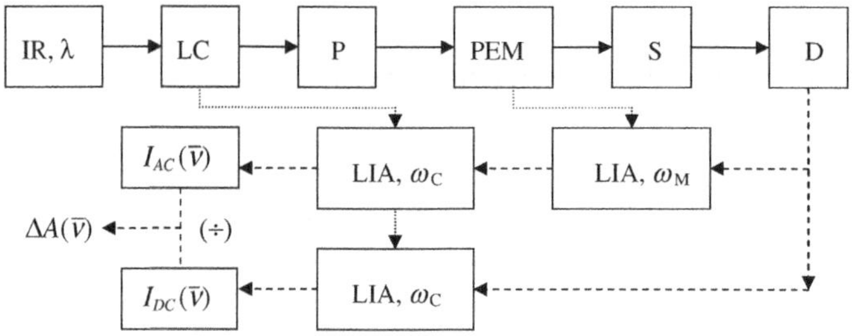

Figure 6.1 *Optical–electronic diagram of dispersive scanning VCD instrumental layout consisting of an optical path (solid line) starting with an infrared source (IR) and wavelength selector (λ), a light chopper (LC) at frequency ω_C, a linear polarizer (P), a photoelastic modulator (PEM) at frequency ω_M, a sample (S) and a detector (D). The electrical signal (dashed lines) starts at the detector and splits into an AC-transmission path through two lock-in amplifiers (LIA) and a DC-transmission path through a single lock-in. The LIA references are indicated by a dotted line. The ratio of the two transmission spectra yields a signal proportional to the VCD spectrum as explained in the text*

two intensity expressions as a function of wavenumber frequency $\bar{\nu}$ across the spectral range of measurement are defined by:

$$I_{AC}(\bar{\nu}) = \frac{1}{2}[I_R(\bar{\nu}) - I_L(\bar{\nu})]\sin[\alpha_M(\bar{\nu})] \quad (6.1)$$

$$I_{DC}(\bar{\nu}) = \frac{1}{2}[I_R(\bar{\nu}) + I_L(\bar{\nu})] \quad (6.2)$$

where the AC-term also depends on how efficiently the PEM modulates between RCP and LCP states through the factor $\sin[\alpha_M(\bar{\nu})]$. For example, the most efficient modulation cycle is a square wave between retardation angles, α_M, of $+90°$ (pure RCP) and $-90°$ (pure LCP) where the factor $\sin[\alpha_M(\bar{\nu})]$ alternates between values of $+1$ and -1. In modern CD (ECD and VCD) instruments, the PEM oscillates in time with a sine-wave cycle such that

$$[\alpha_M(\bar{\nu})] = \alpha_M^{\text{o}}(\bar{\nu})\sin\omega_M t \quad (6.3)$$

where, $\alpha_M^{\text{o}}(\bar{\nu})$ is the maximum value of the sine-wave retardation angle at the wavenumber frequency $\bar{\nu}$. The full sine-wave dependence of the AC intensity term in Equation (6.1) is given by:

$$\sin[\alpha_M(\bar{\nu})] = \sin[\alpha_M^{\text{o}}(\bar{\nu})\sin\omega_M t] = \sum_{n=odd} 2J_n[\alpha_M^{\text{o}}(\bar{\nu})]\sin n\,\omega_M t \quad (6.4)$$

where the second equality is a sum over odd-order Bessel functions, J_n, at the odd harmonics of the PEM frequency, $n\omega_M$, where n in the summation is equal to only odd integers starting at 1. The first term in the odd harmonic expansion above is the main CD signal at the detector. This signal, at the fundamental PEM frequency, is measured by a lock-in amplifier tuned to ω_M as the amplitude of $\sin\omega_M t$, and as a result, Equation (6.1) can be written as:

$$I_{AC}(\bar{\nu}) = \frac{1}{2}[I_R(\bar{\nu}) - I_L(\bar{\nu})]2J_1[\alpha_M^{\text{o}}(\bar{\nu})] \quad (6.5)$$

The measured of VCD spectrum is defined by:

$$\Delta A(\bar{\nu}) = A_L(\bar{\nu}) - A_R(\bar{\nu}) \quad (6.6)$$

which is obtained from the ratio of the AC and DC intensities

$$\begin{aligned}\frac{I_{AC}(\bar{\nu})}{I_{DC}(\bar{\nu})} &= 2J_1[\alpha_M^{\text{o}}(\bar{\nu})]\frac{[I_R(\bar{\nu}) - I_L(\bar{\nu})]}{[I_R(\bar{\nu}) + I_L(\bar{\nu})]} \\ &= 2J_1[\alpha_M^{\text{o}}(\bar{\nu})]\frac{[10^{-A_R(\bar{\nu})} - 10^{-A_L(\bar{\nu})}]}{[10^{-A_R(\bar{\nu})} + 10^{-A_L(\bar{\nu})}]} \\ &= 2J_1[\alpha_M^{\text{o}}(\bar{\nu})]\frac{[e^{-2.303A_R(\bar{\nu})} - e^{-2.303A_L(\bar{\nu})}]}{[e^{-2.303A_R(\bar{\nu})} + e^{-2.303A_L(\bar{\nu})}]}\end{aligned} \quad (6.7)$$

If this expression is multiplied by a particular ratio of exponentials equal to unity, the two arguments of the four exponential terms in this last equation can be shifted to be either the positive or negative value of the *same* exponential argument, namely $1.1513\Delta A(\bar{\nu})$,

$$\frac{I_{AC}(\bar{\nu})}{I_{DC}(\bar{\nu})} = 2J_1\left[\alpha_M^{o}(\bar{\nu})\right]\frac{\left[e^{-2.303A_R(\bar{\nu})} - e^{-2.303A_L(\bar{\nu})}\right]}{\left[e^{-2.303A_R(\bar{\nu})} + e^{-2.303A_L(\bar{\nu})}\right]}\left(\frac{e^{2.303[A_L(\bar{\nu})+A_R(\bar{\nu})]/2}}{e^{2.303[A_L(\bar{\nu})+A_R(\bar{\nu})]/2}}\right)$$

$$= 2J_1\left[\alpha_M^{o}(\bar{\nu})\right]\left[\frac{e^{1.1513\Delta A(\bar{\nu})} - e^{-1.1513\Delta A(\bar{\nu})}}{e^{1.1513\Delta A(\bar{\nu}))} + e^{-1.1513\Delta A(\bar{\nu})}}\right] \tag{6.8}$$

Furthermore, the exponential part of this last equation corresponds to the definition of the hyperbolic tangent $\tanh x = (e^x - e^{-x})/(e^x + e^{-x})$ and thus

$$\frac{I_{AC}(\bar{\nu})}{I_{DC}(\bar{\nu})} = 2J_1\left[\alpha_M^{o}(\bar{\nu})\right]\tanh\left[1.1513\Delta A(\bar{\nu})\right] \tag{6.9}$$

This equation can be reduced further by using the trigonometric identity that the hyperbolic tangent equals its argument for sufficiently small values of the argument, $\tanh x \cong x$ for small x. The correction to this approximation is less than cubic in x and is satisfied to less than 1% error for x as large as 0.1. It is rare that CD intensity is larger than one-tenth of an absorbance unit, especially for VCD, and thus one can write without concern,

$$\frac{I_{AC}(\bar{\nu})}{I_{DC}(\bar{\nu})} = 2J_1\left[\alpha_M^{o}(\bar{\nu})\right]\left[1.1513\Delta A(\bar{\nu})\right] \tag{6.10}$$

Thus the following equation for the measurement of CD intensity is written as:

$$\Delta A(\bar{\nu}) = \frac{1}{2J_1\left[\alpha_M^{o}(\bar{\nu})\right]1.1513}\left[\frac{I_{AC}(\bar{\nu})}{I_{DC}(\bar{\nu})}\right] \tag{6.11}$$

Using definitions provided earlier in Equation (1.3), the corresponding absorbance measurement is given by:

$$A(\bar{\nu}) = -\log\left[\frac{I_{DC}(\bar{\nu})}{I_{DC}^{0}(\bar{\nu})}\right] \tag{6.12}$$

where $I_{DC}^{0}(\bar{\nu})$ is the background, or reference, transmission spectrum of the instrument in the absence of the sample. Notice that the expressions for the measurement of both $\Delta A(\bar{\nu})$ and $A(\bar{\nu})$ involve the ratio of two instrumental transmission measurements. For both $\Delta A(\bar{\nu})$ and $A(\bar{\nu})$, the dependence of the instrument is removed by taking a ratio of transmission intensities. In the case of $A(\bar{\nu})$, the ratio is for the same transmission measurement, *with and without* the sample in place, whereas for $\Delta A(\bar{\nu})$, two different transmission measurements are used, both *with* the sample in place. In the case of $\Delta A(\bar{\nu})$, an additional instrumental calibration measurement is required to remove the spectral dependence of the first-order Bessel function, $J_1\left[\alpha_M^{o}(\bar{\nu})\right]$.

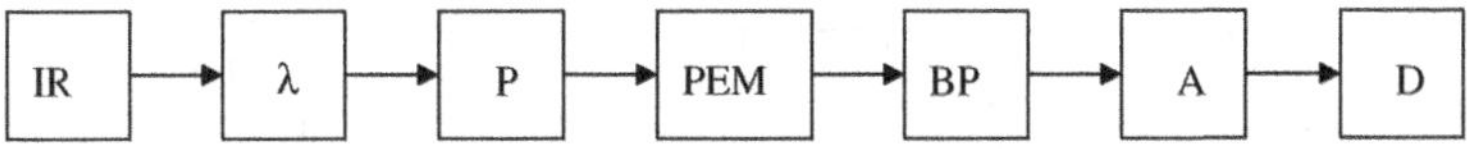

Figure 6.2 Block diagram of the optical layout for calibration of CD intensity consisting of an infrared source (IR), a wavelength selector (λ), a linear polarizer (P), a photoelastic modulator (PEM), a birefringent plate (BP), a linear polarizer that acts as a polarization analyzer (A), and a detector (D)

6.1.2 Calibration of CD Intensities

CD intensities measured with a PEM can be calibrated by using the optical block diagram shown in Figure 6.2.

Instead of a sample S, one places a birefringent plate, BP, with fast and slow axes parallel to the stress axes of the PEM, in this case 45° from vertical, and a polarization analyzer, A, oriented with its polarization axis vertical or horizontal. There are two allowable positions of the BP with its fast axis oriented either ±45° from vertical, and two positions of A, either vertical, 0°, or horizontal, 90°. The CD intensity expression for these four optical configurations are given by:

$$\left[\frac{I_{AC}(\bar{\nu})}{I_{DC}(\bar{\nu})}\right]_{cal} = \frac{\pm 2J_1\left[\alpha_M^{\rm o}(\bar{\nu})\right]\sin[\alpha_B(\bar{\nu})]}{1 \pm J_0[\alpha_M^{\rm o}(\bar{\nu})]\cos[\alpha_B(\bar{\nu})]} \tag{6.13}$$

where $\alpha_B(\bar{\nu})$ is retardation angle of the birefringent plate and $J_0\left[\alpha_M^{\rm o}(\bar{\nu})\right]$ is the zeroth order Bessel function.

The retardation angles of both the PEM, $\alpha_M^{\rm o}(\bar{\nu})$, and the BP, $\alpha_B(\bar{\nu})$, are simple linear functions of the wavenumber frequency,

$$\alpha_M^{\rm o}(\bar{\nu}) = V_{PEM}\bar{\nu} \tag{6.14}$$

$$\alpha_B(\bar{\nu}) = C_{BP}\bar{\nu} \tag{6.15}$$

For the PEM, the factor V_{PEM} is a function of the control voltage of the PEM, while for the BP the proportionality constant is fixed for the plate and carries the information about at which spectral frequencies the plate is a quarter-wave plate or half-wave plate, and so on. In Figure 6.3 the Bessel

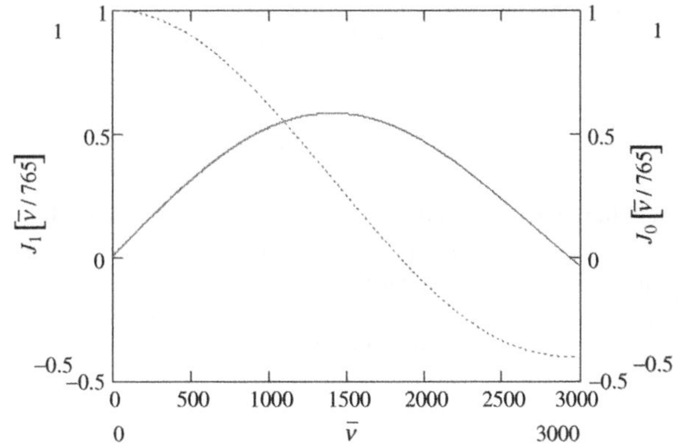

Figure 6.3 Plots of Bessel functions $J_1\left[\alpha_M^{\rm o}(\bar{\nu})\right]$ (solid line) and $J_0\left[\alpha_M^{\rm o}(\bar{\nu})\right]$ (dashed line) where $J_1\left[\alpha_M^{\rm o}(\bar{\nu})\right]$ has a maximum at $\alpha_M^{\rm o}(\bar{\nu}) = \bar{\nu}/765 = 1.83$ rad, or $1400\,\text{cm}^{-1}$, and $J_0\left[\alpha_M^{\rm o}(\bar{\nu})\right] = 0$ at $\alpha_M^{\rm o}(\bar{\nu}) = 2.40$ rad, or $\bar{\nu} = 1865\,\text{cm}^{-1}$

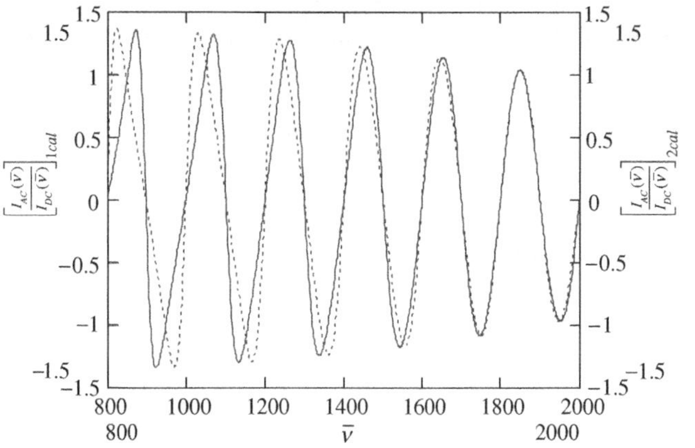

Figure 6.4 Two of the family of four calibration curves defined by Equation (6.13): $I_{1cal}(\nu)$ (solid line) and $I_{2cal}(\nu)$ (dashed line). The value of α_M° has been chosen such that $J_1\left[\alpha_M^{\circ}(\bar{\nu})\right]$ is a maximum (α_M°=1.83 rad) at 1400 cm^{-1} and the value of the α_B corresponds to a half-wave cycle every 100 cm^{-1}

functions $J_1\left[\alpha_M^{\circ}(\bar{\nu})\right]$ and $J_0\left[\alpha_M^{\circ}(\bar{\nu})\right]$ are plotted as a function of $\bar{\nu}$ between 0 and 3000 cm^{-1}. The first maximum of $J_1(x)$ occurs when x equals approximately 1.83 rad (radians). In this plot, the value of V_{PEM} is set to 1/765 such that $\alpha_M^{\circ}(\bar{\nu})$ equals approximately 1.83 at 1400 cm^{-1}. The corresponding plot for $J_0(\bar{\nu}/765)$ is also given where $J_0(x) = 0$ at x equal to approximately 2.40 rad, or 1865 cm^{-1}.

Equation (6.13) describes a family of four curves. Two of these curves with positive numerator and differing signs in the denominator are shown in Figure 6.4. This can be achieved experimentally by rotating the BP by 90°, which changes the sign of the numerator in Equation (6.13), and simultaneously rotating the analyzing polarizer by 90°, which changes the signs of both the numerator and in the denominator of this equation, and thereby only the sign in the denominator changes between the curves. The maximum PEM retardation value, $\alpha_M^{\circ}(\bar{\nu}) = \bar{\nu}/765$, is the same as in Figure 6.3, and the value of C_{BP} is chosen to be $\pi/100$ such that $\alpha_B(\bar{\nu}) = \pi\bar{\nu}/100$ and $\sin[\alpha_B(\bar{\nu})]$ crosses zero at retardation points $n\pi$ every 100 cm^{-1}.

Explicitly, the two curves shown in Figure 6.4 are described by:

$$\left[\frac{I_{AC}(\bar{\nu})}{I_{DC}(\bar{\nu})}\right]_{cal} = \frac{2J_1(\bar{\nu}/765)\sin[\pi\bar{\nu}/100]}{1 \pm J_0(\bar{\nu}/765)\cos[\pi\bar{\nu}/100]} \tag{6.16}$$

It can be seen from both Figure 6.4 and this equation that the retardation value of $\sin[\alpha_B(\bar{\nu})]$ varies through complete 2π cycles every 200 cm^{-1} such that $n\pi$ occurs 12 times across the region displayed. The calibration intensity takes a particularly simple form when $\cos[\alpha_B(\bar{\nu})] = 0$. At these spectral locations, $\alpha_B(\bar{\nu}) = (2n+1)\pi/2$, every 100 cm^{-1} starting at 50 cm^{-1}, the condition $\sin[\alpha_B(\bar{\nu})] = \pm 1$ must also hold, and we therefore have that

$$\left[\frac{I_{AC}(\bar{\nu})}{I_{DC}(\bar{\nu})}\right]_{cal,\cos=0} = \pm 2J_1(\bar{\nu}/765)_{\cos=0} \tag{6.17}$$

where the subscript $\cos = 0$ refers to just the crossing points or a smooth curve drawn through them. In order to draw a smooth curve through all the crossing points in Figure 6.4, one can take the absolute values of the two calibration curves which eliminates all negative values according to

$$\left|\left[\frac{I_{AC}(\bar{\nu})}{I_{DC}(\bar{\nu})}\right]_{cal,\cos=0}\right| = 2J_1(\bar{\nu}/765)_{\cos=0} \tag{6.18}$$

If we plot absolute value of the full frequency dependence

$$\left|\left[\frac{I_{AC}(\bar{\nu})}{I_{DC}(\bar{\nu})}\right]_{cal}\right| = \left|\frac{2J_1(\bar{\nu}/765)\sin[\pi\bar{\nu}/100]}{1 \pm J_0(\bar{\nu}/765)\cos[\pi\bar{\nu}/100]}\right| \tag{6.19}$$

we obtain Figure 6.5 where the crossing points of the two calibration curves are seen to occur every 100 cm^{-1} starting 850 cm^{-1}. These are the same crossing points as in Figure 6.4 where now the two curves cross only with positive CD intensity. The crossing points trace out the function $2J_1\left[\alpha_M^0(\bar{\nu})\right]$ as given by Equation (6.18). As an aside, the crossing points in Figure 6.5 become harder to see toward higher frequencies where the value of $J_0\left[\alpha_M^0(\bar{\nu})\right]$ itself passes through zero at 1865 cm^{-1} and the two curves blend into one another and change places on each side of 1865 cm^{-1}. If a smooth curve is envisioned to pass through the crossing points in Figure 6.5, the maximum of $J_1\left[\alpha_M^0(\bar{\nu})\right]$ near 1400 cm^{-1} can be verified by inspection.

The measured CD spectrum can be calibrated by dividing Equation (6.10) by the positive calibration curve obtained by smoothly connecting the positive values of the crossing points of Equation (6.18) according to:

$$\left[\frac{I_{AC}(\bar{\nu})}{I_{DC}(\bar{\nu})}\right] \Bigg/ \left|\frac{I_{AC}(\bar{\nu})}{I_{DC}(\bar{\nu})}\right|_{cal,\cos=0} = 1.1513\Delta A(\bar{\nu}) \tag{6.20}$$

From this equation, the calibrated CD intensity for any measured spectrum is given by:

$$\Delta A(\bar{\nu}) = \frac{1}{1.1513}\left[\frac{I_{AC}(\bar{\nu})}{I_{DC}(\bar{\nu})}\right] \Bigg/ \left|\frac{I_{AC}(\bar{\nu})}{I_{DC}(\bar{\nu})}\right|_{cal,\cos=0} \tag{6.21}$$

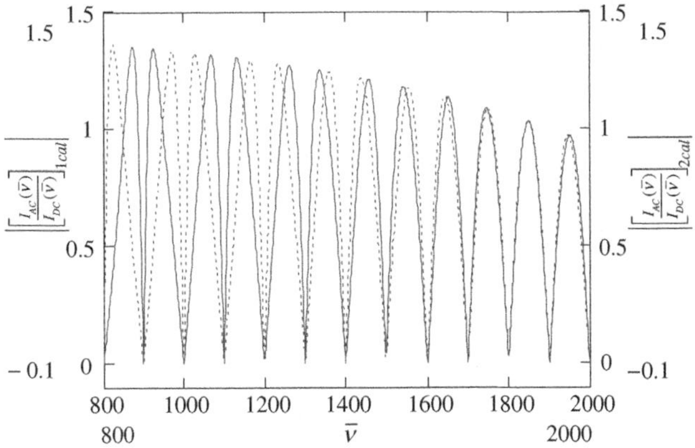

Figure 6.5 *The absolute values of the two calibration curves given by Equation (6.18) where the crossing points occur every 100 cm^{-1} starting at 850 cm^{-1} trace out the function $J_1\left[\alpha_M^0(\bar{\nu})\right]$: $|I_{1cal}(\nu)|$ (solid line), and $|I_{2cal}(\nu)|$ (dashed line)*

6.1.3 Photoelastic Modulator Optimization

The general functional dependence of the Bessel functions $J_1(x)$ and $J_0(x)$ are given in Figure 6.6. When applied to a description of a PEM, as was done in the previous section, the value of x is taken to be the maximum PEM retardation angle in radians, $\alpha_M^0(\bar{\nu})$, at a particular wavenumber frequency, $\bar{\nu}$. As mentioned above, the value of $\alpha_M^0(\bar{\nu})$ that corresponds to the first maximum $J_1\left[\alpha_M^0(\bar{\nu})\right]$ is approximately 1.83 rad as can be seen from Figure 6.6. This corresponds to a retardation angle of 105°, which in turn corresponds to the best maximum retardation angle for sine-wave modulation of the PEM. The most efficient modulation cycle is a square wave between perfect RCP and LCP radiation at retardation angles ±90°. By over modulating to maxima at ±105° with sine-wave modulation cycles, the function $J_1\left[\alpha_M^0(\bar{\nu})\right]$ stays in the vicinity of ±90° retardation for the longest average amount of time. The efficiency of square-wave modulation from Equation (6.1) is $\sin(+90°) = 1$ minus $\sin(-90°) = -1$, which is equal to 2. The corresponding efficiency for sine-wave modulation is obtained from the corresponding factor $2J_1\left[\alpha_M^0(\bar{\nu})\right]$ in Equation (6.5). From Figure 6.6, this factor is approximately 1.2 at 105° compared with 2 from square modulation. Square-wave modulation is not available for high-frequency polarization modulation, particularly in the infrared region, so we must use sine-wave oscillating PEMs.

The maximum value of the function $J_1\left[\alpha_M^0(\bar{\nu})\right]$ can be adjusted to any point in the IR region above say 800 cm^{-1} by varying the function V_{PEM}. As seen in Figure 6.3, the dependence of $J_1\left[\alpha_M^0(\bar{\nu})\right]$ on $\bar{\nu}$ is mild near the maximum value of $\bar{\nu}$. For example, for a maximum value of approximately 0.6 at 1400 cm^{-1}, $J_1\left[\alpha_M^0(\bar{\nu})\right]$ remains above 0.5 from 1000 to 2000 cm^{-1}. Some advantage is gained by maximizing $J_1\left[\alpha_M^0(\bar{\nu})\right]$ in the middle of the spectral range of interest, but adjustment of the PEM optimum voltage is not critical for mid-IR spectral measurements if the PEM maximum is set to 1400 cm^{-1} as shown in Figure 6.3.

Other points of interest in the general plots of $J_1(x)$ and $J_0(x)$ in Figure 6.6 are the value of $J_0(x) = 0$ at $x = 2.40$ and $J_1(x) = 0$ at $x = 3.83$. The former point is especially important when optimizing the PEM settings for dual PEM operation, to be described later in the chapter, and the zero crossing of $J_1(x)$ is useful for easily locating the maximum of $J_1(x)$. This zero crossing of $J_1(x)$ can be seen in the quarter-wave calibration plots carried out further in the spectrum as shown below. The maximum of $J_1(x)$ occurs at a wavenumber frequency that is a factor of (1.83/3.83) times the location of inversion

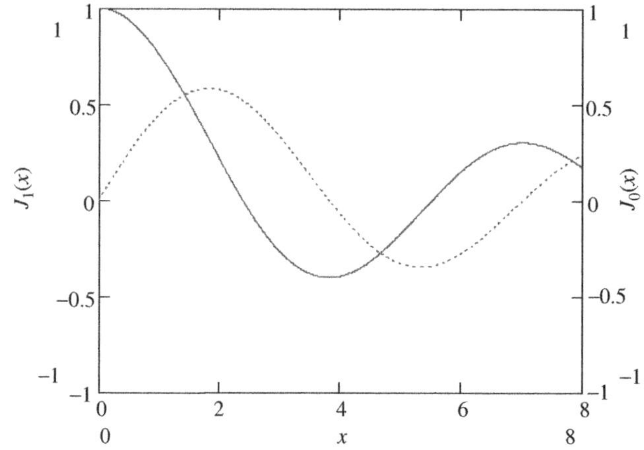

Figure 6.6 Dependence of Bessel function $J_1(x)$ (solid line) and $J_0(x)$ (dashed line) as a function of the general variable x

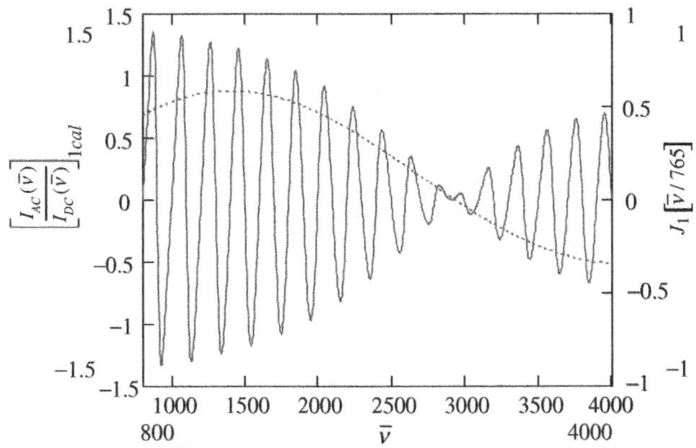

Figure 6.7 *Plot of one of the calibration curves in Equation (6.13) (solid line) and $J_1\left[\alpha_M^0(\bar{\nu})\right]$ (dashed line) as a function of wavenumber frequency showing the inversion point in the calibration curve when $J_1\left[\alpha_M^0(\bar{\nu})\right]$ passes through zero*

point. In Figure 6.7, one of the calibration curves from Figure 6.3 is plotted along with $J_1\left[\alpha_M^0(\bar{\nu})\right]$. The inversion point of the calibration curves occurs at 2930 cm^{-1} where $J_1\left[\alpha_M^0(\bar{\nu})\right]$ passes through zero and where the maximum of $J_1\left[\alpha_M^0(\bar{\nu})\right]$ is still located at 1400 cm^{-1}. Beyond the inversion point, for any CD intensities measured with this PEM setting, for example hydrogen stretching modes, the signs will appear opposite until a proper calibration correction is carried out. In actual practice, if VCD spectra were desired in the hydrogen stretching region, the PEM setting should be changed such that the maximum of $J_1\left[\alpha_M^0(\bar{\nu})\right]$ is set near to 3000 cm^{-1}.

6.2 Stokes–Mueller Optical Analysis

In the first section of this chapter, we derived CD intensity expressions for the measurement and calibration of CD spectra. In this section, the elements of the Stokes–Mueller analysis are presented as a means of easily determining these same intensity expressions. The power of this analysis is that it facilitates the description of more complex optical arrangements including the description of birefringence in the optical components, more than one PEM in the beam, or the measurement of other polarization modulation spectra such as linear dichroism (LD) or circular birefringence (CB), where the latter is also known as optical rotation or optical rotatory dispersion (ORD), described in Chapter 3.

6.2.1 Basic Stokes–Mueller Formalism

The Stokes–Mueller formalism involves the definition of Stokes vectors and Mueller matrices. A Stokes vector is a four-dimensional vector that completely represents the polarization state of any radiation beam. It is defined by the follow expression:

$$\boldsymbol{S} = \begin{pmatrix} S_0 \\ S_1 \\ S_2 \\ S_3 \end{pmatrix} = \begin{pmatrix} I_{Total} \\ I_0 - I_{90} \\ I_{45} - I_{135} \\ I_R - I_L \end{pmatrix} \tag{6.22}$$

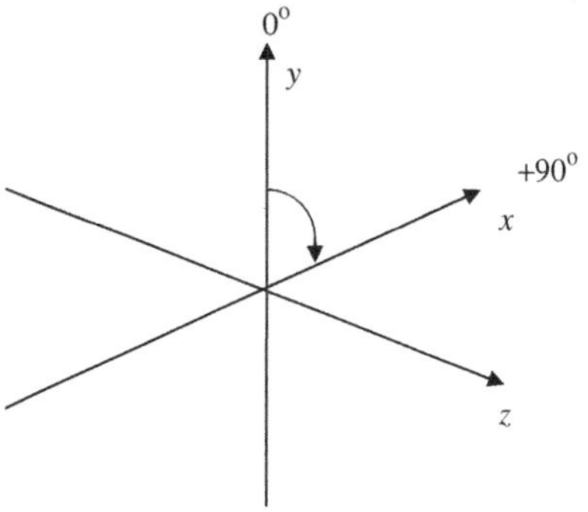

Figure 6.8 Right-handed Cartesian coordinate system used to describe the polarization state of radiation using Stokes vectors. Radiation propagates in the positive z-direction (arrow) from the source to the observer with polarization states specified in the xy-plane

The first element of the Stokes vector is the total intensity of the beam. This intensity is the sum of both the polarized and unpolarized intensities. The following inequality holds $S_0^2 \geq S_1^2 + S_2^2 + S_3^2$ for the squares of the intensities defined in any Stokes vector. The equality holds if there is no unpolarized intensity present, in which case the second, third, and fourth Stokes vector components describe how the polarized light is divided among two orthogonal linear polarization differences and the circular polarization difference. In our analysis, we usually begin with light in a single pure polarization state with no unpolarized light present.

The coordinate system used to describe a Stokes vector is illustrated in Figure 6.8.

Radiation propagates along the z-axis from source to observer in the direction of the arrow. The vertical axis is taken to be the positive y-axis at 0°, and a clockwise rotation of this axis through $+90°$, when viewed by an observer back toward the source of radiation, is the positive x-axis, as illustrated in Figure 6.8. Right circularly polarized (RCP) radiation corresponds to the rotation of the polarization state of light in a clockwise sense in time as the radiation passes through an xy-plane between the viewer and the radiation source.

Some simple examples will illustrate the use of Stokes vectors. Linear polarization states that are polarized at 0°, 90°, 45°, and 135° (equal to $-45°$) are described by:

$$\boldsymbol{S}_0 = \begin{pmatrix} 1 \\ 1 \\ 0 \\ 0 \end{pmatrix} \quad \boldsymbol{S}_{90} = \begin{pmatrix} 1 \\ -1 \\ 0 \\ 0 \end{pmatrix} \quad \boldsymbol{S}_{45} = \begin{pmatrix} 1 \\ 0 \\ 1 \\ 0 \end{pmatrix} \quad \boldsymbol{S}_{135} = \begin{pmatrix} 1 \\ 0 \\ -1 \\ 0 \end{pmatrix} \tag{6.23}$$

Similarly RCP and LCP states are described as:

$$\boldsymbol{S}_R = \begin{pmatrix} 1 \\ 0 \\ 0 \\ 1 \end{pmatrix} \quad \boldsymbol{S}_L = \begin{pmatrix} 1 \\ 0 \\ 0 \\ -1 \end{pmatrix} \tag{6.24}$$

A Mueller matrix, $\boldsymbol{M}$, is a 4×4 matrix that transforms the ith Stokes vector, $\boldsymbol{S}_i$, into the jth Stokes vector, $\boldsymbol{S}_j$. The matrix describes the effect of an optical element in the beam on the polarization state of the initial Stokes vector, and this effect results in the Stokes vector of the beam after the optical element. This transformation of matrix multiplication is represented by:

$$\boldsymbol{S}_j = \boldsymbol{M} \cdot \boldsymbol{S}_i = \begin{pmatrix} S_{j,0} \\ S_{j,1} \\ S_{j,2} \\ S_{j,3} \end{pmatrix} = \begin{pmatrix} M_{00} & M_{01} & M_{02} & M_{03} \\ M_{10} & M_{11} & M_{12} & M_{13} \\ M_{20} & M_{21} & M_{22} & M_{23} \\ M_{30} & M_{31} & M_{32} & M_{33} \end{pmatrix} \begin{pmatrix} S_{i,0} \\ S_{i,1} \\ S_{i,2} \\ S_{i,3} \end{pmatrix} \tag{6.25}$$

The different elements of the Mueller matrix can be described by the elements of the two Stokes vectors that they connect. For example, M_{00} represents the effect of the optical element on the total intensity of the new Stokes vector by the total intensity of the original Stokes vector, and M_{31} describes the effect of the optical element on the vertical/horizontal linear polarization preference of the initial polarization state, $S_{i,1}$, that affects the RCP/LCP polarization-state bias in the resulting Stokes vector, $S_{j,3}$. More specifically M_{31} describes the action of linear birefringence in the optical element that causes phases shifts between orthogonal linear polarization states of the incident light that changes the circular polarization content of the emerging beam.

The Mueller matrices of some basic optical elements are given by the following expressions. A linear polarizer with its polarization axis set at an angle θ with respect to the vertical, shown in Figure 6.8, is given by:

$$\boldsymbol{M}_P(\theta) = \frac{1}{2} \begin{pmatrix} 1 & \cos 2\theta & \sin 2\theta & 0 \\ \cos 2\theta & \cos^2 2\theta & \sin 2\theta \cos 2\theta & 0 \\ \sin 2\theta & \sin 2\theta \cos 2\theta & \sin^2 2\theta & 0 \\ 0 & 0 & 0 & 0 \end{pmatrix} \tag{6.26}$$

For example, it is straightforward to show that vertically polarized light, $\boldsymbol{S}_0$, incident on a polarizer with polarization axis set at 45°, $\boldsymbol{M}_P(45°)$, produces linearly polarized light at 45°, $\boldsymbol{S}_{45}$, namely

$$\boldsymbol{S}_f = \boldsymbol{M}_P \boldsymbol{S}_i = \boldsymbol{M}_P(45°)\boldsymbol{S}_0 = \frac{1}{2} \begin{pmatrix} 1 & 0 & 1 & 0 \\ 0 & 0 & 0 & 0 \\ 1 & 0 & 1 & 0 \\ 0 & 0 & 0 & 0 \end{pmatrix} \begin{pmatrix} 1 \\ 1 \\ 0 \\ 0 \end{pmatrix} = \frac{1}{2} \begin{pmatrix} 1 \\ 0 \\ 1 \\ 0 \end{pmatrix} = \frac{1}{2} \boldsymbol{S}_{45} \tag{6.27}$$

The factor of $^1/_2$ arises from the polarizer. Vertically polarized light can be expressed as the sum of two in-phase orthogonally polarized components at ±45°,

$$\boldsymbol{S}_0 = \frac{1}{2}(\boldsymbol{S}_{45} + \boldsymbol{S}_{-45}) \tag{6.28}$$

and the polarizer only transmits the +45° component while rejecting the −45° component. As shown in Equations (6.25) and (6.27), the Mueller matrix of the polarizer operates on the initial Stokes vector from the left to produce the resulting new Stokes vector.

The Mueller matrix of a general linear birefringent (LB) retardation plate is given by:

$$\boldsymbol{M}_{LB}(\theta,\delta) = \begin{pmatrix} 1 & 0 & 0 & 0 \\ 0 & \cos^2(\delta/2) + \cos 4\theta \sin^2(\delta/2) & \sin 4\theta \sin^2(\delta/2) & -\sin 2\theta \sin\delta \\ 0 & \sin 4\theta \sin^2(\delta/2) & \cos^2(\delta/2) - \cos 4\theta \sin^2(\delta/2) & \cos 2\theta \sin\delta \\ 0 & \sin 2\theta \sin\delta & -\cos 2\theta \sin\delta & \cos\delta \end{pmatrix} \tag{6.29}$$

where θ is the angle of the slow axis of the plate from the vertical, and δ is the retardation of the slow axis relative to the fast axis. We can simplify this expression by considering a general birefringent plate with slow and fast axes at $+45°$and $-45°$ from the vertical, respectively. Thus for $\theta = +45° = \pi/4$ and, after applying the trigonometric identities $\cos^2(\delta/2) - \sin^2(\delta/2) = \cos\delta$ and $\cos^2(\delta/2) + \sin^2(\delta/2) = 1$, we obtain

$$M_{LB}(45°,\delta) = \begin{pmatrix} 1 & 0 & 0 & 0 \\ 0 & \cos\delta & 0 & -\sin\delta \\ 0 & 0 & 1 & 0 \\ 0 & \sin\delta & 0 & \cos\delta \end{pmatrix} \tag{6.30}$$

A quarter-wave plate (QWP) retards or advances the phase of the electric field of the radiation by one quarter of a wave ($90° = \pi/2$) upon passage by the radiation through the plate corresponding to $\delta = \pi/2$. The Mueller matrix for a QWP with its slow axis at 45° is therefore given by:

$$M_{LB}(45°, \pi/2) = \begin{pmatrix} 1 & 0 & 0 & 0 \\ 0 & 0 & 0 & -1 \\ 0 & 0 & 1 & 0 \\ 0 & 1 & 0 & 0 \end{pmatrix} \tag{6.31}$$

With these definitions, it is straightforward to demonstrate that a QWP with its slow axis at 45° from vertical transforms vertically polarized radiation to RCP radiation as follows,

$$M_{LB}(45°, \pi/2)S_0 = \begin{pmatrix} 1 & 0 & 0 & 0 \\ 0 & 0 & 0 & -1 \\ 0 & 0 & 1 & 0 \\ 0 & 1 & 0 & 0 \end{pmatrix} \begin{pmatrix} 1 \\ 1 \\ 0 \\ 0 \end{pmatrix} = \begin{pmatrix} 1 \\ 0 \\ 0 \\ 1 \end{pmatrix} = S_R \tag{6.32}$$

This can be visualized by noting that vertically polarized (VP) light has equal in-phase projections onto the fast and slow axes of the QWP as it enters the plate as shown in Figure 6.9. Passage of the beam though the plate then advances the wave projected onto the fast axis by one quarter of a wavelength relative to the wave traveling (more slowly) along the slow axis. After passage through the plate, a recombination of the two waves produces RCP radiation that circulates clockwise through a plane located between the QWP and the observer as the right-hand helical wave form approaches the

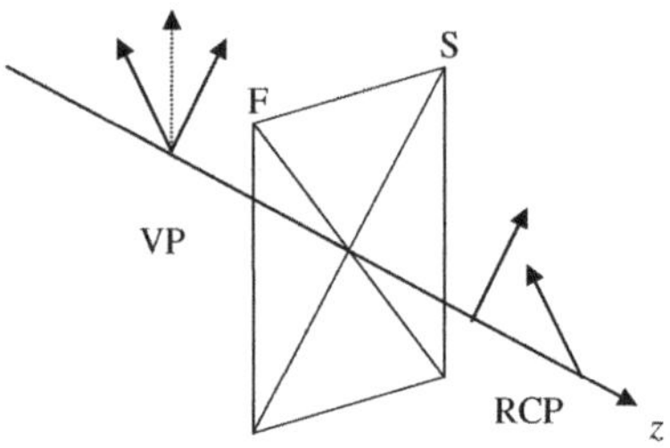

Figure 6.9 *Illustration of the action of a QWP with a slow axis at $+45°$ from vertical on vertical linearly polarized radiation. VP radiation is first resolved into two in-phase linearly polarized components along the fast and slow axes of the QWP. After the plate, the $+45°$ component is retarded by one quarter of a wave relative to the $-45°$ component, yielding RCP radiation*

observer along the positive z-axis. The fast axis component of the radiation arrives at the plane first followed by the slow axis component, and this corresponds to clockwise circulation of the electric field vector of the radiation.

If we drop the restriction of the orientation of the slow and fast axes, an arbitrarily oriented QWP is obtained by setting $\delta = \pi/2$ in Equation (6.29) and applying the half-angle trigonometric identities $(1+\cos 4\theta)/2 = \cos^2 2\theta$, $(1-\cos 4\theta)/2 = \sin^2 2\theta$ and $\sin 4\theta = 2\sin 2\theta \cos 2\theta$, giving

$$\boldsymbol{M}_{QWP}(\theta, \pi/2) = \begin{pmatrix} 1 & 0 & 0 & 0 \\ 0 & \cos^2 2\theta & \sin 2\theta \cos 2\theta & -\sin 2\theta \\ 0 & \sin 2\theta \cos 2\theta & \sin^2 2\theta & \cos 2\theta \\ 0 & \sin 2\theta & -\cos 2\theta & 0 \end{pmatrix} \tag{6.33}$$

Another useful limit is a half-wave plate (HWP) ($\delta = \pi$) with its slow axis oriented at $+45°$. This can be obtained from Equations (6.29) or (6.30) and is:

$$\boldsymbol{M}_{HWP}(45°, \pi) = \begin{pmatrix} 1 & 0 & 0 & 0 \\ 0 & -1 & 0 & 0 \\ 0 & 0 & 1 & 0 \\ 0 & 0 & 0 & -1 \end{pmatrix} \tag{6.34}$$

This plate transforms vertically polarized radiation into horizontally polarized radiation, or vice versa, by retarding radiation projected along the slow axes by half a wavelength relative to the wave projected along the fast axis. It has no effect on incident linearly polarized light oriented at $\pm 45°$, but interconverts LCP and RCP polarization states. If the restriction on the orientation of the HWP is eliminated, the expression for an HWP at any angle is obtained from Equation (6.29) as:

$$\boldsymbol{M}_{HWP}(\theta, \pi) = \begin{pmatrix} 1 & 0 & 0 & 0 \\ 0 & \cos 4\theta & \sin 4\theta & 0 \\ 0 & \sin 4\theta & -\cos 4\theta & 0 \\ 0 & 0 & 0 & -1 \end{pmatrix} \tag{6.35}$$

The general HWP interconverts LCP and RCP at any orientation angle and rotates the plane of LP states at a rate twice the angle change of the HWP, but because LP light oriented at 180° is the same as that at 0°, the LP state is averaged for all angles four times for every rotation of the HWP by 360°.

One additional Mueller matrix associated with linear birefringence is the case of small birefringence. The matrix is used to describe optical imperfections in lenses and windows. It can be derived from the general expression for a birefringent plate in Equation (6.29) by the using the expansions $\sin\delta = \delta - \delta^3/3! \ldots$ and $\cos\delta = 1 - \delta^2/2 + \ldots$ and keeping only terms to first order in δ for small values of LB. This yields

$$\boldsymbol{M}_{LB}(\theta, \delta) = \begin{pmatrix} 1 & 0 & 0 & 0 \\ 0 & 1 & 0 & -\delta \sin 2\theta \\ 0 & 0 & 1 & \delta \cos 2\theta \\ 0 & \delta \sin 2\theta & -\delta \cos 2\theta & 1 \end{pmatrix} \tag{6.36}$$

The $\delta \sin 2\theta$ terms represent small linear birefringence with slow and fast axes at $+45°$ and $-45°$, respectively, whereas the $\delta \cos 2\theta$ terms represent small LB with slow and fast axes at 0° and $+90°$, respectively. This matrix can be simplified in appearance by writing $\delta = \delta \sin 2\theta$ and $\delta' = \delta \cos 2\theta$ where δ and δ' now represent the sum of the two types of LB ($\pm 45°$ and $0°/90°$) that might occur as different types of optical imperfections in various locations with different intensities and orientations across an optical element. This simple but important Mueller matrix is given by:

$$\boldsymbol{M}_{LB}(\delta,\delta') = \begin{pmatrix} 1 & 0 & 0 & 0 \\ 0 & 1 & 0 & -\delta \\ 0 & 0 & 1 & \delta' \\ 0 & \delta & -\delta' & 1 \end{pmatrix} \tag{6.37}$$

The general Mueller matrix of sample through first order can be represented by:

$$\boldsymbol{M}_S = 10^{-A}\begin{pmatrix} 1 & -LD & -LD' & CD \\ -LD & 1 & CB & -LB \\ -LD' & -CB & 1 & LB' \\ CD & LB & -LB' & 1 \end{pmatrix} \tag{6.38}$$

Here A is the decadic absorbance of the sample, LD is the vertical–horizontal linear dichroism, $(\ln 10/2)(A_0 - A_{90})$, LD' is linear dichroism at 45° from vertical–horizontal, $(\ln 10/2)(A_{45} - A_{135})$, CD is the circular dichroism, $(\ln 10/2)(A_L - A_R)$, and $(\ln 10/2) = 1.1513$. The linear birefringence entries, LB and LB', and the circular birefringence, CB, are the corresponding birefringence differences, $n_0 - n_{90}$, $n_{45} - n_{135}$, and $n_L - n_R$, respectively, where n is the real part of the index of refraction of the sample.

Finally, we present the effective Mueller matrix of a detector that has a different response, p_x and p_y, to radiation polarized along its local x- and y-axes, respectively, and where the these local axes are oriented at an angle α with respect to the laboratory x-axis defined in Figure 6.8.

$$\boldsymbol{D}(\alpha) = \begin{bmatrix} 1 & \left(p_x^2 - p_y^2\right)\cos 2\alpha & \left(p_x^2 - p_y^2\right)\sin 2\alpha & 0 \end{bmatrix} \tag{6.39}$$

Normally, a Mueller matrix is a 4×4 matrix, but in the case of the detector, only the total intensity of the final Stokes vector, S_{f0}, is measured. The first entry, M_{00} in Equation (6.39) is 1 for simplicity. It could also be represented by $\left(p_x^2 + p_y^2\right)$, but we take the sum of the x and y detector responses to be normalized and the detector response to be nearly the same, but not exactly the same, in the x and y directions. In the case of the detector, only a 4×1 top row of the matrix is needed to covert a final Stokes vector $\boldsymbol{S}_f$ in the optical train into a scalar intensity measured by detector as:

$$\begin{aligned} I_D = \mathrm{D}(\alpha)\cdot \mathrm{S}_f &= \begin{pmatrix} 1 & \left(p_x^2 - p_y^2\right)\cos 2\alpha & \left(p_x^2 - p_y^2\right)\sin 2\alpha & 0 \end{pmatrix}\begin{pmatrix} S_{f0} \\ S_{f1} \\ S_{f2} \\ S_{f3} \end{pmatrix} \\ &= S_{f0} + S_{f1}\left(p_x^2 - p_y^2\right)\cos 2\alpha + S_{f2}\left(p_x^2 - p_y^2\right)\sin 2\alpha \end{aligned} \tag{6.40}$$

Using the basic Stokes vectors and Mueller matrices presented in this section, it is possible to determine any of the polarization states and the final detector intensity along the complete optical pathway between the source and the detector. In the following sections, three examples are presented.

6.2.2 Stokes–Mueller Derivation of Circular Dichroism Measurement

We now use this formalism to re-derive, more directly, the expression of the measurement of CD intensity in terms of intensities $I_{AC}(\bar{\nu})$ and $I_{DC}(\bar{\nu})$ as in Equation (6.10). The intensity at the detector is given by the following expression for a measurement of CD corresponding to the optical elements in Figure 6.1

$$I_D(\bar{\nu}) = \boldsymbol{D}(\alpha) \cdot \boldsymbol{M}_S(\bar{\nu}) \cdot \boldsymbol{M}_{PEM}(\bar{\nu}) \cdot \boldsymbol{M}_P(0^\circ) \cdot \boldsymbol{S}_0(\bar{\nu}) \tag{6.41}$$

The initially unpolarized beam from the IR spectrometer with spectral distribution $I_0(\bar{\nu})$ passes through a polarizer at an angle of 0° resulting in vertically polarized light with half the original unpolarized intensity,

$$\boldsymbol{S}_1(\bar{\nu}) = \boldsymbol{M}_P(0^\circ) \cdot \boldsymbol{S}_0(\bar{\nu}) = \frac{1}{2}\begin{pmatrix} 1 & 1 & 0 & 0 \\ 1 & 1 & 0 & 0 \\ 0 & 0 & 0 & 0 \\ 0 & 0 & 0 & 0 \end{pmatrix} I_0(\bar{\nu}) \begin{pmatrix} 1 \\ 0 \\ 0 \\ 0 \end{pmatrix} = \frac{I_0(\bar{\nu})}{2} \begin{pmatrix} 1 \\ 1 \\ 0 \\ 0 \end{pmatrix} \tag{6.42}$$

This beam then passes through the PEM with stress axes as 45°, obtained from Equation (6.30), resulting in the Stokes vector, $\boldsymbol{S}_2(\bar{\nu})$,

$$\begin{aligned} \boldsymbol{S}_2(\bar{\nu}) &= \boldsymbol{M}_{PEM}[45^\circ, \alpha_M(\bar{\nu})] \cdot \boldsymbol{S}_1(\bar{\nu}) \\ &= \begin{pmatrix} 1 & 0 & 0 & 0 \\ 0 & \cos\alpha_M(\bar{\nu}) & 0 & -\sin\alpha_M(\bar{\nu}) \\ 0 & 0 & 1 & 0 \\ 0 & \sin\alpha_M(\bar{\nu}) & 0 & \cos\alpha_M(\bar{\nu}) \end{pmatrix} \frac{I_0(\bar{\nu})}{2} \begin{pmatrix} 1 \\ 1 \\ 0 \\ 0 \end{pmatrix} = \frac{I_0(\bar{\nu})}{2} \begin{pmatrix} 1 \\ \cos\alpha_M(\bar{\nu}) \\ 0 \\ \sin\alpha_M(\bar{\nu}) \end{pmatrix} \end{aligned} \tag{6.43}$$

Finally, the PEM-modulated beam passes through the sample, from Equation (6.38), resulting in the last Stoke vector before the detector,

$$\begin{aligned} \boldsymbol{S}_3(\bar{\nu}) &= \boldsymbol{M}_S(\bar{\nu}) \cdot \boldsymbol{S}_2(\bar{\nu}) \\ &= 10^{-A(\bar{\nu})} \begin{pmatrix} 1 & 0 & 0 & CD(\bar{\nu}) \\ 0 & 1 & CB(\bar{\nu}) & -LB(\bar{\nu}) \\ 0 & -CB(\bar{\nu}) & 1 & LB'(\bar{\nu}) \\ CD(\bar{\nu}) & LB(\bar{\nu}) & -LB'(\bar{\nu}) & 1 \end{pmatrix} \frac{I_0(\bar{\nu})}{2} \begin{pmatrix} 1 \\ \cos\alpha_M(\bar{\nu}) \\ 0 \\ \sin\alpha_M(\bar{\nu}) \end{pmatrix} \\ &= \frac{I_{DC}(\bar{\nu})}{2} \begin{pmatrix} 1 + CD(\bar{\nu})\sin\alpha_M(\bar{\nu}) \\ \cos\alpha_M(\bar{\nu}) - LB(\bar{\nu})\sin\alpha_M(\bar{\nu}) \\ -CB(\bar{\nu})\cos\alpha_M(\bar{\nu}) + LB'(\bar{\nu})\sin\alpha_M(\bar{\nu}) \\ CD(\bar{\nu}) + LB(\bar{\nu})\cos\alpha_M(\bar{\nu}) + \sin\alpha_M(\bar{\nu}) \end{pmatrix} \end{aligned} \tag{6.44}$$

For simplicity, we have assumed the sample to be a liquid, solution, or disordered solid, thus eliminating all linear dichroism (*LD*) terms from Equation (6.38). The *LB* terms have been retained to represent linear birefringence in the transparent sample cell windows. Finally, this last Stokes vector is converted into a scalar intensity by the Mueller matrix row vector of the detector as discussed above.

$$\begin{aligned} I_D(\bar{\nu}) &= \boldsymbol{D}(\alpha)\cdot \boldsymbol{S}_3(\bar{\nu}) \\ &= \frac{I_{DC}(\bar{\nu})}{2}\Big\{[1+CD(\bar{\nu})\sin\alpha_M(\bar{\nu})] + (p_X^2-p_Y^2)\cos 2\alpha[\cos\alpha_M(\bar{\nu}) - LB(\bar{\nu})\sin\alpha_M(\bar{\nu})] \\ &\quad + (p_X^2-p_Y^2)\sin 2\alpha[-CB(\bar{\nu})\cos\alpha_M(\bar{\nu}) + LB'(\bar{\nu})\sin\alpha_M(\bar{\nu})]\Big\} \end{aligned} \tag{6.45}$$

If the detector has no linear polarization sensitivity, $p_x^2 - p_y^2 = 0$, only the first of the three detector terms is non-zero, and we can write

$$I_D(\bar{\nu}) = I_{DC}(\bar{\nu}) + I_{AC}(\bar{\nu}) = \frac{I_{DC}(\bar{\nu})}{2}[1 + CD(\bar{\nu})\sin\alpha_M(\bar{\nu})] \tag{6.46}$$

Using CD equal to $(1/2)\ln 10[A_L(\bar{\nu}) - A_R(\bar{\nu})] = 1.1513\Delta A(\bar{\nu})$ and the expression for the lowest order term in the expansion of $\sin\alpha_M(\bar{\nu})$, namely $2J_1\left[\alpha_M^o(\bar{\nu})\right]\sin\omega_M t$ from Equation (6.4), the ratio of the expressions for $I_{AC}(\bar{\nu})$ and $I_{DC}(\bar{\nu})$is given by:

$$\frac{I_{AC}(\bar{\nu})}{I_{DC}(\bar{\nu})} = 2J_1\left[\alpha_M^o(\bar{\nu})\right][1.1513\Delta A(\bar{\nu})]\sin\omega_M t \tag{6.47}$$

This is the same expression as given in Equation (6.10) after the lock-in amplifier measures the amplitude of the sine-wave PEM modulation signal, $\sin\omega_M t$.

6.2.3 Stokes–Mueller Derivation of the CD Calibration

If instead of the Mueller matrix of the sample, the matrix for a combination of a linear birefringent plate at ±45°, given in Equation (6.30), followed by a vertical (upper sign) or horizontal (lower sign) polarizer (analyzer), is used, namely,

$$\begin{aligned} \boldsymbol{M}_{cal}(\bar{\nu}) &= \boldsymbol{M}_A(0°/90°)\boldsymbol{M}_{LB}(\pm 45°, \delta_B(\bar{\nu})) \\ &= \frac{1}{2}\begin{pmatrix} 1 & \pm 1 & 0 & 0 \\ \pm 1 & 1 & 0 & 0 \\ 0 & 0 & 0 & 0 \\ 0 & 0 & 0 & 0 \end{pmatrix}\begin{pmatrix} 1 & 0 & 0 & 0 \\ 0 & \cos\delta_B(\bar{\nu}) & 0 & \mp\sin\delta_B(\bar{\nu}) \\ 0 & 0 & 1 & 0 \\ 0 & \pm\sin\delta_B(\bar{\nu}) & 0 & \cos\delta_B(\bar{\nu}) \end{pmatrix} \\ &= \frac{1}{2}\begin{pmatrix} 1 & \pm\cos\delta_B(\bar{\nu}) & 0 & \pm\mp\sin\delta_B(\bar{\nu}) \\ \pm 1 & \cos\delta_B(\bar{\nu}) & 0 & \mp\sin\delta_B(\bar{\nu}) \\ 0 & 0 & 0 & 0 \\ 0 & 0 & 0 & 0 \end{pmatrix} \end{aligned} \tag{6.48}$$

The optical setup for the CD calibration intensity described by Equation (6.13) can be derived from the following Stokes–Mueller intensity expression at the detector,

$$\begin{aligned} I_D(\bar{\nu}) &= \boldsymbol{D}(\alpha)\cdot \boldsymbol{M}_{cal}(\bar{\nu})\cdot \boldsymbol{M}_{PEM}(\bar{\nu})\cdot \boldsymbol{M}_{\mathrm{p}}(0^0)\cdot \boldsymbol{S}_0(\bar{\nu}) \\ &= \frac{I_{DC}(\bar{\nu})}{4}[1 \pm \cos\alpha_B(\bar{\nu})\cos\alpha_M(\bar{\nu}) \pm \mp \sin\alpha_B(\bar{\nu})\sin\alpha_M(\bar{\nu})] \end{aligned} \quad (6.49)$$

For the product of sign choices, $\pm\mp$, $\pm$ corresponds to the polarizer choice indicated above, while $\mp$ refers to the wave plate angle setting. We have previously evaluated the PEM expression for $\sin[\alpha_M(\bar{\nu})]$ in Equation (6.4), and the corresponding expression for $\cos[\alpha_M(\bar{\nu})]$ is given by the expansion

$$\begin{aligned} \cos[\alpha_M(\bar{\nu})] &= \cos\left[\alpha_M^{\mathrm{o}}(\bar{\nu})\sin\omega_M t\right] \\ &= J_0\left[\alpha_M^{\mathrm{o}}(\bar{\nu})\right] + \sum_{n=even} 2J_n\left[\alpha_M^{\mathrm{o}}(\bar{\nu})\right]\cos n\omega_M t \end{aligned} \quad (6.50)$$

Keeping terms through twice the PEM frequency, these two PEM functions are given by:

$$\sin[\alpha_M(\bar{\nu})] = 2J_1\left[\alpha_M^{\mathrm{o}}(\bar{\nu})\right]\sin\omega_M t \quad (6.51)$$

$$\cos[\alpha_M(\bar{\nu})] = J_0\left[\alpha_M^{\mathrm{o}}(\bar{\nu})\right] + 2J_2\left[\alpha_M^{\mathrm{o}}(\bar{\nu})\right]\cos 2\,\omega_M t \quad (6.52)$$

CD intensity arises from $\sin[\alpha_M(\bar{\nu})]$ at the frequency ω_M, whereas $\cos[\alpha_M(\bar{\nu})]$ contributes a constant DC-term and a $2\omega_M$ term that provides LD intensities. Through first order in the PEM frequency, the CD calibration intensity given in Equation (6.49) becomes

$$I_D(\bar{\nu}) = \frac{I_0(\bar{\nu})}{4}[1 \pm J_0(\bar{\nu})\cos\alpha_B(\bar{\nu}) \pm \mp 2J_1(\bar{\nu})\sin\alpha_B(\bar{\nu})] \quad (6.53)$$

Identifying the first two terms as DC terms and the last term as the AC term, the measured CD calibration intensity is given by:

$$\left[\frac{I_{AC}(\bar{\nu})}{I_{DC}(\bar{\nu})}\right]_{cal,\omega_M} = \frac{\pm\mp 2J_1\left[\alpha_M^{\mathrm{o}}(\bar{\nu})\right]\sin[\alpha_B(\bar{\nu})]}{1 \pm J_0[\alpha_M^{\mathrm{o}}(\bar{\nu})]\cos[\alpha_B(\bar{\nu})]} \quad (6.54)$$

This is the same as Equation (6.13) but with the inclusion of an additional sign choice to indicate the two ways in which the sign of $I_{AC}(\bar{\nu})$ can be changed, whereas the expression for $I_{DC}(\bar{\nu})$ is sensitive only to whether the polarizer and analyzer are parallel (both vertical, upper sign) or perpendicular (lower sign).

6.2.4 Measurement of Circular Birefringence

This same formalism can now be used to develop the expression for the measurement of the CB spectrum, also known as the optical rotatory dispersion (ORD) spectrum, which is the Kramers–Kronig transform of the corresponding CD spectrum (Lombardi and Nafie, 2009). CB intensity can be seen in the Stokes vector representation of the beam emerging from the sample, $S_3(\bar{\nu})$, in

Equation (6.44). To measure CB intensity, a polarization analyzer at 45° is inserted into the beam after the sample, represented by $M_A(45°)$, such that the detector intensity is given by:

$$I_D(\bar{\nu}) = D(\alpha) \cdot M_A(45°)M_S(\bar{\nu}) \cdot M_{PEM}(\bar{\nu}) \cdot M_P(0°) \cdot S_0(\bar{\nu}) = D(\alpha) \cdot M_A(45°)S_3(\bar{\nu})$$

$$= D(\alpha) \cdot \frac{1}{2}\begin{pmatrix} 1 & 0 & 1 & 0 \\ 0 & 0 & 0 & 0 \\ 1 & 0 & 1 & 0 \\ 0 & 0 & 0 & 0 \end{pmatrix} \frac{I_{DC}(\bar{\nu})}{2}\begin{pmatrix} 1 + CD(\bar{\nu})\sin\alpha_M(\bar{\nu}) \\ \cos\alpha_M(\bar{\nu}) - LB(\bar{\nu})\sin\alpha_M(\bar{\nu}) \\ -CB(\bar{\nu})\cos\alpha_M(\bar{\nu}) + LB'(\bar{\nu})\sin\alpha_M(\bar{\nu}) \\ CD(\bar{\nu}) + LB(\bar{\nu})\cos\alpha_M(\bar{\nu}) + \sin\alpha_M(\bar{\nu}) \end{pmatrix} \tag{6.55}$$

Evaluating this expressions yields,

$$I_D(\bar{\nu}) = \frac{I_0(\bar{\nu})}{4}[1 + CD(\bar{\nu})\sin\alpha_M(\bar{\nu}) - CB(\bar{\nu})\cos\alpha_M(\bar{\nu}) + LB'(\bar{\nu})\sin\alpha_M(\bar{\nu})] \tag{6.56}$$

If the AC part of this signal is detected with a lock-in amplifier synchronized with the twice the value of the PEM frequency, $2\omega_M$, the AC and DC detector intensities are:

$$I_D(\bar{\nu}) = I_{DC}(\bar{\nu}) + I_{AC}(\bar{\nu}) = \frac{I_{DC}(\bar{\nu})}{4}\left[1 - CB(\bar{\nu})2J_2[\alpha_M^0(\bar{\nu})]\cos 2\omega_M t\right] \tag{6.57}$$

The CB spectrum, $\Delta n_{LR}(\bar{\nu})$, can be obtained from the ratio

$$\frac{I_{AC}(\bar{\nu})}{I_{DC}(\bar{\nu})} = -2J_2\left[\alpha_M^{\circ}(\bar{\nu})\right][\Delta n_{LR}(\bar{\nu})] \tag{6.58}$$

The spectrum of $J_2\left[\alpha_M^{\circ}(\bar{\nu})\right]$ can be obtained from a calibration spectrum of the optical setup with only the initial vertical polarizer, PEM at 45°, and the analyzing polarizer at two different positions, vertical and horizontal, $M_A(0°/90°)$. The detector signal can be obtained starting from $S_2(\bar{\nu})$ in Equation (6.43) as:

$$I_D(\bar{\nu}) = D(\alpha) \cdot M_A(0°/90°) \cdot M_{PEM}(\bar{\nu}) \cdot M_P(0°) \cdot S_0(\bar{\nu}) = D(\alpha) \cdot M_A(0°/90°) \cdot S_2(\bar{\nu})$$

$$= D(\alpha) \cdot \frac{1}{2}\begin{pmatrix} 1 & \pm 1 & 0 & 0 \\ \pm 1 & 1 & 0 & 0 \\ 0 & 0 & 0 & 0 \\ 0 & 0 & 0 & 0 \end{pmatrix} \frac{I_0(\bar{\nu})}{2}\begin{pmatrix} 1 \\ \cos\alpha_M(\bar{\nu}) \\ 0 \\ \sin\alpha_M(\bar{\nu}) \end{pmatrix}$$

$$= \frac{I_0(\bar{\nu})}{4}[1 \pm \cos\alpha_M(\bar{\nu})]$$

$$= \frac{I_0(\bar{\nu})}{4}\left[1 \pm J_0\left[\alpha_M^{\circ}(\bar{\nu})\right] \pm 2J_2\left[\alpha_M^{\circ}(\bar{\nu})\right]\cos 2\omega_M t\right] \tag{6.59}$$

If we define $I_{AC}(\bar{\nu})$ to be the intensity associated with the $2\omega_M$ lock-in signal, and define $I_{DC,av}(\bar{\nu})$ to be the average of the DC intensities, we have

$$I_{DC,av}(\bar{\nu}) = \frac{I_0(\bar{\nu})}{8}\left[\left\{1 + J_0\left[\alpha_M^{\circ}(\bar{\nu})\right]\right\} + \left\{1 - J_0\left[\alpha_M^{\circ}(\bar{\nu})\right]\right\}\right] = \frac{I_0(\bar{\nu})}{4} J_0\left[\alpha_M^{\circ}(\bar{\nu})\right] \tag{6.60}$$

The ratio of the negative choice for $I_{AC}(\bar{\nu})$, corresponding to a horizontal analyzer, to the average DC intensity $I_{DC,av}(\bar{\nu})$, yields

$$\left[\frac{I_{AC}(\bar{\nu})}{I_{DC,av}(\bar{\nu})}\right]_{cal,2\omega_M} = -2J_2\left[\alpha_M^{\circ}(\bar{\nu})\right] \tag{6.61}$$

This is precisely the intensity needed to calibrate the measurement of CB given in Equation (6.58).

6.3 Fourier Transform VCD Measurement

The description of VCD instrumentation is now extended to the case of FT-VCD measurement. There are several conceivable ways one could measure VCD spectra using an FT-IR spectrometer. Two early attempts that did not yield successful methods involved subtraction of infrared absorbance spectra measured with left and right circularly polarized radiation, either sequentially as two long-term high signal-to-noise ratio measurements or simultaneously using two optical paths with opposite Fourier phase as a true double-beam FT-IR circular polarization difference measurement (Yang *et al.*, 1984). In first case, the dynamic range of the digitized detector signal, limited to 16-bits of data per digitization, prevented the ordinary transmission intensity from being measured at the same time as the much smaller VCD transmission spectrum that is approximately four to six orders of magnitude smaller. An analogue to digital converter of more than 20 bits would be necessary to capture the difference, and, furthermore, instrumental variations over long time periods between LCP and RCP measurements prevented any meaningful VCD spectrum from being obtained. For the second case of simultaneous measurement of RCP and LCP transmission intensity, difficulties with the precise cancellation of the two IR beams with opposite Fourier phase on the same detector element prevented the achievement of a robust measurement method, even though some VCD spectral results were obtained (Yang *et al.*, 1984).

The first successful extension of VCD measurement methodology to FT-IR spectrometers was achieved using the so-called double modulation method (Nafie and Diem, 1979; Nafie *et al.*, 1979; Lipp *et al.*, 1982). In this approach, the FT-IR beam is linearly polarized and passed through a PEM in the same way as a dispersive VCD measurement. The detector then records two Fourier interferogram intensities, a normal DC-interferogram, $I_{DC}(\delta)$, and a polarization modulated AC-interferogram, $I_{AC}(\delta)$. These interferograms are then Fourier transformed in a manner directly analogous to the measurement of VCD using dispersive instrumentation as described above, the details of which will be described in subsequent sections.

A number of years prior to the discovery of VCD, a method to measure CD using a PEM and a step-scanning interferometer was described (Russel *et al.*, 1972). The step-scan approach to VCD measurement will be described further below. The double modulation method of VCD measurement is the first such method to combine a rapid-scan FT-IR spectrometer with a lock-in amplifier that produces an AC interferogram as its output. The double modulation method can be applied to any high-frequency modulation difference FT-IR measurement, and has been used to measure vibrational linear dichroism (VLD), infrared reflection-absorption spectra (IRRAS), and recently vibrational circular birefringence (VCB).

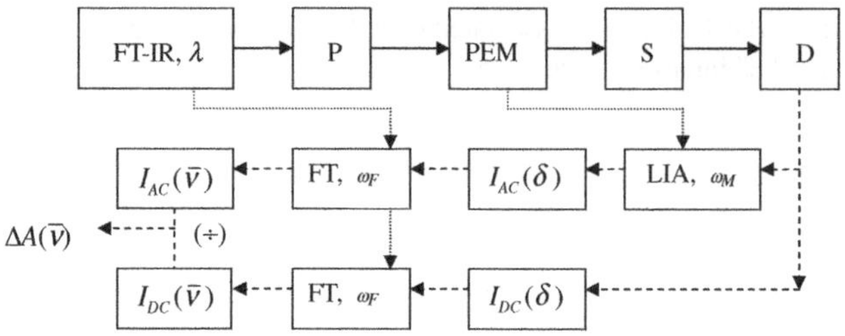

Figure 6.10 *FT-VCD optical block diagram and light path (solid lines) consisting of an FT-IR spectrometer with broadband infrared radiation (λ), a linear polarizer (P), and photoelastic modulator (PEM), a sample (S), and a detector (D). The electronic pathway (dashed lines) with LIA references (dotted lines) is similar to that of the dispersive instrument in Figure 6.1, where here Fourier transformation of the interferogram replaces the lock-ins synchronized to a light chopper. The Fourier transforms of the AC and the DC interferograms are divided (÷) to produce the VCD spectrum* $\Delta A(\bar{\nu})$

6.3.1 Double-Modulation Instrumental Setup and Block Diagram

The double modulation approach to FT-VCD measurement is carried out by replacing an infrared source and dispersive grating monochromator (wavelength selector), illustrated in Figure 6.1, with the output beam of an FT-IR spectrometer, shown in Figure 6.10. Each wavelength, λ, of the FT-IR spectrometer is encoded with a sinusoidal intensity modulation at a frequency called the Fourier frequency, ω_F, determined by the scanning of the dual-arm interferometer equipped with a beamsplitter, according to the usual design of an FT-IR spectrometer. In the double modulation method, two interferograms are simultaneously present at the detector. The first is the standard FT-IR detector signal, the DC interferogram, $I_{DC}(\delta)$, modulated at the Fourier frequencies. The second interferogram is modulated at both the Fourier frequencies and the PEM modulation frequency, ω_M. The signal at the PEM frequency is first demodulated by a lock-in amplifier referenced to the PEM frequency, (LIA, ω_M). The output of the LIA is the AC interferogram, $I_{AC}(\delta)$, which is now modulated only at the Fourier frequencies. After both AC and DC interferograms are Fourier transformed (FT), the AC and DC transmission intensities are ratioed to yield the FT-VCD spectrum, $\Delta A(\bar{\nu})$.

A more detailed mathematical description of the measurement and phase correction of the DC and AC interferograms is provided in the following two sections.

6.3.2 DC Interferogram and Phase Correction

The DC detector signal of the FT-IR spectrometer is a complex intensity pattern of Fourier modulated IR radiation known as an interferogram. The interferogram arises from the interference of two beams of infrared radiation as illustrated in Figure 6.11. Infrared radiation I_0 from a source S is directed to a beamsplitter (BS) creating two beams, one that is reflected at the beamsplitter to become I_R, and the other that is transmitted through it to become I_T. After reflection at mirror surfaces, which are moving relative to each other, the beams recombine at BS where they interfere according to their relative path difference. For different wavelengths, this path difference corresponds to different relative phases of the two beams. Two beams emerge from the BS, one of which returns to the source and is normally not used. This is the combination of the beams that either have been reflected twice at the BS or transmitted twice, I_{RR+TT}. The other beam is a combination of the two beams that each have one reflection and one transmission at the beamsplitter, I_{RT+TR}. These two combined beams are interferograms that have complementary interference patterns, as the sum of these two beams is the original beam from

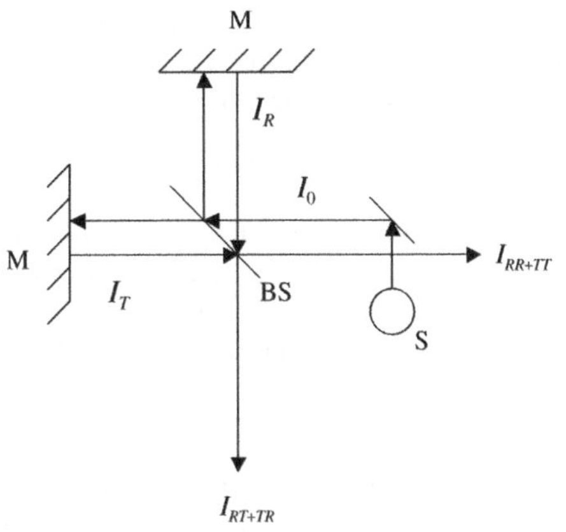

Figure 6.11 *Optical layout of a basic FT-IR spectrometer. The beam from the source, S, is divided into two beams at the beamsplitter (BS). These beams reflect at mirror surfaces and return to the beamsplitter and recombine. The subscripts R and T refer to reflection and transmission at the BS as described in the text. Reproduced with permission from the Society for Applied Spectroscopy (Nafie et al., 2004)*

the source, namely $I_{RR+TT} + I_{RT+TR} = I_0$. I_{RT+TR} is the beam that normally goes to the sample and here is called the DC interferogram, $I_{DC}(\delta) = I_{RT+TR}(\delta)$.

The mathematical expression for the DC interferogram is:

$$I_{DC}(\delta) = I_{RT+TR}(\delta) = \int_0^\infty I_{DC}(\bar{\nu})\cos[2\pi\delta\bar{\nu} + \theta_{DC}(\bar{\nu})]d\bar{\nu} \tag{6.62}$$

$I_{DC}(\delta)$ can be seen to be an integral superposition of the transmission intensities of all the spectral frequencies in the spectrum times a cosine factor that depends on the wavenumber frequency and the retardation, δ, the optical path difference of the interferometer from the equal mirror position. The units of δ are typically in centimeters. For a rapid-scan interferometer operating with constant mirror velocity V_m in cm s^{-1}, the retardation distance is related to the time-based Fourier frequency $f_{F,\bar{\nu}} = \omega_F/2\pi$ in cycles per second, or Hertz (Hz), for each wavenumber $\bar{\nu}$ by the relationships

$$2\pi\delta\bar{\nu} = 2\pi(2V_M t)\bar{\nu} = 2\pi(2V_M\bar{\nu})t = 2\pi f_{F,\bar{\nu}}t = \omega_F t \tag{6.63}$$

This equation exhibits the equivalence of expressing interferograms in terms of a spatial coordinate δ, with inverse spatial frequencies $\bar{\nu}$ or in terms of a time coordinate t, and inverse time frequencies, $f_{F,\bar{\nu}}$. The factor two associated with the mirror velocity arises from the fact that the change in distance between the beamsplitter and moving mirror is only half of the optical path difference of the interferometer as the mirror moves. Finally, in Equation (6.62) the function $\theta_{DC}(\bar{\nu})$ is the phase shift that may be present between Fourier cosine waves of different wavenumber frequency. These phase shifts arise from the various speeds at which electronic signals of different Fourier frequency pass through the processing electronics, such as filters and amplifiers.

A common time-based Fourier frequency is 16 kHz (16 000 Hz) for the HeNe reference laser frequency. From Equation (6.63) this Fourier frequency corresponds to a retardation velocity $2V_M$of $1\,\mathrm{cm\,s^{-1}}$ as the frequency $\bar{\nu}$ of an HeNe laser at 623.8 nm is approximately $16\,000\,\mathrm{cm^{-1}}$.

$$f_{F,\bar{\nu}} = 2V_M\bar{\nu} \tag{6.64}$$

This is a convenient mirror velocity as the Fourier frequency corresponding to any wavenumber frequency is very close to the numerical value of that frequency. Thus the band of Fourier frequencies associated with the mid-IR region from 1000 to $2000\,\mathrm{cm^{-1}}$ have values between 1 and 2 kHz.

In order to obtain the DC transmission spectrum, $I_{DC}(\bar{\nu})$, in the integral of Equation (6.62), the interferogram must be Fourier transformed. This is represented by the following expression,

$$I_{DC}(\bar{\nu}) = FT[I_{DC}(\delta)] = \frac{1}{2\pi}\int_0^\infty I_{DC}(\delta)\cos(2\pi\delta\bar{\nu})d\delta \tag{6.65}$$

Before the Fourier transform can be performed mathematically, the phase correction spectrum, $\theta_{DC}(\bar{\nu})$, must be measured and removed from the expression for $I_{DC}(\delta)$ in Equation (6.62). This is an operation that is automatically performed by software that controls the operation of the FT-IR spectrometer and several types of phase-correction algorithms are available for this purpose.

6.3.3 AC Interferogram and Phase Correction

In addition to the DC interferogram, there is also an AC interferogram present at the detector that carries the spectral information associated with the VCD spectrum. This interferogram initially has two modulation frequencies, the Fourier frequency for each radiation wavenumber, as in the DC interferogram, but in addition a higher frequency polarization modulation, typically in the range of tens of kilohertz. Each of these modulations must be demodulated to retrieve the desired VCD transmission spectrum, $I_{AC}(\bar{\nu})$, given in Equation (6.1). The first signal to be demodulated is that of the high-frequency polarization modulation of the PEM. This is achieved by passing the detector signal through an LIA, shown in Figure 6.10, the output of which is the AC interferogram given by:

$$I_{AC}(\delta) = \int_0^\infty I_{AC}(\bar{\nu})\cos[2\pi\delta\bar{\nu} + \theta_{AC}(\bar{\nu})]d\bar{\nu} \tag{6.66}$$

This interferogram is identical in form to $I_{DC}(\delta)$ in Equation (6.62); however, the phase shift function $\theta_{AC}(\bar{\nu})$ differs in general from $\theta_{DC}(\bar{\nu})$. The reason for this difference is that these two interferograms pass through different electronic pathways where the Fourier components of the two interferograms experience different phase shifts for the same Fourier frequency. The algorithm that determines this phase correction function automatically assumes that all detector intensities are positive. This is the case for $I_{DC}(\bar{\nu})$, but the intensities of $I_{AC}(\bar{\nu})$ can be either positive or negative across the spectrum depending on whether the VCD intensity is negative or positive. Consequently, a method of measuring $I_{AC}(\bar{\nu})$ when only positive intensities are present is needed, after which that phase correction function can be transferred to a measurement of $I_{AC}(\bar{\nu})$ when both positive and negative intensities are present.

A VCD spectrum with only positive intensities can be achieved by inserting into the sample position of the spectrometer an optical plate with a controlled but relatively small amount of birefringence followed by a polarizer. The induced birefringent plate is called a stress birefringent plate (SBP) because the birefringence is created by squeezing a plate with initially zero birefringence along a direction parallel to the stress axes of the PEM. An analyzing polarizer is placed after the SBP oriented to pass horizontally polarized radiation. The Mueller matrix of the SBP can be taken from Equations (6.30) and (6.37) under the assumption of small birefringence is:

$$\boldsymbol{M}_{SBP}(\pi/4, \delta_{SBP}) = \begin{pmatrix} 1 & 0 & 0 & 0 \\ 0 & 1 & 0 & -\delta_{SBP}(\bar{\nu}) \\ 0 & 0 & 1 & 0 \\ 0 & \delta_{SBP}(\bar{\nu}) & 0 & 1 \end{pmatrix} \tag{6.67}$$

Here $\delta_{SBP}(\bar{\nu})$ is the value of a small amount of birefringence across the region of the spectrum of interest. The signal for each wavenumber frequency at the detector for the full optical setup is then given by the expression

$$I_D(\bar{\nu}) = \boldsymbol{D}(\alpha)\cdot\boldsymbol{M}_{\mathrm{A}}(90^\circ)\boldsymbol{M}_{\mathrm{SBP}}(\bar{\nu})\cdot\boldsymbol{M}_{\mathrm{PEM}}(\bar{\nu})\cdot\boldsymbol{M}_{\mathrm{P}}(0^\circ)\cdot\boldsymbol{S}_0(\bar{\nu})$$

$$= \boldsymbol{D}(\alpha)\frac{I_0(\bar{\nu})}{2}\begin{pmatrix} 1 & -1 & 0 & 0 \\ -1 & 1 & 0 & 0 \\ 0 & 0 & 0 & 0 \\ 0 & 0 & 0 & 0 \end{pmatrix}\begin{pmatrix} 1 & 0 & 1 & 0 \\ 0 & 1 & 0 & -\delta_{SBP}(\bar{\nu}) \\ 0 & 0 & 1 & 0 \\ 0 & \delta_{SBP}(\bar{\nu}) & 0 & 1 \end{pmatrix}\begin{pmatrix} 1 \\ \cos\alpha_M(\bar{\nu}) \\ 0 \\ \sin\alpha_M(\bar{\nu}) \end{pmatrix}$$

$$= \frac{I_0(\bar{\nu})}{2}\left\{1 + J_0\left[\alpha_M^{\mathrm{o}}(\bar{\nu})\right] + 2J_1\left[\alpha_M^{\mathrm{o}}(\bar{\nu})\right]\delta_{SBP}(\bar{\nu})\right\} \tag{6.68}$$

The first two terms in this expression are DC-terms associated with the average transmission, and the last term is the AC-term associated with the SBP intensity. Both are positive intensity spectra and their interferograms are:

$$I_{DC}(\delta) = \int_0^\infty \frac{I_0(\bar{\nu})}{2}\left\{1 + J_0\left[\alpha_M^{\mathrm{o}}(\bar{\nu})\right]\right\}\cos[2\pi\delta\bar{\nu} + \theta_{DC}(\bar{\nu})]\mathrm{d}\bar{\nu} \tag{6.69}$$

$$I_{AC}(\delta) = \int_0^\infty I_0(\bar{\nu})J_1\left[\alpha_M^{\mathrm{o}}(\bar{\nu})\right]\delta_{SBP}(\bar{\nu})\cos[2\pi\delta\bar{\nu} + \theta_{AC}(\bar{\nu})]\mathrm{d}\bar{\nu} \tag{6.70}$$

Each of these interferograms may be phase corrected using the assumption that the spectral transmission intensity for all frequencies is positive. Once the AC phase function, $\theta_{AC}(\bar{\nu})$, has been determined with this optical setup, this phase function can be used to Fourier transform any AC interferogram of any VCD transmission spectrum regardless of the signs of the intensities at various frequencies. With the phase correction in hand, the AC analogue of Equation (6.65) is written as:

$$I_{AC}(\bar{\nu}) = FT[I_{AC}(\delta)] = \frac{1}{2\pi}\int_0^\infty I_{AC}(\delta)\cos(2\pi\delta\bar{\nu})\mathrm{d}\delta \tag{6.71}$$

The final calibrated VCD spectrum is obtained from the ratio of the Fourier transforms of the AC and DC interferogram,

$$\Delta A(\bar{\nu}) = \frac{1}{2J_1\left[\alpha_M^{\mathrm{o}}(\bar{\nu})\right]1.1513}\left[\frac{FT[I_{AC}(\delta)]}{FT[I_{DC}(\delta)]}\right] \tag{6.72}$$

where the function Bessel function $2J_1\left[\alpha_M^{\text{o}}(\bar{\nu})\right]$ is obtained from the VCD calibration spectrum given in Equations (6.16) and (6.18). In particular the calibration of VCD using an FT-IR spectrometer follows the same procedure detailed in Section 6.1 for dispersive VCD instrumentation.

6.3.4 Polarization Division FT-VCD Measurement

A second proven method of FT-VCD measurement was achieved a number of years after the establishment of the double modulation method (Ragunathan *et al.*, 1990; Polavarapu *et al.*, 1990). This was based on a modified Martin–Puplett polarization division interferometer (Polavarapu, 1988; Nafie *et al.*, 1988). The method circumvents the use of a PEM by creating circular polarization modulation at the Fourier frequencies. This is achieved by use of a polarization beamsplitter instead of a standard beamsplitter. The infrared beam from the source is first polarized at 45° relative to the grid lines of a polarizing beamsplitter. The divided beams have orthogonal linear polarization (polarization division) and their recombination at the beamsplitter with one mirror in motion creates a beam directed at the sample that goes through complete cycles of polarization phase retardation (vertical 0°, right CP 90°, horizontal 180°, left CP 270°) for each wavelength at its Fourier frequency. Because the recombining beams are orthogonally polarized, they do not interfere with one another, and hence there are only complete cycles of polarization modulation without intensity modulation.

The usual intensity modulated DC interferogram can be obtained by inserting a vertical polarizer between the sample position and the detector. From the polarization modulation cycle, full throughput is obtained at the zero degree position of the polarization modulation cycle, and the radiation is completely blocked when the light is horizontally polarized half-way through its modulation cycle. This measurement and its subsequent phase correction defines the cosine transform, or real part of the Fourier transform. If a sample is linearly dichroic its spectrum is available in the real part of the Fourier transform of the complex interferogram, and no polarizer is required for the VLD measurement. Again with no polarizer but a chiral sample in place, the VCD spectrum is available from the sine or imaginary part of the Fourier transform of the complex interferogram.

In the mid-IR region where successful measurements of VCD were achieved (Ragunathan *et al.*, 1990), the polarization division method proved to be less efficient than the double modulation method, but the polarization division method may well prove to be the best method for future extensions of VCD measurements into the far-IR region where PEMs are not commercially available.

6.3.5 Step-Scan FT-VCD Measurement

A third method for the measurement of FT-VCD is based on a step-scan rather than a rapid scan FT-IR spectrometer (Wang and Keiderling, 1995; Long *et al.*, 1997a; Long *et al.*, 1997b). In this approach, the moving mirror in the interferometer is advanced in discrete steps to determine over time the interferogram after all positions of the moving mirror have been sampled. At each step the DC and AC transmission values are determined by signal averaging. The step-scan measurement of the interferograms may be repeated one or more times to improve the signal-to-noise ratio. The arrangement of optical elements is the same as in the rapid-scan double modulation method. As there are no Fourier frequencies, the step-scan method has some similarities to dispersive VCD measurements where the interferometer rather that the grating spectrometer is slowly advanced over a full range of positions to complete the spectral measurement. After measurements have been completed, the measured AC and DC interferograms still need to be Fourier transformed, but there is no concern about the closeness of Fourier frequencies to PEM frequencies or to differences in phase correction between AC and DC interferograms. Only marginal improvements in signal quality relative to the rapid-scan double modulation were seen in some cases, although more recent advances in electronics for rapid-scan instruments have eliminated these advantages. The use of step-scan interferometers for FT-VCD measurements has not been reported in recent years, in part because these instruments are more expensive, without additional benefit, than their rapid-scan counterparts.

6.4 Commercial Instrumentation for VCD Measurement

Instrumentation for VCD measurement has become commercially available in several relatively distinct stages over the past 30 years. The earliest VCD spectrometers were based on dispersive scanning monochromators and required extensive electronics construction, including at first the PEMs. To date all dispersive VCD spectrometers are constructed from isolated components and essentially are not commercially available. The advent of rapid-scan FT-VCD measurements in 1979 using the double-modulation method described in the previous section simplified considerably the construction of a VCD spectrometer. This is because an FT-IR spectrometer necessarily comes with computer-controlled electronic processing, spectral ratio and calibration subroutines, and with no need to be concerned about a wavelength scanning control mechanism. An FT-IR VCD spectrometer still required assembly of a VCD side-bench with the PEM, as well as spectral collection algorithms, but this was far simpler than building a complete dispersive-scanning VCD spectrometer from individual components.

6.4.1 VCD Side-Bench Accessories

The first FT-VCD spectrometer assembled was based on a Nicolet 7199 FT-IR spectrometer. This was followed a few years later by a similar spectrometer based on a Digilab FT-IR spectrometer. These instruments were used principally in the spectral region below 2000 cm^{-1} with HgCdTe detectors as their performance to higher frequencies with InSb detectors, where hydrogen stretching bands occur, did not produce spectra of quality higher than could be obtained from the original dispersive VCD instruments. The principal reason for this result was the higher Fourier frequencies in this region compared with the mid-IR, which reduced the separation of modulation frequencies between the AC and DC interferograms. This issue was addressed by the development of step-scan FT-VCD spectrometers, where it was found that better VCD spectra could be obtained in the hydrogen stretching region, but mid-IR step-scan VCD did not produce an advance in sensitivity, or overall performance, relative to the double modulation rapid-scan approach.

The advent of step-scan FT-IR instrumentation spurred interest in several instrument companies to offer VCD side-bench accessories and with them software for obtaining calibrated VCD spectra by the ratio of AC and DC transmission intensities followed by a suitable intensity calibration. These companies were Digilab/BioRad, Nicolet/Thermo, and Bruker. Users could chose between step-scan and rapid-scan operation, but by the mid-1990s most applications were carried out with rapid-scan, double-modulation operation as the perceived benefits of step-scan waned.

The major benefit of the side-bench VCD accessory is that the original instrument remains available for normal operation and other types of experiments, such as IRRAS, can performed as alternative applications of the side-bench. The major disadvantage of the side-bench accessory is that the instrument baseline is not established permanently at the point of manufacture, but rather at the point installation at the customer site or after reconfiguration of the instrument for VCD operation following an alternative measurement configuration. The VCD baseline is the most important instrumental feature associated with the quality and reliability of VCD measurement, and setting the baseline close to zero on VCD intensity scales across the spectrum requires sensitive, sometimes lengthy, optical adjustment. Further, as the main-bench and side-bench are on separate optical platforms, the VCD baseline in general changes for small changes in the relative positions of these benches.

6.4.2 Dedicated VCD Spectrometers

In 1997 the first fully-dedicated VCD spectrometer on a single optical bench with a factory-aligned, permanent baseline became commercially available as the Chiral*IR* FT-VCD spectrometer. This instrument was originally produced as a joint venture between Bomem Inc. and BioTools Inc., but now the base system, including the interferometer and detector is supplied to BioTools from ABB Quebec (formerly Bomem) and all VCD assembly and testing is carried out by BioTools. Manufacture of the Chiral*IR* is followed by approximately two weeks of baseline and stability tests, VCD intensity tests, and a set of calibrated reference spectra that provide an extensive library of test data against which the performance of the instrument can be compared. Owing to a thick solid-aluminum baseplate and secure optical mounts, including the mounting and alignment of the detector, the detector-focusing lens and PEM, the baseline is maintained through the shipping process and no further alignment is needed during the installation at the customer site. As a result, final VCD spectra are usually available immediately upon setting up the instrument in its intended location.

Another vital advance brought about by dedicated VCD instruments is the software that allows complete processing of the raw, un-calibrated VCD spectrum with every scan of the Fourier interferometer. For home-built VCD instruments based on a side-bench assembly, blocks of AC and DC scans used to be measured for periods of hours, and, even with VCD macro programs for averaging, dividing and calibrating the raw data files, the first glimpse of the VCD data would not be available until all these human-driven, processing steps were complete. For VCD measurements with the Chiral*IR*, for example, properly ratioed VCD spectra can be seen to be improving with time as *each* scan is measured, processed and co-added to the previous VCD scans. In this way it is seen immediately if the measurement is going well or whether, for some reason, it needs to be stopped and changed before the final spectrum is measured.

Within the past several years, a second dedicated VCD spectrometer has become commercially available from Jasco, Inc., the FVS-6000 Vibrational Circular Dichroism spectrometer. This instrument is also built on a single fixed baseplate and does not use the side-bench approach. Nevertheless, this instrument does not have one important intensity enhancement feature found in the standard Chiral*IR*, namely dual-source operation (Nafie *et al.*, 2004). In addition, the Chiral*IR* is the only commercial VCD instrument available with dual-PEM operation (Nafie, 2000). Finally, a new version of the Chiral*IR*, the Chiral*IR-2X*, became available in March of 2009 in which dual-source and dual-PEM operation are offered on a single fixed baseplate without any external lock-in amplifiers or filters. All processing after the preamplifier is carried out with software using a single computer card inserted into a PC, thus reducing significantly the size and complexity of single- and dual-PEM VCD instrumentation. Each of these advances will be described in the following sections.

6.5 Advanced VCD Instrumentation

In this section we describe the fundamentals of several advanced VCD improvements that, in combination, provide unprecedented VCD sensitivity and routine, near artifact-free VCD measurement for any sample in any sample cell. Further details about the measurement of VCD spectra will be addressed in Chapter 8.

6.5.1 Dual Source Intensity Enhancement and Detector Saturation Suppression

All Chiral*IR* FT-VCD spectrometers employ dual-source operation as a standard feature (Nafie *et al.*, 2004). Unlike for single-source operation, the mirrors in a dual-source FT-IR spectrometer must be cube-corner mirrors (CCMs). A CCM has the remarkable property that any ray of light reflected from a CCM does so along a line exactly parallel to the incident ray after having been reflected off all

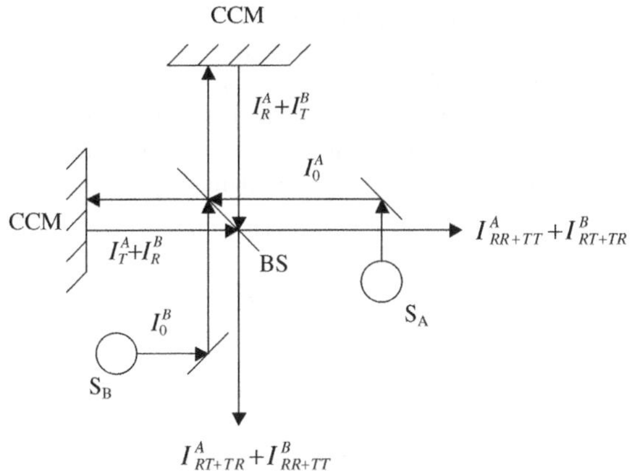

Figure 6.12 *Optical layout of a dual-source FT-IR spectrometer. The beam from sources S_A and S_B are each divided into two beams at the beamsplitter (BS). These four beams reflect at cube-corner mirrors (CCM) and recombine with interference at the beamsplitter. The subscripts R refer to reflection and T to transmission as described in the text. Reproduced with permission from the Society for Applied Spectroscopy (Nafie et al., 2004)*

three cube-corner mirror surfaces. Rays entering a CCM in the lower half of the interferometer horizontal plane are reflected into the upper half plane. The optical diagram associated the dual-source operation of the Chiral*IR* is illustrated in Figure 6.12. The beams from the two sources, S_A and S_B, are directed to the lower half of the beam splitter and recombine in the upper half. Source S_A corresponds to the normal position of a single source interferometer while source S_B is a new source located beneath this beam, which is normally directed toward the standard sample position. The beam directed to the sample for dual-source operation contains two interferograms, one from each source. The beam from S_A is the standard interferogram intensity described above, I_{RT+TR}^A while that from source S_B corresponds to the intensity that usually returns to the source, namely I_{RR+TT}^B. As indicated above, these two sources have opposite Fourier phase, and if they came from the same source, the Fourier modulations would cancel. The expression for the DC interferogram analogous to Equation (6.62) is given by:

$$I_{DC}^{AB}(\delta) = I_{RT+TR}^{A,DC}(\delta) + I_{RR+TT}^{B,DC}(\delta) = \int_0^\infty \left(I_{DC}^A(\bar{\nu}) - I_{DC}^B(\bar{\nu})\right)\cos[2\pi\delta\bar{\nu} + \theta_{DC}(\bar{\nu})]d\bar{\nu} \tag{6.73}$$

For dual-source VCD operation, the polarizer before the PEM is replaced by two polarizers, one in front of each of the two sources with orthogonal orientation to each other, one vertical and the other horizontal. Although beams from S_A and S_B are nearly the same, their different polarization states with respect to reflection and transmission at the beamsplitter leads to different intensities for $I_{DC}^A(\bar{\nu})$ and $I_{DC}^B(\bar{\nu})$, for the outgoing beams. As a result, even though the beams $I_{RT+TR}^A(\delta)$ and $I_{RR+TT}^A(\delta)$ have opposite Fourier phase and opposite signs in Equation (6.73), their Fourier modulations do not cancel but rather their ratio, $I_{DC}^A(\bar{\nu})/I_{DC}^B(\bar{\nu})$, is found experimentally to be about 1.6. If the weaker source $I_{DC}^B(\bar{\nu})$ is normalized to unity, then the difference intensity, $I_{DC}^A(\bar{\nu}) - I_{DC}^B(\bar{\nu})$, in Equation (6.73) is about 60% of $I_{DC}^B(\bar{\nu})$ and 40% of $I_{DC}^A(\bar{\nu})$. Thus, even if either or both of these interferograms saturate the detector, their combined reduced intensity, $I_{DC}^A(\bar{\nu}) - I_{DC}^B(\bar{\nu})$, is found not to saturate the detector. *This is the first major advantage of dual source operation.*

The measurement of FT-VCD with orthogonal polarizers in front of the two sources leads to *the second major advantage of dual-source operation.* Here, the small AC interferograms associated with the two sources add at the detector rather than subtract, as is the case for the DC interferograms described above. This is because the interferograms from the two sources have both opposite Fourier phase and opposite VCD phase. These two opposite phases, each represented by a minus sign, combine to produce constructive addition of the two AC interferograms. The expression of the dual-source interferogram is given by:

$$I_{AC}^{AB}(\delta) = I_{RT+TR}^{A,AC}(\delta) + I_{RR+TT}^{B,AC}(\delta) = \int_0^\infty \left(I_{AC}^{A}(\bar{\nu}) + I_{AC}^{B}(\bar{\nu})\right)\cos[2\pi\delta\bar{\nu} + \theta_{AC}(\bar{\nu})]\mathrm{d}\bar{\nu} \tag{6.74}$$

Experimentally, the ratio $I_{AC}^{A}(\bar{\nu})/I_{AC}^{B}(\bar{\nu})$ is again found to be approximately 1.6, but now the combined intensity $I_{AC}^{A}(\bar{\nu}) + I_{AC}^{B}(\bar{\nu})$ is 260% of $I_{AC}^{B}(\bar{\nu})$ and 160% of $I_{AC}^{A}(\bar{\nu})$. These are intensity gains at the detector against the same noise background that would be present for single-source operation and hence represent a signal-to-noise ratio (SNR) advantage of approximately 2.6 relative to the weaker source and 1.6 relative to the stronger source for dual source operation. The savings in scanning time to achieve the same SNR as single sources is either 6.8 or 2.6.

For dual-source operation, VCD intensity is proportional to the ratio of the combined AC and DC interferograms. This ratio relative to the stronger single-source operation is found experimentally to be approximately

$$\left|\frac{I_{AC}^{A}(\bar{\nu}) + I_{AC}^{B}(\bar{\nu})}{I_{DC}^{A}(\bar{\nu}) - I_{DC}^{B}(\bar{\nu})}\right| = \left|\frac{1+1.6}{1-1.6}\right| \approx 4.3 \tag{6.75}$$

This represents an increase in the uncalibrated dual-source VCD intensity relative to the corresponding uncalibrated single-source VCD intensity. This factor represents the gain in AC transmission intensity relative to the reduction in DC transmission intensity, and this scale factor is removed, but not the associated reduction in VCD noise, when the final intensity-calibrated, dual-source VCD spectrum is determined.

6.5.2 Dual-PEM Theory of Artifact Suppression

The use of two PEMs, one before the sample and one after the sample, dates back to the first measurements of VCD at the University of Southern California in the laboratory of Philip Stephens (Nafie *et al.*, 1975; Nafie *et al.*, 1976). This method of artifact reduction was then called polarization scrambling and was implemented such that the stress level of the second PEM was varied linearly with the monochromator drive to keep the PEM near to the optimum point of artifact cancellation (Cheng *et al.*, 1975). The extension of this method to FT-VCD measurement was problematic because all wavelengths were measured at the same time and the second PEM could only be set optimally for one wavelength. This problem was solved in 2000 by simultaneously recording the VCD interferograms associated with each of the two PEM and subtracting these VCD spectra in real time with each interferometric scan (Nafie, 2000). In the FT dual-PEM method, the signal from the first PEM reports the VCD spectrum plus the birefringent baseline while the second PEM reports only a birefringent baseline. By varying the voltage of the second PEM the net VCD baseline can be systematically brought close to zero with very little slope. This enormously simplified the task of optical alignment and eliminated baseline drift during measurement as any thermal changes in the instrument baseline were dramatically reduced by the subtraction of two interferograms from the PEMs, each reporting the birefringent state of the same optical path.

The optical diagram for the dual-PEM setup is given in Figure 6.13. It is simply an extension of the single PEM FT-VCD optical–electronic diagram in Figure 6.10. Here there are two PEMs and two lock-in

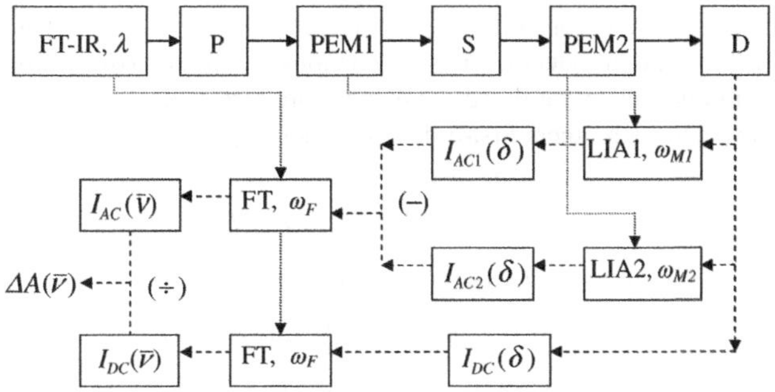

Figure 6.13 *Optical–electronic diagram for dual-PEM FT-VCD with light path (solid lines) consisting of an FT-IR spectrometer with broadband infrared radiation (λ), a linear polarizer (P), PEM1, a sample (S), PEM2, and a detector (D). The electronic pathway (dashed lines) and two LIA references (dotted lines) are analogous to the FT-VCD optical–electronic diagram in Figure 6.10. After the real time subtraction (−) of the two AC-interferogram outputs of the two LIAs, the resulting AC interferogram and the DC interferogram are Fourier transformed and divided as usual to produce the baseline-corrected VCD spectrum, $\Delta A(\bar{\nu})$*

amplifiers (LIA1 and LIA2) with separate LIA reference signals. The outputs of the two LIAs are two AC interferograms, $I_{AC1}(\delta)$ and $I_{AC2}(\delta)$. These are subtracted to produce a single AC interferogram that is Fourier transformed along with the FT of the DC interferogram. The AC and DC transmission spectra, $I_{AC}(\bar{\nu})$ and $I_{DC}(\bar{\nu})$, are divided and calibrated to yield the final VCD spectrum.

The theory of dual-PEM operation is easily described using the Stokes–Mueller formalism developed earlier in this chapter. We start with the Stokes vector $\boldsymbol{S}_3(\bar{\nu})$ from Equation (6.44) that represents the state of the light beam after the polarizer, the first PEM and the sample. The Stokes vector after the second PEM, $\boldsymbol{S}_4(\bar{\nu})$, can be obtained by operating on $\boldsymbol{S}_3(\bar{\nu})$, with the Mueller matrix of the second PEM as:

$$\begin{aligned}
\boldsymbol{S}_4(\bar{\nu}) &= \boldsymbol{M}_{PEM2}[45^\circ, \alpha_{M2}(\bar{\nu})]\cdot \boldsymbol{S}_3(\bar{\nu}) \\
&= \boldsymbol{M}_{PEM2}[45^\circ, \alpha_{M2}(\bar{\nu})]\cdot \{\boldsymbol{M}_S(\bar{\nu})\cdot \boldsymbol{M}_{PEM1}[45^\circ, \alpha_{M1}(\bar{\nu})]\cdot \boldsymbol{M}_P(0^\circ)\cdot \boldsymbol{S}_0(\bar{\nu})\} \\
&= \begin{pmatrix} 1 & 0 & 0 & 0 \\ 0 & \cos\alpha_{M2}(\bar{\nu}) & 0 & -\sin\alpha_{M2}(\bar{\nu}) \\ 0 & 0 & 1 & 0 \\ 0 & \sin\alpha_{M2}(\bar{\nu}) & 0 & \cos\alpha_{M2}(\bar{\nu}) \end{pmatrix} \frac{I_{DC}(\bar{\nu})}{2} \begin{pmatrix} 1 + CD(\bar{\nu})\sin\alpha_{M1}(\bar{\nu}) \\ \cos\alpha_{M1}(\bar{\nu}) - LB(\bar{\nu})\sin\alpha_{M1}(\bar{\nu}) \\ -CB(\bar{\nu})\cos\alpha_{M1}(\bar{\nu}) + LB'(\bar{\nu})\sin\alpha_{M1}(\bar{\nu}) \\ CD(\bar{\nu}) + LB(\bar{\nu})\cos\alpha_{M1}(\bar{\nu}) + \sin\alpha_{M1}(\bar{\nu}) \end{pmatrix} \\
&= \frac{I_{DC}}{2} \begin{pmatrix} 1 + CD\sin\alpha_{M1} \\ \cos\alpha_{M1}\cos\alpha_{M2} - LB\sin\alpha_{M1}\cos\alpha_{M2} - CD\sin\alpha_{M2} - LB\cos\alpha_{M1}\sin\alpha_{M2} - \sin\alpha_{M1}\sin\alpha_{M2} \\ -CB\cos\alpha_{M1} + LB'(\bar{\nu})\sin\alpha_{M1} \\ \cos\alpha_{M1}\sin\alpha_{M2} - LB\sin\alpha_{M1}\sin\alpha_{M2} + CD\cos\alpha_{M2} + LB\cos\alpha_{M1}\cos\alpha_{M2} + \sin\alpha_{M1}\cos\alpha_{M2} \end{pmatrix}
\end{aligned} \tag{6.76}$$

Here, for simplicity, we have dropped the explicit wavenumber frequency dependence in the last expression for $\boldsymbol{S}_4(\bar{\nu})$. We next introduce the Bessel functions $J_0(\alpha_M)$ and $J_1(\alpha_M)$ from Equations (6.4) and (6.50)–(6.52), eliminate the terms that depend on the product $\sin\alpha_{M1}\sin\alpha_{M2}$ as the PEMs are not synchronized and average each others first-order frequency dependence to zero. This gives

$$\boldsymbol{S}_4(\bar{\nu}) = \frac{I_{DC}}{2}\begin{pmatrix} 1+2J_1(\alpha_{M1})CD \\ J_0(\alpha_{M1})J_0(\alpha_{M2}) - 2J_1(\alpha_{M1})J_0(\alpha_{M2})LB - 2J_1(\alpha_{M2})CD - 2J_0(\alpha_{M1})J_1(\alpha_{M2})LB \\ -J_0(\alpha_{M1})CB + 2J_1(\alpha_{M1})LB' \\ 2J_0(\alpha_{M1})J_1(\alpha_{M2}) + J_0(\alpha_{M2})CD + J_0(\alpha_{M1})J_0(\alpha_{M2})LB + 2J_1(\alpha_{M1})J_0(\alpha_{M2}) \end{pmatrix} \tag{6.77}$$

The intensity at the detector is obtained by use of the Mueller matrix for the detector given by Equations (6.39) and (6.40) yielding,

$$\begin{aligned} I_D = \boldsymbol{D}(\alpha)\cdot\boldsymbol{S}_4 &= \begin{pmatrix} 1 & (p_X^2 - p_Y^2)\cos 2\alpha & (p_X^2 - p_Y^2)\sin 2\alpha & 0 \end{pmatrix}\begin{pmatrix} S_{4,0} \\ S_{4,1} \\ S_{4,2} \\ S_{4,3} \end{pmatrix} \\ &= \frac{I_{DC}}{2}\{1 + 2J_1(\alpha_{M1})CD + (p_X^2 - p_Y^2)\cos 2\alpha[-2J_1(\alpha_{M1})J_0(\alpha_{M2})LB \\ &\quad - 2J_1(\alpha_{M2})CD - 2J_0(\alpha_{M1})J_1(\alpha_{M2})LB] + (p_X^2 - p_Y^2)\sin 2\alpha[2J_1(\alpha_{M1})LB']\} \end{aligned} \tag{6.78}$$

Here we have eliminated $J_0(\alpha_{M1})J_0(\alpha_{M2})$ and $-J_0(\alpha_{M1})CB$ as constant, non-modulated terms that are small and insignificant relative to unity (1) in the first term of this equation. Recall that the polarization dependence of the detector is typically on the order of 1%, so the CD intensity term $2J_1(\alpha_{M1})CD$ from the first PEM is two orders of magnitude larger than the corresponding term $-2J_1(\alpha_{M2})CD$ associated with the second PEM. The terms $-2J_1(\alpha_{M1})J_0(\alpha_{M2})LB$ and $2J_1(\alpha_{M1})LB'$ determine the VCD baseline due to birefringence as detected by the first PEM while the term $-2J_0(\alpha_{M1})J_1(\alpha_{M2})LB$ represents the same baseline as seen by the second PEM.

Recall that LB is linear birefringence in the optical train with fast and slow axes parallel to the PEM axes, $(+45^\circ, -45^\circ)$ whereas LB' is linear birefringence with fast and slow axes vertical and horizontal, $(0^\circ, 90^\circ)$. It can be seen that the terms depending on LB can be made to be equal in magnitude if the two PEMs have the same modulation strength such that $J_0(\alpha_{M1}) = J_0(\alpha_{M2})$, and then necessarily $J_1(\alpha_{M1}) = J_1(\alpha_{M2})$. Hence, if the AC interferograms obtained from the lock-ins tuned to PEM1 and PEM2 are subtracted, with PEMs at the same retardation value, the two LB terms cancel. The resulting intensity at the detector is given by:

$$I_D = \frac{I_{DC}}{2}\{1 + 2J_1(\alpha_{M1})CD + (p_X^2 - p_Y^2)\sin 2\alpha[2J_1(\alpha_{M1})LB']\} \tag{6.79}$$

It has been shown that the lone, uncompensated LB' term appears only for LB' present *between* the two PEMs as derived here (Cao *et al.*, 2008). All sources of LB and LB' before or after the PEMs can be canceled in pairs in the same way as the LB terms are canceled for LB occurring between the two PEMs. If the VCD baseline is adjusted with no sample or sample cell between the two PEMs, then the uncompensated LB' term in Equation (6.79) cannot appear and a zero baseline can be obtained provided the two PEMs can be set to be equal to each other. In practice, with a sample with windows in the beam between the two PEMs, the LB term from the PEM2 can be adjusted if desired so that both the

LB term and the *LB*′ term from the PEM1 are approximately canceled. In the following section, we discuss two ways to eliminate the *LB*′ term when a sample cell with birefringent windows, or a solid sample, is in place.

6.5.3 Rotating Achromatic Half-Wave Plate

It has been demonstrated, both theoretically and experimentally, that placing a rotating achromatic half-wave plate (HWP) after the second PEM in a dual-PEM FT-VCD instrument eliminates all sources of artifacts throughout the optical train (Cao *et al.*, 2008). We assume that the plate rotates slowly on the of the order of ten revolutions per minute asynchronously with respect to all other modulation frequencies in the instrument. After adding an additional source of linear birefringence, LB2, after the rotating HWP (described further below), the optical setup for the dual-PEM rotating HWP is as shown in Figure 6.14.

The Mueller matrix for an HWP as a function of its orientation angle is given by Equation (6.35). For a rotating HWP, the plate can be averaged over all angles by integration between the angles 0 and 2π, as:

$$\boldsymbol{M}_{RHWP} = \frac{1}{2\pi}\int_0^{2\pi}\begin{pmatrix} 1 & 0 & 0 & 0 \\ 0 & \cos 4\theta & \sin 4\theta & 0 \\ 0 & \sin 4\theta & -\cos 4\theta & 0 \\ 0 & 0 & 0 & -1 \end{pmatrix} d\theta = \begin{pmatrix} 1 & 0 & 0 & 0 \\ 0 & 0 & 0 & 0 \\ 0 & 0 & 0 & 0 \\ 0 & 0 & 0 & -1 \end{pmatrix} \tag{6.80}$$

This Mueller matrix averages to zero all states of linear polarization and simultaneously converts any component of circular polarization from right to left or *vice versa.* If a rotating HWP is placed after the second PEM in a dual PEM optical train the resulting Stokes vector $\boldsymbol{S}_5(\bar{\nu})$ can be obtained starting from Equation (6.77) by:

$$\boldsymbol{S}_5(\bar{\nu}) = \boldsymbol{M}_{RHWP}\cdot\boldsymbol{S}_4(\bar{\nu}) = \frac{I_{DC}}{2}\begin{pmatrix} 1+2J_1(\alpha_{M1})CD \\ 0 \\ 0 \\ -2J_0(\alpha_{M1})J_1(\alpha_{M2}) - 2J_1(\alpha_{M1})J_0(\alpha_{M2}) \end{pmatrix} \tag{6.81}$$

Here we have eliminated small non-modulation terms $J_0(\alpha_{M2})CD$ and $J_0(\alpha_{M1})J_0(\alpha_{M2})LB$ from the last element of the Stokes vector describing the state of circular polarization. If a second source of small linear birefringence LB2 occurs after the rotating HWP it can be described by the Mueller matrix from Equations (6.37) and (6.38),

$$\boldsymbol{M}_{LB2} = \begin{pmatrix} 1 & 0 & 0 & 0 \\ 0 & 1 & 0 & -LB2 \\ 0 & 0 & 1 & LB2' \\ 0 & LB2 & -LB2' & 1 \end{pmatrix} \tag{6.82}$$

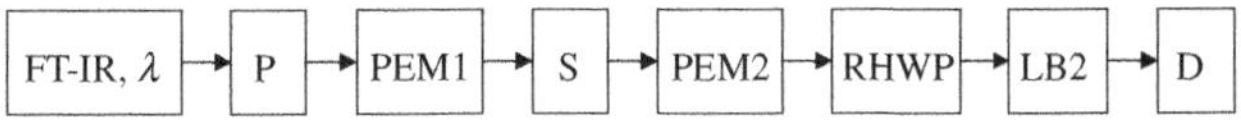

Figure 6.14 *Optical diagram for dual-PEM rotating HWP FT-VCD with light path (solid lines) consisting of an FT-IR spectrometer with broadband infrared radiation (λ), a linear polarizer (P), PEM1, a sample (S), PEM2, rotating HWP (RHWP), an additional source of linear birefringence (LB2), and a detector (D)*

This birefringence can be due to birefringence in the detector focusing lens or window. The resulting Stokes vector after this post rotating HWP LB is given by:

$$\begin{aligned} \boldsymbol{S}_6(\bar{\nu}) &= \boldsymbol{M}_{LB2} \cdot \boldsymbol{S}_5(\bar{\nu}) \\ &= \frac{I_{DC}}{2} \begin{pmatrix} 1 + 2J_1(\alpha_{M1})CD \\ LB2[2J_0(\alpha_{M1})J_1(\alpha_{M2}) + 2J_1(\alpha_{M1})J_0(\alpha_{M2})] \\ -LB2'[2J_0(\alpha_{M1})J_1(\alpha_{M2}) + 2J_1(\alpha_{M1})J_0(\alpha_{M2})] \\ -(1 + LB2 - LB2')[2J_0(\alpha_{M1})J_1(\alpha_{M2}) + 2J_1(\alpha_{M1})J_0(\alpha_{M2})] \end{pmatrix} \end{aligned} \tag{6.83}$$

The terms in square brackets, which are simply the circular polarization term of the Stokes vector in Equation (6.81), become zero if the two PEMs are set to equal retardation and the LIA1 and LIA2 signals are subtracted the way they are in the previous section describing dual-PEM artifact subtraction. If this is done, the last entry in Equation (6.81) or all the square bracket term expressions in Equation (6.83) are zero and then all *LB*and *LB*′ terms become zero and the final Stokes vector before the detector is given by:

$$\boldsymbol{S}_6(\bar{\nu}) = \frac{I_{DC}}{2} \begin{pmatrix} 1 + 2J_1(\alpha_{M1})CD \\ 0 \\ 0 \\ 0 \end{pmatrix} \tag{6.84}$$

The signal at the detector, with or without polarization sensitivity at the detector, is given by:

$$\begin{aligned} I_D(\bar{\nu}) &= \boldsymbol{D}(\alpha) \cdot \boldsymbol{M}_{LB2}(\bar{\nu}) \cdot \boldsymbol{M}_{RHWP} \cdot \boldsymbol{M}_{PEM2}(\bar{\nu}) \cdot \boldsymbol{M}_S(\bar{\nu}) \cdot \boldsymbol{M}_{PEM1}(\bar{\nu}) \cdot \boldsymbol{M}_P(0^\circ) \cdot \boldsymbol{S}_0(\bar{\nu}) \\ &= \boldsymbol{D}(\alpha) \cdot \boldsymbol{S}_6(\bar{\nu}) = \frac{I_{DC}}{2}[1 + 2J_1(\alpha_{M1})CD(\bar{\nu})] \end{aligned} \tag{6.85}$$

This represents a birefringent artifact-free VCD spectrum. The dual-PEM setup eliminates all sources of birefringence except the *LB*′ contributions located between the two PEMs, primarily due to the sample cell windows in the case of liquid or solution sample, and also orientational effects in a solid-phase sample. By contrast, the rotating HWP eliminates all sources of birefringence before it in the optical train but has no effect on birefringence that occurs after it, such as in the detector focusing lens or dewar windows. Together, and only together, they eliminate all birefringent artifact signals.

The principal drawback of the rotating HWP is the cost of acquisition of an achromatic HWP in the IR region. The fabrication is fairly straightforward. For the near-infrared region down to 2000 cm^{-1} the cost of acquisition in 2004 of a super achromatic plate from B. Halle Nachfolger in Berlin, Germany, made from six optically contacting layers of MgF_2 was over \$10 000 US. Extension beyond this limit into the mid-IR is currently untested optical technology and the cost of fabrication is likely to be significantly higher. To date, this extension has not been pursued.

6.5.4 Rotating Sample Cell

An alternative complete solution to the birefringence artifact problem that remains after incorporation of the dual-PEM setup is rotation of the sample about the optic axis of the IR beam. With this

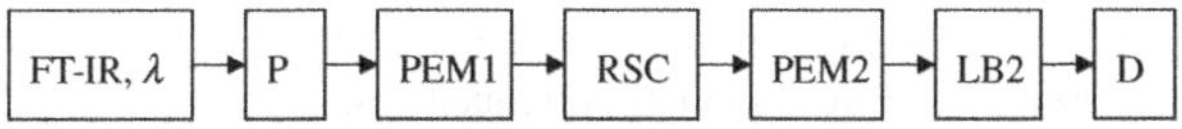

Figure 6.15 *Optical diagram for dual-PEM rotating sample cell FT-VCD with light path (solid lines) consisting of an FT-IR spectrometer with broadband infrared radiation (λ), a linear polarizer (P), PEM1, a rotating sample cell (RSC), PEM2, an second source of linear birefringence (LB2), and a detector (D)*

setup, both sources of linear birefringence, $LB(\bar{\nu})$ and $LB'(\bar{\nu})$, associated with the sample cell are averaged to zero over the course of the measurement, and this yields a VCD baseline that is identical with the baseline of the spectrometer with the same optical aperture and no optical elements between the two PEMs. As mentioned above, the dual-PEM FT-VCD spectrometer, both theoretically and experimentally, can be set to eliminate all remaining sources of LB and LB' both before and after the two PEMs. The optical arrangement for the Mueller analysis to be described below is given in Figure 6.15.

The proof of this method can derived directly from a Mueller matrix analysis that includes a description of the sample matrix in Equation (6.38) as a function of orientation angle,

$$\boldsymbol{M}_S(\theta) = 10^{-A}\begin{pmatrix} 1 & 0 & 0 & CD \\ 0 & 1 & CB & -LB\cos 2\theta - LB'\sin 2\theta \\ 0 & -CB & 1 & LB'\cos 2\theta + LB\sin 2\theta \\ CD & LB\cos 2\theta + LB'\sin 2\theta & -LB'\cos 2\theta + LB\sin 2\theta & 1 \end{pmatrix} \tag{6.86}$$

Equation (6.38) is recovered if the angle $\theta = 0°$. If this matrix is then averaged over all orientation angles as was done for the rotation HWP in Equation (6.79), one can write

$$\boldsymbol{M}_{RS} = \int_0^{2\pi} \boldsymbol{M}_S(\theta)\mathrm{d}\theta = 10^{-A}\begin{pmatrix} 1 & 0 & 0 & CD \\ 0 & 1 & CB & 0 \\ 0 & -CB & 1 & 0 \\ CD & 0 & 0 & 1 \end{pmatrix} \tag{6.87}$$

This eliminates through first order all artifacts arising from linear birefringence from the sample. The action of the dual-PEM then eliminates all remaining linear birefringent sources and the VCD intensity at the detector is again given by the final VCD intensity in Equation (6.85). The detector signal has only pure VCD intensity at the fundamental PEM frequency with no surviving artifact terms due to linear birefringence that may be present in the sample and throughout the optical train,

$$\begin{aligned} I_D(\bar{\nu}) &= \boldsymbol{D}(\alpha)\cdot \boldsymbol{M}_{LB2}(\bar{\nu})\cdot \boldsymbol{M}_{PEM2}(\bar{\nu})\cdot \boldsymbol{M}_{RS}(\bar{\nu})\cdot \boldsymbol{M}_{PEM1}(\bar{\nu})\cdot \boldsymbol{M}_P(0°)\cdot \boldsymbol{S}_0(\bar{\nu}) \\ &= \boldsymbol{D}(\alpha)\cdot \boldsymbol{S}_6(\bar{\nu}) = \frac{I_{DC}(\bar{\nu})}{2}\{1 + 2J_1[\alpha^0_{M1}(\bar{\nu})]CD(\bar{\nu}) - 2J_2[\alpha^0_{M2}(\bar{\nu})]CB(\bar{\nu})\} \end{aligned} \tag{6.88}$$

The circular birefringence (VCB) can be measured at twice the PEM frequency and is proportional to $2J_2(\alpha_{M1})$ as described in Section 6.2.4.

6.5.5 Direct All-Digital VCD Measurement and Noise Improvement

Recently, it has become possible to measure dual-PEM FT-VCD without any conventional lock-in amplifiers, as described above for the Chiral*IR-2X*. Taking advantage of high-speed digital signal processing, the signal from the preamplifier is directly digitized and subsequently the DC, AC1, and AC2 interferograms are isolated using software. Numerical filtering further reduces out-of-band noise and extraneous harmonics of the modulation frequencies.

6.5.6 Femtosecond-IR Laser-Pulse VOA Measurements

Early in 2009, two new techniques for measuring VCD and also VCB (VORD) were reported using heterodyned femtosecond IR laser pulses. These new methods of VCD/VCB take measurements that occur within the femtosecond time window of the IR pulse and can either measure time-independent spectra or time-resolve spectra using a gated pump pulse to prepare the sample in a suitable pre-kinetic state, such as an excited electronic state of some type. In one approach, laser heterodyning is achieved by self-heterodyning orthogonally polarized components of elliptical polarization modulation of the IR laser pulses (Helbing and Bonmarin, 2009). Here, the minor axis component of the elliptically modulated radiation is heterodyned with the chiral depolarization of the major component of the elliptically polarized component of the IR beam. The principal advantage of this method is that the two orthogonally polarized components of the laser beam both follow the same optical path and are locked in-phase with one another. A minor change in the optical arrangement allows conversion from VCD to

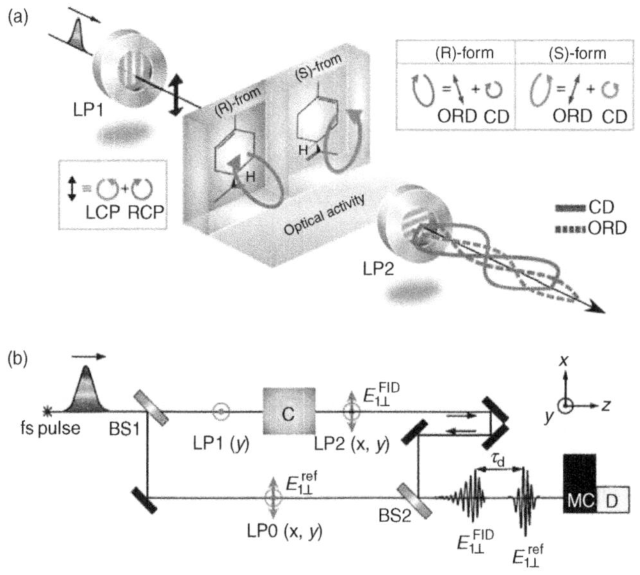

Figure 6.16 *(a) The effect of mirror-image chiral molecular samples on a state of pure linearly polarized radiation creating both VCD and VORD (VCB) of opposite signs that are 90° out of phase with respect to one another. (b) The laser optical heterodyning method of simultaneous VCD and VORD measurement where an initial femtosecond IR broadband pulse is first divided at BS1 into reference and sample beams that are linearly polarized by LP0 and LP1. An additional polarizer placed after the sample allows isolation of the chiral responses of the sample shown in (a) when the sample and reference beams are time delayed relative to each other and combined as time-dependent waveforms by the beamsplitter BS2. The waveforms interfere and are dispersed in frequency by a monochromator (MC) and detected by an IR detector D. The detector signal is time-base Fourier transformed where the real and imaginary parts are the VORD and VCD spectra, respectively. Reproduced with permission from the Nature Publishing Group (Rhee et al., 2009b)*

VCB measurement. The VCD or VCB spectra are obtained as time-base Fourier transforms of the heterodyned intensity at the detector, which is measured one wavenumber band at a time by scanning a monochromator. Preliminary VCD measurements in the CH stretching were reported.

In a different approach, there is no polarization modulation. Instead the IR laser pulse is first divided into two beams. One beam acts as a reference and is directed around the sample through a polarizer while the signal beam passes through a linear polarizer, the sample, and a second polarizer aligned either parallel or perpendicular to the first polarizer. The chiral sample creates a small orthogonally polarized component that is time-delayed and heterodyned with the reference beam (Rhee *et al.*, 2009b). The basic concepts underlying this measurement method are illustrated in Figure 6.16.

In this situation the VCD and VCB are simultaneously available as the real and imaginary parts of the heterodyned detector intensity. As with the self-heterodyning method, the setup allows for either time-independent measurements of VCD and VCB or the optical setup is easily modified to initiate a photochemical event with a pump laser and to then probe the system with femtosecond time resolution using the same detection scheme as that for static measurements. The quality of the VOA measurement depends heavily on the quality of the polarizers used in initial polarization division of the beam. The first simultaneous measurements of VCD and VCB using this technique were reported for the CH-stretching region with calcite polarizers. Subsequently, a new design of reflective Brewster's angle germanium polarizers permitted extension of measurements into the mid-IR region (Rhee *et al.*, 2009a). A further improvement of the method was recently reported in which a multi-channel IR detector eliminated the need to scan the heterodyne detector signal with a monochromator over wavenumber spectral locations for several different regions of the IR (Rhee *et al.*, 2010).

References

Cao, X., Dukor, R.K., and Nafie, L.A. (2008) Reduction of linear birefringence in vibrational circular dichroism measurement: use of a rotating half-wave plate. *Theor. Chem. Acc.*, **119**, 69–79.

Cheng, J.C., Nafie, L.A., and Stephens, P.J. (1975) Polarization scrambling using a photoelastic modulator: Application to circular dichroism measurement. *J. Opt. Soc. Am.*, **65**, 1031–1035.

Helbing, J., and Bonmarin, M. (2009) Vibrational dichroism signal enhancement using self- heterodyning with elliptically polarized laser pulses. *J. Chem. Phys.*, **1131**, 174507.

Lipp, E.D., Zimba, C.G., and Nafie, L.A. (1982) Vibrational circular dichroism in the mid-infrared using Fourier transform spectroscopy. *Chem. Phys. Lett.*, **90**, 1–5.

Lombardi, R.A., and Nafie, L.A. (2009) Observation and calculation of a new form of vibrational optical activity: Vibrational circular birefringence. *Chirality*, **21**, E277–E286.

Long, F., Freedman, T.B., Hapanowicz, R., and Nafie, L.A. (1997a) Comparison of step-scan and rapid-scan approaches to the measurement of mid-infrared Fourier transform vibrational circular dichroism. *Appl. Spectrosc.*, **51**, 504–507.

Long, F., Freedman, T.B., Tague, T.J., and Nafie, L.A. (1997b) Step-scan Fourier transform vibrational circular dichroism measurements in the vibrational region above $2000\,cm^{-1}$. *Appl. Spectrosc.*, **51**, 508–511.

Nafie, L.A. (2000) Dual polarization modulation: a real-time, spectral-multiplex separation of circular dichroism from linear birefringence spectral intensities. *Appl. Spectrosc.*, **54**, 1634–1645.

Nafie, L.A., Cheng, J.C., and Stephens, P.J. (1975) Vibrational circular dichroism of 2,2, 2-trifluoro-1-phenylethanol. *J. Am. Chem. Soc.*, **97**, 3842–3843.

Nafie, L.A., Keiderling, T.A., and Stephens, P.J. (1976) Vibrational circular dichroism. *J. Am. Chem. Soc.*, **98**, 2715–2723.

Nafie, L.A., and Diem, M. (1979) Theory of high frequency differential interferometry: Application to infrared circular and linear dichroism via Fourier transform spectroscopy. *Appl. Spectrosc.*, **33**, 130–135.

Nafie, L.A., Diem, M., and Vidrine, D.W. (1979) Fourier transform infrared vibrational circular dichroism. *J. Am. Chem. Soc.*, **101**, 496–498.

Nafie, L.A., Lee, N.S., Paterlini, G., and Freedman, T.B. (1988) Polarization modulation Fourier transform infrared spectroscopy. *Microchim. Acta*, 1987(III) 93–104.

Nafie, L.A., Buijs, H., Rilling, A., Cao, X., and Dukor, R.K. (2004) Dual source Fourier transform polarization modulation spectroscopy: An improved method for the measurement of circular and linear dichroism. *Appl Spectrosc.*, **58**, 647–654.

Polavarapu, P.L. (1988) Far infrared circular dichroism measurements with Martin-Puplett interferometer: Methods and analysis. *Infrared Phys.*, **28**, 109.

Polavarapu, P.L., Quincey, P.G., and Birch, J.R. (1990) Circular dichroism in the far infrared and millimeter wavelength regions: Preliminary measurements. *Infrared Phys.*, **30**, 175–179.

Ragunathan, N., Lee, N.-S., Freedman, T.B., Nafie, L.A., Tripp, C., and Buijs, H. (1990) Measurement of vibrational circular dichroism using a polarizing Michelson interferometer. *Appl. Spectrosc.*, **44**, 5–7.

Rhee, H., Kim, S.-S., Jeon, S.-J., and Cho, M. (2009a) Femtosecond measurements of vibrational circular dichroism and optical rotatory dispersion spectra. *ChemPhysChem*, **10**, 2209–2211.

Rhee, H., Young-Gun, J., Lee, J.-S. *et al.* (2009b) Femtosecond characterization of vibrational optical activity of chiral molecules. *Nature*, **458**, 310–313.

Rhee, H., Kim, S.-S., and Cho, M. (2010) Multichannel array detection of vibrational otpical activity free-induction-decay. *J. Anal. Sci. Technol.*, **1**, 147–151.

Russel, M.I., Billardon, M., and Badoz, J.P. (1972) *Appl. Opt.*, **11**, 2375–2378.

Wang, B., and Keiderling, T.A. (1995) Observations on the measurement of vibrational circular dichroism with rapid-scan and step-scan FT-IR techniques. *Appl. Spectrosc.*, **49**, 1347–1355.

Yang, W.-J., Griffiths, P.R., and Kemeny, G.J. (1984) Vibrational circular dichroism measurements by optical subtraction FT-IR spectrometry. *Appl. Spectrosc.*, **38**, 337–343.

7

Instrumentation for Raman Optical Activity

Instrumentation for the measurement of ROA is more sophisticated, more varied, and generally more expensive than instrumentation for VCD. Most of the additional expense comes from the more expensive light source, a high-powered laser versus a black-body glower, and a more expensive detector, a cooled multi-channel, charge-coupled device (CCD) detector with an electronic interface versus a cooled single-element detector with a preamplifier. Because of the more complex theory of ROA compared with VCD, ROA invites a much wider variation in instrumental setup to measure the numerous theoretically distinct forms of ROA intensity. In Chapter 5 on the theory of ROA, it was explained that there are eight distinct forms of conventional, linear (versus non-linear) ROA compared with one invariant and two forms of infrared VOA, the rotational strength associated with both VCD and VCB. In this chapter, we describe the instrumental setups for measuring all four forms of circular polarization (CP) ROA and discuss briefly the prospects for the measurement of any of the four forms of linear polarization (LP) ROA.

7.1 Incident Circular Polarization ROA

Incident circular polarization ROA, or ICP-ROA, is the original form of ROA first measured in 1973 (Barron *et al.*, 1973). The basic idea is to modulate the incident laser radiation between right and left circular polarization states in a square-wave duty cycle and accumulate Raman scattering counts separately in two digital storage channels, one for RCP and the other for LCP. The sum of both channels is the ICP-Raman spectrum and the difference of RCP minus LCP counts is the ICP-ROA spectrum. The energy-level diagram and defining equations for the measurement of ICP-Raman and ICP-ROA, summarized from Chapter 1, are presented in Figure 7.1.

Comparison of measured and calculated ROA must take into account the enantiomeric excess (*ee*) of the sample defined in Chapters 1 and 3. For most natural biological samples, such as proteins, nucleic acids, and carbohydrates, this is not an issue due to the homo-chirality of our biosphere, in

Vibrational Optical Activity: Principles and Applications, First Edition. Laurence A. Nafie.

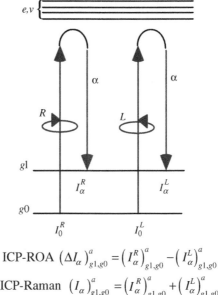

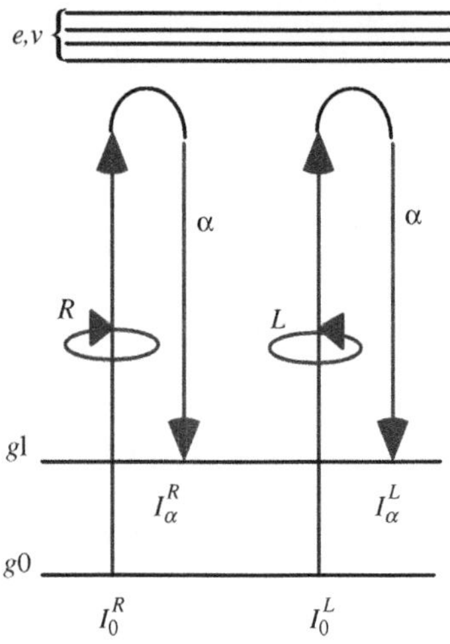

$$\text{ICP-ROA } (\Delta I_\alpha)^a_{g1,g0} = (I^R_\alpha)^a_{g1,g0} - (I^L_\alpha)^a_{g1,g0}$$

$$\text{ICP-Raman } (I_\alpha)^a_{g1,g0} = (I^R_\alpha)^a_{g1,g0} + (I^L_\alpha)^a_{g1,g0}$$

Figure 7.1 *Energy-level diagram and basic definitions of ICP-Raman and -ROA for a Raman transition between vibrational states 0 and 1 of the ground electronic state, g, for normal mode a. The scattered light is unpolarized or in a fixed state of linear polarization, α, having no circular polarization content*

which case the *ee* equals 1. The molecular property associated with Raman and ROA intensity is expressed in terms of the differential Raman and ROA cross-sections, $\mathrm{d}\sigma_\alpha(\theta)/\mathrm{d}\Omega$ and $\Delta\mathrm{d}\sigma_\alpha(\theta)/\mathrm{d}\Omega$, respectively, defined in Chapter 1. The cross-sections have units of area per solid angle Ω and are defined for light collected in a cone of angle θ about a specified direction, such as 90° right-angle scattering or 180° backscattering. These differential cross-sections can be related to the experimentally measured intensities and their corresponding theoretical expressions developed in Chapter 5 by the relationships

$$(\mathrm{d}\sigma_\alpha(\theta)/\mathrm{d}\Omega)^a_{ev',ev} = \frac{1}{I_0 NCV}(I_\alpha)^a_{ev',ev} \tag{7.1}$$

$$[\Delta\mathrm{d}\sigma_\alpha(\theta)/\mathrm{d}\Omega]^a_{ev',ev} = \frac{1}{I_0 NCV(ee)}(\Delta I_\alpha)^a_{ev',ev} \tag{7.2}$$

where I_0 is the incident laser intensity, N is Avagadro's number, C is the molar concentration of the sample, and V is the laser illumination volume in the sample. The total Raman or ROA cross-section for light scattered in all directions can be obtained by integrating the differential cross-section over the 4π stereradians of a sphere as:

$$(\sigma_\alpha)^a_{ev',ev} = \int_0^\pi\int_0^{2\pi} [\mathrm{d}\sigma_\alpha(\theta)/\mathrm{d}\Omega]^a_{ev',ev}\mathrm{d}\Omega \tag{7.3}$$

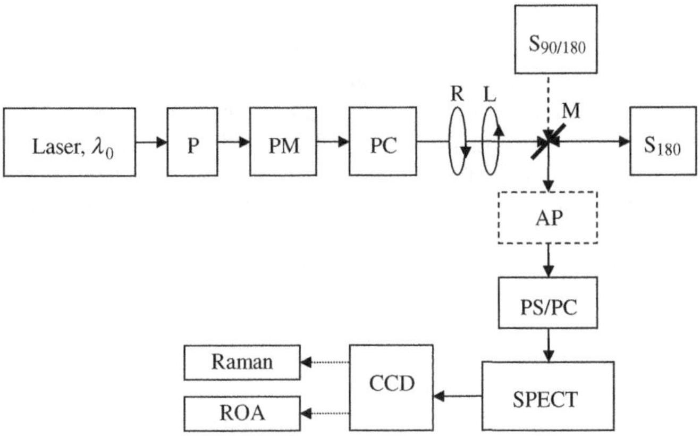

Figure 7.2 *Instrument block diagram for the measurement of ICP-Raman and -ROA showing the optical components, polarizer (P), polarization modulator (PM), polarization conditioning optics (PC), sample positions (S), mirror used for backscattering (M), optional analyzing polarizer (AP), polarization scrambling or conditioning optics (PS/PC), spectrograph (SPECT), and charge-coupled device (CCD) detector*

$$(\Delta\sigma_\alpha)^a_{ev',ev} = \int_0^{\pi}\int_0^{2\pi} \left[\Delta d\sigma_\alpha(\theta)/d\Omega\right]^a_{ev',ev} d\Omega \tag{7.4}$$

where the solid angle, as defined conventionally, is $d\Omega = \sin\theta d\theta d\phi$.

7.1.1 Optical Block Diagram for ICP-Raman and ROA Scattering

The measurement of ICP-Raman and ROA is illustrated in Figure 7.2. This diagram is kept as general as possible to encompass the various designs for the implementation ICP-ROA measurement that have taken place since the early 1970s. Light from a laser is linearly polarized with a polarizer (P). In many instances the output of the laser is already highly polarized, in which case the polarizer simply cleans up the polarization to ensure that the beam is as close to 100% linearly polarized as possible. Next the laser light is polarization modulated (PM) in a square-wave fashion between right (R) and left (L) circular polarization states. This is typically accomplished with an electro-optic modulator where the retardation voltages can be adjusted independently to achieve RCP and LCP states of near perfect circularity, or $\pm 90°$ phase retardation of the linearly polarized beam. Polarization modulation is followed by polarization conditioning (PC), to be discussed further below, that achieves in a time-averaged sense equal intensities of incident RCP and LCP radiation free of any linear polarization components. Because ROA is measured in the visible region of the spectrum, the major source of noise is so-called photon noise or shot noise. This noise for any signal is equal to the square root of the number of photo-electron counts measured. As such, there is no noise advantage to rapid polarization modulation, as is the case of infrared VCD where the major source of noise is background detector noise, independent of the radiation intensity. The square-wave polarization modulation in an ICP-ROA measurement is the in range of ten to a few hundred Hz (cycles per second) at most.

Once polarization modulation has been achieved, the beam is directed to the sample for Raman scattering. In Figure 7.2, two possible sample positions, backscattering (S_{180}) or right-angle (S_{90}) scattering are illustrated. In the case of backscattering, two further strategies have been implemented. The first is illustrated where the laser beam passes through a small hole in the back side of a 45°-angle mirror (M), and the back-scattered light is reflected at right angles by the mirror and sent down the optical rail for the scattered light (Hecht *et al.*, 1992a). The other backscattering strategy is to use a small mirror or reflecting right-angle prism at M to direct the laser beam to the sample position currently labeled $S_{90/180}$ (Nafie and Che, 1994; Che *et al.*, 1991). Nearly all of the backscattered light passes around the small mirror or prism, and then continues along the scattered-light rail. The final strategy illustrated is right-angle scattering. For this measurement, the mirror or prism is removed, and the sample $S_{90/180}$ is moved along the dashed line in the figure such that laser light enters directly into the sample cell and is scattered at right angles along the scattered-light rail (Hecht *et al.*, 1992b).

For all ICP measurements, one seeks to measure the scattered Raman intensities with either no polarization or with some specified state of linear polarization without any circular or elliptical polarization content. In the case of unpolarized ICP scattering intensity, the polarization state of the scattered light needs to be scrambled (PS) or averaged over time to zero in a manner similar to the polarization conditioning (PC) of the laser radiation described above. PS can be achieved with a Lyot depolarizer that spatially scrambles the polarization (Hecht *et al.*, 1992a), while PC is carried out by a rotating half-wave plate that averages linear polarizations states to zero (Hug, 2003) as demonstrated previously by the expression for the Mueller matrix for a rotating half-wave plate given in Equation (6.80). If desired both strategies can be implemented.

For the measurement of the vertically or horizontally polarized components of the scattered light, one must include an analyzing polarizer (AP) before the PS or PC step. Once these polarization conditioning steps have been completed, the scattered light may be brought to the spectrograph (SPECT), dispersed by a grating onto a charge-coupled device (CCD) detector, and processed into ICP-Raman and -ROA spectra, as indicated in Figure 7.2. The most efficient spectrographs possess a holographically ruled transmission grating, obtained from Kaiser Optical Systems, Inc., that achieves close to 80% average quantum efficiency over all polarization states, and a back-thinned, cooled CCD, which also achieves well above 80% quantum efficiency in the visible region of the spectrum.

7.1.2 Intensity Expressions

Intensity expressions of ROA for common experimental setups were given previously in Chapter 5. Here we discuss these intensity expressions for ICP-Raman and -ROA scattering in the context of the instrumental setup used for their measurement. The original form of ROA, used in the experimental discovery of ROA, is depolarized right-angle scattering,

$$\text{ICP}_D\text{-ROA}\,(90^\circ): \qquad I_z^R(90^\circ) - I_z^L(90^\circ) = \frac{4K}{c}\left[6\beta(G')^2 - 2\beta(A)^2\right] \tag{7.5}$$

$$\text{ICP}_D\text{-Raman}\,(90^\circ): \qquad I_z^R(90^\circ) + I_z^L(90^\circ) = 2K\left[6\beta(\alpha)^2\right] \tag{7.6}$$

To measure this form of ROA, the laser light is sent directly to the sample cell without mirror reflection, as discussed above for Figure 7.2. One then inserts an analyzing polarizer (AP) into the scattered light beam immediately after the sample in the z-direction that is oriented parallel to the yz-plane of scattering, where the geometry of the scattering experiment is illustrated in Figure 5.1 in Chapter 5. The direction of this polarization state is perpendicular to the xy-plane of the circular polarization state of the incident light and is a completely depolarized form of Raman scattering, as can be seen by the absence of any isotropic ROA and Raman invariants, as described in Chapter 5.

A closely related form of ROA is right-angle polarized ICP. This is measured simply by rotating the AP by 90° such that the scattered light collected is polarized perpendicular to the plane of scattering. The intensity expressions for this form of ROA are:

$$\text{ICP}_P\text{-ROA }(90°):\qquad I_x^R(90°) - I_x^L(90°) = \frac{4K}{c}\left[45\alpha G' + 7\beta(G')^2 + \beta(A)^2\right] \tag{7.7}$$

$$\text{ICP}_P\text{-Raman }(90°):\qquad I_X^R(90°) + I_X^L(90°) = 2K\left[45\alpha^2 + 7\beta(\tilde{\alpha})^2\right] \tag{7.8}$$

where the polarized nature of the scattered light is seen most clearly in Equation (7.8) Here the isotropic Raman invariant is present and the depolarization ratio, defined in Chapter 2, is 6/7 for depolarized vibrational bands. This form of ROA has been measured only a few times due to its higher susceptibility to polarization artifacts (Hug and Surbeck, 1979).

Following the demonstration of backscattering ROA, the form of ICP scattering that has been more recently employed almost exclusively is unpolarized backscattering. To measure this form of ROA, a backscattering geometry is selected in Figure 7.2, either the one with a hole in a right-angle mirror or the one with a small reflecting right-angle prism. No analyzing polarizer is used, and the polarization content of all the scattered light is removed by PS, PC, or both. The intensity expressions for this form of ICP scattering are given by:

$$\text{ICP}_U\text{-ROA }(180°):\qquad I_U^R(180°) - I_U^L(180°) = \frac{8K}{c}\left[12\beta(G')^2 + 4\beta(A)^2\right] \tag{7.9}$$

$$\text{ICP}_U\text{-Raman }(180°):\qquad I_U^R(180^{\circ}) + I_U^L(180°) = 4K\left[45\alpha^2 + 7\beta(\alpha)^2\right] \tag{7.10}$$

The unpolarized ICP Raman scattered intensity here is the same, but larger by a factor of two, compared with that obtained with right-angle *polarized* ICP scattering in Equation (7.8). On the other hand, aside from a much larger magnitude, the backscattered ROA intensity is close to, but somewhat different from, that obtained with *depolarized* ICP-ROA in Equation (7.5). These similarities and differences will be described briefly in the following section.

7.1.3 Advantages of Backscattering

The first backscattering ROA spectrometer was built and described in the early 1980s (Hug, 1982), however this instrument was destroyed by fire and never used for routine application. The idea of backscattering ROA was resurrected a number of years later by collaboration between Hug and Barron where the advantages of backscattering ROA, along with multi-channel detection with a CCD detector, were brought to fruition (Barron *et al.*, 1989). Ever since this second publication of backscattering ROA, the vast majority of ROA measurements have been carried out in backscattering geometry.

The principal advantage of backscattering is one of intensity. Comparison of Equations (7.10) and (7.8) shows an advantage by a factor of two favoring backscattering over polarized right-angle scattering, all other variables being equal, such as the cone-angle of light collection and incident laser intensity. Thus, twice as many photons are collected in backscattering compared with the same spectrum in right-angle scattering during the same period of measurement. In addition, if one compares the two ROA invariants appearing in backscattering ROA in Equation (7.9) with the corresponding invariants in depolarized right-angle ROA in Equation (7.5), one sees that backscattering is favored by a factor of four. Therefore, in backscattering not only is the parent Raman intensity twice as intense as right-angle scattering, but the ROA invariants are twice as intense relative to the Raman invariants in backscattering, making the net effect a factor of four. While the relative sign of the magnetic dipole and electric quadrupole ROA invariants for a given ROA band are not known, simple models, and most often in quantum calculations, predict they have the same sign. If this is true, then backscattering has even a third advantage. The magnetic dipole and electric quadrupole ROA invariants appear with the same sign in backscattering ROA and opposite (canceling) signs in right-angle ROA scattering.

Backscattering ROA carries yet another type of advantage. This is associated with artifacts arising from the incomplete control of circular polarization. If for right-angle scattering the RCP and LCP components of the incident laser radiation have small undesired components of orthogonal linear polarization states, then a large artifact intensity is present in the measured ROA due to the ordinary Raman depolarization ratio, even if the scattered light has no linear or circular polarization intensity. This is because one of the two small orthogonal linear polarization states of the incident radiation may have a different orientation relative to the plane of scattering, which is the basis of a depolarization ratio measurement. On the other hand, if the experimental geometry is backscattering, artifacts arising from the depolarization ratio are not encountered as easily. In particular, if all polarization is removed from the scattered radiation in backscattering any pair of orthogonal linear polarization states associated with the circular polarization modulation have no difference in their contribution to the scattered intensity and hence cancel in the ROA measurement. This because in backscattering there is no plane of polarization, and the only way to measure a depolarization ratio is by the difference in intensity for orthogonal linear polarization states relative to a fixed linear polarization discrimination in the scattered beam that favors one incident linear polarization state over the other. If there is no linear polarization (or circular polarization bias) in the optical components of the scattered beam before the CCD detector, then no depolarization artifacts can occur. Nevertheless, complete elimination of linear polarization bias is difficult to achieve completely and for strongly polarized bands, artifact control in any form of ROA is still the most challenging part of the measurement.

7.1.4 Artifact Suppression

For ICP-ROA measurement, artifact suppression involves control of the circular polarization balance between RCP and LCP states of the incident radiation and elimination of all polarization states in the scattering beam prior to detection.

Control of the balance between RCP and LCP states is most directly accomplished by adjusting the voltages of the polarization modulator, PM in Figure 7.2, so that circular polarization states are created with as small a linear polarization content as possible. While this can be achieved in principle, the tolerances for departure from the pure circular state are very small before artifacts begin appearing in the ROA spectrum. In addition, small optical changes in the PM or other optical components between the PM and sample, due temperature or other intangible causes, can occur within days or even hours, requiring re-adjustment of the PM settings.

An alternative approach is to carry out an initial optimization of the alignment and PM settings after which a stage of polarization conditioning, indicated by PC in Figure 7.2, is carried out. The conditioning involves a two-stage use of half-wave plates to first eliminate linear polarization content from both RCP and LCP states followed by use of a second half-wave plate to interchange the RCP and LCP polarizations states, such that any difference in intensity between them is time-averaged to zero (Hug, 2003). The first step is achieved by a rotating half-wave plate, which, as we have seen from its Mueller matrix in Chapter 6, averages to zero all linear polarization states over time and also interchanges LCP and RCP states. The second step involves carrying out ICP modulation cycles first with a half-wave plate out of the beam followed by the half-wave plate inserted in the beam.

The computer controlling the instrument and its various optical functions and detector read-outs keeps track of whether RCP or LCP is present at the sample and stores the Raman counts in an appropriate register, such that at the conclusion of the measurement all the counts for LCP incident radiation can be subtracted from those with RCP incident on the sample. In this way any linear polarization content or subsequent inequality of the intensity of the laser light reaching the sample is automatically compensated, even if changes in optical conditions are gradual over the course of the ROA spectral collection.

The elimination of polarization content of the scattered light can be addressed by two methods, both of which are illustrated in Figure 7.2. In the first, a polarization scrambling device called a Lyot depolarizer is inserted in the scattered beam. Radiation that passes through the Lyot depolarizer at slightly different angles experiences sharply different degrees of polarization retardation. The scattered light is collected over a wide cone angle and hence any polarization content of the beam as a whole is scrambled spatially to a very large degree. The details of the operation of this device for ROA measurements have not been published other than it is claimed to be effective in reducing artifacts due to polarization effects in the scattered beam (Hecht *et al.*, 1992a).

An alternative strategy to eliminate the polarization states of the scattered beam is similar to the PC operations used to balance the CP state of the incident beam. In this method, a rotating half-wave plate is inserted into the beam to average to zero over time any linear polarization content of the scattered light. Next, a half-wave plate is used that is either in or out of the beam during equal time periods of the LCP and RCP illumination cycles. The purpose of the alternating half-wave plate is to remove completely any circular polarization bias that may exist in the scattered light collection optics between the sample and detector. It is very important that the true Raman intensities scattered with RCP and the LCP light incident on the sample are detected at the CCD without any CP distortion.

The net effect of the use of half-wave plates in the incident and scattered beam is to make the entire ROA instrument from laser to CCD automatically CP neutral so that the true ROA intensity originating at the molecular level within the sample is measured by the spectrometer without any distortion of the ROA spectrum that originates at the sample. Without some form of automated self-correcting optical routines, the typical measurement of ROA without distortion is difficult to achieve. For the methods described above, modulation cycles for the rotation and circulation of the half-wave plates must be devised that are both efficient and non-interactive between components.

7.2 Scattered Circular Polarization ROA

Scattered circular polarization (SCP) ROA, is the second form of ROA to be measured (Spencer *et al.*, 1988), and is the form of ROA used in the first and to date only commercially available ROA spectrometer. The basic idea is to use unpolarized or linearly polarized incident laser radiation that is completely devoid of any circular polarization component. Right and left circular polarization states of the scattered light are separately measured and either summed to produce the SCP-Raman spectrum or subtracted (RCP minus LCP) to produced the SCP-ROA spectrum. SCP-ROA had been predicted before the experimental discovery of ROA, when it was determined theoretically that an ROA spectrum is also present as the degree of circularity in the scattered radiation, where it was correlated to a single right or left elliptical polarization state of the radiation at each point in the Raman spectrum rather than as a difference between RCP and LCP intensity of the scattered beam (Barron and Buckingham, 1971). Of course, these are equivalent ways of thinking about SCP-ROA, but the second concept carries with it a clear procedure for its measurement. The energy-level diagram and equations defining the measurement of SCP-Raman and SCP-ROA analogous to Figure 7.1 for ICP are given below in Figure 7.3.

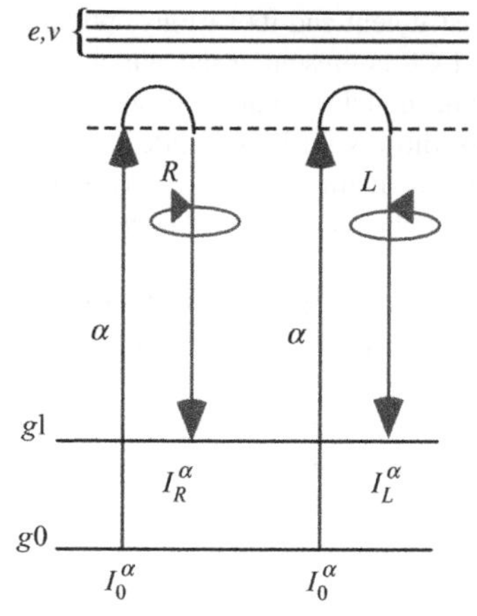

$$\text{SCP-ROA}\ \left(\Delta I^{\alpha}\right)^{a}_{g1,g0} = \left(I^{\alpha}_{R}\right)^{a}_{g1,g0} - \left(I^{\alpha}_{L}\right)^{a}_{g1,g0}$$

$$\text{SCP-Raman}\ \left(I^{\alpha}\right)^{a}_{g1,g0} = \left(I^{\alpha}_{R}\right)^{a}_{g1,g0} + \left(I^{\alpha}_{L}\right)^{a}_{g1,g0}$$

Figure 7.3 Energy-level diagram and basic definitions of SCP Raman and ROA for a Raman transition between vibrational states 0 and 1 of the ground electronic state, g, for normal mode a. The incident light is unpolarized or is in a fixed state of linear polarization, α, having no circular polarization content

7.2.1 Measurement of SCP-ROA and Raman Scattering

The measurement of SCP-ROA was first achieved using a right-angle scattering geometry and in-plane linearly polarized incident radiation, a depolarized Raman measurement (Spencer *et al.*, 1988). In this experiment, incident radiation propagating in the z-direction was polarized along the y-axis, and scattered light was collected at 90° along the y-axis thus forming an yz-plane of scattering and an xz-plane of scattered circular polarization, as can be visualized in reference to Figure 5.1 in Chapter 5. Thus the incident polarization state is perpendicular to any instantaneous polarization state of the scattered light and the scattering is therefore depolarized. The theoretical expressions for SCP-ROA and SCP-Raman intensity in the far-from-resonance (FFR) approximation are given by:

$$\text{SCP}_D\text{-ROA}\ (90°):\quad I_R^y(90°) - I_L^y(90°) = \frac{4K}{c}\left[6\beta(G')^2 - 2\beta(A)^2\right] \tag{7.11}$$

$$\text{SCP}_D\text{-Raman}\ (90°):\quad I_R^y(90°) + I_L^y(90°) = 2K\left[6\beta(\alpha)^2\right] \tag{7.12}$$

This is the same combination of ROA and Raman invariants as given for depolarized right-angle ICP-ROA and Raman given in Equations (7.5) and (7.6).

The SCP-ROA spectrum was measured, in direct analogy to ICP-ROA, as the difference in the alternately measured intensity of RCP and LCP Raman scattered radiation. The orthogonal RCP and LCP states of the scattered radiation were first converted into orthogonal linearly polarized states by a zeroth-order quarter-wave plate (QWP) that was optimized for quarter-wave retardation in the middle of the Raman spectrum. One of the linear polarization states was selected by an analyzing polarizer

(AP), thereby isolating intensity originating in a particular state of circular polarization, such as RCP. The opposite circular polarization, LCP, was then selected by rotating the QWP by 90° while keeping the orientation of the AP fixed. Thus, RCP and LCP intensities of the scattered radiation were measured sequentially and separately using the same state of linear polarization incident on the spectrograph and detector.

Approximately a decade later, Hug proposed a method for measuring the various forms of CP-ROA that included the measurement of SCP-ROA with simultaneous collection of RCP and LCP scattered radiation (Hug and Hangartner, 1999). This was accomplished by first converting orthogonal CP states into orthogonal LP states, as in the original SCP-ROA measurements, followed by a polarization beamsplitting cube (BSC) to spatially separate the two LP states. These two beams were then focused into separate fiber-optic bundles each consisting of 31 fibers in a 1–6–12–12 concentric honeycomb pattern with a central fiber, surrounded by 6 fibers, surrounded by 12 fibers and a final set of 12 more fibers. The two sets of 31 fibers from the two beams were transformed into a single curved linear array of 62 fibers that formed effectively the entrance slit of the ROA spectrograph. The upper 31 fibers carried the intensity of one sense of CP while the lower 31 fibers carried the intensities of the opposite CP state. Finally, the radiation from the linear array of fibers was dispersed by the spectrograph and imaged on to the CCD detector. In this way RCP and LCP scattered radiation were simultaneously recorded without the need to modulate alternately between the measurement of RCP and LCP intensities. The curvature of the linear array of 61 input fibers compensated for a natural curvature associated with the spatial dispersion of the spectrograph, such that each wavelength is imaged at the CCD as a straight vertical column of a single narrow band of wavelengths.

7.2.2 Optical Block Diagram for SCP-Raman and ROA Measurement

A general optical diagram illustrating the measurement of SCP-ROA is given in Figure 7.4. Radiation from the laser is first polarized (P) and sent through a set of polarization conditioning (PC) optics.

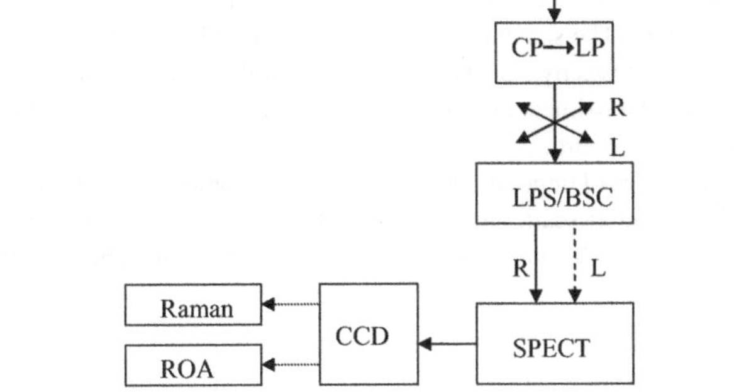

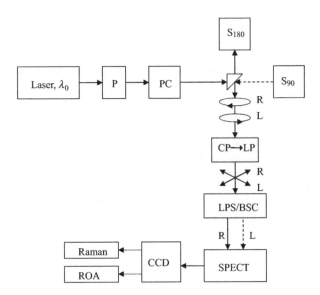

Figure 7.4 *Instrument block diagram for the measurement of SCP-Raman and SCP-ROA. CP to LP refers to the conversion of circular polarization into linear polarization and LPS/BSC is either a linear polarization selector or a linear polarization beam splitting cube*

These consist of a rotating half-wave plate (HWP) to eliminate over time all linear polarization states followed by an HWP that is in the beam half the time and out of the beam the other half. Any residual circular polarization present on the beam is reduced to zero by the action of this second step of PC. Thus the radiation arriving at the sample has no net linear or circular polarization when averaged over time, and hence the beam is effectively unpolarized. If a particular polarization state is desired for the incident radiation beam, the PC optical stage can be removed or modified. Next, the incident laser radiation is directed to the sample as illustrated in either the right-angle or the backscattering geometry. For the backscattering geometry option in Figure 7.4, the incident radiation is reflected off the diagonal surface of a small right-angle prism. The backscattered radiation passes around the prism, with only a few per cent loss due to blockage, and is directed down the scattered-radiation optical rail. In the case of right-angle scattering, the prism in removed and the sample is moved to the prism position where the light scattered at right-angles is directed down the optical rail. The orthogonal CP states of the scattered radiation are converted into orthogonal LP states by a quarter-wave plate. These LP states are either selected (LPS) by an analyzing polarizer for separate measurement of the intensities for RCP and LCP radiation, or they are separated by a beam splitting cube (BSC) for simultaneous measurement of RCP and LCP intensities as described above. Once spectral accumulation of RCP and LCP scattered radiation is achieved, the corresponding Raman and ROA intensities can be obtained by adding or subtracting these two spectral accumulations

7.2.3 Comparison of ICP- and SCP-ROA

We have already seen that in the FFR approximation the intensities for ICP and SCP-ROA are the same. Initially, however, SCP-ROA had some practical disadvantages relative to ICP-ROA. These were that: (i), the incident laser radiation is typically highly polarized making unpolarized SCP measurements difficult to achieve, whereas the measurement of unpolarized ICP-ROA is straightforward; (ii) the separation of the scattered light into orthogonal circular polarization states can be done exactly only at one scattered radiation frequency and only close to exact elsewhere in the spectrum, whereas CP modulation of the incident radiation in ICP is exact as it takes place at the single laser radiation frequency; and (iii) if the selection of RCP and LCP scattered radiation is carried out in square-wave modulation form, then one rejects approximately half (the unmeasured orthogonal CP state) of the scattered light at all times during this process. The advent of the Hug design for ROA instrumentation, discussed in the previous section, overcomes the first and third of these concerns for SCP-ROA measurement. A direct and efficient method of eliminating over time all polarization modulation on the incident radiation beam is included in a polarization condition stage (PC) and both RCP and LCP scattered radiation intensities are measured simultaneously using the dual fiber optic collection system. The second concern is relatively minor and can be corrected by a mild weighting function that increases by less than a few per cent the measured ROA intensity away from the frequency of exact circular to linear polarization conversion by a zeroth-order QWP optimized near the middle of the Raman spectrum.

With the technical advances of unpolarized incident laser radiation and simultaneous collection and measurement of RCP and LCP scattered intensities, the SCP analogue of unpolarized ICP backscattering, given in Equations (7.9) and (7.10), can be measured with the same overall efficiency as ICP and is given by:

$$\text{SCP}_U\text{-ROA}\,(180^\circ):\quad I_R^U(180^\circ) - I_L^U(180^\circ) = \frac{8K}{c}\left[12\beta(G')^2 + 4\beta(A)^2\right] \tag{7.13}$$

$$\text{SCP}_U\text{-Raman}\,(180^\circ):\quad I_R^U(180^\circ) + I_L^U(180^\circ) = 4K\left[45\alpha^2 + 7\beta(\alpha)^2\right] \tag{7.14}$$

As in ICP scattering, the backscattered Raman intensity is the same as the standard polarized right-angle Raman spectrum using unpolarized or circularly polarized incident radiation, and the ROA spectrum involves only depolarized ROA invariants.

With the new SCP-ROA design of simultaneous measurement of RCP and LCP scattered intensities, SCP enjoy two closely related advantages not present in ICP-ROA measurement. The first is the absence of a need to modulate the polarization state between RCP and LCP. Thus, there are no electronics required to switch the CP states. The ROA spectrum simply accumulates in time without modulation. The absence of a polarization modulation mechanism simplifies somewhat the construction of the modern SCP-ROA spectrometer. Secondly, and perhaps more importantly, because both RCP and LCP scattered radiation are measured simultaneously from the same laser focal volume at the same instance of time, fluctuations in the laser intensity and any other transient fluctuation in the sample volume, for example a dust particle in the laser beam, are canceled upon subtraction of LCP intensity from RCP intensity to form the SCP-ROA spectrum.

7.2.4 Artifact Reduction in SCP-ROA Measurement

Essentially, automated artifact reduction in SCP-ROA is similar to that for ICP-ROA although there are some differences arising from the simultaneous detection of RCP and LCP scattered intensities. The artifact reduction occurs in two stages. Rotating and alternating insertion and removal of half-wave plates first eliminate, respectively, the linear and circular polarization content of the laser beam, as was carried out for the scattered beam in ICP-ROA. For the scattered beam in the SCP measurement, the true ROA intensity is preserved while artifacts from optical sources are eliminated. Initially any linear polarization components on the scattered beam are eliminated by a rotating half-wave plate. Net linear polarization induced in the scattered radiation may be present to different extents in the RCP and LCP polarization states. These are then eliminated by inserting, for half of the collection time, a half-wave plate, thereby measuring the RCP radiation either as RCP with the HWP out or as LCP with the HWP in the beam. This neutralizes any CP bias of the instrument detection optics beyond the optical location of the HWP insertions. The conversion from CP to LP radiation is achieved effectively by a QWP device. Because there are different throughput efficiencies for vertical and horizontal polarization states in the polarization beam splitting cube, these LP states are switched for RCP and LCP by interchanging the fast and slow axes of the QWP. As in the case of ICP-ROA, these rotating and position-modulated HWPs serve first to remove all polarization content in the incident beam and neutralize as much as possible any CP bias that may exist along the optical and detection pathway of the scattered beam.

7.3 Dual Circular Polarization ROA

There are two forms of dual circular polarization (DCP) ROA, the in-phase or DCP_I-ROA and the out-of-phase or DCP_{II}-ROA. The energy-level diagrams and definitions of these two forms of ROA are illustrated in Figure 7.5. Because Raman scattering is a single two-step quantum process with correlation between the incident and scattered photons, DCP_I- and DCP_{II}-ROA spectra are in general not simply the sum and difference of the corresponding ICP- and SCP-ROA spectra. Considering all four forms of CP-ROA, one can conclude that DCP_I-ROA has by far the highest intensity of ROA relative to its associated Raman intensity. Furthermore, DCP_{II}-ROA typically has the least ROA intensity relative to its parent Raman intensity, but is extremely sensitive to the breakdown of the FFR approximation as DCP_{II}-ROA vanishes in the FFR limit.

In most respects, DCP_I-ROA backscattering is the ultimate of all ROA measurements. In the FFR approximation, backscattering DCP_I yields the full intensity of ICP or SCP unpolarized backscattering ROA, but with far fewer photons, as the simultaneous selection of the CP states of the incident and

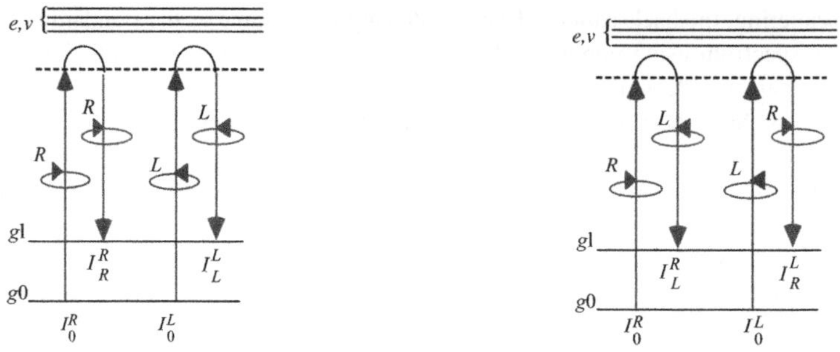

$$\text{DCP}_{\text{I}}\text{-ROA} \quad (\Delta I_I)^a_{gl,g0} = \left(I^R_R\right)^a_{gl,g0} - \left(I^L_L\right)^a_{gl,g0}$$

$$\text{DCP}_{\text{I}}\text{-Raman} \quad (I_I)^a_{gl,g0} = \left(I^R_R\right)^a_{gl,g0} + \left(I^L_L\right)^a_{gl,g0}$$

$$\text{DCP}_{\text{II}}\text{-ROA} \quad (\Delta I_{II})^a_{gl,g0} = \left(I^R_L\right)^a_{gl,g0} - \left(I^L_R\right)^a_{gl,g0}$$

$$\text{DCP}_{\text{II}}\text{-Raman} \quad (I_{II})^a_{gl,g0} = \left(I^R_L\right)^a_{gl,g0} + \left(I^L_R\right)^a_{gl,g0}$$

Figure 7.5 *Energy-level diagram and basic definitions of DCP_I- and DCP_{II}-Raman and -ROA for a Raman transition between vibrational states 0 and 1 of the ground electronic state, g, for normal mode a*

scattered beams allows only depolarized scattering to be detected. As seen in Equations (7.15) and (7.16) only one anisotropic Raman invariant is needed to describe the Raman scattering and two anisotropic ROA invariants are needed for the ROA,

$$\text{DCP}_{\text{I}}\text{-ROA}\,(180^\circ): \qquad I^R_R(180^\circ) - I^L_L(180^\circ) = \frac{8K}{c}\left[12\beta(G')^2 + 4\beta(A)^2\right] \tag{7.15}$$

$$\text{DCP}_{\text{I}}\text{ Raman}\,(180^\circ): \quad I^R_R(180^\circ) + I^L_L(180^\circ) = 4K\left[6\beta(\alpha)^2\right] \tag{7.16}$$

The absence of the isotropic Raman invariant dramatically reduces the susceptibility of the ROA to optical artifacts. The only drawback of DCP-ROA measurement is the additional instrumental complexity, compared with either ICP- or SCP-ROA, needed to measure the spectra, as described in the next section. For completeness, we also provide the corresponding DCP_{II}-ROA and Raman backscattering intensities in the FFR limit:

$$\text{DCP}_{\text{II}}\text{-ROA}\,(180^\circ): \qquad I^R_L(180^\circ) - I^R_L(180^\circ) = 0 \tag{7.17}$$

$$\text{DCP}_{\text{II}}\text{-Raman}\,(180^\circ): \quad I^R_L(180^\circ) + I^R_L(180^\circ) = 4K\left[45\alpha^2 + \beta(\alpha)^2\right] \tag{7.18}$$

DCP_{II}-Raman backscattering is an extreme form of polarized Raman scattering dominated by the isotropic Raman invariant with only a very small contribution from the anisotropic invariant. For backscattering ROA, ICP- and SCP-Raman and ROA, unpolarized intensities are simply the sum and difference of the corresponding DCP_I and DCP_{II} intensities. In the FFR approximation, the DCP_{II}-Raman intensity is the additional empty (no ROA) Raman intensity that must be measured in ICP or SCP unpolarized backscattering that is not needed, and hence does not contribute to additional noise, artifacts, and polarized fluorescent background in DCP_I-ROA measurements.

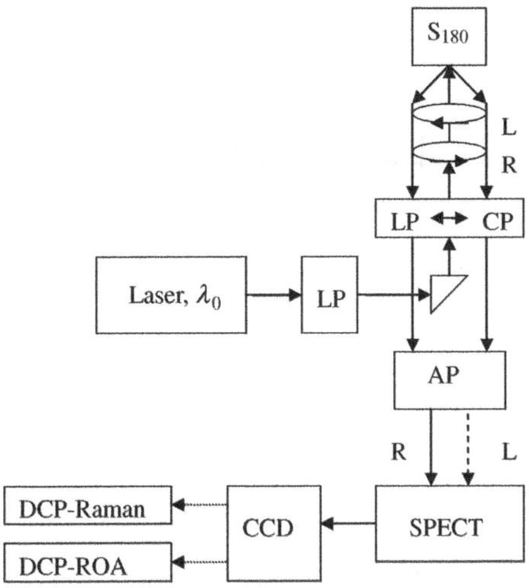

Figure 7.6 *Instrument block diagram for the measurement of backscattering* DCP_I *and* DCP_{II} *forms of Raman and ROA in which the same optical element LP ↔ CP is used to create the incident CP modulation and simultaneously convert CP Raman backscattered radiation into LP where the appropriate polarization state can be selected by an analyzing polarizer (AP)*

7.3.1 Optical Setups for DCP-ROA Measurement

There are two distinct optical block diagrams for the measurement of DCP-ROA. The simplest and the first used for DCP-ROA measurements is configured such that the same optical elements are used to create the incident circular polarization modulation and to select the desired circular polarization state of the scattered radiation. The setup can only be used for a backscattering geometry. For setups with other scattering geometries, the synchronous modulation between RCP and LCP in the incident and scattered beams must be carried out with separate optics.

This first backscattering setup is illustrated in Figure 7.6. Here the laser radiation is placed in a pure vertical or horizontal linear polarization state and directed to a reflection off a small right-angle prism as described above for ICP and SCP backscattering ROA. After the prism, the radiation passes through a quarter-wave plate, with fast and slow axes at $\pm 45^\circ$ that can be rotated between positions that differ by 90° to produce either RCP or LCP. This step is represented by the conversion between LP and CP states of light by LP ↔ CP in Figure 7.6. The backscattered radiation is collected by a lens and returned through the QWP converting RCP or LCP to horizontally or vertically polarized light. If an analyzing polarizer, AP, is kept fixed, say in the horizontal position, then when RCP is incident on the sample, the AP will pass RCP. Rotation of the QWP then produces LCP at the sample and only LCP will pass through the horizontal polarizer. In general DCP_I will be measured with this setup when the incident LP and AP in the scattered beam are set to pass orthogonal linear polarization states, whereas DCP_{II} is measured when they pass parallel polarization states. Because the diffraction grating used in the spectrograph strongly favors light polarized parallel to the diffraction plane, the AP was always maintained in the horizontal position when the LP was in the vertical position. To measure DCP_{II}-ROA, the incident polarization state is rotated to horizontal while the AP remains horizontal. The reason why orthogonal polarization states for the incident and scattered beams is required for DCP_I-ROA measurement is because RCP reflected back from a surface changes to LCP, and RCP traveling in opposite directions represents light with opposite senses of circulation of the polarization vector of the light when seen by a single observer.

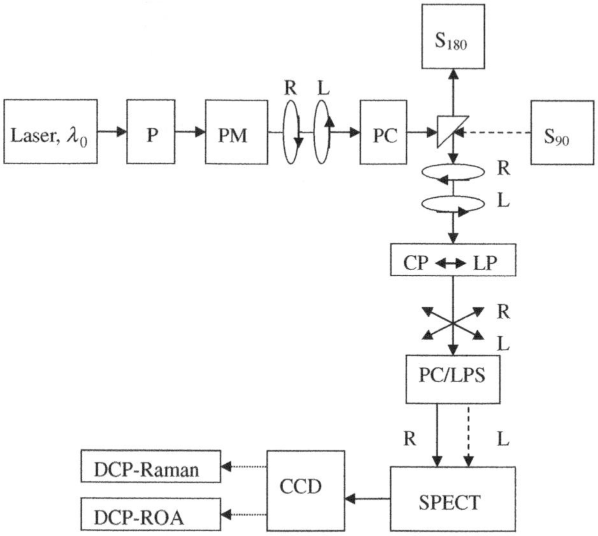

Figure 7.7 *Instrument block diagram for the measurement of DCP_I and DCP_{II} forms of Raman and ROA as described in the text*

An alternative for the measurement of DCP-ROA is illustrated in Figure 7.7. Radiation from the laser is first linearly polarized (P) and the polarization modulated (PM) between RCP and LCP states before polarization conditioning (PC), as described above. The incident beam is then directed to a sample either by a right-angle reflection off a prism mirror for backscattering at the sample (S_{180}), or without reflection inserting the sample (alternate position) for the collection of right-angle (S_{90}) scattering. For backscattering, the analysis along the scattered light rail is the same as in Figure 7.6 except that additional options are introduced. Here PC/LPS represents not only an AP but also additional polarization conditioning followed by a beam splitting cube focusing two beams simultaneously into two fiber-optic bundles as described above for SCP-ROA measurement. In the case of a beamsplitting cube, however, a rotating sector wheel is introduced to alternately block one fiber-optic bundle or the other to avoid the simultaneous measurement of intensities associated with both DCP_I and DCP_{II} as the incident beam is modulated, possibly at relatively high modulation rates, between RCP and LCP states. The phase of the sector wheel relative to the incident beam polarization modulation determines whether DCP_I or DCP_{II} intensities are measured.

7.3.2 Comparison of ICP-, SCP-, and DCP_I-ROA

A simple comparison of two closely related ROA experiments is to consider depolarized right-angle ICP scattering and (depolarized) backscattering DCP_I scattering (Che and Nafie, 1992). As both Raman spectra are the same, collection times can be adjusted such that the two spectra have the same Raman intensity for all vibrational frequencies. At the same time the ROA for the two different scattering experiments can be displayed above the common Raman spectrum, as in Figure 7.8 for the molecule (+)-*trans*-pinane. In order to equate the intensities of the Raman spectra, given for ICP right-angle scattering in Equation (7.6) and DCP_I scattering in Equation (7.16), we multiply the ICP intensities by two (corresponding to measurement times twice as long) such that both Raman spectra are represented by the expression $24\beta(\alpha)^2$. Then the ICP-ROA is given by $16K[3\beta(G')^2 - \beta(A)^2]/c$ while the DCP_I-ROA intensity is $16K[6\beta(G')^2 + 2\beta(A)^2]/c$. If the electric-quadrupole ROA invariant

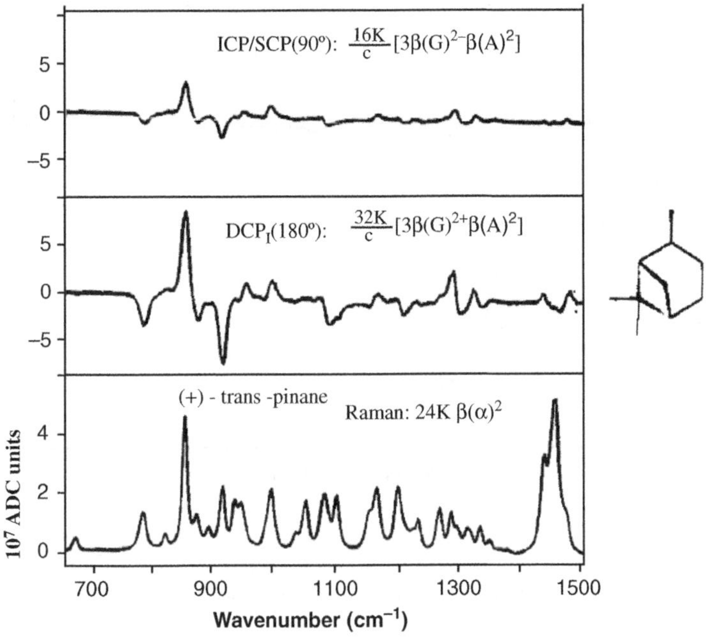

Figure 7.8 *Comparison of right-angle depolarized ICP-ROA with (depolarized) DCP_I-ROA for (+)-trans pinane showing clearly the advantage of backscattering DCP_I-ROA scattering. In total, this is an enormous advantage and illustrates why depolarized right-angle ICP- or SCP-ROA is no longer used for routine ROA measurements. Adapted with permission from Elsevier Publishing Company (Che and Nafie, 1992)*

$\beta(A)^2$ was zero, then these two Raman spectra in Figure 7.8 would differ by a factor of two. If the two Raman invariants were equal in magnitude and the same sign, then the two ROA spectra would differ by a factor of four. If the invariants were equal and opposite in sign the two ROA spectra would have the same intensity. Actually, the two ROA spectra appear to differ by a factor of between two and three, which leads to the conclusion that both invariants are non-zero and the same sign, but that $\beta(G')^2$ is greater than $\beta(A)^2$. From the comparison of ROA in this figure, the clear advantage of backscattering DCP_I-ROA over depolarized right-angle ICP-ROA can be seen, and this advantage is in addition to the fact that the DCP_I-Raman spectrum is twice as strong as the corresponding ICP-Raman spectrum, or equivalently that it took twice as long to collect the ICP-ROA spectrum.

7.3.3 Isolation of ROA Invariants

In the FFR approximation, it is possible to combine different measured ROA spectra, properly normalized with their associated Raman spectra, to isolate the three ROA invariants (Che and Nafie, 1992). In the more general theory, only six linear combinations of the ten ROA invariants can be isolated. Isolating the ROA spectra of individual invariants provides an additional level of comparison between calculated and measured ROA spectra as ROA invariants are calculated individually before combining them in appropriate ways to match a given experimental measurement. The example given in the previous section of (+)-*trans* pinane can be used directly to isolate the ROA spectra of $16K[3\beta(G')^2]/c$ and $16K[\beta(A)^2]/c$ by first multiplying the depolarized right-angle ICP-ROA spectrum by two, and then adding or subtracting the backscattering DCP_I-ROA spectrum. These spectra are

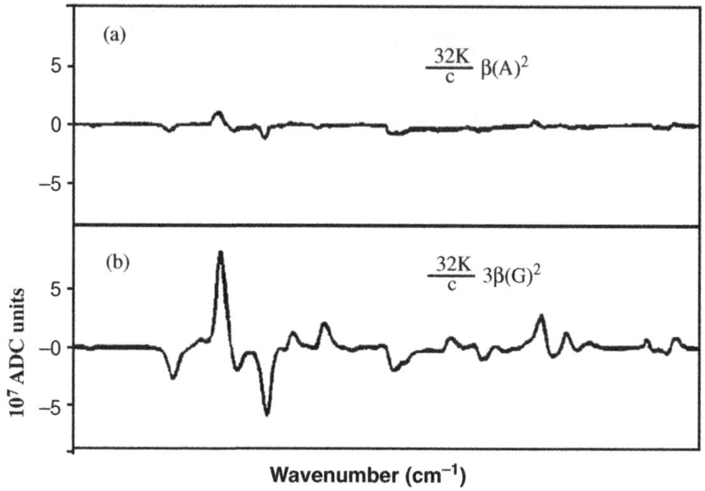

Figure 7.9 *Isolation of invariants from right-angle depolarized ICP-ROA with backscattered (depolarized) DCP$_I$-ROA for (+)-trans pinane showing in this case the dominance of the magnetic-dipole (b) over the electric-quadrupole (a) contributions to ROA intensity. Adapted with permission from Elsevier Publishing Company (Che and Nafie, 1992)*

displayed in Figure 7.9. This choice of display represents visually the relative importance, in this case, of the magnetic-dipole ROA invariant $\beta(G')^2$ to the corresponding electric-quadrupole invariant $\beta(A)^2$ as the factor of three that appears with $\beta(G')^2$ in both original ROA intensity expressions is retained. This figure demonstrates that the electric-quadrupole mechanism of ROA provides only a minor intensity contribution relative to the much more dominant magnetic-dipole contribution. If the ROA intensity of the $16K[3\beta(G')^2]/c$ spectrum is reduced by a factor of three, one can see that this contribution is still larger than the ROA spectrum $16K[\beta(A)^2]/c$ by somewhat less than a factor of two, as concluded by a different line of reasoning in the previous section. At least for the case of *trans*-pinane, the two invariant ROA spectra appear to have features of nearly the same relative magnitude and intensity. From other experiments, for instance the comparison of ICP and SCP of *trans*-pinane, it is known that for *trans*-pinane, a molecule without functional groups or double bonds, the FFR approximation is obeyed to within experimental accuracy. To date, this is the only such molecule for which satisfaction of the FFR limit has been demonstrated. All other molecules starting with α-pinene show some degree of breakdown of the FFR approximation as described in the following section.

7.3.4 DCP$_{II}$-ROA and the Onset of Pre-resonance Raman Scattering

A direct measurement of DCP$_{II}$-ROA has not yet been reported; however, the DCP$_{II}$-ROA spectra of several molecules have been isolated by the subtraction of two ROA measurements that equals DCP$_{II}$-ROA (Yu and Nafie, 1994). In particular, the subtraction of properly normalized backscattering DCP$_I$-ROA from unpolarized ICP-ROA is DCP$_{II}$-ROA at any level of approximation. The DCP$_{II}$-ROA spectra of the four molecules (−)-*trans* pinane, (−)-α-pinene, (−)-verbenone, and (+)-quinidine solution, which have increasing levels of functionality and increasing wavelength for the onset of electronic absorption bands, are shown in Figure 7.10. From these spectra it is clear, that the degree

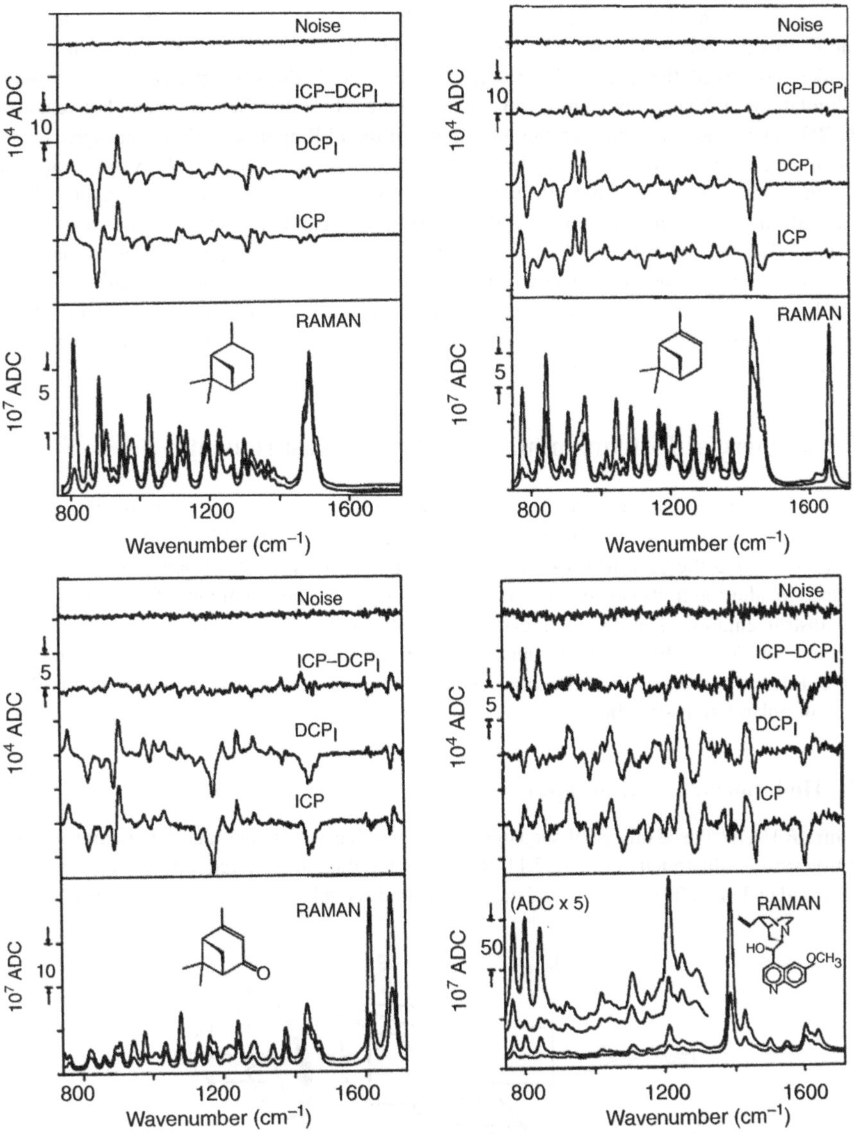

Figure 7.10 *Isolation of DCP$_{II}$-ROA spectra from the subtraction of properly normalized backscattering DCP$_I$-ROA from unpolarized ICP-ROA for (−)-trans pinane, (−)-α-pinene, (−)-verbenone, and (+)-quinidine solution, which have increasing levels of functionality and increasing wavelength for the onset of electronic absorption bands and correspondingly greater departure from the FFR resonance approximation. Note that backscattering unpolarized ICP-Raman (polarized) and backscattering DCP$_I$-Raman spectra (depolarized) properly normalized differ in intensity. Adapted with permission from Elsevier Publishing Company (Yu and Nafie, 1994)*

functionality correlates with the magnitude of the DCP_{II}-ROA spectrum and with the breakdown of the FFR approximation due to the lowering of the energy of the first excited electronic state with respect to the incident laser excitation energy. The results show that the DCP_{II}-ROA spectrum is zero within the noise limitations of the instrument for *trans*-pinane, whereas the other three compounds give non-zero DCP_{II}-ROA and hence have measurable breakdown of the FFR approximation. This agrees with an earlier publication of the comparison of depolarized right-angle ICP- and SCP-ROA for *cis*- and *trans*-pinane and α- and β-pinene where it was found that the pinanes, with no functionality, show no measureable difference between ICP- and SCP-ROA whereas both pinenes showed measurable differences (Hecht *et al.*, 1992b). These examples show the sensitivity of various forms of ROA intensities, or combination of ROA intensities, to the breakdown of FFR approximation that can be achieved at a *single* excitation wavelength. Other methods of detecting the breakdown of the FFR approximation rely on the observation of changes in the Raman spectrum as a function of excitation wavelength as resonance is approached.

7.4 Commercial Instrumentation for ROA Measurement

In contrast to instrumentation for VCD, there is only one source of commercial instrumentation for ROA, a backscattering SCP-ROA spectrometer called the Chiral*RAMAN*, now available in an improved version as the Chiral*RAMAN-2X*, from BioTools, Inc. This instrument possesses a number of features in addition to its capability to measure ROA that are not commonly found in commercial Raman instrumentation. These special features derive from the novel use of a dual fiber-optic bundle light collection design that results in unusually high throughput and the ability to simultaneously measure Raman spectra with orthogonal polarization states of the scattered radiation free of instrument polarization-state bias.

7.4.1 High Spectral Throughput

The Chiral*RAMAN*, based on a published optical design (Hug and Hangartner, 1999), uses a dual arm collection system illustrated in Figure 7.11. RCP and LCP Raman scattered light from a sample focus is collimated by lens (G) and transmitted through a liquid crystal retarded (LCR) that acts as a variable

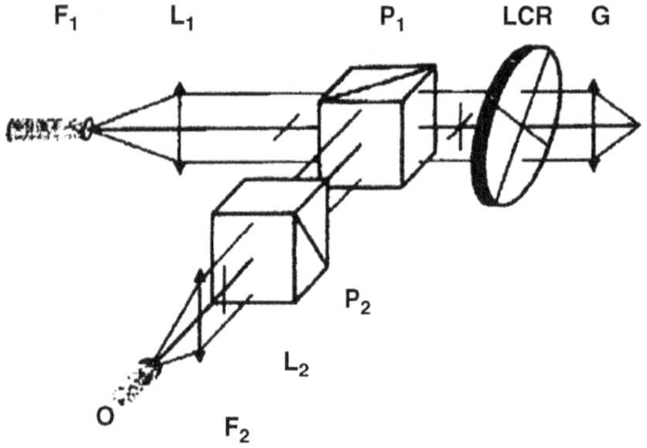

Figure 7.11 Optical diagram of the conversion of a collimated beam (G) of Raman scattered orthogonal CP states to orthogonal LP states by a liquid crystal retarder (LCR) followed by spatial separation of the LP states by beamsplitting cubes (P) and focusing of the two output beams to fiber-optic bundles (F) by lenses (L) to form a dual-arm fiber-optic image transformation optic (Hug and Hangartner, 1999)

quarter-wave plate converting orthogonal circular polarization states into orthogonal linear polarization states. The light is then directed to a polarizing beam splitting cube (P_1).

The cube is configured to transmit horizontally polarized (p-polarized) light and to reflect at right angles the vertically polarized light (s-polarized). The p-polarized component is transmitted without contamination by s-polarized light, but not vice versa, hence a second beamsplitting cube (P_2) is positioned to transmit the s-polarized component as p-polarized (uncontaminated) light by the second cube. Each of the two polarized beams is focused by identical lenses (L_1 and L_2) to a bundle (F_1 and F_2) of 31 quartz fibers in a hexagonal arrangement starting from the center fiber as 1:6:12:12. The two fiber-optic bundles are joined to form a single curved line of 62 fibers that becomes the effective entrance slit of the spectrograph. The curve compensates for the natural curvature produced by a diffraction grating for resolution elements of the light imaged above and below the plane of diffraction. The upper 31 fibers originate in one of the two clusters and passes scattered intensity from one direction of polarization while the lower 31 fibers transmit the scattered light with the orthogonal polarization state. The 75 micron (μm) width of the individual fibers results in a spectral resolution at the CCD covering approximately three pixels spanning about 7 cm^{-1}.

The dual-arm fiber collection optic produces an image transformation that creates two polarization divided images of the same scattering volume at the sample, and distributes the scattered light across the entire height of the 256 × 1024 element CCD array. Normally, scattered radiation is imaged directly at the spectrograph slit concentrating most the scattered light over a limited number of pixels at the center of each vertical column of CCD pixels. This leads to low saturation tolerance by the CCD for high-throughput scattering samples.

The dual-arm fiber transformation optic allows much higher scattering intensity, and hence higher effective throughput, to be imaged on the CCD before the onset of saturation. For example, shown in Figure 7.12 is a Raman spectrum of a protein in water at a concentration of 50 mg mL^{-1} that has been measured with 100:1 signal-to-noise ratio in less than 150 ms, the minimum camera exposure time for the Chiral*RAMAN* spectrometer. It should be noted that if the LCR is modified to operate at zero and 180° retardation, instead of ±90° retardation, then the instrument is able to measure simultaneously the difference between orthogonal linear polarization states scattered by the sample rather than orthogonal circular polarization states.

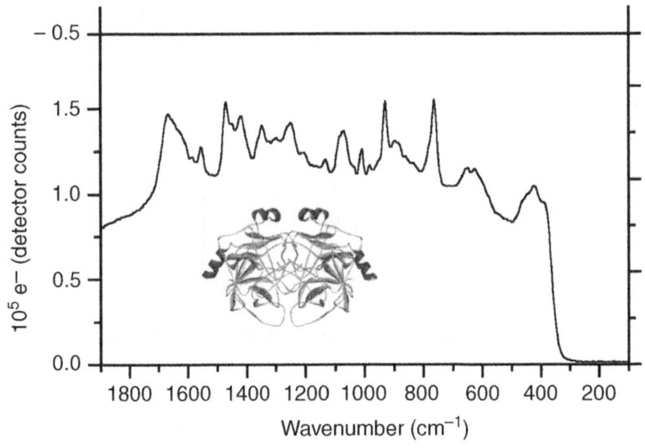

Figure 7.12 *Raman spectrum of a protein in water at concentration 50 mg mL*$^{-1}$ *with exposure time of less than 150 ms illustrating the very high throughput of the ChiralRAMAN spectrometer*

7.4.2 Artifact Suppression and the Virtual Enantiomer

The suppression of artifacts in the Chiral*RAMAN* ROA spectrometer is achieved by use of multiple half-wave plates in the incident and scattered light beams (Hug, 2003). In the previous section, we saw that a half-wave plate interconverts CP light between LCP and RCP states and, properly oriented, switches LP light between vertical and horizontal states. From Equation (6.34), the Mueller matrix for a half-wave plate (180° retardation) with fast and slow axes at 45° from vertical is given by:

$$\boldsymbol{M}_{HWP}(45^\circ) = \begin{pmatrix} 1 & 0 & 0 & 0 \\ 0 & -1 & 0 & 0 \\ 0 & 0 & 1 & 0 \\ 0 & 0 & 0 & -1 \end{pmatrix} \tag{7.19}$$

The action of this plate on RCP light converts it into LCP light as:

$$\boldsymbol{M}_{HWP}(45^\circ)\boldsymbol{S}_{RCP} = \begin{pmatrix} 1 & 0 & 0 & 0 \\ 0 & -1 & 0 & 0 \\ 0 & 0 & 1 & 0 \\ 0 & 0 & 0 & -1 \end{pmatrix}\begin{pmatrix} 1 \\ 0 \\ 0 \\ 1 \end{pmatrix} = \begin{pmatrix} 1 \\ 0 \\ 0 \\ -1 \end{pmatrix} = S_{LCP} \tag{7.20}$$

Similarly, it coverts vertical into horizontal LP light as:

$$\boldsymbol{M}_{HWP}(45^\circ)\boldsymbol{S}_{VLP} = \begin{pmatrix} 1 & 0 & 0 & 0 \\ 0 & -1 & 0 & 0 \\ 0 & 0 & 1 & 0 \\ 0 & 0 & 0 & -1 \end{pmatrix}\begin{pmatrix} 1 \\ 1 \\ 0 \\ 0 \end{pmatrix} = \begin{pmatrix} 1 \\ -1 \\ 0 \\ 0 \end{pmatrix} = S_{HLP} \tag{7.21}$$

In addition, a half-wave plate at an arbitrary orientation, from Equation (6.35) is given by:

$$\boldsymbol{M}_{HWP}(\theta) = \begin{pmatrix} 1 & 0 & 0 & 0 \\ 0 & \cos 4\theta & \sin 4\theta & 0 \\ 0 & \sin 4\theta & -\cos 4\theta & 0 \\ 0 & 0 & 0 & -1 \end{pmatrix} \tag{7.22}$$

When averaged over all angles of orientation, the Mueller matrix for a rotating HWP, also called a linear rotator (LR), becomes,

$$\boldsymbol{M}_{RHWP} = \begin{pmatrix} 1 & 0 & 0 & 0 \\ 0 & 0 & 0 & 0 \\ 0 & 0 & 0 & 0 \\ 0 & 0 & 0 & -1 \end{pmatrix} \tag{7.23}$$

The action of the LR on any linear polarization state of arbitrary orientation is to completely eliminate the LP content of the state as:

$$\boldsymbol{M}_{RHWP}\boldsymbol{S}_{LP}(\theta) = \begin{pmatrix} 1 & 0 & 0 & 0 \\ 0 & 0 & 0 & 0 \\ 0 & 0 & 0 & 0 \\ 0 & 0 & 0 & -1 \end{pmatrix}\begin{pmatrix} 1 \\ \cos\theta \\ \sin\theta \\ \varepsilon \end{pmatrix} = \begin{pmatrix} 1 \\ 0 \\ 0 \\ -\varepsilon \end{pmatrix} \tag{7.24}$$

Any residual circular polarization content (ε) is reversed to the opposite CP state as in Equation (7.20).

Artifacts in SCP-ROA can be dramatically reduced as follows. Laser radiation is first linearly polarized and then sent to a fast LR that, as in Equation (7.24), eliminates by averaging over time all LP states of the light. Any remnant CP content (ε) can be averaged to zero by use of a circularity converter (CC), which is simply an HWP that is in the beam for half the collection time and out of the beam the other half. After a complete cycle of averaging, the incident radiation has no LP or CP content, as desired for unpolarized backscattering SCP-ROA.

A similar procedure is used to eliminate polarization distortions or polarization biases in the scattered radiation. This is a bit more difficult, because the true intensities for scattered RCP and LCP states must be preserved until the intensities are recorded and digitized. The process of preserving the separate intensities of RCP and LCP radiation involves the concept of a virtual enantiomer. A real enantiomer interchanges the intensity responses of the chiral molecule for RCP and LCP. For example, scattered RCP intensity from a chiral molecule and its enantiomer are different and the corresponding LCP intensities are different in the reverse sense. If now an HWP plate is inserted in the beam, the response of the enantiomer of the chiral molecule is simulated virtually in that the intensities of RCP and LCP light to be measured are now the same as if the actual enantiomer of the chiral molecule had been used.

In general, there is a bias in an ROA instrument for the measurement of RCP versus LCP scattered light. To overcome this bias, a large CC is used. The large size is needed to accommodate the large size of the collimated scattered light beam. When the CC is in the beam, the HWP interconverts the CP states of the light, creating a virtual enantiomer of the sample such that the intensity of scattered RCP light is measured instead as LCP light and the scattered LCP light is measured as RCP light. If these intensities are properly averaged with the intensities when the CC is out of the beam by the instrument control electronics, any CP bias of the instrument is averaged in first-order to zero.

Prior to the use of the CC, a large LR is placed in the beam. The function of the LR is to remove any LP content of the scattered light due to birefringence in the sample cell or light collection lens. As the incident light is unpolarized, and because for backscattering geometry there is no scattering plane, there is no molecular source of LP content to the scattered light. The only source is imperfections in the sample cell and light collection optics for the CP polarized scattered light. After any such spurious LP content of the scattered light is removed by the LR, the CC is used to equalize the instrument response to pure LCP and RCP scattered light.

After the CC, the orthogonal CP states of the scattered beam are converted into orthogonal LP states by the QWP action of the LCR as described in Section 7.4.1. Because the beam path and the light efficiency through the beamsplitting cube are measurably different for s-polarized and p-polarized radiation, the voltage of the LCR is alternately switched between positive and negative states thereby interchanging whether RCP light is converted into s-polarized or p-polarized radiation and oppositely for LCP light.

A field programmable array (FPGA) computer chip controls all modulations cycles and collection logic such that the incident radiation has no LP or CP polarization bias and that the original RCP and LCP scattered light intensities are stored in the proper registers despite the different types of CP and LP interconversions that occur in the light collection modulation cycles. All cycles of interconversions must be precisely controlled and all modulation cycles must be properly averaged with respect to each other.

7.5 Advanced ROA Instrumentation

Raman scattering is one of the most fascinating forms of the interaction of light with matter. It is directly analogous with vibrational IR absorption in that there is a linear response between the incident radiation and measured transmitted or scattered radiation. At the same time, there is an enormous difference between IR and Raman spectroscopies in that Raman scattering is fundamentally a two-photon quantum mechanical effect. As such, there is a depolarization ratio between incident and scattered radiation even for randomly oriented sample molecules. Raman scattering is also balanced more directly between vibrational spectroscopy and electronic absorption or fluorescence spectroscopy. The richness of Raman scattering makes possible a wide variety experiments that extend well beyond conventional Raman scattering for samples far-from-resonance with electronic transitions, the simplest form of Raman scattering. Below, we briefly describe instrumentation for some departures from conventional ROA which may someday, possibly soon, become standard methods for research and applications of ROA.

7.5.1 Resonance ROA (RROA)

Resonance ROA (RROA) occurs when the frequency of the incident radiation in an ROA measurement coincides with an allowed electronic transition of the molecule. The far-from-resonance (FFR) theory of ROA breaks down when the frequency of the incident radiation begins to approach resonance with an allowed excited electronic state. The regime between the FFR theory and the strong resonance (SR) theory is called the near-resonance (NR) theory (Nafie, 2008), and all three of these resonance regimes were described in Chapter 5 on the theory of ROA, where the SR regime subdivides into the single-electronic-state (SES) theory (Nafie, 1996), the two-electronic-state (TES) theory, and beyond.

No special instrumentation is required, beyond that described previously in this chapter, for the measurement of ROA in the FFR or NR regimes. The same is true for SR-ROA (or more simply RROA) if resonance occurs in the visible region of the spectrum, such as near readily available lasers sources at 532, 514 or 488 nm, and where significant sample absorption does not occur during the scattering process. If RROA is to be measured in the UV or near-IR regions, then modified instrumentation is needed, as described in the next two sub-sections of this chapter. If significant sample absorption is present, corrections are needed both for the parent RR spectrum as well as the RROA spectrum. Corrections to the RR spectrum follow standard procedures for correcting Raman spectra, typically as a function of incident laser wavelength, for sample absorption of both the incident radiation and the scattered radiation as a function of scattered wavelength.

In the case of RROA, the sample absorption may result in a background ROA offset from zero due electronic circular dichroism (ECD) associated potentially with both the incident radiation and across the spectrum of the scattered radiation. In addition, the effect of ECD on the incident radiation is to cause a molecule induced CP bias that can add significant depolarization ratio effects to the ROA spectrum. Typically, instead of measuring just unpolarized SCP backscattering as $I_R^U(180°) - I_L^U(180°)$, the incident polarized beam may carry, for example, a slight RCP bias as $U + \delta R$ from positive ECD in the sample where LCP is absorbed more strongly than RCP. This positive bias developing in the sample as the incident radiation penetrates deeper into the sample before scattering results in an additional term in the RROA spectrum as $\delta\left[I_R^R(180°) - I_L^R(180°)\right]$ where this spectrum is not ROA but a difference between co-rotating and contra-rotating Raman backscattering, an intrinsically large spectrum. The influence on this term and on other baseline offset effects due to ECD depend on the magnitude of sample absorption of the Raman incident and scattered light beams. The effects can be minimized by focusing incident radiation, and therefore collecting scattered radiation, just inside the front surface of the cell containing the sample solution.

7.5.2 Near-Infrared Excitation ROA

In order to reduce the effect of interference from sample fluorescence in the measurement of Raman and ROA spectra, usually, but not always, due to impurities in the sample, one can increase the wavelength of the incident radiation from the visible region to the near-IR (NIR) region. Here sample absorption is typically much weaker and fluorescence is greatly reduced from corresponding levels in the visible region. NIR-excited ROA has been achieved for excitation at 780 nm by exchanging wavelength-sensitive visible optical elements in a Chiral*RAMAN* SCP-ROA spectrometer with the corresponding NIR elements (Nafie *et al.*, 2007).

Changing an ROA spectrometer from operation in the visible to the NIR involves changing the following items: (i) laser, (ii) all half-wave plates, (iii) Rayleigh line rejection filter, and (iv) diffraction grating. In addition, one must check that all other optical components still function well in the NIR, possibly by selecting long-wavelength enhanced components. These include: (i) all optical reflection surfaces, (ii) incident radiation polarizer, (iii) sample focusing and collection lens, (iv) liquid crystal retarder, (v) beamsplitting cube, (vi) dual-arm, fiber-optic image transformer, and (vii) CCD detector. All optical components that were changed for the conversion of a Chiral*RAMAN* SCP ROA spectrometer from operation at 532 nm to operation at 780 nm are shown in Figure 7.13.

After the conversion, the performance of the instrument in these two regions was compared for the test molecule α-pinene, as shown in Figure 7.14. There are distinct advantages for 532 nm operation involving detector sensitivity and also the usual ν^4-dependence that gives approximately a factor of four increase in scattering intensity for the same total incident energy (power in watts times duration of exposure). Beyond that, ROA intensity has a ν^5-dependence on the incident laser frequency. This frequency dependence results in a factor of approximately 1.5 decrease in ROA intensity relative to the parent Raman intensity for 780 versus 532 nm operation. In Figure 7.14 it is easy to see that the ROA spectrum is less intense for 780 nm excitation compared with 532 nm excitation for the same level of Raman scattering.

7.5.3 Ultraviolet Excitation ROA

Extending the laser excitation in the opposite spectral direction to the UV region carries mostly relative advantages rather than disadvantages. Doubling the incident laser frequency, such as from 488 to 244 nm excitation using an argon ion laser, increases the Raman intensity by a factor of 16 and the ROA by a factor of 32 for the same incident beam intensity. In addition, the FFR approximation certainly breaks down in this region and the scattering is either near-resonance or strongly in resonance, adding even more intensity per unit laser excitation energy. Reports of UV ROA have yet to be published, but it has been at least partially developed in the laboratories of Barron and Hecht at Glasgow University.

The achievement of deep UV resonance ROA (DUV-RROA), particularly for biological molecules such as proteins or nucleic acids, would combine the powerful incisive advantages of DUV-RR scattering with the stereo-selectivity of ROA to isolate particular structural regions for analysis. ROA excitation profiles (ROA intensity versus excitation frequency) would provide an understanding of a typical UV electronic CD spectrum at the level of its underlying vibronic contributions and would allow, for example, the overlapping ECD intensity from backbone peptide contributions and aromatic amino acid side chains to be analyzed and resolved into separate spectral contributions.

DUV-RROA spectra would need to be interpreted in terms of resonance ROA theory, for which the NR theory (Nafie, 2008) and the strong resonance theory from a single resonant electronic state (SES) (Nafie, 1996) has been described in Chapter 5. Extension of the strong resonance theory beyond the SES limit to involve two or more electronic states and other RR scattering mechanisms has also been described earlier. *Ab initio* calculations have been performed for molecules in the SES limit confirming this theory (Jensen *et al.*, 2007).

7.5.4 Linear Polarization ROA

Linear polarization ROA (LP-ROA) has yet to be observed even though it was predicted theoretically nearly 20 years ago (Hecht and Nafie, 1990). As describe previously, one measures the intensity

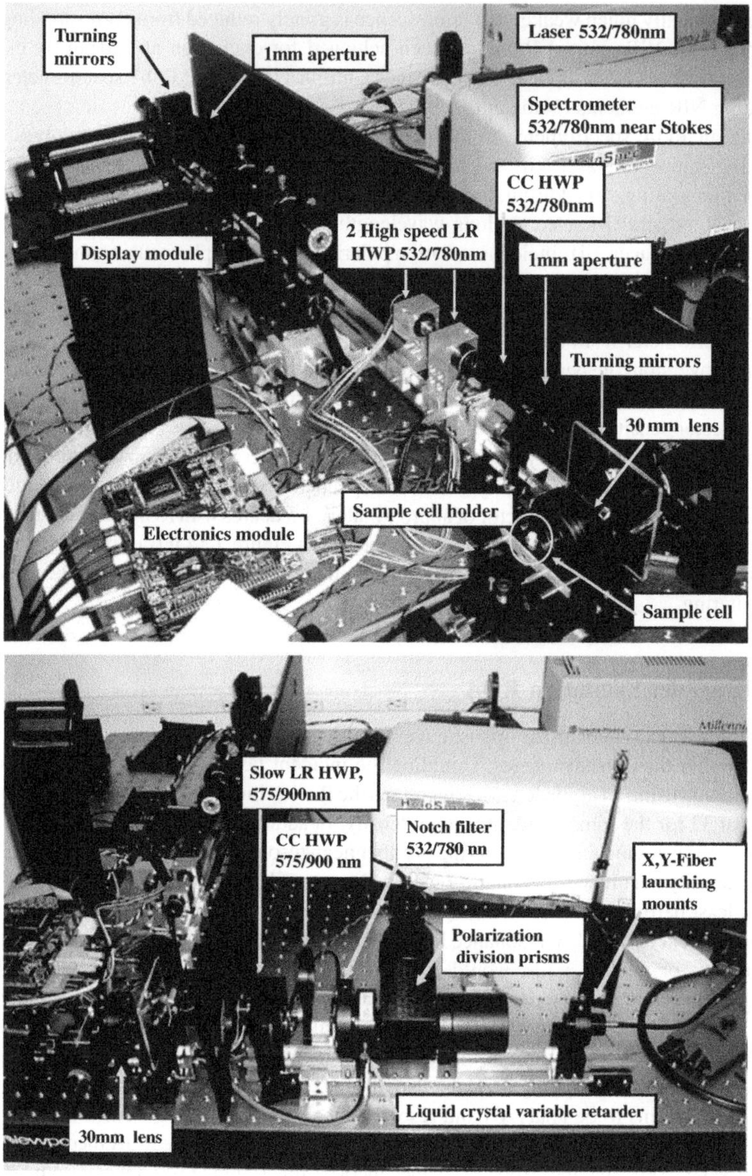

Figure 7.13 Optical layout for the incident rail (upper) and scattered rail (lower) showing the principal optical components and indicating the wavelength changes for conversion from visible operation with 532 nm excitation to NIR operation with excitation at 780 nm

difference for Raman scattering between states of linear polarization that are at ±45° relative to the vertical or horizontal direction. As with CP-ROA there are four forms of LP-ROA depending on whether the linear polarization difference is carried out for the incident, scattered, or both incident and scattered light beams in-phase or out-of-phase. To measure LP-ROA, therefore, one needs only convert CP modulation into the corresponding LP modulation.

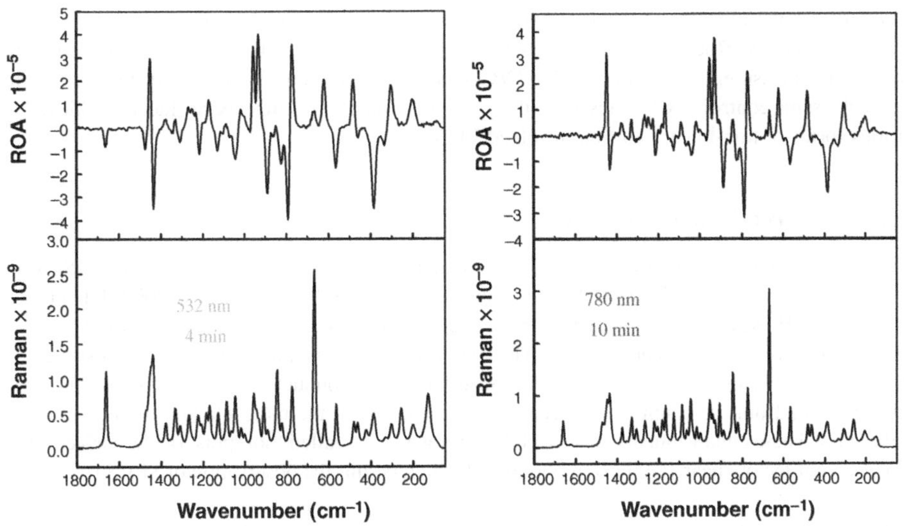

Figure 7.14 *Comparison of SCP-ROA and Raman spectra of (−)-α-pinene for excitation at 532 and 780 nm. Adapted with permission from the Society of Applied Spectroscopy (Nafie et al., 2007)*

There are a variety of instrumental methods for achieving LP modulation starting from a CP modulation instrument. The simplest method might be to measure backscattering SLP-ROA by polarizing the beam in the vertical direction and using a CC plate to eliminate any CP bias. Then, one needs to remove the LR in the scattered beam to maintain the original LP states scattered by the sample. The CC, acting as a linear converter (LC) in this circumstance, can be retained because, properly oriented, it interchanges the two scattered LP (or CP) states. The LCR likewise can be retained but it needs to operate between zero and half-wave retardation, rather than between opposite QWP states, in order to maintain LP states and not convert LP into CP states. Finally, the polarization beamsplitter needs to be rotated by 45° about the scattered beam direction so that it separates two orthogonal LP polarization states that are ±45° relative to the vertical direction.

One reason that LP-ROA has yet to be measured is that, like DCP_{II}-ROA, it is predicted to be zero in the FFR approximation (Hecht and Nafie, 1990; Hecht and Barron, 1994). Unless one is in, or close to, resonance, LP-ROA is a small effect close to the noise level for measurement times of hours in duration. On the other hand, LP-ROA is a new form of ROA and has at least conceptual links to the phenomenon of optical rotation as one can imagine, for example in the case of SLP-ROA, that the linear polarization bias of the scattered beam is rotated from the vertical in opposite directions for enantiomers, such that the intensities for light polarized at ±45° relative to the vertical direction are different from one another.

7.5.5 Non-Linear and Time-Resolved ROA

ROA has not been measured for any non-linear forms of ROA. A theoretical description of coherent anti-Stokes Raman scattering (CARS) ROA has been published, but control of polarization properties of the light beam for short pulses needed for all non-linear forms of Raman scattering has not been achieved. On the other hand, VCD measurements using femtosecond IR laser pulses have been successful recently (Rhee *et al.*, 2009b; Rhee *et al.*, 2009a; Helbing and Bonmarin, 2009), setting the stage for the measurement of time-resolved VCD measurement. Thus, it seems only a matter of time before the same control of CP states and sufficient reduction of birefringence artifacts is achieved for various forms of non-linear and time-resolved ROA.

7.5.6 Surfaced-Enhanced ROA

A few reports of either the observation or theoretical prediction of surface-enhanced ROA (SEROA) have been published (Abdali, 2006; Johannessen and Abdali, 2007; Abdali *et al.*, 2007; Johannessen *et al.*, 2007). However, successful comparisons between observed and calculated SEROA have not yet been achieved. Surface-enhanced Raman scattering (SERS) is a long-established sensitive form of Raman scattering in which molecules adsorbed onto conducting surfaces show enormous enhancements from the coupling between the polarizability of the molecule and plasmonic modes in the solid. These enhancements are especially strong near so-called 'hot spots' on the surface where enhancement factors near 10^{14} have reported and which permit the observation of SERS from single molecules.

The attempt to measure SEROA is difficult for a variety of reasons. One is sample, or hot spot, stability over the collection times required to obtain sufficient photon counts for ROA to be observed. SERS spectra are rarely recorded beyond a signal-to-noise ratio (SNR) of approximately 100, whereas ROA as large as 10^{-3} compared with the parent SERS spectrum requires an SNR of approximately 10^4 for a well-defined SERS spectrum, which in turn requires collecting 10^8 SERS photon counts, or at least 1 h of collection time.

A second reason is the possibility that SERS hot spots may differ in the local chirality. This may not lead to much variation between the SERS from different hot spots, but it could lead to very large differences between SEROA spectra. It is well known that a major problem of the quantitative use of SERS is the variability between SERS intensities from even the same sample. This variability is amplified to the point where individual hot spots might even have opposing or canceling contributions. What is needed to solve this problem is the fabrication of nano-surface structures that have identical local morphology and identical chirality. This has not yet been achieved.

A practical solution to this problem, and to the stability problem is to design SERS surfaces that provide a buffer from surface hot spots. This would lower the overall SERS enhancement, but may allow a much larger ensemble of chiral molecules to be seen, and averaged over, in the SERS and SEROA spectra. This so-called near-SERS and near-SEROA would be a compromise between conventional ROA and the averages for direct-surface hot-spot SEROA. Theoretical descriptions of orientationally averaged-SERS and -SEROA have been proposed (Janesko and Scuseria, 2009; Janesko and Scuseria, 2006).

7.5.7 Rayleigh Optical Activity

If ROA is measured near the Rayleigh line but below the frequencies of the normal vibrational modes of the molecule, one can measure instead Rayleigh optical activity (RayOA) (Barron and Buckingham, 1971). RayOA spectra have not yet been reported. The principal reason for this is that Rayleigh scattering is dominated by the polarized part of the scattering. Most of this polarized scattering is merely redirected incident radiation with little or no change in wavelength. The weaker depolarized part, by contrast, contains information about the hydrodynamic diffusion and reorientational motion of the molecules in the scattering medium. This depolarized scattering component, also known as the Rayleigh wing, extends much further from zero frequency shift, as far as several hundred cm^{-1}, and underlies the low frequency Raman bands.

Most ROA instruments are now set up as backscattering spectrometers, but in backscattering both unpolarized ICP and SCP forms of Raman/ROA are polarized scattering. On the other hand, DCP_I backscattering is a completely depolarized form of scattering, and is ideal for the measurement of RayOA. Furthermore, theoretical simulations of RayOA find that depolarized RayOA is much more intense, relative to the parent depolarized Rayleigh scattering, than the corresponding polarized RayOA is to its parent polarized Rayleigh scattering (Zuber *et al.*, 2005). Efforts to measure RayOA should therefore focus on measuring depolarized RayOA using backscattering DCP_I Rayleigh scattering.

References

Abdali, S. (2006) Observation of SERS effect in Raman optical activity, a new tool for chiral vibrational spectroscopy. *J. Raman Spectrosc.*, **37**, 1341.

Abdali, S., Johannessen, C., Nygaard, J., and Norbygaard, T. (2007) Resonance surface enhanced Raman optical activity of myoglobin as a result of optimized resonance surface enhanced Raman scattering conditions. *J. Phys. Condens. Mattter*, **19**, 285205.

Barron, L.D., and Buckingham, A.D. (1971) Rayleigh and Raman scattering from optically active molecules. *Mol. Phys.*, **20**, 1111–1119.

Barron, L.D., Bogaard, M.P., and Buckingham, A.D. (1973) Raman scattering of circularly polarized light by optically active molecules. *J. Am. Chem. Soc.*, **95**, 603–605.

Barron, L.D., Hecht, L., Hug, W., and MacIntosh, M.J. (1989) Backscattered Raman optical activity with CCD detector. *J. Am. Chem. Soc.*, **111**, 8731–8732.

Che, D., Hecht, L., and Nafie, L.A. (1991) Dual and incident circular polarization Raman optical activity backscattering of (–)-*trans*-pinane. *Chem. Phys. Lett.*, **180**, 182–190.

Che, D., and Nafie, L.A. (1992) Isolation of Raman optical activity invariants. *Chem. Phys. Lett.*, **189**, 35–42.

Hecht, L., and Nafie, L.A. (1990) Linear polarization Raman optical activity: A new form of natural optical activity. *Chem. Phys. Lett.*, **174**, 575–582.

Hecht, L., Barron, L.D., Gargaro, A.R. et al. (1992a) Raman optical activity instrument for biochemical studies. *J. Raman Spectrosc.*, **23**, 401–411.

Hecht, L., Che, D., and Nafie, L.A. (1992b) Experimental comparison of scattered and incident circular polarization Raman optical activity in pinanes and pinenes. *J. Phys. Chem.*, **96**, 4266–4270.

Hecht, L., and Barron, L.D. (1994) Linear polarization Raman optical activity: The importance of the non-resonant term in the Kramers-Heisenberg-Dirac dispersion formula under resonance conditions. *Chem. Phys. Lett.*, **225**, 519–524.

Helbing, J., and Bonmarin, M. (2009) Vibrational dichroism signal enhancement using self-heterodyning with elliptically polarized laser pulses. *J. Chem. Phys.*, **1131**, 174507.

Hug, W. (1982). Instrumental and theoretical advances in Raman optical activity. In: *Raman Spectroscopy* (ed. J. Lascombe), pp. 3–12.

Hug, W. (2003) Virtual enantiomers as the solution of optical activity's deterministic offset problem. *Appl. Spectrosc.*, **57**, 1–13.

Hug, W., and Surbeck, H. (1979) Vibrational Raman optical activity spectra recorded in perpendicular polarization. *Chem. Phys. Lett.*, **60**, 186–192.

Hug, W., and Hangartner, G. (1999) A very high throughput Raman and Raman optical activity spectrometer. *J. Raman Spectrosc.*, **30**, 841–852.

Janesko, B.G., and Scuseria, G.E. (2006) Surface enhanced Raman optical activity of molecules on orientationally averaged substrates: Theory of electromagnetic effects. *J. Chem. Phys.*, **125**, 124704.

Janesko, B.G., and Scuseria, G.E. (2009) Molecule-surface orientational averaging in surface enhanced Raman optical activity spectrosocpy. *J. Phys. Chem. C*, **113**, 9445.

Jensen, L., Autschbach, J., Krykunov, M., and Schatz, G.C. (2007) Resonance vibrational Raman optical activity: A time-dependent density functional theory approach. *J. Chem. Phys.*, **127**, 134101–134111.

Johannessen, C., and Abdali, S. (2007) Surface enhanced Raman optical activity as an ultra sensitive tool for ligand binding analysis. *Spectrosocpy*, **21**, 143–149.

Johannessen, C., White, P.C., and Abdali, S. (2007) Resonance Raman optical activity and surface enhanced resonance Raman optical activity analysis of cytochrome c. *J. Phys. Chem. A*, **111**, 7771–7776.

Nafie, L.A. (1996) Theory of resonance Raman optical activity: The single-electronic state limit. *Chem. Phys.*, **205**, 309–322.

Nafie, L.A. (2008) Theory of Raman scattering and Raman optical activity. Near resonance theory and levels of approximation. *Theor. Chem. Acc.*, **119**, 39–55.

Nafie, L.A., and Che, D. (1994). Theory and measurement of Raman optical activity. In: *Modern Nonlinear Optics, Part 3* (eds M. Evans, and S. Kielich) John Wiley & Sons, Inc., New York, pp. 105–149.

Nafie, L.A., Brinson, B.E., Cao, X. et al. (2007) Near-infrared excited Raman optical activity. *Appl. Spectrosc.*, **61**, 1103–1106.

Rhee, H., Kim, S.-S., Jeon, S.-J., and Cho, M. (2009a) Femtosecond measurements of vibrational circular dichroism and optical rotatory dispersion spectra. *ChemPhysChem*, **10**, 2209–2211.

Rhee, H., Young-Gun, J., Lee, J.-S. et al. (2009b) Femtosecond characterization of vibrational optical activity of chiral molecules. *Nature*, **458**, 310–313.

Spencer, K.M., Freedman, T.B., and Nafie, L.A. (1988) Scattered circular polarization Raman optical activity. *Chem. Phys. Lett.*, **149**, 367–374.

Yu, G.-S., and Nafie, L.A. (1994) Isolation of preresonance and out-of-phase dual circular polarization Raman optical activity. *Chem. Phys. Lett.*, **222**, 403–410.

Zuber, G., Goldsmith, M.-R., Beratan, D.N., and Wipf, P. (2005) Towards Raman optical activity calculations of large molecules. *ChemPhysChem*, **6**, 595–597.

8

Measurement of Vibrational Optical Activity

The measurement of vibrational optical activity follows primarily the same procedures as those employed for the measurement of the parent infrared or Raman spectra. As these methods can be found in numerous source materials (Griffiths and de Haseth, 2007; Lewis and Edwards, 2001), the emphasis in this chapter will be on those concepts that involve the optimization of the VOA spectrum given that the parent IR or Raman spectrum is already optimized. For example, there are well-known trade-offs between signal-to-noise ratio, spectral resolution, and measurement time, and while these will be mentioned briefly, extensive discussions of these trade-offs will not be undertaken. Instead, issues involving VOA intensity calibration, baseline correction, verification of instrument performance, and intensity standards, will be described in detail. In Chapter 9, standard procedures for the calculation of VOA spectra will be described, and in the final chapter, applications of VOA involving measured spectra, or the comparison of measured with calculated spectra, will be described.

8.1 VOA Spectral Measurement

The measurement of a VOA spectrum is accompanied by the simultaneous measurement of the parent IR or Raman spectrum. As any VOA spectrum is several orders of magnitude weaker than its parent spectrum, the sampling conditions and measurement parameters for the VOA spectrum must be amenable to the measurement of a high-quality parent spectrum. The measurement of such a spectrum involves different issues for IR absorption compared with Raman scattering due to differences in the sources of both signal and noise in these two types of spectra. For IR absorption spectra, the information contained in the transmission signal arises from the spectrum of radiation *removed* from the beam. The dominant source of noise for IR measurements is from the background radiation seen by the detector. In absolute terms, this noise level for a given optical setup depends only on measurement time and is independent of the sampling conditions. By contrast, the information content of a Raman spectrum is due to the *presence*, not the absence, of photons detected, and for Raman scattering the

Vibrational Optical Activity: Principles and Applications, First Edition. Laurence A. Nafie.

dominant source of noise is proportional the square root of the total number of scattered photons detected and not background radiation (after stray light is eliminated). As a result, the approaches to optimize the measurement conditions for IR and Raman scattering are fairly different.

Once the conditions for the measurement of the parent IR or Raman spectrum have been optimized, an additional level of optimization is needed for obtaining a high-quality VCD or ROA spectrum. At this secondary level, the considerations needed for the optimization of VCD and ROA spectral measurement are more closely related. The issues involved include verification of instrument performance using standard samples, verification of the sign of the VOA spectrum, determination and correction of the zero baseline, determination of the optimum measurement time, removal of spectral artifacts dependent on instrumental and sampling conditions, isolation of the VOA noise spectrum, calibration of the VOA intensity, and finally presentation of the parent, VOA, and VOA noise spectra for spectral interpretation and application. Because of fundamental differences between the measurement of IR absorption and VCD spectra compared with Raman scattering and ROA spectra, the measurement methodology for these two forms of VOA will be described separately.

8.2 Measurement of IR and VCD Spectra

The measurement of high-quality IR and VCD spectra requires following a number of specific steps. The first of these is the selection of the spectral region to be covered, which includes the choice of IR detector and optical components needed. This is followed by sampling considerations such as the choice of the solvent, the type of sample cell to use, and the optimization of pathlength and concentration. Once an optimized IR spectrum is obtained, which may include more than one combination of pathlength and concentration for particular sub-spectral regions, one must decide on spectral resolution and collection time to achieve the desired VCD spectrum. Finally, the format for the presentation of VCD, IR, and VCD noise spectra should be selected. Examples of a preferred format from each of the major spectral regions for which VCD spectra have been measured will be presented.

8.2.1 Selection of Frequency Range, Detector and Optical Components

VCD spectra has been measured with an FT-IR spectrum across a wide range of frequencies from 800 to 14 000 cm^{-1}. For VCD measurements, a semiconductor detector sensitive to the generation of signals by individual photons is required. Such detectors have a response time in the micro-second range, which is required for following the high-frequency (tens of kilohertz) polarization modulation of the photoelastic modulators (PEMs) used for the measurement of VCD spectra.

8.2.1.1 Mid-Infrared Spectral Region

The most common spectral region for VCD measurement is the mid-IR from 800 to 2000 cm^{-1}, often called the fingerprint region due to overlap of many types of vibrational modes that are highly characteristic of the absorbing molecule. Here, a liquid-nitrogen cooled mercury–cadmium–telluride, [$(Hg_xCd_{1-x})Te$ or MCT] photoconductive (PC) detector is typically employed where the relative composition of Hg and Cd, designated by x and $1-x$, can be adjusted to achieve different semiconductor band-gap energies. The most common band-gap, as noted above, corresponds to a low-energy photon cut-off of approximately 800 cm^{-1}. Photons with lower energy, and hence lower wavenumber frequency, have insufficient energy to promote an electron in the MCT semiconductor element from the valence to the conduction band, and thereby activate the response of the detector. An 800 cm^{-1} MCT cut-off is referred to as a type-A MCT detector. The normalized response profile of this detector is plotted in Figure 8.1 along with the profiles of several other detectors to be discussed below in subsequent subsections (Cao *et al.*, 2004). An alterative type of MCT (type-B) can be fabricated with a

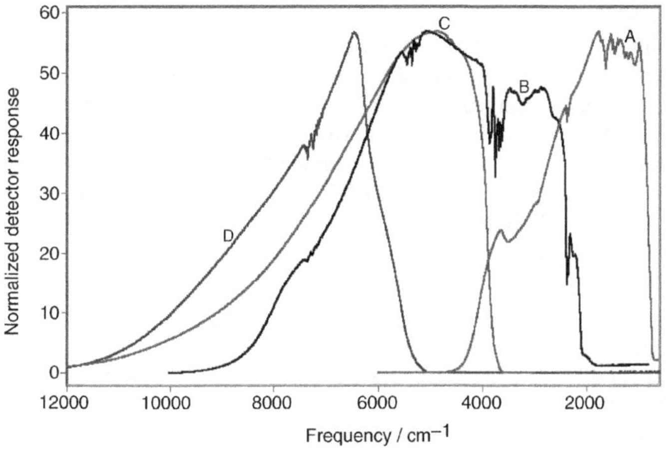

Figure 8.1 *Response curves for four IR detectors, normalized to the same maximum value and covering the frequency range from 800 to 12 000 cm^{-1}, exhibit their relative overlap of spectral coverage. The relative sensitivities of the detectors is not shown but sensitivities increase with increasing spectral frequency coverage. A, PC-MCT-LN2; B, PV-MCT-TE; C, InGaAs; and D, Ge. Adpated with permission from the Elsevier Publishing Company (Cao et al., 2004)*

low-frequency cut-off of 400 cm^{-1}, but this type of detector has a lower intrinsic signal-to-noise ratio (SNR) response due to its wider spectral range of response and associated higher background noise level.

The optical components used for mid-IR VCD measurement are usually compatible with the spectral range from 800 to 2000 cm^{-1}. The most important of these is the photoelastic modulator (PEM), which usually possesses a ZnSe optical element with a low-frequency cut-off between 500 and 600 cm^{-1} and a non-absorbing high-frequency transmission extending to the mid-visible region of the spectrum. The principal drawback for ZnSe as an optical element is its high index of refraction and hence high reflection losses at air–ZnSe interfaces. This is ameliorated by the use of anti-reflection coatings that can be tailored for relatively broad spectral ranges which include the entire mid-IR range. The remaining optical elements are sample cell windows and lenses. The preferred optical material for mid-IR VCD measurements is BaF_2, which has low reflection losses and a low-frequency cut-off near 800 cm^{-1}, which nicely matches the cut-off of the type-A MCT detector. An alternative optical material for the PEM, cell windows, and lenses is CaF_2. The optical properties of CaF_2 are similar to those of BaF_2 except that the low-frequency cut-off is 1200 instead of 800 cm^{-1}. This choice of optical material significantly reduces or eliminates spectral coverage below 1200 cm^{-1}; however, in the case of biological molecules in aqueous solutions with high levels of water absorption below 1000 cm^{-1}, use of a CaF_2 cell does little to reduce the mid-IR coverage available for these applications.

8.2.1.2 Hydrogen-Stretching Region

The region between 2000 and 4000 cm^{-1} includes primarily fundamental vibrations associated with hydrogen stretching. Two types of semiconductor detector span this region with high-efficiency; these are indium antinomide (InSb) and photovoltaic (PV) MCT. The latter is in contrast with the mid-IR photoconductive (PC) MCT detector described above. InSb is the more traditional detector in the hydrogen-stretching region and requires liquid-nitrogen for operation. When cooled, InSb has a low-energy cut-off near 2000 cm^{-1}. By contrast PV-MCT is also available with a low-frequency cut-off near 2000 cm^{-1} but requires only thermoelectric (TE) cooling, thus simplifying its use, particularly for long scans when the reservoir of liquid nitrogen typically needs to be refilled after 4–6 h (or longer with long-hold-time dewars). The spectral response for the PV-MCT-TE cooled detector is shown in Figure 8.1 and is similar in profile to the InSb detector. The preferred optical material for the hydrogen stretching region is CaF_2. This material is readily available and is transparent from the UV region to 1200 cm^{-1} in the mid-IR. The PEM and cell windows are typically comprised of CaF_2. Focusing lenses, where used, can be made either of CaF_2, BaF_2, or anti-reflection coated ZnSe.

8.2.1.3 First Overtone and Combination-Band Region

The region between 4000 and 6500 cm^{-1} is associated primarily with two-quantum vibrational transitions, either as the first overtone or the combination band of two fundamentals. This region is the first of a broader region know as the near-IR region, which extends from the hydrogen stretching region to the visible region. Although the PV-MCT detector has good coverage in this region, somewhat better coverage is provided by an InGaAs detector, which can be operated at room temperature or with TE cooling. The spectral response profile of an InGaAs detector is shown in Figure 8.1. The optical material used for this and subsequent regions in the near-IR is typically CaF_2 for the PEM, lenses and windows can be comprised of CaF_2, BaF_2, or quartz.

8.2.1.4 Second Overtone and Second Combination Band Region

In this region, transitions intensities are somewhat weaker than the first overtone and combination-band region, but aside from this difference most of the instrumental components can remain the same. The InGaAs detector maintains coverage to 8000 cm^{-1} but within this region performance of a germanium (Ge) detector surpasses that of InGaAs, as shown in Figure 8.1. Replacement becomes highly desirable at vibrational frequencies above 7000 cm^{-1} up to 10 000 cm^{-1}.

8.2.1.5 Third Overtone and Combination Band Region and Beyond

For spectral coverage from 10 000 to 14 000 cm^{-1}, one can use either A Ge detector or a silicon (Si) detector. Alternatively, photomultiplier detectors can be used in this region, which overlaps with the visible region of the spectrum at 700 nm.

8.2.2 Choice of IR Solvents

Once the spectral region and instrumental configuration have been determined, the focus of measurement shifts to how the sample will be measured. Most applications of VCD involve solution-state sampling, and hence the most important consideration for sampling is the choice of solvent. The preferred solvents have a large mid-IR window and relatively small interaction of the molecule with solvent. Deuterium substitution typically enlarges the spectral window in the mid-IR by elimination of solvent absorption in the CH-bending region from 1500 to 1200 cm^{-1}. Some solvents typically used for VCD measurements, in order from non-polar to polar, are: CCl_4, CS_2, $CDCl_3$, CD_2Cl_2, DMSO-d_6, CD_3OD, CD_3CN, CD_3CD_3CO, H_2O, and D_2O. Spectra of selected solvents of various pathlengths are presented in Figures 8.2 and 8.3.

Although not shown in Figure 8.2, carbon tetrachloride, CCl_4, is an ideal solvent with no significant absorbance between 2000 and 800 cm^{-1}; however, its use is limited as a solvent to the most non-polar of sample molecules. The more polar molecule, chloroform, is a widely-used solvent, and deuterated chloroform ($CDCl_3$) is almost ideal for VCD measurement. For organic molecules that are more difficult to dissolve, deuterated dimethylsulfoxide (DMSO-d_6) is often useful although it tends to

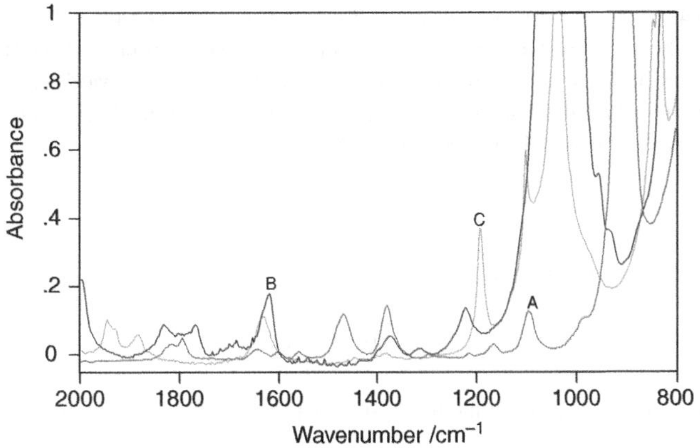

Figure 8.2 Mid-IR absorbance spectra for a pathlength of 75 microns (μm): A, $CDCl_3$; B, DMSO-d_6; and C, CD_3CN. $CDCl_3$ provides spectral coverage to almost 900 cm^{-1} without significant spectral interference whereas DMSO-d_6 and CD_3CN coverage is to only 1100 cm^{-1}

interact with solute molecules more strongly than chloroform. Another solvent for more polar organic molecules is deuterated acetonitrile, shown in Figure 8.2.

The polar solvents illustrated in Figure 8.3 are used for biological molecules, such as proteins and nucleic acids that are only appreciably soluble in aqueous solvents. D_2O can be used with higher concentrations than H_2O, but D_2O suffers from incompleteness of dynamics of deuterium–hydrogen exchange both in sample preparation and during the course of measurement.

8.2.3 Optimization of Concentration, Pathlength, and Spectral Resolution

In general, the concentration and pathlength of a sample for VCD should be adjusted such that the absorbance bands of interest fall in the range of 0.1–1.0 absorbance units. IR detectors have a constant noise from background radiation, as discussed above, and as a result the optimum absorbance level is

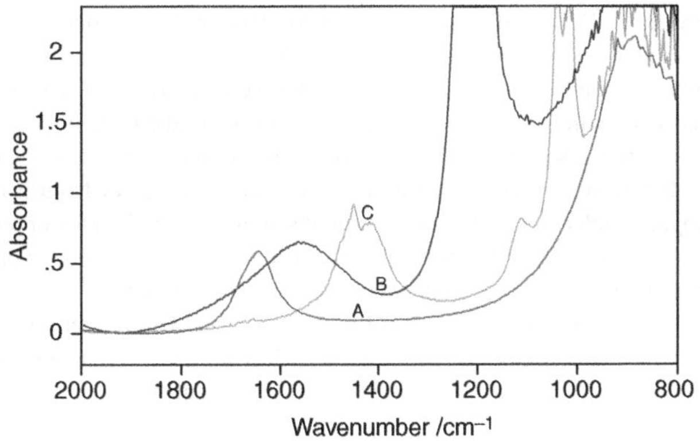

Figure 8.3 Mid-IR absorbance spectra of: A, H_2O (6 μm); B, D_2O (80 μm); and C, CD_3OD (20 μm) at typical pathlengths for measurements between 1400 and 1800 cm^{-1}. Use of D_2O instead of H_2O allows use of lower concentrations due to the increased pathlengths that can be used with D_2O

approximately 0.4. This can be shown in the following way. We write the signal intensity as the absorbance A, and the noise level relative to this intensity at the detector is proportional to the inverse of the sample transmission $1/T$. In other words, as the transmission decreases toward zero (infinite relative noise), the signal intensity at the detector decreases relative to the constant detector noise background. We can therefore write the signal-to-noise ratio (SNR) as:

$$SNR = A/(1/T) = AT \tag{8.1}$$

If we now express the transmission in terms of absorbance as:

$$T = I/I_0 = 10^{-A} = \mathrm{e}^{-\ln 10A} \tag{8.2}$$

we can find the maximum SNR as a function of A by setting the derivative of SNR with respect to A equal to zero as:

$$\frac{\partial(SNR)}{\partial A} = \frac{\partial\left(A\mathrm{e}^{-\ln 10A}\right)}{\partial A} = 0 \tag{8.3}$$

Differentiation yields the equation

$$(1 - A\ln 10)\mathrm{e}^{-\ln 10A} = 0 \tag{8.4}$$

This holds in general when the expression in parentheses vanishes which yields,

$$A = 1/\ln 10 \cong 0.43 \tag{8.5}$$

The basic idea is that if the transmission is too high not enough of the light beam is used to provide information about the sample, and if the transmission is too low, there are not enough photons at the detector, relative to the constant level of background noise, to give a good SNR ratio. For samples with wide ranges of absorbance bands of interest, it may be desired to make separate measurements of the high and low intensity bands at different pathlengths, thereby optimizing each region or sets of bands, separately.

Pathlengths in the mid-IR typically range from 5 to 10 microns for aqueous samples to 50 to 100 microns for non-aqueous samples. In the near-IR, pathlengths are in the millimeter range because overtone and combination band spectra of the solutes and solvents are two to three orders of magnitude weaker than their corresponding fundamental transitions in the mid-IR region.

Selection of the spectral resolution in the mid-IR region is typically 4 cm^{-1} as the bandwidth of spectral features for most common-sized molecules is in the range of from 5 to 10 cm^{-1}. For more complex biological molecules, for example proteins or nucleic acids, a lower resolution, such as 8 cm^{-1}, may be desired due to the higher density of different normal modes contributing to the observed spectrum and hence overall broader shaped overlapping bands.

In the near-IR regions, spectral features are significantly broader and more overlapped than in the mid-IR. As a result, lower spectral resolution is adequate to record all the spectral detail that is needed. Typical spectral resolution is 16 cm^{-1}, and in some cases 32 cm^{-1} is adequate for particular applications.

8.2.4 Measurement and Optimization of VCD Spectra

Given that the measurement of the parent IR spectrum has been optimized, one can proceed with the various steps needed for the correct measurement of a VCD spectrum. In Chapter 6, the instrumental aspects of a VCD spectrometer were described. Here we detail the steps needed for the measurement of a high-quality Fourier transform VCD spectrum. For a VCD spectrometer with a single photoelastic modulator (PEM), these steps include setting the retardation value of the PEM, determining the Fourier phase for the VCD interferogram, setting the phase of the lock-in amplifier (LIA), determining the sign of the VCD spectrum, obtaining a VCD calibration spectrum, and calibrating the VCD spectrum. For dual-PEM operation, additional steps are needed to determine the phase of the second LIA and to optimize the retardation of the second PEM, in order to bring the VCD baseline as close to zero as possible.

8.2.4.1 Fourier Phase Correction for the VCD Interferogram

The first step for VCD measurement is to make sure that a correct AC Fourier phase, $\theta_{AC}(\bar{\nu})$, is available for the phase correction of the VCD transmission spectrum as defined in Equation (6.66). This can be obtained by measuring the VCD transmission spectrum using a so-called stressed birefringent plate where the stress is well below quarter-wave retardation across the entire spectrum. This plate is used with a second polarizer oriented parallel or perpendicular to the first polarizer (located before the PEM) in the optical train. In order to measure the VCD phase spectrum, $\theta_{AC}(\bar{\nu})$, with the stress plate, the phase of the lock-in amplifier must be properly set. This is typically accomplished with the auto-phase option of the lock-in amplifier. One must also chose the spectral resolution and the value of PEM retardation (typically the desired wavenumber frequency of the quarter-wave maximum) that will be used for the VCD measurement. Separate phase-correction spectra must be obtained for each choice of spectral resolution and PEM retardation used for VCD measurement. The use of the stress plate setup yields a strong, monosignate VCD spectrum across the range of VCD measurement that can be phase corrected in the same way that an ordinary IR transmission spectrum can be Fourier phase corrected. Once $\theta_{AC}(\bar{\nu})$ is obtained, it can be used, by transfer, to phase correct any VCD transmission interferogram without loss of VCD sign information.

8.2.4.2 Setting the Retardation Value of the First PEM

As explained in Chapter 6, the retardation value of the PEM determines the Bessel function, $J_1(\bar{\nu})$, which expresses the efficiency as a function of wavenumber frequency for the measurement of a VCD spectrum. Typically one sets the maximum of $J_1(\bar{\nu})$ to be centered near the middle of the spectral region of interest. The PEM will maintain an efficiency of greater than 90% over the range from 2/3 to 4/3 of the frequency of maximum efficiency. For example, for optimization at 1200 cm^{-1}$J_1(\bar{\nu})$ is within 90% of its maximum value from 800 to 1600 cm^{-1}. This corresponds close to the range of spectral coverage commonly used for mid-IR VCD measurement using an MCT detector.

8.2.4.3 Calibration of the Intensity and Sign of the VCD Spectrum

With the Fourier phase, lock-in phase, spectral resolution, and PEM retardation set, one can proceed to the measurement of the uncalibrated VCD spectrum as the ratio of the Fourier transformed VCD to IR transmission spectra. In order to convert the uncalibrated VCD spectrum into a calibrated VCD spectrum, one needs a calibration spectrum as described in Chapter 6. This is obtained from a pair of VCD measurements using a multiple-wave plate together with a second polarizer. The resulting calibration function $2(1.1513)J_1(\bar{\nu})$ can be divided into the uncalibrated VCD spectrum, taking into account any changes in gain used to measure the calibration function, to obtain the final calibrated VCD spectrum.

If a sample, such as (+)- or (−)-α-pinene is used, the accuracy of the calibration procedure can be checked against published standards and minor discrepancies remedied by scale adjustment. A calibrated intensity of neat (−)-α-pinene with a 75 micron pathlength is presented in Figure 8.4

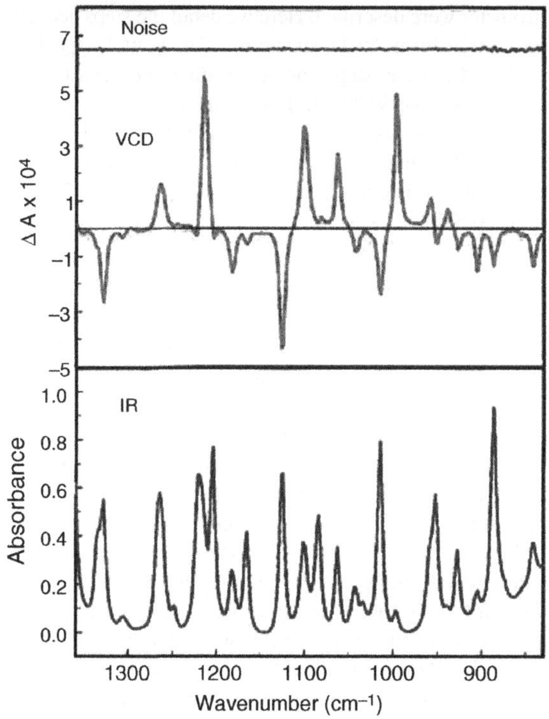

Figure 8.4 *VCD, infrared absorbance, and VCD noise spectrum (top) for neat (−)-α-pinene measured in a cell with a 75 micron pathlength*

against which a newly measured VCD spectrum can be compared. At the same time as the intensities are compared, the sign of the VCD spectrometer can be verified. If the signs of the bands are the opposite of those shown here, and if one is sure that the sample is (−)-α-pinene, and not (+)-α-pinene, then the opposite VCD sign can be selected by changing the phase of the PEM lock-in amplifer by 180°. This changes the signs of all VCD bands by −1 and should bring the measured and reference VCD bands of (−)-α-pinene into exact agreement. It is good practice for the maintenance of a VCD spectrometer to measure regularly, if only briefly, the VCD of (−)-α-pinene, or some suitable reference spectrum, to verify that the performance of the instrument in terms of VCD sign, intensity, and signal-to-noise ratio is what it is supposed to be. Most important of these is to check the sign to avoid a sign error in VCD measurement that could occur if the LIA is reset by, for example, a power failure.

Figure 8.4 also illustrates a preferred format for the presentation of VCD spectra. VCD and IR spectra are presented in a vertical stack so that the origin of VCD features can be visually correlated with the corresponding parent IR features. Typically the VCD are collected in two successive blocks of co-added spectra. If these two blocks are added and divided by two, the average VCD of the two blocks is obtained. If the two VCD blocks are subtracted and divided by two, a corresponding VCD noise spectrum is obtained. Such a spectrum is shown directly above the VCD spectrum in Figure 8.4. A VCD noise spectrum is an important reference of signal reliability as the level of noise may be difficult to discern in a VCD spectrum that is small and exhibits genuine VCD features that are close to the noise level.

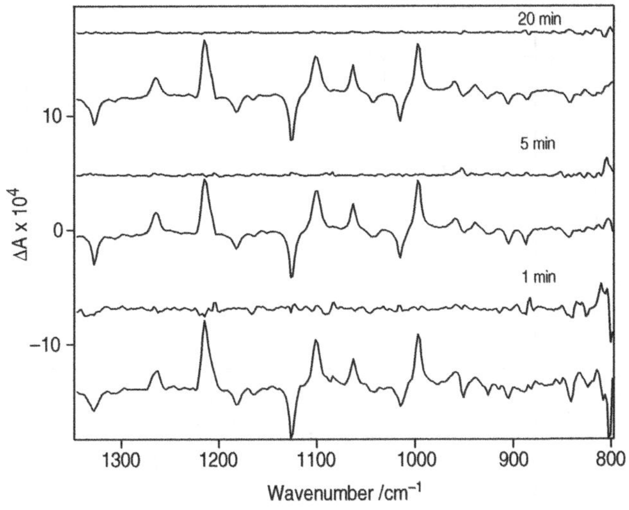

Figure 8.5 *VCD and VCD noise spectra of neat (−)-α-pinene for collection times of 1, 5, and 20 min showing a noise improvement equal to the square root of the increase in spectral measurement time*

8.2.4.4 Check of Signal-Averaging Improvement

Once a correct, calibrated VCD spectrum has been obtained for a given resolution and spectral region, one can decide how much time should be devoted to the measurement of the VCD spectrum. Generally, a signal-to-noise ratio of ten or more is desired for the major VCD bands. For weak spectra, long averaging times may be necessary. For IR and VCD spectra, the SNR increases with the square root of the increase in measurement time. This is illustrated for (−)-α-pinene in Figure 8.5, where VCD and their noise spectra are compared for collection times of 1, 5, and 20 min. While more difficult to discern from the VCD spectrum themselves, it is clear from the VCD noise spectra that the SNR approximately doubles between the 1 and 5 min collection times (theoretically $\sqrt{5}$) and between the spectra with 5 and 20 min collection times. For weak spectra it is not unusual to collect for 4 or 8 h, or even overnight with a 12–18 h measurement for long hold-time dewars. The time one devotes to obtaining a VCD spectrum is a matter of choice, but given that the interpretation or the theoretical calculation of VCD spectra may require many hours or days, it is reasonable that one should devote a significant amount of time to the measurement of the VCD to obtain as high a signal quality as possible under any particular circumstances.

8.2.4.5 VCD Baseline Correction and Artifact Elimination

After measurement of a VCD spectrum, a baseline correction is applied to compensate for deviations of the VCD instrument baseline from true zero. This can be achieved in a number of ways. The most accurate is to measure under identical conditions (including concentration, cell pathlength, and cell window orientation) the pure enantiomers of the sample. Subtraction of the VCD spectrum of a sample of the opposite enantiomer and division by two yields a baseline and artifact corrected VCD spectrum. If the opposite enantiomer is not available, the racemic mixture can be used instead, but this option adds noise to the corrected VCD spectrum, because after subtraction of the two spectra, which results in a $\sqrt{2}$ increase in the noise level, one does not divide by two to obtain the final spectrum. If neither the opposite enantiomer nor the racemic mixture is available (or indeed possible, as in the case of biological molecules such as proteins), the VCD spectrum must be baseline corrected by use of the solvent VCD spectrum measured under the same conditions as those for the solution of the chiral

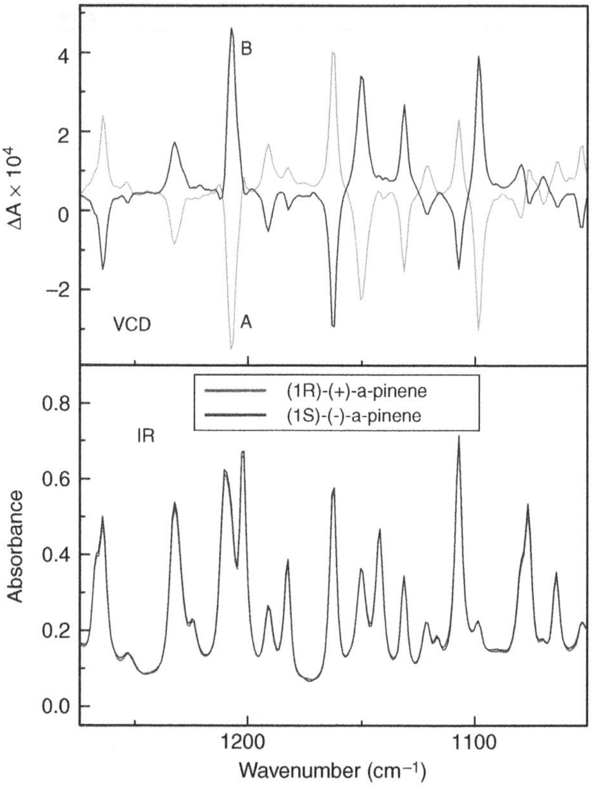

Figure 8.6 *The VCD spectra of neat (−)- (B) and (+)-α-pinene (A) for 1 h collection times that show a slight baseline offset and deviation from perfect mirror symmetry of the opposite VCD spectra due to either small levels of artifacts or VCD noise*

sample molecule. While the VCD spectrum of the solvent has a zero true VCD spectrum, the absorbance spectrum is not the same as the sample, and in some cases, for baselines significantly displaced from zero, artifacts can arise in the sample associated with its absorbance spectrum that are not present in the baseline VCD spectrum of the solvent. This problem is addressed below when the use of two PEMs is discussed. In Figure 8.6 the near mirror-image VCD spectra of (−)- and (+)-α-pinene are presented showing the small offset of the baseline from zero and small deviations from perfect mirror symmetry of the enantiomeric pair of VCD spectra due to VCD baseline artifacts or noise.

8.2.4.6 Dual PEM with Rotating Sample Cell and Artifact Reduction

The theory of dual-PEM measurement of VCD spectra was presented in Chapter 6 (Nafie, 2000). By adding a second PEM after the sample and with a sample cell rotating about the axis of light beam propagation (Lombardi, 2011), one is able to measure dynamically the VCD baseline with the second PEM at the same time the VCD of the sample with the same baseline is measured with the first PEM. The principal requirement to achieve optically is that the IR beam profile should be close to the same for each of the two PEMs. The VCD spectrum with baseline and the VCD baseline measured by the two associated lock-in amplifiers can be subtracted in real time to yield the baseline-corrected VCD spectrum. The correction process is illustrated in Figure 8.7. The advantage of dynamic subtraction is

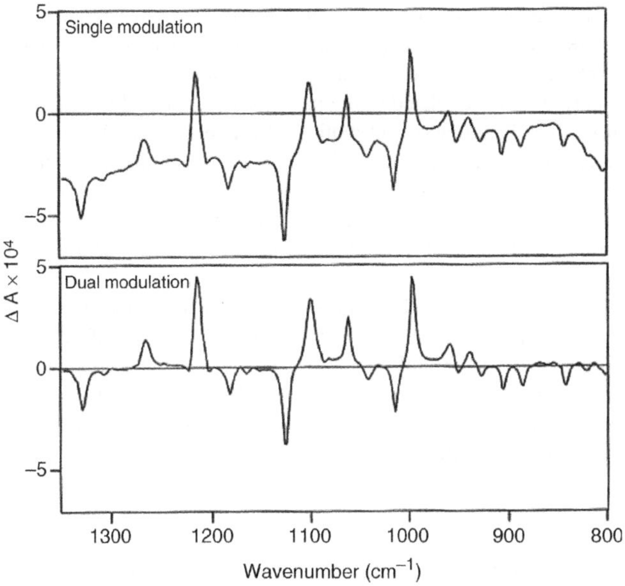

Figure 8.7 *The VCD spectra of neat (−)- and (+)-α-pinene with a dual PEM optical setup where in the upper spectrum only the first PEM was used, with the second PEM turned off, while in the lower spectrum, both PEMs are operating and their corresponding VCD spectra (the second measuring only the VCD baseline) are dynamically subtracted*

that it facilitates optical alignment of the VCD spectrometer as the baseline can be adjusted simply by changing the retardation voltage of the second PEM. In addition, such undesirable effects as baseline drift and absorption artifacts that are a problem for large deviations of the baseline from zero, are either eliminated or greatly reduced with dual PEM operation. Noise is increased in the baseline-corrected VCD spectrum by a factor of $\sqrt{2}$ compared with single PEM operation, but this is more than compensated by long-term instrument stability that permits long block averaging without baseline drift, as any drift is canceled by subtraction of the VCD baselines measured simultaneously by the two PEMs.

8.2.5 Solid-Phase VCD Sampling

The most practiced medium for the measurement of VCD spectra is the liquid phase, either neat liquids or solutions. A few publications featuring VCD of films or mulls of microcrystalline samples were published at an early stage in the development of VCD (Diem *et al.*, 1979; Narayanan *et al.*, 1985; Sen and Keiderling, 1984), but these sampling methods have been largely abandoned due to undesired variations in the measured VCD spectrum with sample preparation. More recently, several publications have appeared in which films were cast of chiral molecules, most notably proteins, with carbohydrates acting as a host medium (Shanmugam and Polavarapu, 2005). The VCD from these films appear to be artifact free, but they are not pure samples and involve a chiral host requiring the subtraction of a VCD, and also an IR, background. Artifacts in solid-phase VCD, or ECD, arise from two sources: one is scattering from particles comparable in size to the wavelength of the light and the other is linear birefringence from the solid-phase medium, particles, or film.

Scattering in the IR region is associated with the Christiansen effect (Laufer *et al.*, 1980) where the symmetry of the IR band is distorted by the addition of negative intensity on the high-frequency side of the absorption band and additional positive intensity on the low-frequency side. These scattering contributions are related to the index of refraction of the solid sample and have a dispersive (derivative-like) bandshape contribution. As a result, the IR bands no longer have a symmetric lineshape, a distinctive feature of the presence of significant scattering contributions. Correspondingly, positive VCD bands have distortions of the same shape as the parent IR bands while negative VCD bands have the opposite signed distortion, that is the addition of positive intensity on the high-frequency side of the band and additional negative intensity on the low-frequency side of the band.

It can be shown theoretically that the Christiansen effect depends on the third power of the ratio of the average particle size to the wavelength of the light. As a consequence, if a film can be cast without small particle constituents or if the mull of a polycrystalline sample contains (after grinding) an average particle size smaller than the wavelength of the light, then scattering is essentially eliminated as a problem and the observed IR bands have symmetric bandshapes. Then the only remaining sources of artifacts are those arising from linear birefringence in the solid-state sample.

Linear birefringence is associated with any linear alignment of the structure of the sample molecules that could lead to differences in the index refraction for orthogonal states of linear polarization. One can test for this dependence by rotating the sample about the axis of light propagation. Any systematic variation in the measured VCD spectrum with angle of rotation is most likely due to this source. Such variations for a liquid or solution spectrum cannot arise for the sample itself (neglecting liquid crystals), but could be due to the windows of the sample cell. As nearly all VCD measurements involve solid-phase windows (except unsupported films or isolated crystals), linear birefringence is a universal source of potential artifacts in VCD measurements unless special precautions are taken, as described below and in the previous chapter. An important word of caution must be stated here. The absence of any variation in a VCD spectrum with rotation of a sample (and cell) about the propagation axis does not mean that there are no artifacts due to linear birefringence. To demonstrate this, the Mueller matrix formalism of Chapter 6 is extended to second order in the effect that an optical element, such as a solid-phase sample or cell window, can have on the optical beam. A higher-order sample Mueller matrix can be written as (Lombardi, 2011):

$$\boldsymbol{M}_{SF}(\bar{\nu}) = 10^{-A(\bar{\nu})}\left(1 - \boldsymbol{F}(\bar{\nu}) + \frac{1}{2}\boldsymbol{F}^2(\bar{\nu}) + \ldots\right) \tag{8.6}$$

$$\boldsymbol{F}(\bar{\nu}) = \begin{pmatrix} 0 & LD(\bar{\nu}) & LD'(\bar{\nu}) & -CD(\bar{\nu}) \\ LD(\bar{\nu}) & 0 & CB(\bar{\nu}) & LB(\bar{\nu}) \\ LD'(\bar{\nu}) & -CB(\bar{\nu}) & 0 & -LB'(\bar{\nu}) \\ -CD(\bar{\nu}) & -LB(\bar{\nu}) & LB'(\bar{\nu}) & 0 \end{pmatrix} \tag{8.7}$$

If only the first two terms are considered, then one obtains the first-order sample Mueller matrix given previously in Equation 6.38 but without the wavenumber dependence included here,

$$\boldsymbol{M}_{S1}(\bar{\nu}) = 10^{-A(\bar{\nu})}\begin{pmatrix} 1 & -LD(\bar{\nu}) & -LD'(\bar{\nu}) & CD(\bar{\nu}) \\ -LD(\bar{\nu}) & 1 & -CB(\bar{\nu}) & -LB(\bar{\nu}) \\ -LD'(\bar{\nu}) & CB(\bar{\nu}) & 1 & LB'(\bar{\nu}) \\ CD(\bar{\nu}) & LB(\bar{\nu}) & -LB'(\bar{\nu}) & 1 \end{pmatrix} \tag{8.8}$$

For simplicity, we drop the wavenumber dependence and then insert the full angular dependence of the linear dichroism and linear birefringence terms. This gives

$$M_{S1}(\theta) = 10^{-A}\begin{pmatrix} 1 & -LD\text{c}2\theta + LD'\text{s}2\theta & -LD'\text{c}2\theta - LD\text{s}2\theta & CD \\ -LD\text{c}2\theta + LD'\text{s}2\theta & 1 & -CB & -LB\text{c}2\theta - LB'\text{s}2\theta \\ -LD'\text{c}2\theta - LD\text{s}2\theta & CB & 1 & LB'\text{c}2\theta - LB\text{s}2\theta \\ CD & LB\text{c}2\theta + LB'\text{s}2\theta & -LB'\text{c}2\theta + LB\text{s}2\theta & 1 \end{pmatrix} \tag{8.9}$$

where we have abbreviated $\cos 2\theta$ as $\text{c}2\theta$ and $\sin 2\theta$ as $\text{s}2\theta$. This equation is an extension of Equation (6.86) where we have now included the *LD* terms. It can be seen that rotating the sample about the optical axis by an angle θ mixes the primed and unprimed *LB* and *LD* terms from their original values when $\theta = 0$ and therefore $\cos 2\theta = 1$ and $\sin 2\theta = 0$. If as in Equation (6.87), the sample Mueller matrix is rotated uniformly during sample measurement such that all angles are evenly sampled, all θ dependence vanishes, along with all contributions to the VCD from linear birefringence.

If, on the other hand, the Mueller matrix of the sample is extended to second order using Equation (8.6), the following angle dependent expression is obtained,

$$M_{S2}(\theta) = \begin{pmatrix} 1+\frac{1}{2}(LD^2+LD'^2) & -LD\text{c}2\theta + LD'\text{s}2\theta & -LD'\text{c}2\theta - LD\text{s}2\theta & CD+\frac{1}{2}(LBLD-LBLD) \\ -LD\text{c}2\theta + LD'\text{s}2\theta & \begin{array}{l} 1+\frac{1}{2}(LD^2-LB^2)\text{c}^2 2\theta \\ +\frac{1}{2}\left(LD'^2-LB'^2\right)\text{s}^2 2\theta \\ -\frac{1}{2}(LD'LD+LBLB')\text{s}4\theta \end{array} & \begin{array}{l} -CB+\frac{1}{2}(LD'LD+LBLB')\text{c}4\theta \\ +\frac{1}{4}(LD^2-LD'^2+LB'^2-LB^2)\text{s}4\theta \end{array} & -LB\text{c}2\theta - LB'\text{s}2\theta \\ -LD'\text{c}2\theta - LD\text{s}2\theta & \begin{array}{l} CB+\frac{1}{2}(LD'LD+LBLB')\text{c}4\theta \\ +\frac{1}{4}(LD^2-LD'^2+LB'^2-LB^2)\text{s}4\theta \end{array} & \begin{array}{l} 1+\frac{1}{2}(LD'^2-LB'^2)\text{c}^2 2\theta \\ +\frac{1}{2}(LD^2-LB^2)\text{s}^2 2\theta \\ +\frac{1}{2}(LD'LD+LBLB')\text{s}4\theta \end{array} & LB'\text{c}2\theta - LB\text{s}2\theta \\ CD-\frac{1}{2}(LBLD-LBLD) & LB\text{c}2\theta + LB'\text{s}2\theta & -LB'\text{c}2\theta + LB\text{s}2\theta & 1+\frac{1}{2}\left(LB^2+LB'^2\right) \end{pmatrix} \tag{8.10}$$

Here, terms involving products second order with circular properties, such as CD^2, *CDCB*, *CDLD*, and *CDLB*, are ignored as being very small. Only second-order terms in linear properties are retained as linear difference properties tend to be larger than circular difference properties, except for chiral liquid crystals and similar supramolecular chiral structures. If the orientation of the sample cell is averaged over all angles, such as with a uniformly rotating cell, then all terms dependent on θ vanish leaving the following expression for the Mueller matrix,

$$M_{RS2} = 10^{-A}\begin{pmatrix} 1+\frac{1}{2}(LD^2+LD'^2) & 0 & 0 & CD-\frac{1}{2}(LBLD-LB'LD') \\ 0 & 1+\frac{1}{4}(LD^2-LB^2+LD'^2-LB'^2) & -CB & 0 \\ 0 & CB & 1+\frac{1}{4}(LD^2-LB^2+LD'^2-LB'^2) & 0 \\ CD-\frac{1}{2}(LBLD-LB'LD') & 0 & 0 & 1+\frac{1}{2}(LB^2+LB'^2) \end{pmatrix} \tag{8.11}$$

The quadratic terms along the diagonal are very small relative to unity and therefore can be deleted. This gives an even simpler expression for the second-order rotating-sample Mueller matrix,

$$M_{RS2} = 10^{-A}\begin{pmatrix} 1 & 0 & 0 & CD - \frac{1}{2}(LBLD - LB'LD') \\ 0 & 1 & -CB & 0 \\ 0 & CB & 1 & 0 \\ CD - \frac{1}{2}(LBLD - LB'LD') & 0 & 0 & 1 \end{pmatrix} \tag{8.12}$$

Comparing Equation (8.12) with Equation (6.87), the *CD* matrix elements have an additional orientation-independent term. The generalization of Equation (6.88) that includes the orientation-independent second-order contributions is:

$$I_D = \frac{I_{DC}}{2}\left\{1 + 2J_1(\alpha_{M1}^0)\left[CD + \frac{1}{2}(LBLD - LD'LB')\right] - 2J_2(\alpha_{M1}^0)CB\right\} \tag{8.13}$$

The additional terms due to the product of *LB* and *LD* can produce an orientation independent artifact for a solid phase sample and also from the cell windows of a solution-phase sample. The fast and slow axes of the *LB* term are defined as aligned with those of the PEMs while the axes of linear dichroism for the *LD* term are vertical and horizontal. One can think of this CD intensity mechanism as one in which *LB* shifts the natural balance between RCP and LCP states of the radiation from the PEM that results in a small additional vertical component for RCP radiation and a corresponding horizontal component for LCP, or vice versa. The *LD* of the sample then converts the small vertical–horizontal polarization modulation at the PEM frequency into intensity modulation at the PEM frequency, and hence an artifact contribution to the CD spectrum. The *LD′LB′* term acts in the same way except both the *LB′* axes and the *LD′* axes differ by 45° from those of the *LDLB* term. The difference between the *LDLB* term and the *LD′LB′* term may not be large. Most likely this solid-phase artifact will be significant only for highly aligned samples or chiral single crystals with both *LB* and *LD* axes appropriately aligned. With polycrystalline mulls, which have no overall spatial alignment, there is a lower likelihood for this source of artifact to be significant.

In Figure 8.8 we present a comparison of the IR and VCD spectra of L-alanine as a solution and hydrocarbon (nujol) oil mull (Lombardi, 2011). Alanine is the simplest of the chiral amino acids. It has a methyl group at the asymmetric carbon instead of the hydrogen found on the non-chiral glycine molecule, the simplest amino acid. The solution spectra are generally broader and represent a composition of the conformers of alanine in solution. The composition includes numerous configurations of the L-alanine molecule with hydrogen bonded water molecules. The solid-phase mull spectra on the other hand possess bands that are narrower and represent the more limited and specific configuration of alanine molecules in an alanine crystal. Alanine crystallizes with two alanine molecules per unit cell and the effects of the splitting of pairs of vibrational modes from the interaction of associated pairs of alanine molecules can be seen in the spectra. The most dominant VCD bands in the solution spectra arise from a (+, −, +) triplet of modes near 1410, 1360, and 1300 cm^{-1} representing the symmetric stretch of the CO_2^- group and a pair of orthogonal methine CH bending modes at the asymmetric chiral carbon. These same three VCD bands with the same intensities relative to each other and their parent IR bands can be seen in the solid-phase mull spectrum, but in the latter spectrum these VCD bands are no longer the dominant features. Instead the vibrations in the region of the anti-symmetric stretch of CO_2^- centered at 1600 cm^{-1} exhibit very large VCD. The reason these bands are large in the mull VCD but not in the solution VCD is that in the crystal, the average torsional-angle orientation of the CO_2^- is fixed in the crystal but varies widely in the solution

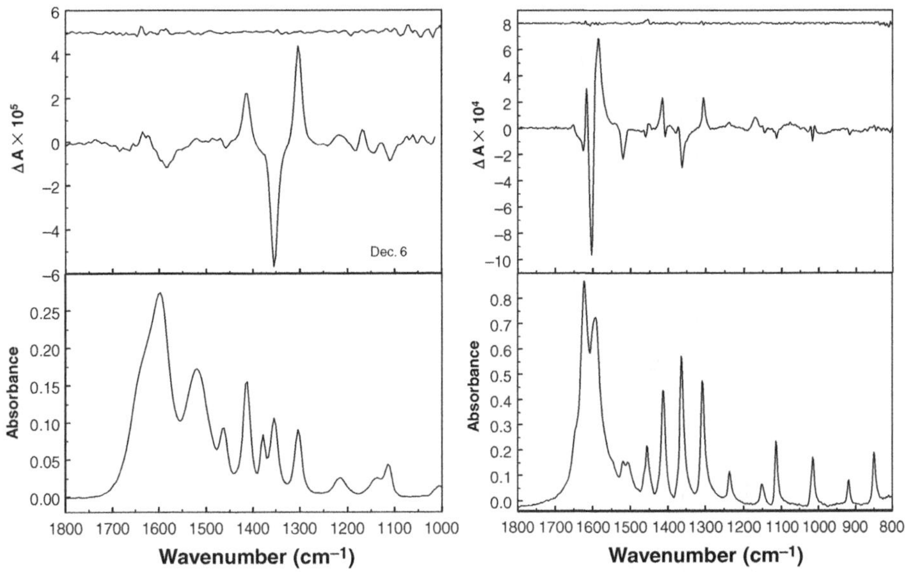

Figure 8.8 *IR and VCD spectra of L-alanine as an aqueous solution (left) and a mull in nujol oil (right). Both IR and VCD spectra are background subtracted with a water spectra for the solution and spectra of nujol oil for the mull (Lombardi, 2011)*

phase. The VCD intensities from the different torsional angle of the CO_2^- group possess different signs that heavily cancel in the observed solution-phase VCD spectrum.

If one inspects the solid-phase IR and VCD spectra for artifacts, the parent IR bands are very close to symmetric in shape indicating only minor contributions, if any, from particle scattering. The angular dependence of the mull spectra about the transmission direction of the IR beam show very small variations of intensity, and therefore one can conclude that contributions from linear birefringence are minor. This can be confirmed by comparing the VCD of mulls of L- and D-alanine and looking for departures from mirror symmetry, which again is a signature of LB-artifact contributions to the VCD spectra as mull VCD of solids of mirror-image molecules should display mirror symmetry to within the noise level of the spectra. In Figure 8.9, the IR and VCD spectra of mulls of L- and D-alanine are compared. Here it can be seen that a high degree of mirror symmetry is found for the VCD of these two enantiomorphic solids. Furthermore, the IR spectra are nearly identical in both magnitude and shape except for some variation in intensity near 1600 cm^{-1}, most likely due to differences in the water content of the L- and D-alanine crystalline samples.

We conclude this section by discussing briefly additional methods of solid-phase sampling beyond those of films and mulls discussed above. The first is spray-dried films (Lombardi, 2011). These are produced by spraying a solution of a sample, typically aqueous, through a small nozzle that produces a fine mist. If the nozzle is placed a few centimeters from a heated IR window, such as a CaF_2 plate, the solvent is rapidly evaporated leaving a spray-dried film that has crystallized predominantly in a plane perpendicular to the direction of the spray. Spray-dried films of relatively small molecules that are either zwitterions or charged chiral species with a counter ion have been shown to exhibit unusually intense VCD compared with their parent IR intensities. Examples of two VCD spectra of spray-dried films are given in Figure 8.10. In the first (left), the VCD spectrum of L-alanine as a spray-dried film is compared with the corresponding IR and VCD spectra of a nujol mull. The IR spectra have been matched to have nearly the same intensities and show close to identical spectra for the spray film and

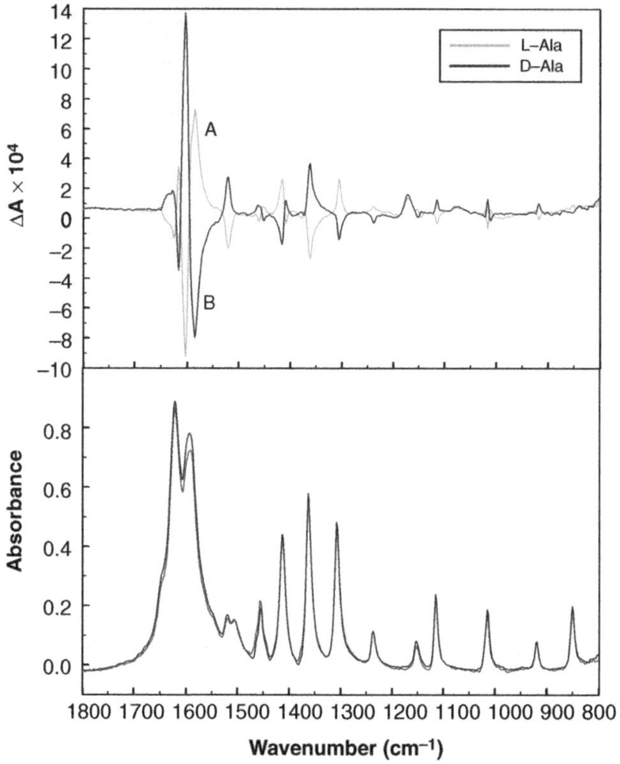

Figure 8.9 *IR and VCD spectra of L- (A) and D-alanine (B) as nujol mulls. The experimental conditions are 20 min of measurement time at 4 cm^{-1} spectral resolution (Lombardi, 2011)*

the mull. The VCD spectra, on the other hand, are completely different, both in magnitude and shape. In particular, the bands of the spray-dried film VCD spectra are one to two orders of magnitude larger than those of the mull. Such a large increase can only arise from an additional degree of chiral order existing in the film beyond that found in the mull. It can be shown that the unit cell geometry of the alanine crystals in both the spray film and mull are the same, in agreement with only one known form of L- (or D-) alanine crystal structure. The difference between these two solid-phase samples must be due to some chiral arrangement of two-dimensionally grown micro-crystals of alanine. The exact source of this large VCD enhancement is presently unknown and is currently beyond present computational models of VCD intensity.

Another interesting example of spray-dried VCD is that of glycine (Lombardi, 2011). As mentioned above, glycine is both the simplest amino acid and the only naturally occurring achiral amino acid. Despite this, glycine has three known crystal forms, α, β, and γ, the last two of which are chiral crystals. Mulls of glycine typically show VCD originating from one or both chiral crystal forms even though the solution of glycine naturally shows no VCD. The ability to observe VCD from chiral crystals of achiral molecules is another interesting application of the VCD of solids. VCD of spray-dried films of from pure glycine solutions show enhanced VCD that can either be predominantly one sign or another, depending on the chirality of the initial crystals formed in the spray process or possibly from any chiral contamination on the surface of the IR window prior to the initiation of the spray process. As shown in on the right side of Figure 8.10, the chirality of the glycine spray-dried film

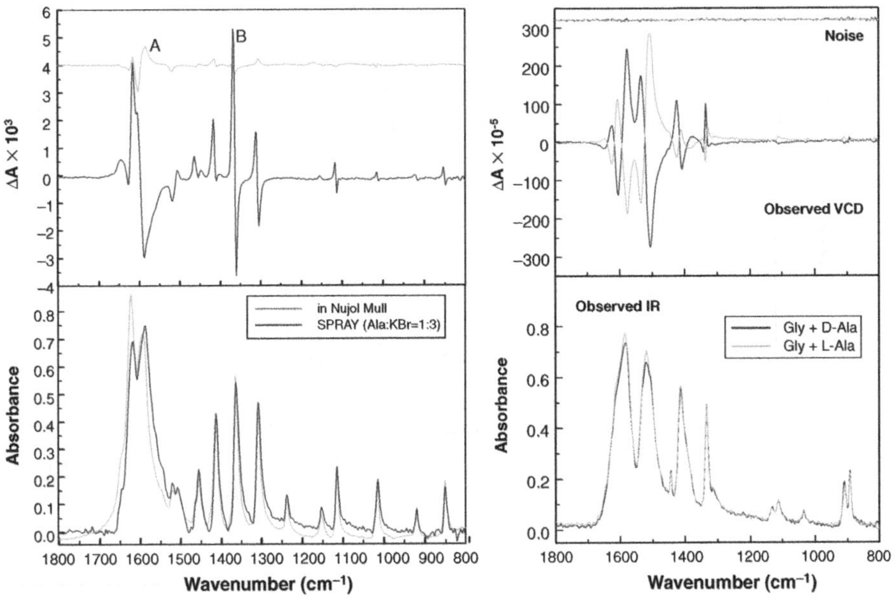

Figure 8.10 *IR and VCD spectra of L-alanine as a spray-dried film are compared with the corresponding mull IR and VCD spectra (left). IR and VCD spectra of glycine spray-dried films (right) formed from solutions of achiral glycine seeded with 0.5% L-alanine or 0.5% D-alanine (Lombardi, 2011)*

can be controlled by the addition of a small amount of either L- or D-alanine. The alanine molecules direct the formation of the glycine film to the chiral sense of the alanine molecules, while the presence of alanine bands in either the parent IR or VCD spectra of the glycine is not visually detectable. The level of VCD enhancement relative to the parent IR spectrum in the spray-dried film of glycine is on the same scale as the intense VCD exhibited by spray-dried alanine films.

Finally, we note that VCD can be measured from KBr pellets provided again that the crystal particle size of both the KBr and included micro-crystals is small relative to the wavelength of the IR radiation such that scattering contributions are small. To date no publications of VCD spectra using attenuated total reflection (ATR) have been published. Such publications must await the development of a theoretical and practical methodology for the control of artifacts arising from the ATR beam path that heavily depends on off-axis reflections which badly distort the balance between RCP and LCP states of radiation.

8.2.6 Presentation of IR and VCD Spectra with Noise Spectra

We conclude this section on IR and VCD measurement by presenting in Figure 8.11 the IR, VCD, and VCD noise spectra of the molecule 2,2-dimethyl-dioxolane-4-methanol (DDM) as a neat liquid from the mid- IR region through to most of the near-IR region.

The instrumentation used in these various spectral regions is described in Section 8.2.1. The mid-IR region features a rich array of fundamental vibrational modes. The hydrogen-stretching region exhibits CH-stretching bands centered near 2900 cm^{-1} and an OH-stretching band near 3400 cm^{-1} that is broadened by hydrogen bonding. The spectra from 3800 to 10 000 cm^{-1} feature overtone and combination bands of predominantly hydrogen-stretching and bending vibrations.

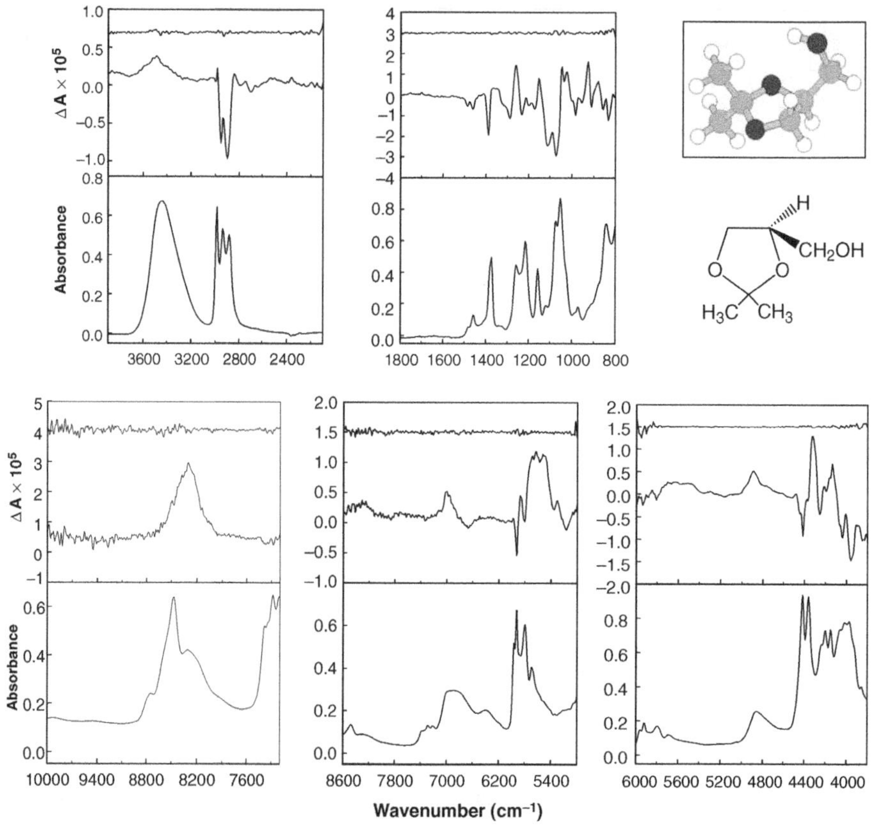

Figure 8.11 The IR, VCD and VCD noise spectra of neat (−)-R-2, 2-dimethyl-dioxolane-4-methanol (DDM) from 800 to 10 000 cm^{-1} with a dual PEM optical setup using various, optical components and detectors as described in the text. Adpated with permission from the Society of Applied Spectroscopy (Cao et al., 2004)

The combination band region extends from 3800 to 4500 cm^{-1} for the CH modes and near 4800 cm^{-1} for the OH modes. The overtone of the CH stretching region occurs near 5800 cm^{-1} while the overtone of the OH stretching mode is centered near 6800 cm^{-1}. At a weaker level of intensity are the second combination bands (two stretches and one bend) of the CH modes near 7200 and the OH modes near 8400 cm^{-1}. Closer to 8700 cm^{-1} is the weak feature of the second overtone of the CH-stretching mode, but interestingly it shows little VCD intensity compared with that of the second OH combination band at 8400 cm^{-1}. Overall, these spectra demonstrate that even though the intensity of near-IR absorption decreases significantly with higher levels of overtone and combination band modes, the VCD remains proportionately the same size and therefore equally measurable compared with the corresponding spectra in the mid-IR regions where bands from primarily fundamental transitions are observed.

8.3 Measurement of Raman and ROA Spectra

The measurement of Raman and ROA involves many differences from the measurement of IR and VCD spectra described above. One major difference is the preferred solvents; aqueous solvents versus non-aqueous are preferred for Raman and ROA measurement as water has a relatively weak Raman scattering spectrum. This opens up the full range of biomolecules, such as proteins, nucleic acids, carbohydrates, glycoproteins, and viruses for direct sampling as aqueous solution with ROA. The principal concerns for obtaining quality of ROA spectra involve reduction of background fluorescence, sufficient sample concentration, and sample stability under extended laser radiation. These concerns are related to the need to obtain close to 10^{10} Raman counts for the major bands in the Raman spectrum as the dominant noise source at each point in the spectrum is the square root of the total Raman counts. Because ROA intensities are 3–5 orders of magnitude smaller than the parent Raman intensities, the noise level needs to be approximately five orders of magnitude smaller than the parent Raman bands of interest in order to see all the feature in the ROA spectrum well. To generate this level of Raman counts, relatively high laser powers and relatively long accumulation times are necessary. Background fluorescence, usually originating in impurities, diminishes the signal-to-noise ratio in the Raman and ROA spectra, and hence must be reduced as much as possible. Thus, obtaining a high-quality ROA usually means taking steps to make sure the sample is free of dust and impurities, and stable over one or more hours of collection time, except for simple samples, such as neat liquids of rigid chiral molecules, which yield quality ROA spectra in tens of minutes.

8.3.1 Choice of Form of ROA and Scattering Geometry

Within the past two decades, the scattering geometry of choice for the measurement of ROA has been 180° backscattering. Prior to that nearly all experiments were performed in right-angle scattering geometry, usually as incident circular polarization (ICP) ROA. During the 1990s, two forms of backscattering ROA were often performed, unpolarized ICP (ICP_U) ROA from Barron's laboratory in Glasgow (Hecht *et al.*, 1992; Hecht and Barron, 1995) and in-phase dual circular polarization (DCP_I) ROA in Nafie's laboratory in Syracuse (Nafie *et al.*, 1995; Vargek *et al.*, 1997). With the advent of the commercial ROA spectrometer, using backscattering scattered circular polarization (SCP) ROA (Spencer *et al.*, 1988), and an automated artifact-correction design with dual fiber optic collection optics (Hug and Hangartner, 1999; Hug, 2003), almost all ROA measurement are now backscattering SCP_U-ROA. Future developments, however, may offer all four forms of ROA (ICP, SCP, DCP_I, and DCP_{II}) as described further below. Once these options become available, then consideration of which form of ROA to use, depending on the goal of the measurements will become an important issue.

As described theoretically above, both ICP_U and SCP_U Raman spectra are polarized Raman scattering intensities. In the far-from-resonance (FFR) approximation, which holds well for most colorless samples, the ROA for these two forms of ROA are the same. Artifacts are most susceptible from highly polarized Raman bands, and hence ROA from these bands may not always be reliable compared with ROA from Raman bands that are predominantly depolarized. By contrast, backscattering DCP_I-Raman is a strictly depolarized form of Raman scattering, and yet in the FFR approximation DCP_I-ROA is identical with that of ICP_U- and SCP_U-ROA. This is because in the FFR, DCP_{II}-ROA is zero and DCP_I-Raman and -ROA differ from ICP_U- and SCP_U-Raman and -ROA by the elimination of DCP_{II}-Raman and -ROA. On the other hand, DCP_{II}-Raman is a highly polarized form of ROA, which contributes polarized bands and ROA artifact susceptibility to the measurement of ICP_U- and SCP_U-Raman and -ROA. In the future, when automated artifact reduction instrumentation with the capability to measure DCP_I-Raman and -ROA, in most cases, and particularly for biological samples, DCP_I will become the method of choice for ROA measurements.

8.3.2 Raman and ROA Sampling Methods

The active region of a sample undergoing Raman and ROA measurement is confined to the highly focused volume of the incident laser intensity. For standard focusing lenses and incident laser beams, this region can be measured in fractions of a microliter. As a result sampling volumes in the absence of laser absorption can be very small. This currently is a major difference between ROA and VCD measurements, where for the latter, the IR beam at the sample is not highly focused and sample volumes must be significantly larger. Development of infrared microscope imaging for VCD in the future, however, may equalize this difference.

8.3.2.1 Sample Cells and Accessories

Two types of samples cells have been used for backscattering ROA measurements. One is a small quartz cuvette with flat thin windows for passage of the incident laser radiation and the backscattering of the Raman radiation. The other, requiring even smaller sample volumes, is a thin quartz capillary tube where the Raman sample need only be present for a short section along the tube where the laser focus occurs. Here, the surface of the tube is cylindrical in shape. This has some advantage compared with a flat surface due to an increase in the solid angle of radiation collected from the laser focal volume for a given solid angle subtended by the collection lens. For backscattering the lens used to focus the incident laser radiation is the same as that used to collect the backscattered Raman radiation. ROA measurements using a thin capillary tube significantly reduces the sample volume required for the measurement of an ROA spectrum but introduces a greater challenge in positioning the sample in the laser beam due to the higher level of spatial precision required.

8.3.2.2 Sample Purification and Fluorescence Reduction

Most measurements of ROA to date have involved either neat liquids or aqueous solutions. In the case of neat liquids, the samples may be used as available from a commercial source provided there has not been chemical decomposition due to temperature or exposure to ambient radiation, that is, photochemical damage. Typically, fluorescence is not a problem for neat liquids such as chiral monoterpenes.

For aqueous solutions of biological samples, such as proteins, greater care is needed in sample preparation to avoid unwanted fluorescence. Firstly, the water used to dissolve the sample should be de-ionized and distilled. The biological solution should then be passed through a micron-sized millipore filter to removed dust particles and other small particulate impurities. Another useful step is to add activated charcoal to the solution, shake, and let the charcoal settle to the bottom of the solution. The solution may also be centrifuged and the supernatant only used as the ROA sample. Finally, if a fluorescence background larger than the Raman bands is encountered, or if any undesired fluorescence is present with the Raman spectrum, the sample may be exposed to the incident laser radiation to burn off the fluorescence. The duration of the laser fluorescence burn-off depends on the severity of the fluorescence and the desired level of background fluorescence. Decreasing the background noise from fluorescence photon counts using laser burn-off is usually well worth the time in achieving a desired level of ROA spectral quality. Because biological samples, even ones commercially obtained, are isolated from living organisms, there is almost always a background level of fluorescence from organic compounds left over after the biochemical isolation and purification steps. The amount of background fluorescence present for different types, or even lots, of biological sample can vary significantly. Further purification of the biological sample may be necessary if fluorescence levels are encountered that exceed what can be tolerated in the measurement of the ROA spectrum of the sample.

8.3.3 Instrument Laser Alignment

Commercial ROA spectrometers are pre-aligned and reasonably stable; nevertheless, in some cases refinement of the alignment of the ROA spectrometer is necessary to achieve a spectrum with sufficiently low interfering artifacts, mostly from the polarized Raman bands as discussed above. The basic idea with ROA instrument alignment is to keep the incident and scattered radiation beams centered in the optical elements along the incident and scattered radiation rails of the spectrometer. Centering the beam in the optical elements ensures that distortions of the polarization states of the incident and scattered radiation are kept to a minimum, which in turn allows the spectrometer to maintain the desired balance between the delivery and detection of RCP and LCP states of the light.

8.3.4 ROA Artifact Suppression

Artifacts arise in ROA from two principal sources. If there is an overall imbalance of either the delivery or detection of RCP and LCP radiation, an excess Raman spectrum or combination of Raman spectra will be present in addition to the desired ROA spectrum when the difference between Raman scattering between designated RCP and LCP radiation states is taken. A related effect is if the modulation between RCP and LCP is not between perfect CP states, such that small net linear polarization components of the light are delivered to the sample or the detector after scattering. Such small linear polarization components create in effect a modulation between elliptical polarization states instead of pure circular polarization states. As Raman scattering is sensitive to differences in orthogonal linear polarization states at the level of the depolarization ratio, any net linear polarization content of the incident and/or scattered radiation can lead to artifacts related to the depolarization ratio of the sample. Because strong polarized bands give rise to the largest intensity differences between parallel and perpendicular polarization states for the incident and scattered radiation, strongly polarized Raman bands are most susceptible to the occurrence of artifacts dependent on the optical alignment and CP modulation of the spectrometer.

8.3.4.1 Artifact Reduction Scheme of Hug

In 1999 Hug completed construction of a new backscattering Raman spectrometer that included an automated ROA artifact reduction scheme, which was explained in more detail in a paper published in 2003 (Hug and Hangartner, 1999; Hug, 2003). The idea behind the artifact reduction scheme is to use a number of half-wave plates (HWPs) in both the incident and scattered optical rails to create, in a time-average sense, equal CP balance as desired for both the incident and scattered radiation. The exact method is somewhat dependent on the form of ROA measured, whether it be ICP, SCP, or one of the two DCP forms of ROA. There is more than one way to explain the artifact reduction scheme. One way is to say that by insertion of two HWPs, one before and one after the sample, one can create in effect a virtual (optically created) enantiomeric sample compared with the original one measured when the HWPs were not present. Hug called this the method of the virtual enantiomer. For example, if a difference in RCP minus LCP is being measured for the chiral sample, insertion of an HWP before the sample changes RCP to LCP and LCP to RCP for measurement at the sample, and the HWP after the sample restores the CP states to their original states. Insertion of two HWPs, one before and one after the sample, is equivalent to exchanging the original chiral sample with its mirror-image (enantiomeric) form. This automatically cancels any CP bias that may be present in the beam as insertion of HWP plates allows the measurement of LCP intensity minus RCP intensity, as opposed to the original intensity difference for RCP minus LCP, without any change in the actual polarization states, first RCP and then LCP, being measured by the instrument. This focuses the measurement only on the actual desired intensity difference to be measured without a concern for differences or biases in the responses of the instrument for RCP versus LCP to enter into the measurement. Such biases cancel out when the two RCP minus LCP differences, one for the chiral sample and one for its virtual enantiomer, are combined by subtraction.

8.3.4.2 Artifact Suppression for Backscattered SCP_U Measurement

With these basic concepts in mind, the way in which an SCP_U-ROA instrument in general, and the commercial Chiral*RAMAN* ROA spectrometer in particular, can be configured to reduce to first order CP biases in the incident and scattered radiation is as follows. The laser radiation passes through a linear polarizer to reduce the levels of circular polarization that might be present. The beam then passes through a linear rotator (LR) comprised of a rotating HWP optimized for the incident laser wavelength. The LR rotates the plane of polarization of the light beam through 360° twice per rotation of the HWP. In this way, the LP polarization content of the beam is eliminated in a time-averaged sense during the course of the ROA measurement. After the LR the beam passes through a circular converter (CC), which consists of another HWP that for exactly equal time periods is either in the beam path or out of the path. When in the beam path, the CC reverses whatever CP bias might be present on the beam. Following the CC the incident laser beam has, in a time-average sense, no LP or CP bias and hence is effectively completely unpolarized, as desired for SCP_U intensities.

For the scattered beam containing different intensities of the RCP and LCP light, the beam first passes through an LR with half-wave retardation set to approximately the wavelength corresponding to the middle of the Raman scattering or roughly 1000 cm^{-1} from the Rayleigh line. It is important that HWPs used for the scattered radiation are zeroth-order so that the half-wave dependence of the plate about the optimum wavelength is as slow as possible. This LR removes any stray LP components from the scattered beam so that only differences between CP states of radiation are measured. This is followed by a CC that for half of the collection time is in the beam and for the other half is out of the beam. When the CC HWP is out of the beam, LCP and RCP intensities are measured as their own CP states, and when the HWP is in the beam, LCP and RCP intensities are interchanged and measured as their opposite CP states. In this way, LCP and RCP intensities are measured without any bias the instrument may have for the measurement of RCP versus LCP radiation. In an SCP-ROA measurement, orthogonal CP states are converted into orthogonal LP states by a liquid crystal retarder acting as a quarter-wave plate. These LP states are then spatially divided by a polarization beamsplitter and after several optical steps imaged simultaneously on the upper and lower halves of a CCD detector for the measurement of the SCP_U-Raman (sum) and -ROA (difference) spectra.

8.3.5 Forms of Backscattering ROA and their Artifacts

In this section we look a bit further into sources of artifacts in ROA backscattering for all four forms of ROA. We do this by assuming the presence of undesired imbalances for the intensity of RCP versus LCP in the delivery of incident radiation to the sample or undesired imbalances in the measurement of RCP versus LCP intensities in the scattered beam. At the level of Raman spectra, ignoring effects of chirality, and hence ROA, there are two main forms of CP backscattered Raman spectra, namely co-rotating or contra-rotating CP states of the incident and scattered radiation. From Equations (2.118) and (2.119), these Raman intensities are given by:

$$I_{\text{co}} = I_R^R(180°) = I_L^L(180°) = 2K\left[6\beta_S(\tilde{\alpha})^2\right] = 2K\left[6\beta(\tilde{\alpha})^2\right] \tag{8.14}$$

$$I_{\text{contra}} = I_L^R(180°) = I_R^L(180°) = 2K\left[45\alpha^2 + \beta_S(\tilde{\alpha})^2 + 5\beta_A(\tilde{\alpha})^2\right] = 2K\left[45\alpha^2 + \beta(\tilde{\alpha})^2\right] \tag{8.15}$$

where for simplicity we have reduced these expressions to the far-from-resonance (FFR) approximation using $\beta_S(\tilde{\alpha})^2 = \beta(\tilde{\alpha})^2$ and $\beta_A(\tilde{\alpha})^2 = 0$. The co-rotating intensities, I_R^R and I_L^L, are pure depolarized Raman intensities, while the contra-rotating intensities, I_L^R and I_R^L, are highly polarized Raman intensities. These two classes of Raman intensities can be measured by using first RCP and then LCP for the polarization of the incident beam of a backscattering SCP-ROA spectrometer in which case one first measures simultaneously I_R^R and I_L^R, and then simultaneously I_R^L and I_L^L.

8.3.5.1 Direct Measurement of all Four Forms of ROA Intensities

If one now includes contributions from chirality, the four measured CP Raman intensities, I_R^R, I_L^L, I_L^R, and I_R^L can be grouped into the four distinct forms of CP ROA as:

$$\text{SCP-ROA:} \quad \Delta I^U = (I_R^R + I_R^L) - (I_L^R + I_L^L) = I_R^U - I_L^U \tag{8.16}$$

$$\text{ICP-ROA:} \quad \Delta I_U = (I_R^R + I_L^R) - (I_R^L + I_L^L) = I_U^R - I_U^L \tag{8.17}$$

$$\text{DCP}_\text{I}\text{-ROA:} \quad \Delta I_I = I_R^R - I_L^L \tag{8.18}$$

$$\text{DCP}_\text{II}\text{-ROA:} \quad \Delta I_{II} = I_L^R - I_R^L \tag{8.19}$$

where in the first two equations we use the fact that the sum of equal amounts of RCP and LCP intensities, either in the incident or scattered radiation, is equivalent to unpolarized (U) intensity. Here, it can be seen that if one has measured the four Raman intensities, I_R^R, I_L^L, I_L^R, and I_R^L, the two forms of DCP-ROA are the easiest forms of ROA to construct. These four DCP Raman measurements are illustrated in Figure 8.12 (Li and Nafie, 2011) for neat (+)-α-pinene where the highly polarized form of contra-rotating Raman intensity, associated with DCP$_\text{II}$-ROA measurement in Equation (8.19) can be compared with the pure depolarized co-rotating Raman intensity associated with DCP$_\text{I}$-ROA measurement in Equation (8.18).

The corresponding four forms of ROA constructed from the four DCP Raman spectra shown in Figure 8.12 using Equations (8.16)–(8.19) are shown in Figure 8.13 for neat (+)-α-pinene (Li and Nafie, 2011). Here we can see that DCP$_\text{II}$-ROA is close to zero within the noise level and that the other three forms of ROA are essentially equal to one another. From Equations (8.16) and (8.17), it can be seen that ICP- and SCP-ROA differ from DCP$_\text{I}$-ROA by the addition or subtraction of a zero plus noise DCP$_\text{II}$-ROA spectrum. By inspection of Figure 8.13 it can be seen that the DCP$_\text{I}$-ROA spectrum, although essentially the same as the ICP- and SCP-ROA spectra, has distinctly less noise. Another way of stating this idea is that backscattering DCP$_\text{I}$-ROA arises in this FFR case from pure co-rotating depolarized Raman spectra and that *all the ROA intensity* in an unpolarized backscattering ICP- or SCP-ROA *originates from* DCP$_\text{I}$-ROA but with extra 'empty' Raman intensity from DCP$_\text{II}$ contra-rotating Raman spectra.

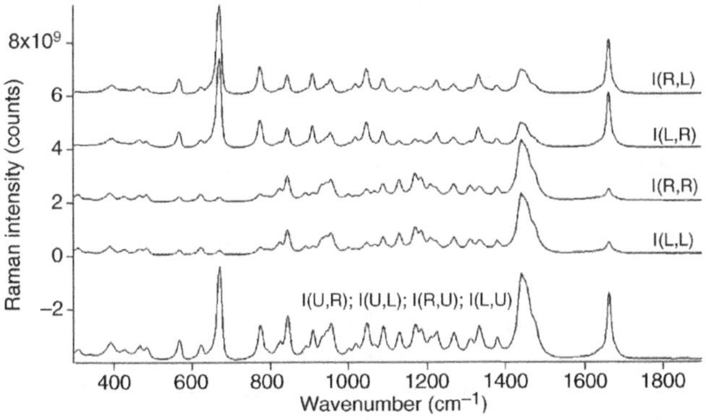

Figure 8.12 *Pure dual circular polarization (DCP) backscattering Raman intensities of neat (+)-α-pinene. The two uppermost spectra, I_R^L [I(L,R)], and I_L^R [I(R,L)], are polarized contra-rotating Raman spectra while the next two, I_R^R [I(R,R)], and I_L^L [I(R,R)], are depolarized co-rotating Raman spectra. The bottom spectrum represents the four different ways of forming unpolarized backscattering Raman spectra associated backscattering ICP- and SCP-ROA measurements. Reproduced with permission from John Wiley & Sons (Li and Nafie, 2011)*

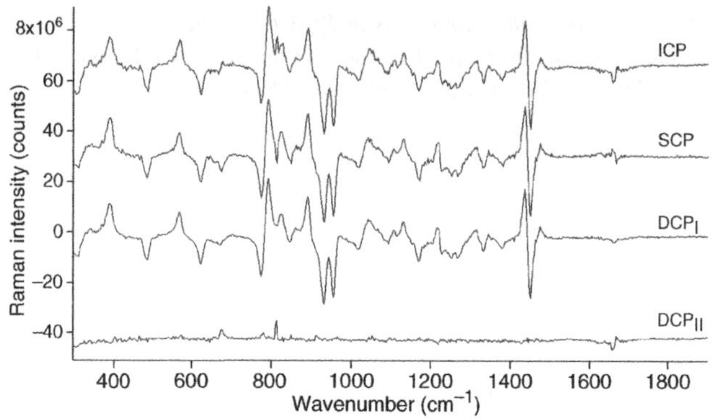

Figure 8.13 *Four forms of ROA constructed from the four DCP-ROA measurements for neat (+)-α-pinene shown in Figure 8.12 according the expressions: ICP-ROA* $(I_R^R + I_L^R) - (I_R^L + I_L^L)$*; SCP-ROA* $(I_R^R + I_R^L) - (I_L^R + I_L^L)$*; DCP*$_I$*-ROA* $I_R^R - I_L^L$*; and DCP*$_{II}$*-ROA:* $I_L^R - I_R^L$*. Reproduced with permission from John Wiley & Sons (Li and Nafie, 2011)*

8.3.5.2 Artifacts from Imbalance in Incident CP Intensities

If in ROA measurements there is an imbalance, δ, in the amount of RCP and LCP incident radiation intensity, the four different forms of ROA have deviations (artifacts) from their true values defined above by the addition (or subtraction) of small fractions, δ, of the sample's co-rotating, I_{co}, and/or contra-rotating, I_{contra} Raman spectra. This is shown by the following equations where positive δ is defined as an excess of LCP:

$$\text{SCP:}\quad \Delta I^U(\delta) = \left[I_R^R + I_R^L(1+\delta)\right] - \left[I_L^R + I_L^L(1+\delta)\right]$$

$$= \left(I_R^U - I_L^U\right) + \delta\left(I_R^L - I_L^L\right) = \Delta I^U + \delta(I_{\text{contra}} - I_{\text{co}}) \tag{8.20}$$

$$\text{ICP:}\quad \Delta I_U(\delta) = \left[I_R^R + I_L^R\right] - \left[I_R^L(1+\delta) + I_L^L(1+\delta)\right]$$

$$= \left(I_U^R - I_U^L\right) - \delta\left(I_R^L + I_L^L\right) = \Delta I_U - \delta(I_{\text{contra}} + I_{\text{co}}) \tag{8.21}$$

$$\text{DCP}_{\text{I}}:\quad \Delta I_I(\delta) = I_R^R - I_L^L(1+\delta) = \Delta I_I - \delta I_{\text{co}} \tag{8.22}$$

$$\text{DCP}_{\text{II}}:\quad \Delta I_{II}(\delta) = I_L^R - I_R^L(1+\delta) = \Delta I_{II} - \delta I_{\text{contra}} \tag{8.23}$$

For this source of ROA artifact, each of the four forms of ROA has a different artifact expression. Of these, only the artifact for DCP$_{\text{I}}$-ROA has no contribution from I_{contra} that has strong contributions from polarized bands, which are the most serious spectral locations of artifacts in ICP- and SCP-ROA spectra. Each of the artifact expressions are simply small fractions δ of pure or combined forms of co- and contra-rotating Raman spectra.

8.3.5.3 Artifacts from Imbalance in the Detection of Scattered CP Intensities

We next consider artifacts due to an imbalance, ε, in the detection of RCP and LCP scattered radiation intensity. The four different forms of ROA have artifact expressions given by where positive ε is defined as an excess of LCP:

$$\text{SCP:}\quad \Delta I^U(\varepsilon) = \left[I_R^R + I_R^L\right] - \left[I_L^R(1+\varepsilon) + I_L^L(1+\varepsilon)\right]$$
$$= \left(I_R^U - I_L^U\right) - \varepsilon\left(I_R^L + I_L^L\right) = \Delta I^U - \varepsilon(I_{\text{contra}} + I_{\text{co}}) \tag{8.24}$$

$$\text{ICP:}\quad \Delta I_U(\varepsilon) = \left[I_R^R + I_L^R(1+\varepsilon)\right] - \left[I_R^L + I_L^L(1+\varepsilon)\right]$$
$$= \left(I_U^R - I_U^L\right) + \varepsilon\left(I_R^L - I_L^L\right) = \Delta I_U + \varepsilon(I_{\text{contra}} - I_{\text{co}}) \tag{8.25}$$

$$\text{DCP}_\text{I}:\quad \Delta I_I(\varepsilon) = I_R^R - I_L^L(1+\varepsilon) = \Delta I_I - \varepsilon I_{co} \tag{8.26}$$

$$\text{DCP}_\text{II}:\quad \Delta I_{II}(\varepsilon) = I_L^R(1+\varepsilon) - I_R^L = \Delta I_{II} + \varepsilon I_{\text{contra}} \tag{8.27}$$

where ε is associated with a positive value for an imbalance favoring LCP detection over RCP detection.

8.3.5.4 Artifacts from Imbalance in both Incident and Scattered CP Intensities

If we combine these analyses and assume no second order contributions, say proportional to $\delta\varepsilon$, to observed artifacts we can write the expressions for the four forms of ROA as:

$$\begin{aligned} \text{SCP:}\quad \Delta I^U(\delta,\varepsilon) &= \left[I_R^R + I_R^L(1+\delta)\right] - \left[I_L^R(1+\varepsilon) + I_L^L(1+\delta+\varepsilon)\right] \\ &= \Delta I^U + \delta(I_{\text{contra}} - I_{\text{co}}) - \varepsilon(I_{\text{contra}} + I_{\text{co}}) \\ &= \Delta I^U + (\delta - \varepsilon)I_{\text{contra}} - (\delta + \varepsilon)I_{\text{co}} \end{aligned} \tag{8.28}$$

$$\begin{aligned} \text{ICP:}\quad \Delta I_U(\delta,\varepsilon) &= \left[I_R^R + I_L^R(1+\varepsilon)\right] - \left[I_R^L(1+\delta) + I_L^L(1+\delta+\varepsilon)\right] \\ &= \Delta I_U - \delta(I_{\text{contra}} + I_{\text{co}}) + \varepsilon(I_{\text{contra}} - I_{\text{co}}) \\ &= \Delta I_U - (\delta - \varepsilon)I_{\text{contra}} - (\delta + \varepsilon)I_{\text{co}} \end{aligned} \tag{8.29}$$

$$\text{DCP}_\text{I}:\quad \Delta I_I(\delta,\varepsilon) = I_R^R - I_L^L(1+\delta+\varepsilon) = \Delta I_I - (\delta+\varepsilon)I_{\text{co}} \tag{8.30}$$

$$\text{DCP}_\text{II}:\quad \Delta I_{II}(\delta,\varepsilon) = I_L^R(1+\varepsilon) - I_R^L(1+\delta) = \Delta I_{II} - (\delta-\varepsilon)I_{\text{contra}} \tag{8.31}$$

These expression demonstrate that only two forms of correction terms are needed for all four forms of ROA, namely $(\delta+\varepsilon)I_{\text{co}}$ and $(\delta-\varepsilon)I_{\text{contra}}$. The signs and relative magnitudes of the CP imbalance parameters, δ and ε, for any particular measurement of the four uncorrected CP Raman intensities I_R^R, $I_L^L(1+\delta+\varepsilon)$, $I_L^R(1+\varepsilon)$, and $I_R^L(1+\delta)$ need to be determined by an error analysis procedure. Various approaches are possible, but an obvious one for backscattering geometry is that for a sample satisfying the FFR approximation, such as *trans*-pinane or α-pinene, the DCP$_\text{II}$-ROA spectrum is zero and all three other fours of ROA are equal to one another, as demonstrated in Figure 8.13. Satisfying these conditions would lead to an error analysis that is somewhat over-determined, but satisfaction of these two constraints for all four forms of ROA should provide clear specification of the imbalance parameters, δ and ε, and allow recovery of any form of artifact-free ROA desired. While δ is most likely a constant across all Raman frequencies, ε may vary slowly across the spectrum as a function of the HWP efficiency associated with standard artifact-suppression optics for the SCP-ROA scattered beam.

The artifact analysis just presented assumes an ROA spectrometer equipped with artifact suppression optics. In particular, it assumed that all linear polarization components are eliminated from the incident laser radiation and from the scattered radiation by LR HWPs that average to zero by rotation over time all linear polarization components of the light beams. If this assumption is violated, say by non-optimized HWPs, the artifact analysis becomes much more complex.

8.3.6 Presentation of Raman and ROA Spectra

ROA and Raman spectra are displayed similarly to VCD and IR. It is customary to plot the parent Raman spectrum below and on the same frequency axis as the ROA spectrum as shown in Figure 8.14 for the sample *R*-(+)-α-pinene (Hug, 2003). For Raman and ROA spectra, the dominant noise level at any point in the spectrum is the square root of number of Raman counts. The counts registered on a CCD camera are proportional to the number of Raman photoelectrons registered by the CDD, but as the conversion from electrons to counts is calibrated and strictly proportional with no offset, the square-root rule for the noise level is appropriate. Because the noise level is easily estimated from the Raman spectrum, it is customary not to include a noise spectrum, contrary to the custom for VCD. Instead, as shown in Figure 8.14, the degree of circularity, defined previously in Chapter 2, can be presented above the ROA spectrum. The degree of circularity spectrum is defined for backscattering as

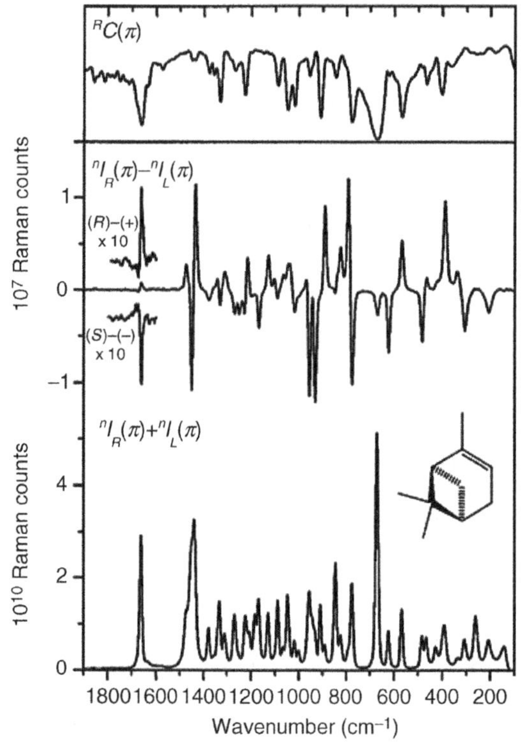

Figure 8.14 *The backscattered unpolarized SCP-Raman, -ROA and degree of circularity (lower to upper) of R-(+)-α-pinene from 35 μL of neat liquid. The laser exposure time was 33 min and the laser power at the sample was 420 mW. Reproduced with permission from the Society of Applied Spectroscopy (Hug, 2003)*

the difference in the co-rotating Raman spectrum minus the contra-rotating Raman spectrum divided by their sum as shown in Equation (8.32):

$$^{R}C(180^\circ) = \frac{I_R^R - I_L^R}{I_R^R + I_L^R} = \frac{-45\alpha^2 + 5\beta(\tilde{\alpha})^2}{45\alpha^2 + 7\beta(\tilde{\alpha})^2} = -{}^{L}C(180^\circ) \tag{8.32}$$

This contains information about the conventional right-angle-scattering depolarization ratio involving unpolarized incident or scattered radiation, $\rho_n(90^\circ)$, by $^{R}C(180^\circ) = 2\rho_n(90^\circ) - 1$. The degree of circularity, $^{R}C(180^\circ)$, varies from $+5/7$ for depolarized bands, where $\alpha^2 = 0$ and $\rho_n(90^\circ) = 6/7$, to -1 for completely polarized bands where $\beta(\tilde{\alpha})^2 = 0$ and $\rho_n(90^\circ) = 0$.

References

Cao, X., Shah, R.D., Dukor, R.K. *et al.*, (2004) Extension of Fourier transform vibrational circular dichroism into the near-infrared region: Continuous spectral coverage from 800 to 10,000 cm^{-1}. *Appl. Spectrosc.*, **58**, 1057–1064.

Diem, M., Photos, E., Khouri, H., and Nafie, L.A. (1979) Vibrational circular dichroism in amino acids and peptides. 3. Solution and solid phase spectra of serine and alanine. *J. Am. Chem. Soc.*, **101**, 6829–6837.

Griffiths, P.R., and de Haseth, J.A. (2007) *Fourier Transform Infrared Interferometry*, 2nd edn, John Wiley & Sons, Ltd, Chichester.

Hecht, L., Barron, L.D., Gargaro, A.R. *et al.* (1992) Raman optical activity instrument for biochemical studies. *J. Raman Spectrosc.*, **23**, 401–411.

Hecht, L., and Barron, L.D. (1995) Instrumentation for Raman optical-activity measurements. *J. Mol. Struct.*, **347**, 449–458.

Hug, W. (2003) Virtual enantiomers as the solution of optical activity's deterministic offset problem. *Appl. Spectrosc.*, **57**, 1–13.

Hug, W., and Hangartner, G. (1999) A very high throughput Raman and Raman optical activity spectrometer. *J. Raman Spectrosc.*, **30**, 841–852.

Laufer, G., Huenke, J.T., Royce, B.S.H., and Teng, Y.C. (1980) Elimination of dispersion-induced distortion of infrared absorption spectra by use of photoacoustic spectroscopy. *Appl. Phys. Lett.*, **37**, 517–519.

Lewis, I.R., and Edwards, H.G.M. (2001) *Handbook of Raman Spectroscopy: From the Research Laboratory to the Process Line*, Marcel Dekker, Inc., New York.

Li, H., and Nafie, L.A. (2011) Simultaneous acquisition and comparison of all forms of Raman optical activity (ROA) for neat alpha-pinene and lysozyme protein solutions. *J. Raman Spectrosc.*, in press.

Lombardi, R.A. (2011) Ph.D. thesis, Syracuse University.

Nafie, L.A. (2000) Dual polarization modulation: A real-time, spectral-multiplex separation of circular dichroism from linear birefringence spectral intensities. *Appl. Spectrosc.*, **54**, 1634–1645.

Nafie, L.A., Yu, G.S., and Freedman, T.B. (1995) Raman optical-activity of biological molecules. *Vib. Spectrosc.*, **8**, 231–239.

Narayanan, U., Keiderling, T.A., Bonora, G.M., and Toniolo, C. (1985) Vibrational circular dichroism of polypeptides. IV. Film studies of L-alanine homo-oligopeptides. *Biopolymers*, **24**, 1257–1263.

Sen, A.C., and Keiderling, T.A. (1984) Vibrational circular dichroism of polypeptides. III. Film studies of several alpha-helical and beta-sheet polypeptides. *Biopolymers*, **23**, 1533–1545.

Shanmugam, G., and Polavarapu, P.L. (2005) Film techniques for vibrational circular dichroism measurements. *Appl. Spectrosc.*, **59**, 673–681.

Spencer, K.M., Freedman, T.B., and Nafie, L.A. (1988) Scattered circular polarization Raman optical activity. *Chem. Phys. Lett.*, **149**, 367–374.

Vargek, M., Freedman, T.B., and Nafie, L.A. (1997) Improved backscattering dual circular polarization Raman optical activity spectrometer with enhanced performance for biological applications. *J. Raman Spectrosc.*, **28**, 627–633.

9

Calculation of Vibrational Optical Activity

The goal of this chapter is to provide an introduction to the basic concepts associated with the calculation of VOA intensities. Combined with the methodology for measuring of VOA spectra, described in Chapter 8, knowing how to understand, execute, and interpret calculations of VOA will enable one to use VOA to obtain stereochemical information and solve structural problems in a variety of ways as described in the final chapter of this book.

We describe here how theoretical expressions for VCD and ROA intensity developed in Chapters 4 and 5, respectively, are used to calculate VOA intensities using modern quantum mechanical methods. The steps required to calculate a VOA spectrum begin with the specification of the molecular structure and a particular choice of absolute configuration. This is followed by a potential energy search to identify the lowest-energy conformational states and the selection of the lowest of these for a full quantum mechanical force field analysis and determination of vibrational frequencies. Finally, a VOA spectrum for each conformer is calculated and the resulting VOA spectra are averaged, usually weighted by a Boltzmann energy distribution, to obtain a final VOA spectrum of the molecule. At this point, the calculated VOA spectrum can be compared with the measured VOA spectrum and analyzed in a manner appropriate to the application.

9.1 Quantum Chemistry Formulations of VOA

There are a variety of ways of thinking about the theoretical formulation of VCD and ROA intensities. At one extreme are the simplest formulations or model expressions that either provide no excited electronic state vibronic detail or provide it with the most stringent restrictions. Examples of these formulations are VCD intensity that has no reference to the electronic states of the molecule or ROA intensity that is independent of the frequency of the incident radiation field. The other extreme of formulation involves explicit complex theoretical expressions summed over all the vibronic states of the molecule that are applicable to any particular circumstance of proximity of vibrational states to electronic transitions. In the case of VCD, this involves coupling between electronic and vibrational

Vibrational Optical Activity: Principles and Applications, First Edition. Laurence A. Nafie.

magnetic dipole transition moments and for ROA any degree of resonance between the incident radiation field and electronic excited states of the molecule.

Historically, the calculation of VCD and ROA started at the very simplest levels of theory in an attempt to understand the measured VOA spectra from an elementary viewpoint. Early calculations of VOA involved conceptual models, such as the coupled oscillator model (Holzwarth and Chabay, 1972) or the fixed partial charge model for VCD (Schellman, 1973), as quantum calculations were initially not practical. In the case of VCD, even if there were practical routes the calculation of VCD intensity using quantum chemistry, the necessary fundamental quantum theory had not yet been found. For ROA, the fundamental quantum theory was known from an early stage (Barron and Buckingham, 1971), before ROA was discovered experimentally, but practical ways of implementing the theory were not available, and so, as for VCD, simple models where used as the first approach to spectral interpretation (Barron and Buckingham, 1975). When finally practical ways were found for handling the sum over all defined excited electronic states of a molecule, as required for both ROA and VCD intensity, the field of VOA was transformed from being concerned about models and model calculations, to being focused on increasingly accurate quantum chemistry methods that could be applied to the calculation of intensities without serious approximations. Nevertheless, to date, the formulations of VOA used for quantum chemistry calculations tend toward the more elementary versions of available VOA formulations that do not include proximity or resonance with particular excited electronic states of the molecule. In this section, we will define the basic formulations of VOA intensity used in currently available quantum chemistry software, and in subsequent sections describe the necessary steps needed to carry out practical calculations of the VOA spectra.

In the following subsections, we extend the theoretical developments of Chapters 2, 4, and 5 to form a basis for describing the dependence of vibrational intensities on perturbations of the Hamiltonian, and hence the total energy, of a molecule by an applied external field. Some of these expressions have been presented in part previously, but we gather them here at the beginning of this chapter so we can gain an overview of the entire formalism and appreciate the degree of unity between the formalisms for vibrational absorption, VCD, Raman scattering, and ROA. In particular all four classes of intensities can be expressed as field-dependent and nuclear-coordinate dependent derivatives of the total energy of the molecule.

9.1.1 Formulation of VA Intensities

We start with the perturbation of a molecule with an applied electric field $\boldsymbol{F}$ and show how the atomic polar tensor (APT) in the position representation can be derived from derivatives of the energy gradient of the molecule with respect to the applied electric field. We consider the adiabatic electronic Hamiltonian operator, $H_E^A(\boldsymbol{R})$, of the molecule with an electric field perturbation term, $-\boldsymbol{\mu}_E \cdot \boldsymbol{F}$, resulting in the field adiabatic (*FA*) Hamiltonian operator,

$$H_E^{FA}(\boldsymbol{R},\boldsymbol{F}) = H_E^A(\boldsymbol{R}) - \boldsymbol{\mu}_E \cdot \boldsymbol{F} \tag{9.1}$$

where $\boldsymbol{\mu}_E$ is the electronic part of the electric-dipole moment operator of the molecule. The field perturbation term describes the interaction energy between the electrons of the molecule and the field. The field-dependent total ground state electronic energy is defined using the field-perturbed Hamiltonian operator and the perturbed electronic wavefunction of the molecule $\psi_g^{FA}(\boldsymbol{r},\boldsymbol{R},\boldsymbol{F})$ as:

$$E_g^{FA}(\boldsymbol{R},\boldsymbol{F}) = \left\langle \psi_g^{FA}(\boldsymbol{r},\boldsymbol{R},\boldsymbol{F}) \middle| H_E^{FA}(\boldsymbol{R},\boldsymbol{F}) \middle| \psi_g^{FA}(\boldsymbol{r},\boldsymbol{R},\boldsymbol{F}) \right\rangle \tag{9.2}$$

We next recall from Equation (2.48) that the adiabatic Hamiltonian operator and the electronic wavefunction, when expanded to first order in the nuclear position coordinates, are given by:

$$H_E^{FA}(\boldsymbol{R},\boldsymbol{F})_{F=0} = H_E^A(\boldsymbol{R}) = H_E^A(\boldsymbol{R}_0) + \left(\frac{\partial H_E^A}{\partial \boldsymbol{R}_J}\right)_{R=0} \cdot \boldsymbol{R}_J \tag{9.3}$$

$$\psi_g^{FA}(\boldsymbol{r},\boldsymbol{R})_{F=0} = \psi_g^A(\boldsymbol{r},\boldsymbol{R}) = \psi_E^A(\boldsymbol{r},\boldsymbol{R}_0) + \sum_{J,\alpha}\left(\frac{\partial \psi_E^A}{\partial \boldsymbol{R}_J}\right)_{R=0} \cdot \boldsymbol{R}_J \tag{9.4}$$

In these equations, we first removed the external field dependence by setting $F = 0$ and then expressing the nuclear coordinate dependence. Thus, the first-order dependence of the field-adiabatic Hamiltonian operator in Equation (9.1) with respect to the applied field, F, and the nuclear coordinate, R, is given by:

$$\frac{\partial}{\partial F_\beta} H_E^{FA}(\boldsymbol{R},\boldsymbol{F}) = -\mu_{E,\beta} \tag{9.5}$$

$$\frac{\partial}{\partial R_\alpha} H_E^{FA}(\boldsymbol{R},\boldsymbol{F}) = \left(\frac{\partial H_E^A}{\partial \boldsymbol{R}_J}\right)_{R=0} \tag{9.6}$$

Recall that the atomic polar tensor, used to express vibrational absorption intensities is defined from Equations (2.72) and (2.73) as the derivative of the total electric dipole moment, $\boldsymbol{\mu}$, of the molecule with respect the Cartesian nuclear coordinate where

$$\boldsymbol{\mu} = \boldsymbol{\mu}_E + \boldsymbol{\mu}_N = -\sum_j e\boldsymbol{r}_j + \sum_J Z_J e \boldsymbol{R}_J \tag{9.7}$$

and

$$P_{r,\alpha\beta}^J = E_{r,\alpha\beta}^J + N_{r,\alpha\beta}^J = \left[\frac{\partial}{\partial R_{J,\alpha}}\left\langle \psi_g^A(\boldsymbol{r},\boldsymbol{R})|\mu_\beta|\psi_g^A(\boldsymbol{r},\boldsymbol{R})\right\rangle\right]_{R=0} \tag{9.8}$$

Furthermoe, the APT is related to the dipole strength of a vibrational transition in the ground electronic state between levels gv and gv' by:

$$D_{r,gv',gv}^a = \left|\left\langle \Psi_{gv'}^A(\boldsymbol{r},Q_a)|\boldsymbol{\mu}|\Psi_{gv}^A(\boldsymbol{r},Q_a)\right\rangle\right|^2 = \left|\sum_J P_{r,\alpha\beta}^J S_{J\alpha,a}\left\langle \phi_{gv'}^a|Q_a|\phi_{gv}^a\right\rangle\right|^2 \tag{9.9}$$

where $S_{J\alpha,a} = (\partial R_{J\alpha}/\partial Q_a)_{Q=0}$ is the transformation between Cartesian and normal coordinates that represents the Cartesian displacement vector of each nucleus J in normal mode a. The nuclear part $N_{r,\alpha\beta}^J$ in Equation (9.8) of the APT is trivial as given in Equation (2.74), and as a result, we focus our attention on the electronic part of the APT, $E_{r,\alpha\beta}^J$, the contribution of the electronic motion to the change in the electronic part of dipole moment of the molecule, $\mu_{E,\beta}$, with respect to the displacement of the nuclear coordinate $R_{J,\alpha}$ from equilibrium:

$$E_{r,\alpha\beta}^J = \left(\frac{\partial}{\partial R_{J,\alpha}}\left\langle \psi_g^A|\mu_{E,\beta}|\psi_g^A\right\rangle\right)_{R=0} = 2\left\langle \psi_g^A|\mu_{E,\beta}|\frac{\partial \psi_g^A}{\partial R_{J,\alpha}}\right\rangle_{R=0} \tag{9.10}$$

When the molecule is exposed, at least theoretically, to an external electric field, $\boldsymbol{F}$, an additional opportunity opens for expressing the electronic APT, namely as the second-order derivative of the electronic energy, E_g, with respect to first the applied electric field and second to the nuclear position. This can be seen from the definition of $E^J_{r,\alpha\beta}$ by converting the dipole moment operator using Equation (9.5) into the derivative of the electric-field perturbed Hamiltonian operator and then continuing on to the second-derivative expressions with respect to the electronic energy as:

$$\begin{aligned} E^J_{r,\alpha\beta} &= \left(\frac{\partial}{\partial R_{J,\alpha}}\langle\psi^A_g|\mu_{E,\beta}|\psi^A_g\rangle\right)_{R=0} = -\left(\frac{\partial}{\partial R_{J,\alpha}}\left(\left\langle\psi^A_g\left|\frac{\partial H^{FA}_E}{\partial F_\beta}\right|\psi^A_g\right\rangle\right)_{F=0}\right)_{R=0} \\ &= -\left(\frac{\partial}{\partial R_{J,\alpha}}\left(\frac{\partial E^{FA}_g}{\partial F_\beta}\right)_{F=0}\right)_{R=0} = -\left(\frac{\partial^2 E^{FA}_g}{\partial R_{J,\alpha}\partial F_\beta}\right)_{F=0,R=0} \end{aligned} \tag{9.11}$$

If now we reverse the order of the differentiation of the second-order derivative, first the nuclear position derivative followed by the electric field derivative, we can work backwards as:

$$\begin{aligned} E^J_{r,\alpha\beta} &= -\left(\frac{\partial^2 E^{FA}_g}{\partial F_\beta\partial R_{J,\alpha}}\right)_{R=0,F=0} = -\left(\frac{\partial}{\partial F_\beta}\left(\frac{\partial E^{FA}_g}{\partial R_{J,\alpha}}\right)_{R=0}\right)_{F=0} \\ &= -\left(\frac{\partial}{\partial F_\beta}\left(\left\langle\psi^{FA}_g\left|\frac{\partial H^A_E}{\partial R_{J,\alpha}}\right|\psi^{FA}_g\right\rangle\right)_{R=0}\right)_{F=0} = -2\left\langle\psi^0_g\left|\frac{\partial H^A_E}{\partial R_{J,\alpha}}\right|\frac{\partial\psi^{FA}_e}{\partial F_\beta}\right\rangle_{R=0,F=0} \end{aligned} \tag{9.12}$$

The advantage of the second approach is that the energy gradient, the derivative of the energy with respect to nuclear positions, $(\partial E^A_g/\partial R_{J,\alpha})_{R=0}$, is already available from the determination of the equilibrium conformation of the molecule, the geometry being optimized when the energy gradients in all nuclear coordinate directions are sufficient small. The second approach then involves only three additional perturbations, F_x, F_y, and F_z, of the matrix element, $\partial/\partial F_\beta$, in Equation (9.12) instead of the $3N$ Cartesian nuclear coordinate derivatives, $\partial/\partial R_{J,\alpha}$, in Equation (9.11).

Analogous expressions can be written for the velocity formulation of infrared vibrational absorption intensity (Nafie, 1992). We present these expressions to show the close relationship of the velocity form of the APT to the magnetic-dipole moment derivative with respect to nuclear velocity that is needed for the atomic axial tensor (AAT) and VCD. Here, as with the formulation of VCD, we must rely on the complete adiabatic (CA) approximation and the correlation of the electronic current density (electron velocity fields) with nuclear velocity instead of just the Born–Oppenheimer adiabatic approximation and correlation of the electron population density with nuclear position. The nuclear position dependence is not needed as a dynamic variable. As a result, two perturbations are included in the CA Hamiltonian operator, the nuclear-velocity perturbation operator from Equation (4.22) and the field perturbation operator of the vector potential, $\boldsymbol{A}$, as described previously in Chapter 4 by Equation (4.94). Thus we write, analogously to Equation (9.1),

$$\tilde{H}^{FCA}_E(\boldsymbol{R},\dot{\boldsymbol{R}},\boldsymbol{A}) = H^A_E(\boldsymbol{R}) - i\hbar\left(\frac{\partial}{\partial\boldsymbol{R}}\right)_E\cdot\dot{\boldsymbol{R}} - \dot{\boldsymbol{\mu}}_E\cdot\boldsymbol{A}/c \tag{9.13}$$

where the superscript FCA refers the field (electric-dipole vector potential $\boldsymbol{A}$) complete adiabatic Hamiltonian, and the subscript E refers to differentiation with respect to $\boldsymbol{R}$ of only the electronic part of the molecular wavefunction, and not the nuclear part. The tilda over the Hamiltonian operator indicates a complex operator due to the imaginary perturbation terms associated with the nuclear velocity perturbation and the velocity dipole operator, $\dot{\boldsymbol{\mu}}_E$. The dipole velocity operator, first defined in Chapter 2, is given by:

$$\dot{\boldsymbol{\mu}} = \dot{\boldsymbol{\mu}}_E + \dot{\boldsymbol{\mu}}_N = -\sum_j e\dot{\boldsymbol{r}}_j + \sum_J Z_J e\dot{\boldsymbol{R}}_J = -\sum_j e\left(\frac{-i\hbar}{m}\right)\frac{\partial}{\partial \boldsymbol{r}_j} + \sum_J Z_J e\left(\frac{-i\hbar}{M_J}\right)\frac{\partial}{\partial \boldsymbol{R}_J} \quad (9.14)$$

Using this perturbed Hamiltonian operator, we can write the derivative of the Hamiltonian with respect to nuclear velocity as:

$$\frac{\partial}{\partial \dot{R}_{J,\alpha}}\tilde{H}_E^{FCA}(\boldsymbol{R},\dot{\boldsymbol{R}},\boldsymbol{A}) = -i\hbar\left(\frac{\partial}{\partial R_{J,\alpha}}\right)_E \quad (9.15)$$

Notice the close connection between the nuclear velocity derivative of the Hamiltonian in Equation (9.15) and the nuclear position operator that remains after the nuclear velocity derivative is taken. The derivative of the field perturbation term is given by:

$$\frac{\partial}{\partial A_\beta}\tilde{H}_E^{FCA}(\boldsymbol{R},\dot{\boldsymbol{R}},\boldsymbol{A}) = -\dot{\mu}_{E,\beta}/c = -i\hbar\sum_j\left(\frac{e}{mc}\right)\frac{\partial}{\partial r_{j\beta}} \quad (9.16)$$

If one compares the form of Equations (9.15) and (9.16) one sees close parallels between these two perturbation derivatives; both are pure imaginary and both generate current density in the motion of electrons in molecules. As previously for Equation (9.2), the electronic energy of the molecule in the ground electronic state corresponding to the Hamiltonian in Equation (9.13) is given at the equilibrium nuclear position, $\boldsymbol{R}_0$, by:

$$E_g^{FCA}(\boldsymbol{R}_0,\dot{\boldsymbol{R}},\boldsymbol{A}) = \left\langle \tilde{\psi}_g^{FCA}(\boldsymbol{r},\boldsymbol{R}_0,\dot{\boldsymbol{R}},\boldsymbol{A})\middle|\tilde{H}_E^{FCA}(\boldsymbol{R}_0,\dot{\boldsymbol{R}},\boldsymbol{A})\middle|\tilde{\psi}_g^{FCA}(\boldsymbol{r},\boldsymbol{R}_0,\dot{\boldsymbol{R}},\boldsymbol{A})\right\rangle \quad (9.17)$$

The velocity formulation of the electronic part of the APT is defined as the derivative of the electronic velocity dipole moment with respect to the nuclear velocity as:

$$E^J_{v,\alpha\beta} = \left(\frac{\partial}{\partial \dot{R}_{J,\alpha}}\left\langle \tilde{\psi}_g^{CA}\middle|\dot{\mu}_{E,\beta}\middle|\tilde{\psi}_g^{CA}\right\rangle\right)_{\substack{R=0\\ \dot{R}=0}} = 2\left\langle \tilde{\psi}_g^{CA}\middle|\dot{\mu}_{E,\beta}\middle|\frac{\partial\tilde{\psi}_g^{CA}}{\partial \dot{R}_{J,\alpha}}\right\rangle_{\substack{R=0\\ \dot{R}=0}} \quad (9.18)$$

The same line of development as in Equations (9.11) and (9.12) leads to the following relationships,

$$\begin{aligned} E^J_{v,\alpha\beta} &= \left(\frac{\partial}{\partial \dot{R}_{J\alpha}}\left\langle \tilde{\psi}_g^{CA}\middle|\dot{\mu}_\beta^E\middle|\tilde{\psi}_g^{CA}\right\rangle\right)_{\substack{R=0\\ \dot{R}=0}} = -c\left(\frac{\partial}{\partial \dot{R}_{J,\alpha}}\left(\left\langle \tilde{\psi}_g^{CA}\middle|\frac{\partial\tilde{H}_E^{FCA}}{\partial A_\beta}\middle|\tilde{\psi}_g^{CA}\right\rangle\right)_{A=0}\right)_{\substack{R=0\\ \dot{R}=0}} \\ &= -c\left(\frac{\partial}{\partial \dot{R}_{J,\alpha}}\left(\frac{\partial E_g^{FCA}}{\partial A_\beta}\right)_{A=0}\right)_{\substack{R=0\\ \dot{R}=0}} = -c\left(\frac{\partial^2 E_g^{FCA}}{\partial \dot{R}_{J,\alpha}\partial A_\beta}\right)_{\substack{R=0\\ A=0\\ \dot{R}=0}} \end{aligned} \quad (9.19)$$

Here a factor c is needed before the second derivative to account for the difference in the form of the field interaction of Equation (9.16) versus Equation (9.5). This last expression is the second-order derivative of the energy with respect to the two perturbations of the Hamiltonian, the nuclear velocity, and the vector potential. Reversing the order of these derivatives allows us to write

$$E^{J}_{v,\alpha\beta} = -c\left(\frac{\partial^2 E_g^{FCA}}{\partial A_\beta \partial \dot{R}_{J,\alpha}}\right)_{\substack{R=0\\A=0\\\dot{R}=0}} = -c\left(\frac{\partial}{\partial A_\beta}\left\langle \tilde{\psi}_g^{FCA}\left|\frac{\partial \tilde{H}_E^{FCA}}{\partial \dot{R}_{J,\alpha}}\right|\tilde{\psi}_g^{FCA}\right\rangle_{\substack{R=0\\\dot{R}=0}}\right)_{A=0}$$

$$= 2c\left\langle \frac{\partial \tilde{\psi}_e^{FA}}{\partial A_\beta}\left| i\hbar\frac{\partial}{\partial R_{J,\alpha}}\right|\psi_e^A\right\rangle_{\substack{A=0\\R=0}} = 2i\hbar c\left\langle \frac{\partial \tilde{\psi}_e^{FA}}{\partial A_\beta}\bigg|\frac{\partial \psi_e^A}{\partial R_{J,\alpha}}\right\rangle_{\substack{A=0\\R=0}}$$

$$= 2\hbar c\,\mathrm{Im}\left\langle \frac{\partial \psi_e^A}{\partial R_{J,\alpha}}\bigg|\frac{\partial \tilde{\psi}_e^{FA}}{\partial A_\beta}\right\rangle_{\substack{A=0\\R=0}} \tag{9.20}$$

The final two expressions above contain the overlap integral of two electronic wavefunction derivatives, one with respect to the applied field, and the other with respect to the nuclear position. There is a sign change for interchanging the order of the two wavefunctions in the overlap matrix element as the left-hand wavefunction in the matrix element involves the complex conjugate wavefunction. This expression for the electronic APT is similar to that of the magnetic field perturbation method used to calculate the AAT where there appears to be no quantum mechanical operator, and the overlap matrix is the not in the form of a conventional quantum mechanical property matrix compared with the nuclear velocity definition of the AAT in Equation (9.18).

9.1.2 Formulation of VCD Intensities

We have already developed the relevant equations for this section when the theory of VCD was presented in Chapter 4. Here we present the equations again in a slightly different form and show the close parallel to the field-dependent formulation of infrared vibrational absorption in general and the velocity-dipole formulation in particular.

The expression for the CA Hamiltonian operator, first derived in Equation (4.22), perturbed by a magnetic field $\boldsymbol{H}$ is given by:

$$\tilde{H}_E^{HCA}(\boldsymbol{R},\dot{\boldsymbol{R}},\boldsymbol{B}) = H_E^A(\boldsymbol{R}) - i\hbar\left(\frac{\partial}{\partial \boldsymbol{R}}\right)_E \cdot \dot{\boldsymbol{R}} - \boldsymbol{m}_E\cdot\boldsymbol{H} \tag{9.21}$$

where, the magnetic-dipole moment operator can be written as:

$$\boldsymbol{m} = \boldsymbol{m}_E + \boldsymbol{m}_N = \frac{e}{2c}\left(-\sum_j \boldsymbol{r}_j\times\dot{\boldsymbol{r}}_j + \sum_J Z_J\boldsymbol{R}_J\times\dot{\boldsymbol{R}}_J\right) \tag{9.22}$$

The derivative of this Hamiltonian operator by the nuclear velocity is the same as that for a vector potential perturbation given in Equation (9.15), whereas the corresponding magnetic field derivative, analogous to Equation (9.16), is given by:

$$\frac{\partial}{\partial H_\beta}\tilde{H}_E^{HCA}(\boldsymbol{R},\dot{\boldsymbol{R}},\boldsymbol{H}) = -m_{E,\beta} = -\sum_j\left(\frac{-e}{2c}\right)(\boldsymbol{r}_j\times\dot{\boldsymbol{r}}_j)_\beta = -i\hbar\sum_j\left(\frac{e}{2mc}\right)\varepsilon_{\beta\gamma\delta}r_{j\gamma}\frac{\partial}{\partial r_{j\delta}} \quad (9.23)$$

VCD intensity is given by the rotational strength, which in turn depends on the imaginary part of the scalar product of the electric- and magnetic-dipole transition moments where the former can be expressed in either the dipole-position or dipole-velocity form as given previously in Equations (4.10) and (4.14):

$$R^a_{r,g1,g0} = \mathrm{Im}\left[\left\langle\Psi^a_{g0}\middle|\mu_{r,\beta}\middle|\Psi^a_{g1}\right\rangle\cdot\left\langle\tilde{\Psi}^a_{g1}\middle|m_\beta\middle|\tilde{\Psi}^a_{g0}\right\rangle\right] \quad (9.24)$$

$$R^a_{v,g1,g0} = \omega_a^{-1}\,\mathrm{Re}\left[\left\langle\tilde{\Psi}^a_{g0}\middle|\dot{\mu}_\beta\middle|\tilde{\Psi}^a_{g1}\right\rangle\cdot\left\langle\tilde{\Psi}^a_{g1}\middle|m_\beta\middle|\tilde{\Psi}^a_{g0}\right\rangle\right] \quad (9.25)$$

As with the dipole strength these intensity expressions can be reduced to products of atomic polar and atomic axial tensors as:

$$R^a_{r,g1,g0} = \mathrm{Im}\left[\sum_J P^J_{r,\alpha\beta}S_{J\alpha,a}\left\langle\phi^a_{gv'}\middle|Q_a\middle|\phi^a_{gv}\right\rangle\sum_{J'}M^{J'}_{\alpha'\beta}S_{J'\alpha',a}\left\langle\phi^a_{gv'}\middle|\dot{Q}_a\middle|\phi^a_{gv}\right\rangle\right] \quad (9.26)$$

$$R^a_{v,g1,g0} = \omega_a^{-1}\,\mathrm{Re}\left[\sum_J P^J_{v,\alpha\beta}S_{J\alpha,a}\left\langle\phi^a_{gv'}\middle|\dot{Q}_a\middle|\phi^a_{gv}\right\rangle\sum_{J'}M^{J'}_{\alpha'\beta}S_{J'\alpha',a}\left\langle\phi^a_{gv'}\middle|\dot{Q}_a\middle|\phi^a_{gv}\right\rangle\right] \quad (9.27)$$

The atomic axial tensor (AAT) is defined as the derivative of the magnetic-dipole moment of the molecule with respect to the nuclear velocity where

$$M^J_{\alpha\beta} = I^J_{\alpha\beta} + J^J_{\alpha\beta} = \left[\frac{\partial}{\partial\dot{R}_{J,\alpha}}\left\langle\tilde{\psi}^{CA}_g(\boldsymbol{r},\boldsymbol{R},\dot{\boldsymbol{R}})\middle|m_\beta\middle|\tilde{\psi}^{CA}_g(\boldsymbol{r},\boldsymbol{R},\dot{\boldsymbol{R}})\right\rangle\right]_{\substack{R=0\\ \dot{R}=0}} \quad (9.28)$$

The electronic part of the atomic axial tensor is defined as:

$$I^J_{\alpha\beta} = \left(\frac{\partial}{\partial\dot{R}_{J,\alpha}}\left\langle\tilde{\psi}^{CA}_g\middle|m_{E,\beta}\middle|\tilde{\psi}^{CA}_g\right\rangle\right)_{\substack{\dot{R}=0\\ R=0}} = 2\left\langle\tilde{\psi}^{CA}_g\middle|m_{E,\beta}\middle|\frac{\partial\tilde{\psi}^{CA}_g}{\partial\dot{R}_{J,\alpha}}\right\rangle_{\substack{\dot{R}=0\\ R=0}} \quad (9.29)$$

As with the electronic APT, the expression of the electronic AAT can be derived from the second derivative of the energy first with respect to nuclear velocity and second with respect to the external magnetic field. Proceeding as before, we can replace the magnetic-dipole moment operator in

Equation (9.29) by the derivative of the perturbed Hamiltonian with respect to the external magnetic field, which gives

$$I^J_{\alpha\beta} = -\left(\frac{\partial}{\partial \dot{R}_{J,\alpha}}\left\langle \tilde{\psi}^{CA}_g \middle| \left(\frac{\partial \tilde{H}^{HCA}_E}{\partial H_\beta}\right)_{H=0} \middle| \tilde{\psi}^{CA}_g \right\rangle\right)_{\substack{R=0\\ \dot{R}=0}} = \left(\frac{\partial}{\partial \dot{R}_{J,\alpha}}\left(\frac{\partial}{\partial H_\beta}\left\langle \tilde{\psi}^{CA}_g \middle| \tilde{H}^{HCA}_E \middle| \tilde{\psi}^{CA}_g \right\rangle\right)_{H=0}\right)_{\substack{R=0\\ \dot{R}=0}}$$

$$= \left(\frac{\partial}{\partial \dot{R}_{J,\alpha}}\left(\frac{\partial E^{HCA}_g}{\partial H_\beta}\right)_{H=0}\right)_{\substack{R=0\\ \dot{R}=0}} = -\left(\frac{\partial^2 E^{HCA}_g}{\partial \dot{R}_{J,\alpha}\partial H_\beta}\right)_{\substack{R=0\\ \dot{R}=0\\ H=0}} \tag{9.30}$$

The AAT has yet to be calculated by this nuclear velocity perturbation (NVP) route, but there are reasons to believe that some theoretical and computational simplifications may be realized by doing so. The AAT is currently calculated by VCD software using the magnetic field perturbation (MFP) approach. This can be formulated directly by reversing the order of differentiation in the last terms of Equation (9.30) giving

$$I^J_{\alpha\beta} = -\left(\frac{\partial^2 E^{HCA}_g}{\partial H_\beta \partial \dot{R}_{J,\alpha}}\right)_{\substack{H=0\\ R=0\\ \dot{R}=0}} = -\left(\frac{\partial}{\partial H_\beta}\left\langle \tilde{\psi}^{HCA}_g \middle| \frac{\partial H^{HCA}_E}{\partial \dot{R}_{J,\alpha}} \middle| \tilde{\psi}^{HCA}_g \right\rangle_{\substack{R=0\\ \dot{R}=0}}\right)_{H=0}$$

$$= 2\left\langle \frac{\partial \tilde{\psi}^{HA}_g}{\partial H_\beta} \middle| i\hbar \frac{\partial}{\partial R_{J,\alpha}} \middle| \psi^A_e \right\rangle_{\substack{H=0\\ R=0}} = 2i\hbar \left\langle \frac{\partial \tilde{\psi}^{HA}_e}{\partial H_\beta} \middle| \frac{\partial \psi^A_e}{\partial R_{J,\alpha}} \right\rangle_{\substack{H=0\\ \dot{R}=0}}$$

$$= 2\hbar\, \mathrm{Im}\left\langle \frac{\partial \psi^A_e}{\partial R_{J,\alpha}} \middle| \frac{\partial \tilde{\psi}^{HA}_e}{\partial H_\beta} \right\rangle_{\substack{H=0\\ \dot{R}=0}} \tag{9.31}$$

This last term is the commonly presented expression (usually without the $2\hbar$ pre-factor) for the AAT in the MFP formalism that is used as the current basis for all calculations of VCD.

9.1.3 Formulation of Raman Scattering

From Chapter 2, we recall from Equation (2.94) that Raman scattering intensity is proportional to the square of the complex *induced* electric-dipole moment of the molecule, $\tilde{\mu}^I_\alpha$, where ω_0 is angular frequency of the incident laser radiation and $\omega_s$the angular frequency of the scattered Raman radiation. The Raman intensity in terms of the induced dipole moment is given by:

$$I(\omega_0, \omega_s) = k'_\omega \omega_s^4 \left|\tilde{\mu}^I_\alpha\right|^2 \sin^2\theta \tag{9.32}$$

In this expression, Cartesian component notation is used, and as the subscript α is repeated in the absolute square of $\tilde{\mu}^I_\alpha$, it is summed over all three Cartesian directions. The induced dipole moment operator arises from the tensor response of the molecule to the external field $\boldsymbol{F}$ of the incident laser radiation by the following relationship where the tensor is the polarizability operator, $\tilde{\boldsymbol{\alpha}}$, of the molecule

$$\tilde{\mu}_{\alpha}^{I} = \left(\tilde{\boldsymbol{\mu}}^{I}\right)_{\alpha} = (\tilde{\boldsymbol{\alpha}} \cdot \boldsymbol{F})_{\alpha} = \tilde{\alpha}_{\alpha\beta} F_{\beta} \tag{9.33}$$

The tilda above these quantities is included to allow for complex expressions that arise, for example, when there is strong resonance between the polarizability and the incident radiation. In the far-from-resonance limit, the complex form of the induced dipole moment and polarizability operators is not needed. For Raman scattering, we need the derivative of the induced dipole moment of the molecule with respect to the normal coordinate of vibration, Q_a, and this derivative can be expressed in terms of Cartesian nuclear coordinates using the usual transformation involving S-vectors as:

$$\left(\frac{\partial\langle\tilde{\mu}_{\alpha}^{I}\rangle}{\partial Q_a}\right)_{Q=0} Q_a = \left(\frac{\partial\langle\tilde{\alpha}_{\alpha\beta} F_{\beta}\rangle}{\partial Q_a}\right)_{Q=0} Q_a = \left(\frac{\partial\langle\tilde{\alpha}_{\alpha\beta}\rangle}{\partial R_{J,\gamma}}\right)_{R=0} F_{\beta} S_{J,\gamma}^{a} Q_a \tag{9.34}$$

The Raman atomic polar tensor (R-APT) can be defined in analogy to infrared APT as:

$$\tilde{P}_{\alpha\beta\gamma}^{J} = \left(\frac{\partial\langle\tilde{\alpha}_{\beta\gamma}\rangle}{\partial R_{J,\alpha}}\right)_{R=0} = \left[\frac{\partial}{\partial R_{J,\alpha}}\left\langle\psi_g^A(\boldsymbol{r},\boldsymbol{R})\middle|\tilde{\alpha}_{\beta\gamma}\middle|\psi_g^A(\boldsymbol{r},\boldsymbol{R})\right\rangle\right]_{R=0} \tag{9.35}$$

Here and in analogous tensors below, we use the Greek subscript α for the Cartesian nuclear coordinate derivative. The R-APT is a third-order tensor that requires three scalar-product operations to reduce it to a Raman intensity component. If now the field perturbed electronic Hamiltonian operator in Equation (9.1) is expanded to include the *induced* electric-dipole moment, $\tilde{\boldsymbol{\mu}}^{I}$, and also the electronic part of the *intrinsic* electric-dipole moment, $\boldsymbol{\mu}_E$, we can write

$$\begin{aligned}\tilde{H}_E^{FA}(\boldsymbol{R},\boldsymbol{F}) &= H_E^A(\boldsymbol{R}) - \boldsymbol{\mu}_E \cdot \boldsymbol{F} - \tilde{\boldsymbol{\mu}}^{I} \cdot \boldsymbol{F} \\ &= H_E^A(\boldsymbol{R}) - \boldsymbol{\mu}_E \cdot \boldsymbol{F} - (\tilde{\boldsymbol{\alpha}} \cdot \boldsymbol{F}) \cdot \boldsymbol{F} = H_E^A(\boldsymbol{R}) - \mu_{E\beta} \cdot F_{\beta} - \left(\tilde{\alpha}_{\beta\gamma} F_{\gamma}\right) F_{\beta} \\ &= H_E^A(\boldsymbol{R}) - \mu_{E\beta} \cdot F_{\beta} - \tilde{\alpha}_{\beta\gamma} F_{\beta} F_{\gamma}\end{aligned} \tag{9.36}$$

As a result, the R-APT can be written as the third derivative with respect to the electronic energy of the molecule with two field derivatives to arrive at the equilibrium polarizability and one nuclear Cartesian derivative for the Raman polarizability tensor. As noted in Chapter 2, the polarizability and all Raman tensors are properties that arise from just the electrons. Therefore, $\tilde{P}_{\alpha\beta\gamma}^{J}$ does not need to be reduced further to electronic and nuclear parts. From Equation (9.36) we can write

$$\tilde{P}_{\alpha\beta\gamma}^{J} = \left(\frac{\partial}{\partial R_{J,\alpha}}\left\langle\psi_g^A\middle|\tilde{\alpha}_{\beta\gamma}\middle|\psi_g^A\right\rangle\right)_{R=0} = -\left(\frac{\partial}{\partial R_{J,\alpha}}\left(\left\langle\psi_g^A\middle|\frac{\partial^2\tilde{H}_E^{FA}}{\partial F_{\beta}\partial F_{\gamma}}\middle|\psi_g^A\right\rangle\right)_{F=0}\right)_{R=0} = -\left(\frac{\partial^3\tilde{E}_g}{\partial R_{\alpha}\partial F_{\beta}\partial F_{\gamma}}\right)_{\substack{F=0\\R=0}} \tag{9.37}$$

The order of differentiation of the fields and the nuclear coordinates can be reversed so that the first quantity encountered is the energy gradient of the molecule with respect to nuclear coordinates, as discussed above for the calculation of vibrational absorption with infrared APTs. In this case we write

$$\tilde{P}^{J}_{\alpha\beta\gamma} = -\left(\frac{\partial^2}{\partial F_\beta \partial F_\gamma}\left\langle \psi_g^A \left| \frac{\partial \tilde{H}_E^{FA}}{\partial R_{J,\alpha}} \right| \psi_g^A \right\rangle_{F=0}\right)_{R=0} = -\left(\frac{\partial^3 \tilde{E}_g}{\partial F_\beta \partial F_\gamma \partial R_\alpha}\right)_{\substack{R=0\\F=0}} \tag{9.38}$$

If one assumes the far-from-resonance (FFR) approximation, and hence symmetric polarizabilities and Raman tensors, then keeping track of the order of the field derivatives with respect to β and γ is not necessary, and all quantities are real. This may not be the case when resonance is approached and the FFR approximation breaks down. Then the order of the perturbing fields as specified above becomes important in correlating the order of field differentiation with the order of the Cartesian subscripts in the Raman polarizability tensor.

9.1.4 Formulation of ROA Intensities

The corresponding formulation for ROA intensities can be obtained by modifying the formulation presented above for Raman intensities. We first present the expressions for the general unrestricted (GU) level of ROA theory and will then present the simplified version, currently available in commercial ROA intensity programs, in the far-from-resonance (FFR) approximation. The formulation for the calculation of Raman intensities can be expanded to include the calculation of ROA intensities by generalizing the molecular polarizability, $\tilde{\alpha}_{\alpha\beta}$, to the scattering tensor, $\tilde{a}_{\alpha\beta}$, which includes the four optical activity tensors, $\tilde{G}_{\alpha\beta}, \tilde{\mathscr{G}}_{\alpha\beta}, \tilde{A}_{\alpha,\beta\gamma}$, and $\tilde{\mathscr{A}}_{\alpha,\beta\gamma}$ according to the relationship given previously in Equation (5.35),

$$\tilde{a}_{\alpha\beta} = \tilde{\alpha}_{\alpha\beta} + \frac{1}{c}\left[\varepsilon_{\gamma\delta\beta} n^i_\delta \tilde{G}_{\alpha\gamma} + \varepsilon_{\gamma\delta\alpha} n^d_\delta \tilde{\mathscr{G}}_{\gamma\beta} + \frac{i}{3}\left(\omega_0 n^i_\gamma \tilde{A}_{\alpha,\gamma\beta} - \omega_R n^d_\gamma \tilde{\mathscr{A}}_{\beta,\gamma\alpha}\right)\right] \tag{9.39}$$

These new tensors can be calculated using field perturbation methods describe above by expanding the field perturbed Hamiltonian operator to include additional field interactions as:

$$\begin{aligned}
\tilde{H}_E^{FA}(\boldsymbol{R},\boldsymbol{F},\boldsymbol{B},\boldsymbol{FF}) &= H_E^A(\boldsymbol{R}) - \tilde{\boldsymbol{\mu}}^I\cdot\boldsymbol{F} - \tilde{\boldsymbol{m}}^I\cdot\boldsymbol{H} - \tilde{\boldsymbol{\Theta}}^I : \boldsymbol{FF} \\
&= H_E^A(\boldsymbol{R}) - \left(\tilde{\mu}^I_\beta\right)F_\beta - \left(\tilde{m}^I_\beta\right)H_\beta - \left(\tilde{\Theta}^I_{\beta\gamma}\right)F_\beta F_\gamma \\
&= H_E^A(\boldsymbol{R}) - \left(\tilde{\alpha}_{\beta\gamma}F_\gamma + \tilde{G}_{\beta\gamma}H_\gamma + \tilde{A}_{\beta,\gamma\delta}F_\gamma F_\delta\right)F_\beta - \left(\tilde{\mathscr{G}}_{\beta\gamma}F_\gamma\right)H_\beta - \left(\tilde{\mathscr{A}}_{\delta,\beta\gamma}F_\delta\right)F_\beta F_\gamma
\end{aligned} \tag{9.40}$$

For ROA intensities we define two Raman atomic axial tensors (R-AAT), $\tilde{M}_{\alpha\beta\gamma}$ and $\tilde{\mathscr{M}}_{\alpha\beta\gamma}$, as:

$$\tilde{M}_{\alpha\beta\gamma} = \left(\frac{\partial\langle \tilde{G}_{\beta\gamma}\rangle}{\partial R_{J,\alpha}}\right)_{R=0} = \left[\frac{\partial}{\partial R_{J,\alpha}}\left\langle \psi_g^A(\boldsymbol{r},\boldsymbol{R})\middle|\tilde{G}_{\beta\gamma}\middle|\psi_g^A(\boldsymbol{r},\boldsymbol{R})\right\rangle\right]_{R=0} \tag{9.41}$$

$$\tilde{\mathscr{M}}_{\alpha\beta\gamma} = \left(\frac{\partial\langle \tilde{\mathscr{G}}_{\beta\gamma}\rangle}{\partial R_{J,\alpha}}\right)_{R=0} = \left[\frac{\partial}{\partial R_{J,\alpha}}\left\langle \psi_g^A(\boldsymbol{r},\boldsymbol{R})\middle|\tilde{\mathscr{G}}_{\beta\gamma}\middle|\psi_g^A(\boldsymbol{r},\boldsymbol{R})\right\rangle\right]_{R=0} \tag{9.42}$$

and two Raman atomic quadrupole tensors (R-AQT), $\tilde{Q}_{\alpha\beta,\gamma\delta}$ and $\tilde{\mathscr{O}}_{\alpha\beta\gamma,\delta}$, as follows:

$$\tilde{Q}_{\alpha\beta,\gamma\delta} = \left(\frac{\partial\langle\tilde{A}_{\beta,\gamma\delta}\rangle}{\partial R_{J,\alpha}}\right)_{R=0} = \left[\frac{\partial}{\partial R_{J,\alpha}}\left\langle\psi_g^A(\boldsymbol{r},\boldsymbol{R})\middle|\tilde{A}_{\beta,\gamma\delta}\middle|\psi_g^A(\boldsymbol{r},\boldsymbol{R})\right\rangle\right]_{R=0} \tag{9.43}$$

$$\tilde{\mathscr{O}}_{\alpha\delta,\gamma\beta} = \left(\frac{\partial\langle\tilde{\mathscr{A}}_{\delta,\beta\gamma}\rangle}{\partial R_{J,\alpha}}\right)_{R=0} = \left[\frac{\partial}{\partial R_{J,\alpha}}\left\langle\psi_g^A(\boldsymbol{r},\boldsymbol{R})\middle|\tilde{\mathscr{A}}_{\delta,\beta\gamma}\middle|\psi_g^A(\boldsymbol{r},\boldsymbol{R})\right\rangle\right]_{R=0} \tag{9.44}$$

We can now introduce the field-dependent formalism, corresponding to Equations (9.37) and (9.38) for Raman scattering, that is needed for computing ROA spectra,

$$\begin{aligned}\tilde{M}_{\alpha\beta\gamma} &= \left(\frac{\partial}{\partial R_{J,\alpha}}\left\langle\psi_g^A\middle|\tilde{G}_{\beta\gamma}\middle|\psi_g^A\right\rangle\right)_{R=0} = -\left(\frac{\partial}{\partial R_{J,\alpha}}\left\langle\psi_g^A\middle|\frac{\partial^2\tilde{H}_E^{FA}}{\partial F_\beta\partial H_\gamma}\middle|\psi_g^A\right\rangle_{\substack{H=0\\F=0}}\right)_{R=0}\\ &= -\left(\frac{\partial^3\tilde{E}_g}{\partial R_\alpha\partial F_\beta\partial H_\gamma}\right)_{\substack{H=0\\F=0\\R=0}} = -\left(\frac{\partial^3\tilde{E}_g}{\partial F_\beta\partial H_\gamma\partial R_\alpha}\right)_{\substack{H=0\\F=0\\R=0}}\end{aligned} \tag{9.45}$$

$$\begin{aligned}\tilde{\mathscr{M}}_{\alpha\beta\gamma} &= \left(\frac{\partial}{\partial R_{J,\alpha}}\left\langle\psi_g^A\middle|\tilde{\mathscr{G}}_{\beta\gamma}\middle|\psi_g^A\right\rangle\right)_{R=0} = -\left(\frac{\partial}{\partial R_{J,\alpha}}\left\langle\psi_g^A\middle|\frac{\partial^2\tilde{H}_E^{FA}}{\partial H_\beta\partial F_\gamma}\middle|\psi_g^A\right\rangle_{\substack{F=0\\H=0}}\right)_{R=0}\\ &= -\left(\frac{\partial^3\tilde{E}_g}{\partial R_\alpha\partial H_\beta\partial F_\gamma}\right)_{\substack{F=0\\H=0\\R=0}} = -\left(\frac{\partial^3\tilde{E}_g}{\partial H_\beta\partial F_\gamma\partial R_\alpha}\right)_{\substack{F=0\\H=0\\R=0}}\end{aligned} \tag{9.46}$$

$$\begin{aligned}\tilde{Q}_{\alpha\beta,\gamma\delta} &= \left(\frac{\partial}{\partial R_{J,\alpha}}\left\langle\psi_g^A\middle|\tilde{A}_{\beta,\gamma\delta}\middle|\psi_g^A\right\rangle\right)_{R=0} = -\left(\frac{\partial}{\partial R_{J,\alpha}}\left\langle\psi_g^A\middle|\frac{\partial^2\tilde{H}_E^{FA}}{\partial F_\beta\partial(F_\gamma F_\delta)}\middle|\psi_g^A\right\rangle_{\substack{FF=0\\F=0}}\right)_{R=0}\\ &= -\left(\frac{\partial^3\tilde{E}_g}{\partial R_\alpha\partial F_\beta\partial(F_\gamma F_\delta)}\right)_{\substack{FF=0\\F=0\\R=0}} = -\left(\frac{\partial^3\tilde{E}_g}{\partial F_\beta\partial(F_\gamma F_\delta)\partial R_\alpha}\right)_{\substack{FF=0\\F=0\\R=0}}\end{aligned} \tag{9.47}$$

$$\begin{aligned}\tilde{\mathscr{O}}_{\alpha\delta,\beta\gamma} &= \left(\frac{\partial}{\partial R_{J,\alpha}}\left\langle\psi_g^A\middle|\tilde{\mathscr{A}}_{\delta,\beta\gamma}\middle|\psi_g^A\right\rangle\right)_{R=0} = -\left(\frac{\partial}{\partial R_{J,\alpha}}\left\langle\psi_g^A\middle|\frac{\partial^2\tilde{H}_E^{FA}}{\partial(F_\beta F_\gamma)\partial F_\delta}\middle|\psi_g^A\right\rangle_{\substack{F=0\\FF=0}}\right)_{R=0}\\ &= -\left(\frac{\partial^3\tilde{E}_g}{\partial R_\alpha\partial(F_\beta F_\gamma)\partial F_\delta}\right)_{\substack{F=0\\FF=0\\R=0}} = -\left(\frac{\partial^3\tilde{E}_g}{\partial(F_\beta F_\gamma)\partial F_\delta\partial R_\alpha}\right)_{\substack{F=0\\FF=0\\R=0}}\end{aligned} \tag{9.48}$$

where the derivative of the quadrupole perturbing field, $F_\beta F_\gamma = F_\gamma F_\beta$ is written as $\partial/\partial(F_\beta F_\gamma) = \partial/\partial(F_\gamma F_\beta)$. In the last energy derivative expression for each of the last four equations, the order of the nuclear coordinate derivative is interchanged with the field derivatives to allow the field dependent derivatives to be taken with respect to the energy gradient of the molecule, as in Equation (9.38) relative to Equation (9.37) for ordinary Raman intensities, if desired.

We conclude this section by reducing the Raman optical activity tensor derivatives just presented to the level of the FFR approximation. At this level, there are only two ROA tensors, $G'_{\alpha\beta}$ and $A_{\alpha,\beta\gamma}$. The corresponding field perturbed Hamiltonian operator is written as:

$$H_E^{FA}(\boldsymbol{R},\boldsymbol{F},\boldsymbol{H},\boldsymbol{FF}) = H_E^A(\boldsymbol{R}) - \left(\alpha_{\beta\gamma}F_\gamma + G'_{\beta\gamma}H_\gamma + A_{\beta,\gamma\delta}F_\gamma F_\delta\right)F_\beta \tag{9.49}$$

The R-APT, R-AAT, and the R-AQT, respectively, needed for ROA calculations in the FFR approximations are given by:

$$P^J_{\alpha\beta\gamma} = \left(\frac{\partial}{\partial R_{J,\alpha}}\left\langle\psi_g^A\middle|\alpha_{\beta\gamma}\middle|\psi_g^A\right\rangle\right)_{R=0} = -\left(\frac{\partial^2}{\partial F_\beta\partial F_\gamma}\left(\left\langle\psi_g^A\middle|\frac{\partial H_E^{FA}}{\partial R_{J,\alpha}}\middle|\psi_g^A\right\rangle\right)_{F=0}\right)_{R=0} = -\left(\frac{\partial^3 E_g}{\partial F_\beta\partial F_\gamma\partial R_\alpha}\right)_{\substack{R=0\\F=0}} \tag{9.50}$$

$$M'_{\alpha\beta\gamma} = \left(\frac{\partial}{\partial R_{J,\alpha}}\left\langle\psi_g^A\middle|G'_{\beta\gamma}\middle|\psi_g^A\right\rangle\right)_{R=0} = -\left(\frac{\partial}{\partial F_\beta\partial H_\gamma}\left\langle\psi_g^A\middle|\frac{\partial^2 H_E^{FA}}{\partial R_{J,\alpha}}\middle|\psi_g^A\right\rangle_{R=0}\right)_{\substack{H=0\\F=0}}$$

$$= -\left(\frac{\partial^3 E_g}{\partial F_\beta\partial H_\gamma\partial R_\alpha}\right)_{\substack{H=0\\F=0\\R=0}} \tag{9.51}$$

$$Q_{\alpha\beta,\gamma\delta} = \left(\frac{\partial}{\partial R_{J,\alpha}}\left\langle\psi_g^A\middle|A_{\beta,\gamma\delta}\middle|\psi_g^A\right\rangle\right)_{R=0}$$

$$= -\left(\left(\frac{\partial}{\partial F_\beta\partial(F_\gamma F_\delta)}\left\langle\psi_g^A\middle|\frac{\partial^2 H_E^{FA}}{\partial R_{J,\alpha}}\middle|\psi_g^A\right\rangle\right)_{\substack{FF=0\\F=0}}\right)_{R=0} = -\left(\frac{\partial^3 E_g}{\partial F_\beta\partial(F_\gamma F_\delta)\partial R_\alpha}\right)_{\substack{FF=0\\F=0\\R=0}} \tag{9.52}$$

These Cartesian tensor derivatives, once calculated, can be assembled in various combinations to produce the Raman and ROA invariants at the appropriate level of approximation given in Chapter 5, and these invariants in turn can be assembled to represent Raman and ROA spectra for the various forms of ROA and scattering geometries.

9.1.5 Additional Aspects of VOA Intensity Formulation

In the preceding sections, we presented a unified description of the computation-ready formulations of both VCD and ROA in terms of field perturbed electronic energies. Field-perturbed formulations are preferred over direct evaluations of the atomic tensor derivatives in both VCD and ROA as they automatically circumvent the need to evaluate cumbersome sums of excited electronic states of the molecule. In the VCD the field-perturbed approach is the well-known MFP formulation, although there is an alternative nuclear velocity perturbation (NVP) formulation that also avoids summation over all excited electronic states of the molecule and has yet to be programmed as a software application. All perturbation formulations of VCD can be expressed as second-order energy derivatives, whereas to access ROA intensities, third-order energy derivatives are needed. This is because VA and VCD intensities arise from vibrational derivatives of dipole moments of the molecule whereas these same moments, and more, considering the electric quadrupole moment, must first be induced by the incident radiation fields, which brings in a third perturbation derivative. Additional complications arise in Raman and ROA due to the dependence of these property derivatives on the frequency of the incident radiation, which leads to various levels of approximation of the theoretical formulations.

9.1.5.1 Analytic Derivatives Versus Finite Difference Derivatives

The first formulations of both VCD and ROA involved the use of finite difference to calculate the field derivative required for the MFP formulation of VCD and the analogous field-dependent formulation of ROA. Numerical values of the perturbing magnetic or electric quadrupole fields were inserted into the calculations to construct the perturbation operators, and differences in the calculated quantities were used to construct the desired molecular property derivatives. At first, a minimum of five such calculations, zero, two positive, and two negative field values, were needed to ensure linear slope for the derivative. Subsequently, over a period of a few years for VCD and only recently for ROA, computer subroutines have incorporated analytical derivatives of all tensor derivative properties. This advance drastically reduced the computation time required for VOA calculations.

9.1.5.2 Gauge-Origin Independent Formulations

Another important advance required for the calculation of VOA intensities is gauge-origin dependence. The problem with origin dependence of both VCD and ROA intensities was discussed in Chapters 4 and 5. Unless specific remedies are included in the formulated expressions, the calculation of VOA intensities will vary depending on the choice of molecular origin. Two principal solutions exist. The most prevalent is to use gauge-invariant atomic orbitals (GIAOs), also called London orbitals, in the basis functions of the electronic wavefunction or density functional used for the calculations. It can be shown both theoretically and in practice that GIAOs remove this origin dependence from the calculated VOA intensities. While the earliest VOA calculations did not include GIAOs, subsequently they were included and are now used in virtually all calculations of VOA intensities. The second solution is to use velocity dependent electric-dipole transition moment operators for the rotational strength and the optical activity tensors. These velocity formulations of VOA possess inherent origin independence as demonstrated for VCD in Chapter 4. The same principle applies to ROA when using velocity formulations of the polarizability and ROA tensors.

9.1.5.3 Incident Frequency Dependence for ROA

In the preceding development we alluded to the frequency dependence of the atomic tensors needed for ROA calculations. The very first calculations of ROA were carried out in the static-limit approximation of zero incident radiation frequency. Subsequently, frequency dependence was introduced, still within the FFR approximation, to allow a more realistic modeling of the Raman and ROA tensors. This is the simplest level of incident radiation frequency dependence. As described in detail in Chapter 5 there are two more significant levels of frequency dependence. The first is the mild breakdown of the FFR approximation that leads to a loss in symmetry of the Raman tensor and four distinct ROA tensors. This level of theory is known as the near-resonance (NR) theory. This level of theory distinguishes between the resonant properties of the atomic tensors for incident versus scattered radiation frequencies. The form of tensor invariants is identical with that of the GU level of theory presented first for ROA in this chapter. Beyond the NR level of theory are tensor descriptions that occur when strong resonance is encountered with one or more electronic states of the molecule. In this regime, simplifications occur for resonance with a single electronic state, but this simplicity is quickly lost to the GU level of theory when more than one electronic state is needed to describe the Raman and ROA intensities.

9.2 Fundamental Steps of VOA Calculations

In this section, we outline the steps that are generally needed to carry out a calculation of VOA. At this level, we can treat VCD and ROA on an equivalent basis. While some differences exist between the execution of VCD and ROA intensities, they are generally equivalent as commercial software is available to calculate both VCD and ROA with the same preliminary steps and similar choices as encountered for the actual calculations.

9.2.1 Choice of Model Quantum Chemistry

Within the past decade the model chemistry known as density functional theory (DFT) has become the dominant quantum chemistry approach for the calculation of both VCD and ROA. The one used earlier is the *ab initio* Hartree–Fock self-consistent field theory (HF-SCF), which features a single determinant formalism and can be improved systematically up to the single-determinant limit by increasing the size and flexibility of the basis set employed. The principal problem with the HF approach is the absence of any effects of electron correlation, which limits the accuracy of all quantities calculated, such as equilibrium geometry, force field, and intensities. Several methods are available for extending VOA calculations beyond the HF limit. These include Moeller–Plessett perturbation theory to various orders of approximation, MP2, MP4, and so on, configuration interaction (CI) involving expansions of multiple determinant wavefunctions as in multi-configuration self-consistent field (MC-SCF) theory, and coupled-cluster (CC) methods. The drawback with all of these post-HF methods of calculation is the increasingly long computation times required and hence current restrictions of the approach to relatively simple chiral molecules.

In the 1990s it was found that the DFT approach, which includes the use of empirical density functionals incorporating some degree of electron correlation, provides remarkably accurate descriptions of the geometry, force fields, and intensities of most small to medium-sized molecules for which VOA calculations are practical. In particular, it was found that, using similar basis sets, DFT calculations required approximately the same computation time as HF calculations, but the with similar accuracy to that found in the lowest-level post-HF *ab initio* methods, such as MP2, which requires much longer calculation times than DFT. The principal drawback of DFT is that required functionals are developed empirically with no systematic way to improve performance. Thus, one choice of functional may perform well in certain types of applications, or for certain molecules, while a different functional may perform better for other applications or for a different set of molecules. In this sense, DFT is not strictly an *ab initio* method of quantum chemistry, but this point is generally overlooked as many common choices of functionals perform much better than *ab initio* HF-SCF calculations. Generally, DFT calculations improve systematically on increasing the size and flexibility of the basis functions used to construct the functionals, but eventually a limit is reached beyond which one needs to use *ab initio* theory including electron correlation if higher levels of accuracy are desired.

9.2.2 Conformational Search

Calculations of VOA begin with building the structure of the molecule and specifying the absolute stereochemistry at all the structural elements of chirality as described in Chapter 3. Once the structure is completely specified, a search is conducted over the conformational space of the molecule. This can be done either systemically by specifying all angular degrees of freedom, including ring puckers, and allowing a conformational search program to explore all these degrees of freedom to find all the local unique minima on the potential energy surface of the molecule, or by using a Monte Carlo method of searching that randomly specifies displacements of nuclei away from the starting structure and then optimizes the geometry of the molecule from the displaced nuclear locations. Most energy minimizing programs employ some level of molecular mechanics (MM) programs containing force-field parameters that allow rapid exploration of the potential energy surface. The output of an MM search is a set of conformers that differ from each other along the conformation degrees of freedom of the molecule.

The most important consideration at this stage of the analysis is that *all* the lowest-energy conformers of the molecule are found. If one or more important conformers are somehow overlooked by the search process, the validity of the entire calculation of the VOA is placed in jeopardy. In the simplest case, there is only one conformer to consider, but in most cases there are many conformers, for example, as many as 50–100 conformers is not unusual for a molecule of medium to large size.

9.2.3 Optimization of Geometries

From this set of MM conformers, one can select the lowest 10–20 for further optimization at a higher level of model chemistry, such as DFT using the basis set for the final VOA calculations. If computer power is not an issue, say with unrestricted access to a large supercomputer or computer cluster, one might simply calculate all MM conformers at the final DFT level. The result of the DFT calculation of all or a selected set of lowest energy MM conformers is a set of conformers of increasing energy starting with the lowest energy conformer. Generally, one to three conformers dominate the Boltzmann distribution. By examining the relative energies of the DFT set of conformers one can decide how many conformers will be carried forward for further intensity analysis. Generally, all conformers lying within 2 kcal (1 cal = 4.184 J) of the lowest energy conformer will capture well over 95% of the conformational population of the entire set of conformers, which is more than enough for the calculation of the vibrational spectra. It is unusual for more than eight conformers to be included in the calculation of spectra to be compared with the corresponding measured spectra.

9.2.4 Solvent Corrections and Modeling

In some cases, consideration of the influence of the solvent on the molecule is important and needs to be taken into account. Calculations with solvent corrections are now available for improving the solution-state calculation of molecular properties at various levels of approximation. One of the simplest levels is the polarizable continuum model (PCM), which surrounds the solute molecule with a conforming shell with the dielectric constant of the solvent. This mimics the effect of the solvent on the dissolved molecules and can alter the vibrational frequencies of some modes and possibly change the energy ordering of the conformers. A second, more realistic model is to use molecular mechanics to describe the distribution of solvent molecules around the chiral solute molecule. Programs are available, for example the *Oniom* subroutine from Gaussian Inc., that permit the description of tens to hundreds of solvent molecules at the level of molecular mechanics, while the chiral solute molecule is described by a higher-level model chemistry such as DFT. By averaging such composite calculations over an ensemble of solute molecule configurations, a realistic description of the effect of solvent molecules can be obtained. In some cases, it is known that a small number of solvent molecules bind to the solute molecule at particular locations. In such cases, the solvent molecules can be included in the DFT calculations with or without other solvent treatments, such as ensembles of MM solvent molecules and beyond that of a PCM description. Finally, for some combinations of solute and solvent molecules, multiple species, such as dimers may form and need to be modeled either as the full dimer or as the monomer and a fragment representing that part of the dimer pair that describes the dimeric interaction site.

9.2.5 Force Fields and Vibrational Frequencies

After the conformers of the molecule with its solvent interactions, if desired, have been described, the vibrational force field and frequencies of each conformer are calculated. As described in Chapter 2, the optimization of the conformer geometries involves minimizing the values of the slopes of the potential surface along all $3N$ Cartesian coordinates below some preset limit. For each such fully optimized conformer, second derivatives, or curvatures of the Cartesian potential energy surface are determined. The Cartesian potential energy is transformed so as to diagonalize the coordinate representation of the potential-energy surface, and the coordinate transformation that achieves this diagonalization then describes the $3N-6$ independent normal modes of motion. The S-vectors, $S_{J\alpha,a}$, and the associated vibrational frequencies of the normal modes are proportional the square roots of the diagonal force constants. These computation steps for each enantiomer occur automatically in most quantum chemistry programs once the initial equilibrium geometry for a conformer is specified.

9.2.6 Vibrational Intensities

In addition to normal modes and vibrational frequencies, typically options to continue the calculation to the desired vibrational intensities, namely VA, VCD, Raman, and ROA, can be selected in which case the quantum chemistry program does not stop until all selected vibrational intensities have been calculated. In this part of the calculation, the analytical integrals representing the derivatives of the atomic polar (APT, R-APT), axial (AAT, R-AAT), and quadrupole (R-AQT) tensors defined in section 9.1 for VCD and ROA are calculated, assembled into Cartesian invariants, and these invariants are appropriately summed over all atomic displacements in each normal mode to obtain the desired vibrational intensities for each conformer.

The output of a VOA intensity calculation is a list of numbers that includes the normal mode number, associated normal frequency, and normal mode intensity for each vibrational spectrum calculated, such as VA, VCD, Raman, and ROA. For a VA-VCD calculation, one obtains the vibrational frequency, $\bar{\nu}_a$, in wavenumber (cm^{-1}), the dipole strength, $D^a_{g1,g0}$ in esu^2 cm^2, and the rotational strength, $R^a_{g1,g0}$, also in esu^2 cm^2, of each of the $3N-6$ normal modes, a, of the molecule. For Raman and ROA calculations, the basic outputs in the FFR approximation are the vibrational frequencies, $\bar{\nu}_a$, the isotropic and anisotropic Raman invariants, $(\alpha^2)^a_{g1,g0}$ and $\left(\beta(\alpha)^2\right)^a_{g1,g0}$, and the three ROA invariants, the isotropic magnetic-dipole ROA invariant,$(\alpha G')^a_{g1,g0}$, and the anisotropic magnetic-dipole and electric-quadrupole ROA invariants, $\left(\beta(G')^2\right)^a_{g1,g0}$ and $\left(\beta(A)^2\right)^a_{g1,g0}$. From these five invariants, the ROA and Raman intensities for any desired form of ROA and scattering angle can be calculated.

9.2.7 Bandshape Presentation of Spectra

While such lists of intensity numbers represent the raw data of the VOA calculation, it is visually convenient to multiply the spectral intensity of each normal mode, a, by a lineshape function, usually a normalized Lorentzian lineshape of the form

$$f'_a(\bar{\nu}) = \left[\frac{\gamma_a/\pi}{(\bar{\nu}_a-\bar{\nu})^2+\gamma_a^2}\right] \tag{9.53}$$

This is a symmetric lineshape centered at the normal mode frequency with half-width at half-maximum of γ_a illustrated in Figure 3.1. As described in Chapters 1 and 3, the VA and VCD spectra as a function of radiation frequency can then be constructed by summing these products of intensity and lineshape over all the normal modes in the molecule to give

$$\Delta\varepsilon(\bar{\nu}) = \frac{\bar{\nu}}{2.236 \times 10^{-39}} \sum_a R^a_{g1,g0} f'_a(\bar{\nu}) \tag{9.54}$$

$$\varepsilon(\bar{\nu}) = \frac{\bar{\nu}}{9.184 \times 10^{-39}} \sum_a D^a_{g1,g0} f'_a(\bar{\nu}) \tag{9.55}$$

These equations were given previously in a more general form for any set of transitions summed over a in a molecule in Equations (3.51) and (3.52).

For ROA and Raman calculations, the process of spectral simulation is a bit more complicated because there is typically more than one invariant contributing to the ROA and Raman intensity for each normal mode. The final ROA and Raman spectra can be simulated by multiplying the calculated ROA and Raman intensity for each normal mode by a lineshape function, as illustrated above for VCD and VA intensities, and summing the resulting individual bandshapes centered at the vibrational frequency for each normal mode over all the modes in the molecule.

9.2.8 Weighting Spectra of Conformers

Vibrational spectra of molecules in the gas, liquid, or solution phase in more than one equilibrium conformation consist of population-weighted linear superpositions of the vibrational spectra of each conformer. The population-weighting factors result from the Boltzmann distribution of the conformers present in equilibrium under the conditions of the spectral measurement. The spectra are linear superpositions because the conformers typically interconvert on the picosecond timescale, which is slower than the femtosecond timescale of the vibrational oscillations of the molecule, and hence each conformer present is seen oscillating many times over its lifetime before interconverting by equilibrium distribution to another conformer. This is in contrast to NMR spectroscopy, which measures spectra on the microsecond timescale. In NMR spectroscopy, only the ensemble average of the conformer population is measured, which gives no information regarding the conformational distribution in solution. The expressions for the Boltzmann-weighted linear superpositions of, for example, VCD and VA conformer spectra, $\Delta\varepsilon_n(\bar{\nu})$ and $\varepsilon_n(\bar{\nu})$, are given by:

$$\Delta\varepsilon(\bar{\nu}) = \sum_n \Delta\varepsilon_n(\bar{\nu}) \mathrm{e}^{-\Delta G_n/RT} \Big/ \sum_n \mathrm{e}^{-\Delta G_n/RT} \tag{9.56}$$

$$\varepsilon(\bar{\nu}) = \sum_n \varepsilon_n(\bar{\nu}) \mathrm{e}^{-\Delta G_n/RT} \Big/ \sum_n \mathrm{e}^{-\Delta G_n/RT} \tag{9.57}$$

where the fractional population p_n of each conformer n at the absolute temperature T is given by:

$$p_n = \mathrm{e}^{-\Delta G_n/RT} \Big/ \sum_n \mathrm{e}^{-\Delta G_n/RT} \tag{9.58}$$

and $\Delta G_n = \Delta H_n - T\Delta S_n$ is the free-energy difference of each conformer from that of the lowest free-energy conformer. For small- to medium-sized molecules the entropy term, $T\Delta S_n$, is small compared with the enthalpy term, ΔH_n, and so enthalpic energies may be used instead of free energies. Both energies are available for each conformer in most quantum chemistry software packages.

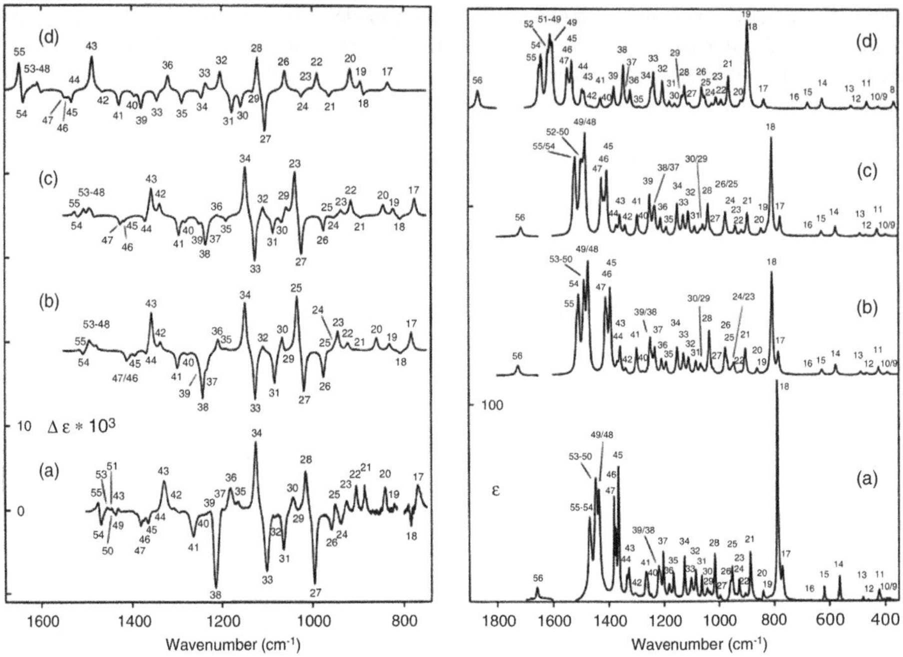

Figure 9.1 *Comparison of measured (a) and calculated (b–d) VCD (left panel) and VA (right panel) for (+)-α-pinene neat liquid in units of molar absorptivity. Calculated normal mode numbers starting from the lowest frequency modes are used for numbering. Calculated spectra used a TZ2P basis set and Hartree–Fock self-consistent field theory (d), and density function theory with B3LYP (c) and B3PW91 (b) functionals. Reproduced with permission from the American Chemical Society (Devlin et al., 1997)*

9.2.9 Comparison of Calculated and Experimental Spectra

Once Boltzmann-weighted vibrational spectra have been calculated and simulated, they can be compared directly with the corresponding measured vibrational spectra. Two methods of spectral analysis can be undertaken. VCD and VA spectra will be featured in the descriptions, but the same analysis applies to ROA and Raman spectra. The first method involves extracting rotational and dipole strengths from the measured VCD and VA spectra, while the second involves comparing the degrees of similarity of the measured and calculated VCD and VA spectra. Each has its advantages and disadvantages as we will see on describing them further.

The most direct, and perhaps theoretically sound, way to compare calculated and measured VCD and VA spectra is accomplished first by correlating each band, to the extent possible, in the measured spectrum with the corresponding band in the calculated spectrum. This method is best carried out for smaller molecules with limited or zero conformational freedom. The basic procedure here is as follows. (i) Plot the measured and calculated VCD and VA spectra in units of molar absorptivity as $\Delta\varepsilon(\bar{\nu})$ and $\varepsilon(\bar{\nu})$, respectively, so that they can be compared on the same intensity scale. (ii) Curve-fit the measured VA spectra to Lorentzian profiles and determine the integrated area, and from that the dipole strength, of each band measured in the VA spectrum. (iii) Correlate each band in the measured

VA spectrum with a corresponding assigned band in the calculated VA spectrum using the frequency and intensity patterns of the two spectra as a guide to their band-by-band correlation. One common method is to use the numbering scheme of the quantum chemistry calculation of the normal modes to label correlated bands in the measured and calculated VA spectra. This scheme is illustrated in Figure 9.1 for (+)-α-pinene neat liquid (Devlin *et al.*, 1997). (iv) Given the set of band frequencies, intensities and bandwidths, the measured VCD spectrum is then curve-fit with the same peak positions and bandwidths while allowing the best fit of the VCD sign and intensity. (v) From the area under each VCD band, determined by the Lorentzian peak height and bandwidth, the measured rotational strength of each vibrational mode is obtained. The reason why the VA spectrum is fitted first, followed by the VCD spectrum with fixed positions and bandwidths, is that VCD features can be obscured in some cases by overlapping bands of opposite sign, which in fact cancels VCD intensity. As a result, VCD peak positions and intensities can be difficult if not impossible to determine without guidance from the corresponding measured VA spectrum where no such cancellation can occur.

Given correlated sets of measured and calculated dipole and rotational strengths, two-dimensional plots of measured versus calculated intensities can be constructed where a perfect fit corresponds to all points lying on one of the diagonal axes of these plots, as illustrated for VCD in Figure 9.2 (Devlin *et al.*, 1997). Statistical analysis, such as r-squared values, can be applied to these plots as a quantitative measure of the degree of agreement between the calculated and measured sets of dipole and rotational strengths. Clearly, this method of spectral analysis is only effective when individual bands from individual vibrational transitions can be discerned in the spectrum, and hence its restriction to simpler chiral molecules. Other drawbacks of this method of analysis are the time-required to carry out careful curve-fitting analyses and the element of human judgment required to correlated bands in the measured spectrum to bands in the calculated spectrum. For example two closely-spaced or partially overlapping modes might interchanged in frequency in the calculated spectrum making the correlation of bands difficult or subject to error.

The other method of spectral analysis is to simply compare the calculated and measured VA and VCD spectra as spectral shapes by an integral overlap procedure that yields a number between 0 and 1 for the degree of similarity of the two spectra being compared, where the value 1 only occurs if the two spectra are identical to within a single scale factor (Debie *et al.*, 2011). This method is particularly useful for more complex, flexible molecules where there is insufficient spectral resolution or too many transitions to be able to assign all bands in the calculated spectrum to corresponding bands in the measured spectrum. The basic idea of spectral similarity is to construct the ratio of the integral overlap of the observed spectrum $g(\nu)$ and calculated spectrum $f(\nu)$ divided by the square root of the product of the integrals of the squares of these two spectra as illustrated below for the similarity, S_{fg}, of the vibrational spectrum:

$$S_{fg} = \frac{\int f(\nu)g(\nu)\mathrm{d}\,\nu}{\sqrt{\int f^2(\nu)\mathrm{d}\nu \int g^2(\nu)\mathrm{d}\,\nu}} \tag{9.59}$$

The denominator in this expression ensures that the range of S_{fg} lies between 0 and 1 where 0 is total dissimilarity and 1 is complete similarity that only occurs if $g(\nu)$ equals $f(\nu)$ to within a constant factor.

In practice, some pretreatment is carried out to account to some degree for frequency shifts between measured and calculated spectra. As it is well known that calculated spectra tend to have frequencies higher than corresponding measured spectra, a scale factor, σ, is applied to calculated spectra such that now $f(\nu) = f(\sigma\nu')$ where ν' is the value of the original calculated frequency. To take into account less regular frequency shifts, due for example to solvent effects or anharmonicity in the measured spectra, a neighborhood weighting factor is applied to the original measured spectra to take into account the

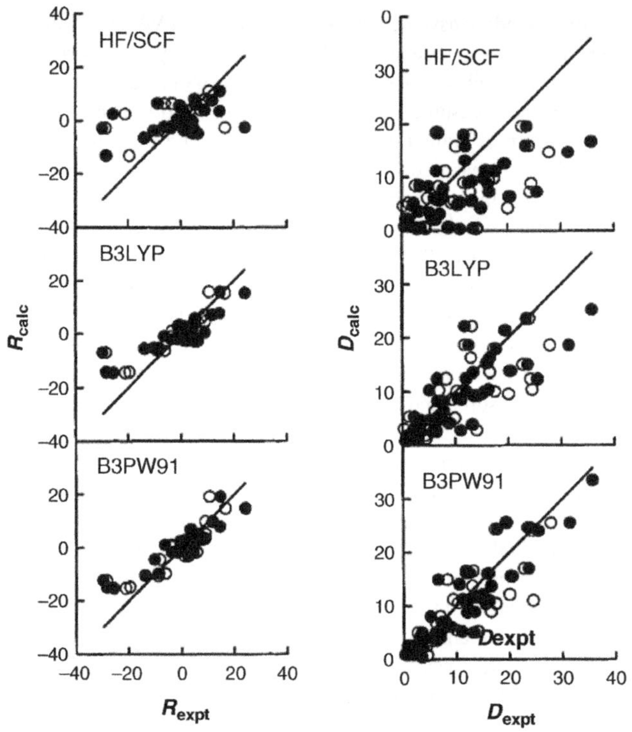

Figure 9.2 Plots of calculated (vertical scale) versus measured (horizontal scale) of rotational strength (left) and dipole strength (right) for (+)-α-pinene neat liquid in units of molar absorptivity for the spectra in Figure 9.1. Two sets of measured data are included as open circles (FT) and filled circles (dispersive) and three calculations using a TZ2P basis set for HF/SCF and DFT with functionals B3LYP and D3PW91. Adapted with permission from the American Chemical Society (Devlin et al., 1997)

spectral intensity in the neighborhood of a given measured frequency. This is accomplished by replacing the original measured spectra $g(\nu')$ by an integral weighted spectrum $g(\nu)$ according to the expression

$$g(\nu) = \int g(\nu')w(\nu' - \nu)\mathrm{d}\nu' \tag{9.60}$$

$$w(\nu' - \nu) = 1 - \frac{|\nu' - \nu|}{l} \quad |\nu' - \nu| \leq l \tag{9.61a}$$

$$w(\nu' - \nu) = 0 \quad |\nu' - \nu| > l \tag{9.61b}$$

Calculating the neighborhood similarity, S_{fg}, for VA or Raman spectra proceeds from this basis, but the calculation of the corresponding factor for VCD or ROA requires more care due to the presence of both positive and negative spectral intensity in the spectra. In particular the normalization integrals in the denominator do not yield the total VOA due to intensity cancellation. Instead, only frequency ranges over which the calculated and measured VOA spectra have the same sign, either positive as S_{fg}^{++} or negative as S_{fg}^{--} are included in the overlap and normalization integrals in Equation (9.59) as:

$$S_{fg}^{++} = \frac{\int\limits_{f,g>0} f(\nu)g(\nu)\mathrm{d}\nu}{\sqrt{\int\limits_{f,g>0} f^2(\nu)\mathrm{d}\nu \int\limits_{f,g>0} g^2(\nu)\mathrm{d}\nu}} \qquad S_{fg}^{--} = \frac{\int\limits_{f,g<0} f(\nu)g(\nu)\mathrm{d}\nu}{\sqrt{\int\limits_{f,g<0} f^2(\nu)\mathrm{d}\nu \int\limits_{f,g<0} g^2(\nu)\mathrm{d}\nu}} \tag{9.62}$$

From these common signed neighborhood similarities a single similarity index, Σ_{fg} can be constructed by weighting the integrated intensities of the positive and negative signed regions as:

$$\Sigma_{fg} = \frac{\Phi^{++} S_{fg}^{++} + \Phi^{--} S_{fg}^{--}}{\Phi^{++} + \Phi^{--}} \tag{9.63}$$

The weight factors Φ^{++} and Φ^{--} are defined as the sum of the measured and calculated VOA intensity of a given sign. For example, Φ^{++} is defined as:

$$\Phi^{++} = \int\limits_{f>0} f(\nu)\mathrm{d}\nu + \int\limits_{g>0} g(\nu)\mathrm{d}\nu \tag{9.64}$$

By taking into account for S_{fg}^{++} and S_{fg}^{--} only those frequency regions where the calculated and measured spectra have the same sign, the regions where there are opposite signs are omitted entirely from consideration. These regions can be reclaimed exclusively by considering calculating the same signed neighborhood similarities, $S_{\bar{f}g}^{++}$ and $S_{\bar{f}g}^{--}$, using the calculated spectrum of the opposite enantiomer, written as $\bar{f}(\nu)$ where the signs of all VOA intensities are reversed. Thus a total similarity measure for the opposite enantiomer can be calculated and designated as $\Sigma_{\bar{f}g}$. From Σ_{fg} and $\Sigma_{\bar{f}g}$ a final similarity measure can be defined called the enantiomeric similarity index, *ESI*, defined as:

$$\Delta = \left|\Sigma_{fg} - \Sigma_{\bar{f}g}\right| = |ESI| \tag{9.65}$$

The *ESI* provides a measure of discriminating power of the similarity measures between a chiral molecule and its enantiomer. It has been found that the *ESI* increases as the value of Σ_{fg} approaches unity. If the *ESI* is negative, then there is a larger similarity for the opposite enantiomer of the originally defined enantiomer. An example of a similarity calculation is shown in Figure 9.3.

Figure 9.3 shows the measured and calculated VCD spectra of (+)-3-methlycyclohexanone (left) and a plot of Δ versus $\Sigma_{fg}^{\max}$ (right) where the location of the values of these measures for (+)-3-methylcyclohexanone (triangle) against a database of previous similarity determinations where agreement (diamonds) and disagreement (squares) with independently determined correct assignments of absolute configuration are shown. $\Sigma_{fg}^{\max}$ is the maximum value of Σ_{fg} with respect to adjustable parameters in the similarity algorithm, such as the scale factor used for the calculated frequencies. From the plot in Figure 9.3, the values of Δ and $\Sigma_{fg}^{\max}$ for (+)-3-methylcyclohexanone are approximately 64 and 68, while the value of $\Sigma_{\bar{f}g}^{\max}$ can be deduced to be only 4. These similarity values represent a high level of discrimination between the two enantiomers of this molecule.

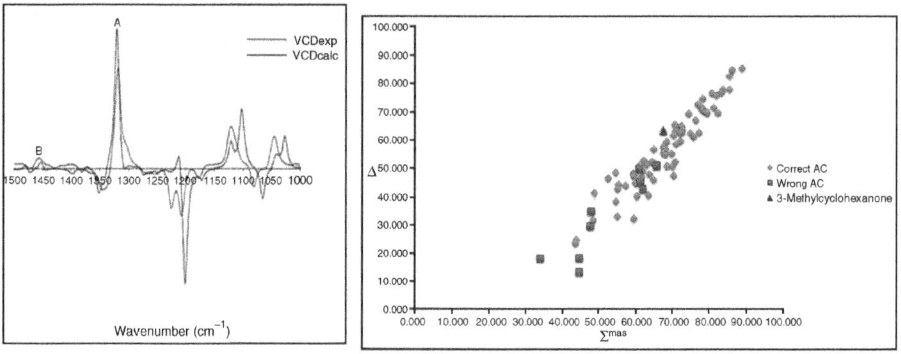

Figure 9.3 *Measured (A) and calculated (B) VCD spectra of (+)-3-methylcyclohexanone (left) and a plot of* Δ *versus* $\Sigma_{fg}^{\max}$ *(right) showing the location of these measures for (+)-3-methylcyclohexanone, indicated by a triangle; diamonds correspond to agreement between the similarity measures and previously assigned absolute configures while squares are instances of disagreement. Reproduced with permission from John Wiley & Sons (Debie et al., 2011)*

The primary use of the similarity measures Δ and Σ_{fg} is to provide an automated user-independent measure of the degree of agreement between calculated and measured spectra. In this way, a statistical measure is available for the quality of agreement between observed and calculated VCD spectra compared with all previous assignments in a database or previous correct determinations of absolute configuration (AC). The database, including points of disagreement between the AC assignment and the sign of the *ESI*, provides the basis for a confidence level (CL) measure in % derived from distances between a given point in the database relative to all other correct and incorrect points. This can be accomplished for a point *a* by summing over all the correct points in the database weight by a Gaussian exponential of their distance divided by the corresponding weighted sum over all points, correct and incorrect, in the database,

$$CL(a) = \frac{\sum_{i}^{N} \mathrm{e}^{-d_{ia}^2} \delta(correct)}{\sum_{i}^{N} \mathrm{e}^{-d_{ia}^2}} \tag{9.66}$$

where, d_{ia} is the Euclidean distance in $\Delta - \Sigma_{fg}$ space between point a and any other point i in the database, and $\delta(correct)$ is unity for a correct point and zero for an incorrect point. For the example given above for (+)-3-methylcyclohexanone, the CL value is approximately 98%.

9.3 Methods and Visualization of VOA Calculations

Although ROA and VCD were discovered experimentally at approximately the same time (Barron *et al.*, 1973; Holzwarth *et al.*, 1974), their theoretical and computational development took place on different timescales. The first rigorous quantum mechanical theory of ROA appeared before its experimental discovery (Barron and Buckingham, 1971) whereas the corresponding level of VCD did not appear until well after VCD measurement became established (Nafie and Freedman, 1983; Nafie, 1983; Galwas, 1983). On the other hand, the first quantum mechanical calculations of VCD appeared shortly thereafter (Stephens and Lowe, 1985; Stephens, 1985) followed five years later by the first such ROA calculations (Polavarapu, 1990). Starting with these first quantum calculations, VCD preceded ROA in its development and general availability for widespread use. The first fully-dedicated VOA instrument became commercially available from BioTools in 1997 for VCD and in 2002 for ROA. Commercially available user-friendly software for the calculation of VOA first became available from Gaussian in 1994 for VCD and 2003 for ROA. Since then both VCD and ROA have been commercially available for applications that combine the power discussed in this chapter of bringing together observed and calculated VOA for detailed analysis at the level of full quantum mechanical theory.

The main reason why VCD preceded ROA for widespread application is the relatively greater simplicity of VCD compared with ROA, both for instrumentation for spectral measurement and software for spectral calculation. As we have seen, VCD requires the calculation of only *one* second-order tensor invariant, the atomic axial tensor (AAT), whereas ROA requires, at a minimum, *three* third-order tensor invariants, the Raman atomic polar, axial, and quadrupole tensors (R-APT, R-AAT, and R-AQT). Nevertheless, current software for VOA has now developed to a point where the incremental computational cost (time of calculation) of continuing from a completed VA/VCD calculation to the corresponding Raman/ROA spectra is comparable to the computational cost of the initial VCD calculation, including the cost required to find the lowest energy conformers and calculate their vibrational force fields and frequencies. We now discuss separately the currently recommended methods for carrying out VCD and ROA calculations and their related visualizations in terms of localized, atomic-level electronic properties.

9.3.1 Recommended Methods for VCD Calculations

Regardless of the quantum mechanical approach chosen for VCD calculations, the minimum recommended basis set for either HF/SCF or DFT is the so-called double-zeta 6-31G(d) basis set, also referred to as the 6-31G* basis set. This basis set contains polarized basis functions, indicated by the d or *. Higher-level basis sets can be used, but for most applications 6-31G(d) has been shown to provide a reasonably reliable VCD calculation. Basis sets such as 3-21G or worse still, STO-3G, provide unreliable VCD spectra of poor quality compared with experiment. Basis sets augmented with diffuse basis functions do not provide any special advantage for VCD calculations, but as we shall the same is not true for ROA calculations. If a higher level of accuracy is desired, then triple-zeta basis sets, such as TZVP or cc-aug-pVTZ are recommended. Of course additional computation cost is required for such calculations.

Beyond the choice of basis set is the choice of functional if a DFT calculation is being performed. Although this point has been discussed previously, we mention here that the most accurate functionals for vibrational intensity calculations are the so-called hybrid density functionals, such as B3LYP and B3PW91. Other functionals besides these can be used, and new functionals continue to be developed, but there is currently no systematic way forward to improve the accuracy of functionals when comparing calculated with observed vibrational spectra, or any other molecular property.

In Figure 9.4 observed and calculated VA, VCD, and VCB spectra of ($-$)-α-pinene are compared for three different DFT calculations, 6-31G(d)/B3LYP, TZVP/B3PW91, and cc-pVTZ/B3PW91. The two DFT calculations with triple-zeta basis sets are seen to agree more closely with the observed

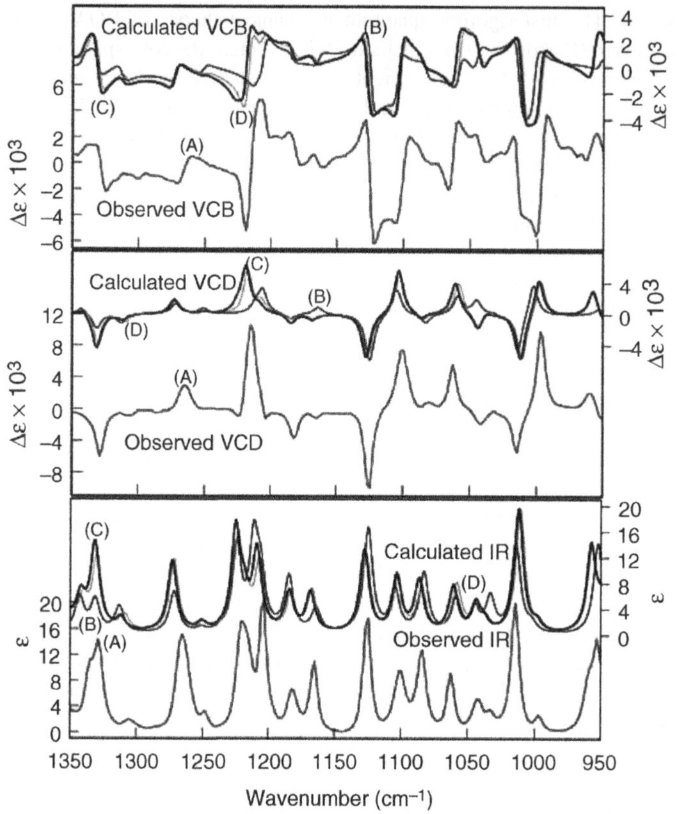

Figure 9.4 *Comparison of measured neat liquid (A) and calculated VCB (upper), VCD (middle), and VA (IR, lower) of (–)-S-α-pinene. The three sets of calculated spectra are DFT 6-31G(d)/B3LYP (B), cc-pVTZ/B3PW91 (C), and TZVP/B3PW91 (D). Reproduced with permission from John Wiley & Sons (Lombardi and Nafie, 2009)*

spectra in all cases, but overall, all three DFT calculations provide reasonably close matches between observed and calculated spectra for VA, VCD, and VCB.

9.3.2 Recommended Methods for ROA Calculations

Raman and ROA spectra arise from a very different type of molecular property than does VA and VCD spectra even though they share the same initial and final states, namely transitions between vibrational states in a fixed electronic state. This difference is the difference between an *intrinsic* electromagnetic moment and an *induced* electromagnetic moment. Intrinsic moments are related to the charge and current density distributions in the molecule in the absence of a perturbing electromagnetic wave, whereas induced moments are generally much weaker and only come into existence in response to the presence of an external electromagnetic wave through polarizabilities (electric-dipole, magnetic-dipole, electric-quadrupole, etc.) of the molecule. The polarizability of a molecule is a property that is more sensitive to the loosely bound electron density than a molecular moment is, which is more of a static property. Loosely held electron density often is located on the surface of a molecule or locations further away from the nuclear centers. Here diffuse basis functions are needed to describe accurately these more loosely bound (diffuse) electron densities in molecules.

It has been found that Raman scattering, and ROA in particular, requires the addition of diffuse basis functions to obtain accurate calculated spectra. In fact, with one exception, basis functions used for ROA calculation must be larger, and hence more expensive than those used for VCD calculations. The one exception is a so-called reduced basis set developed by Hug in which additional diffuse basis functions are added to the hydrogens, and the valence part of the calculation is at the 3-21G level with no additional polarized basis functions in the basis set. This remarkably small, carefully-constructed, basis set, called rDPS for reduced diffuse polarization shell, has been shown to perform as well as the very expensive diffuse augmented triple-zeta basis set aug-cc-pVTZ for the required Raman and ROA tensors. Regardless of what basis set is used, there is a need for diffuse functions in the basis set to accurately describe ROA spectra.

One approach that helps to alleviate the computational burden of the requirement to use diffuse basis functions in ROA calculations is to use one level of basis set for geometry optimization and force-field determination, and to use a second basis set for the polarizability derivatives needed for Raman and ROA intensities. Specification of the model chemistry of such ROA calculations involves listing the functional and basis set for the Raman and ROA polarizability tensors followed by first the functional and the basis set for calculation of the geometry and force field. For example, the following is such a specification: HSEh1PBE/rDPS//B3LYP/6-31G*. A single diagonal slash separates functionals and a double slash separates the two model chemistries. The recommended list of functionals and force fields is the same as that recommended for VA/VCD calculations, namely B3LYP and B3PW91 for the functionals and, in order of increasing sophistication and cost, are the basis sets 6-31G*, DGDZVP, TZVP, 6-311G**, and cc-pVTZ. For Raman/ROA calculations, the functionals are the same, except for the higher addition of HSEh1PBE, while the recommended basis sets in order of increasing cost are rDPS, aug(sp)-cc-pVDZ, aug-cc-pVDZ, 6-311++G**, and aug-cc-pVTZ.

Another important aspect of Raman and ROA calculations is the need to use a time-dependent computational formalism, such as time-dependent Hartree–Fock (TDHF) or time-dependent DFT (TDDFT), in order to capture the incident laser frequency dependence and avoid the so-called static limit approximation discussed above. These only need to be applied to the calculations of the various polarizability derivatives and hence can be restricted to the model chemistry used for those calculations.

An example of a mixed model chemistry is given in Figure 9.5 where the Raman and ROA spectra were calculated using TDHF/rDPS for the calculation of the Raman and ROA tensor derivatives, while the geometry and force field were calculated with DFT/B3LYP/6-311++G** (Haesler and Hug, 2008). Clearly all the principal features of the measured Raman and ROA spectra are well represented by the calculated spectra, demonstrating the ability of these model chemistries to simulate closely the measured spectra. The quality of agreement between measured and calculated ROA spectra is certainly more than sufficient to establish the absolute configuration of this molecule.

9.3.3 Visualization of VCD and VA Spectra

The visualization of normal modes of vibration are almost exclusively restricted to illustrating the nuclear motion, usually with arrows of the appropriate length and direction, attached to each nucleus

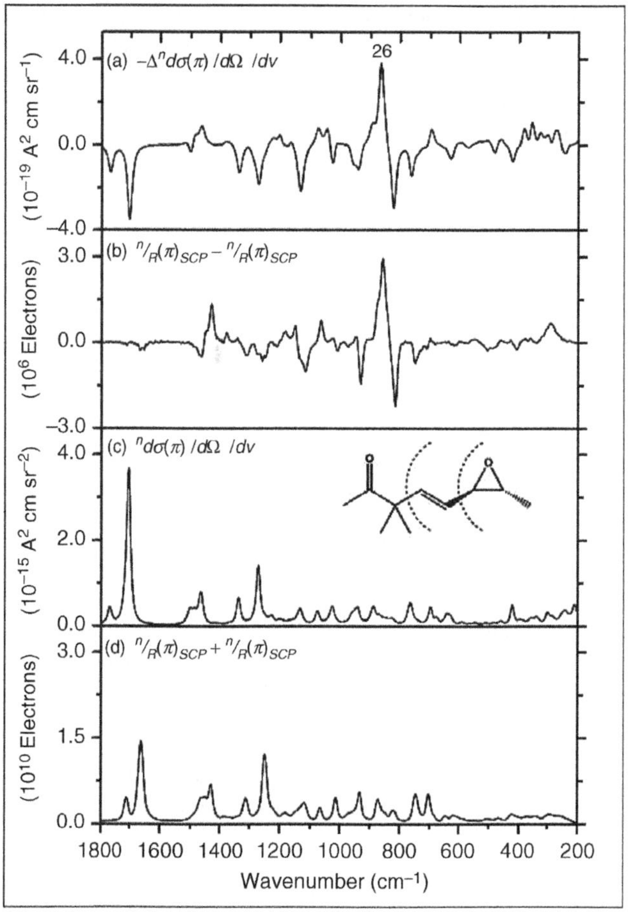

*Figure 9.5 Comparison of measured and calculated unpolarized backscattering SCP-ROA and Raman of the junionone precursor 1 illustrated in panel (c). The exposure time for the measurement was 25 min at a power level at the sample of 500 mW and a sample volume of 35 μL. The model-chemistry used for the calculations is TDHF/rDPS for Raman and ROA tensor derivatives and DFT/B3LYP/6-311++G** for the geometry and force field. Reproduced with permission from the Schweitzerische Chemische Gesellschaft (Haesler and Hug, 2008)*

for each normal mode. Illustrating the accompanying electronic motion has only been published a few times. The problem here is that the accompanying electronic motion needs to be illustrated by a vector field, the origin of which lies beyond the Born–Oppenheimer approximation. The problem was first solved theoretically with the derivation of quantum mechanical expressions of the vibrational electron current density that describes the correlation of electron density to nuclear velocities (Nafie, 1983). The basic formalism of vibrational electron transition current density has been presented earlier in Chapters 2 and 4 (Nafie, 1997; Freedman *et al.*, 1997), but we restate some of the basic ideas for comparison with visualization methods that have been developed primarily for ROA in recent years (Hug, 2001; Hug, 2002).

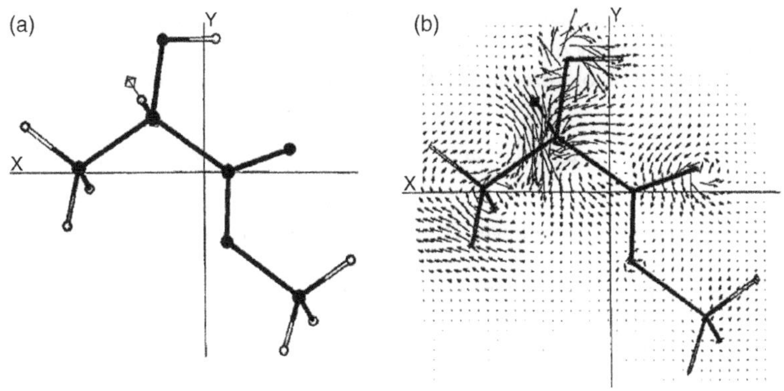

Figure 9.6 *Transition current density (TCD) map for the C* H-stretching vibrational mode of S-methyl-d_3-lactate Cd_3. The structure of the molecule, $CD_3C^*H(OH)(COOCD_3)$ with C and O as filled circles and H and D as open circles, is depicted on the left (a). The only significant atomic displacement is H attached to the chiral center C* of the molecule, the direction and magnitude of which are indicated by a small arrow. The map of the vibrational electron current density is depicted on the right (b) as explained in the text. Reproduced with permission from the American Chemical Society (Freedman et al., 2000)*

Vibrational TCD maps are very detailed showing how electron current density is generated as the result of non-zero nuclear velocities. An example of a vibrational electron transition current density map is presented in Figure 9.6 for the C*H stretching mode of *S*-methyl-d_3-lactate Cd_3 (Freedman *et al.*, 2000). It can be seen that although the nuclear displacement (or velocity) vector for the C*H stretching mode is highly localized, its vibrational motion has a widespread influence on electronic current density well beyond the C*H bond. The perspective in Figure 9.6 is parallel to the axis of the electric dipole transition moment. The current density visualized is that of the entire molecule projected into a plane perpendicular to the electric-dipole transition moment. It is the circulation of current density in this plane that gives rise to the component of the magnetic-dipole transition moment that is parallel or anti-parallel to the electric-dipole transition moment. As a result one is viewing in Figure 9.6 exclusively the vibrationally induced currents that are responsible for the sign and magnitude of the calculated VCD intensity. What is particularly interesting are the strong circulations of electron current density around the two oxygen atomic centers as well as the closer of the two deuterated methyl groups to the C*H stretching motion. These plots were generated by software in the commercial program from Advanced Visual Systems (AVS) that displays electron current density as vector field maps.

We conclude this section by recalling the description of VA and VCD intensity maps in Chapter 4. In Section 4.4.3 we described extension of TCD maps to magnetic-dipole transition current density maps, dipole strength density maps to visualize the spatial location of VA intensity, and rotational strength density maps to view the local contributions, regions of positive and negative value, to VCD intensity. A vibrational magnetic-dipole density map is a simple extension of the vibrational TCD obtained by introducing a moment arm to the vibrational TCD, where the latter, as we have demonstrated previously in Chapters 2 and 4 and Appendix B, is a velocity electric dipole current density map. To obtain three dimensional vibrational intensity maps, one must use the velocity dipole form of the VA and VCD intensities. By withholding a last integration over all space from the dipole and rotational strength expressions, origin-independent maps of VCD and VA intensities are obtained as first described in Equations (4.165) and (4.166). The dipole-strength density, $D^a_{v,g1,g0}(\boldsymbol{r})$, can be obtained by not carrying out the final integration over the velocity dipole transition moment, which results in the total dipole strength, or VA intensity of the transition $g0$ to $g1$ in mode a,

$$D^a_{v,g1,g0} = \omega_a^{-2}\,\mathrm{Re}\left[\dot{\boldsymbol{\mu}}^a_{g0,g1}\cdot\dot{\boldsymbol{\mu}}^a_{g1,g0}\right] = \omega_a^{-2}\int \mathrm{Re}\left[\dot{\boldsymbol{\mu}}^a_{g0,g1}\cdot\dot{\boldsymbol{\mu}}^a_{g1,g0}(\boldsymbol{r})\right]\mathrm{d}\boldsymbol{r} = \int D^a_{v,g1,g0}(\boldsymbol{r})\mathrm{d}\boldsymbol{r} \qquad (9.67)$$

where $\dot{\boldsymbol{\mu}}^a_{g1,g0}(\boldsymbol{r})$ is the sum vibrational TCD plus the nuclear contribution to the transition moment.

The corresponding expression for the rotational-strength density, $R^a_{v,g1,g0}(\boldsymbol{r})$, is given analogously by:

$$\begin{aligned} R^a_{v,g1,g0} &= \omega_a^{-1}\,\mathrm{Re}\left[\dot{\boldsymbol{\mu}}^a_{g0,g1}\cdot\boldsymbol{m}^a_{g1,g0}\right] = \omega_a^{-1}\int \mathrm{Re}\left[\dot{\boldsymbol{\mu}}^a_{g0,g1}(\boldsymbol{r})\cdot\boldsymbol{m}^a_{g1,g0}\right]\mathrm{d}\boldsymbol{r} \\ &= \omega_a^{-1}\int \mathrm{Re}\left[\dot{\boldsymbol{\mu}}^a_{g0,g1}\cdot\boldsymbol{m}^a_{g1,g0}(\boldsymbol{r})\right]\mathrm{d}\boldsymbol{r} = \int R^a_{v,g1,g0}(\boldsymbol{r})\mathrm{d}\boldsymbol{r} \qquad (9.68)\end{aligned}$$

Here the last spatial integration leading to the rotational strength, or VCD intensity, of the transition is withheld either from the velocity electric-dipole transition moment or the magnetic-dipole transition moment. Such maps have not yet been produced, but have the potential to reveal the origin for VA and VCD intensity on whatever scale of local resolution is desired.

9.3.4 Visualization of ROA and Raman Spectra

Within the past decade, a new option for the visualization VOA spectra has been developed and become available for widespread use (Hug, 2001; Hug, 2002). This program was initially used for the visualization of only Raman and ROA spectra; however a recent version of the program called *py Vib* has been published and is freely available for downloading (Zerara, 2008) and can be applied to not only Raman and ROA spectra but also to VA and VCD spectra. There are two basic functions of the program. One is to depict the vibrational motion of the molecule in a three-dimensional view, and the other is to give a local representation of the origin of ROA intensity in the molecule by the use of atomic contribution patterns (ACPs) and group coupling matrices (GCMs). Spheres depicting vibrational motion and the GCM are illustrated in Figure 9.7 for one vibrational mode of the same molecule for which the SCP-Raman and -ROA spectra are presented in Figure 9.5, namely junionone precursor 1 (Haesler and Hug, 2008). The group regions of this molecule, labelled *A*, *B*, and *C*, are illustrated by curved dashed lines in the structure inset of Figure 9.5(c).

The GCMs organize the calculated ROA intensity into regional contributions. The area of the circles is proportional to the scalar magnitude of the ROA contribution while the grey (dark grey/positive and light grey/negative) designates the sign of the contribution. Circles on the diagonal of the GMC matrix correspond to single center contributions while the off-diagonal contributions correspond to intensity generated between the two different groups. The large positive ROA in the band labeled 26 in Figure 9.5 can be seen to arise from the large dominating positive two-group contribution associated with the local regions *A* and *C* in the GCM.

As mentioned above, the program *py Vib* can be applied to either VA/VCD or Raman/ROA calculations and has been programmed to work easily as an interface to the Gaussian program checkpoint file as illustrated in Figure 9.8 (Zerara, 2008). The checkpoint file (*fchk*) provides information about the geometry, vibrational force field, and normal mode nuclear displacement vectors, APT, AAT, R-APT, R-AAT, and R-AQT tensors for calculating VA, VCD, Raman, and ROA spectra followed by the 3D representations of the normal modes, ACPs, and GCMs as illustrated in Figure 9.7.

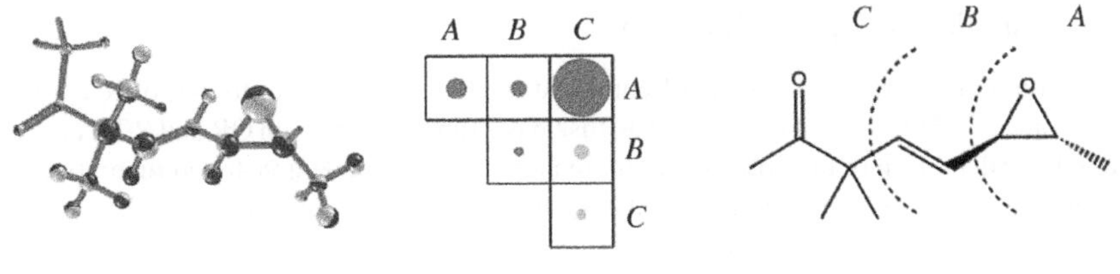

Figure 9.7 Visual representation of the vibrational motion and ROA intensity of mode 26 (the large positive ROA band near 875 cm^{-1} of junionone precursor 1 shown in Figure 9.5. The magnitude and shading of the atomic spheres in the structure of the molecule on the left represents the energy and direction of the nuclear displacements in this vibrational mode. The contributions to the calculated ROA intensity from more local regions of the molecule are given by the group coupling matrix (GCM) in the center where the regions A, B, and C are defined in the structure on the right by dashed lines. In the GCM, dark grey is positive ROA intensity and light grey is negative intensity. Off diagonal boxes represent intensity generated by the interaction of two different groups, whereas diagonal boxes are intensity contributions generated within a local group. Adapted with permission from the Schweitzerische Chemische Gesellschaft (Haesler and Hug, 2008)

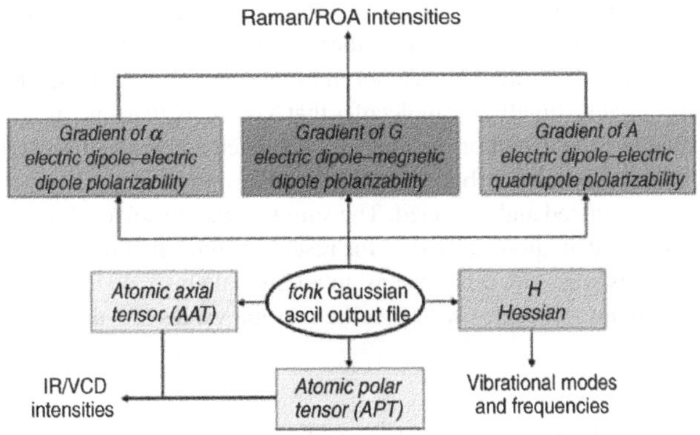

Figure 9.8 Flow diagram for the program py Vib that reads the output of the Gaussian fchk output file for molecular geometry, force field, vibrational modes, APT, and AAT tensors for calculating VA and VCD intensities and Raman R-APT, R-AAT, and R-AQT tensors for calculating Raman and ROA tensors. Reproduced with permission from John Wiley & Sons (Zerara, M. 2008)

9.4 Calculation of Electronic Optical Activity

The focus of this chapter is on the calculation of vibrational optical activity, but in this section we provide a few remarks concerning the calculation of electronic optical activity. Within the past decade, impressive strides have been made in the calculation of both optical rotation (OR) and electronic circular dichroism (ECD). In particular both OR and ECD can now be calculated routinely using commercially available software, and progress in the theory and calculation of EOA, and of VOA, has been reviewed recently (Autschbach, 2009; Polavarapu, 2007). We also briefly mention depolarized Rayleigh OA as a possible additional form of EOA that may prove useful in the future.

9.4.1 Calculation of Optical Rotation

The calculation of OR requires a level of computational analysis similar to that of ROA, namely linear response TD-DFT with at least added diffuse basis functions, as both OR and ROA are closely related to the calculation of the optical activity tensors. It has been demonstrated theoretically that OR is equivalent to Rayleigh optical activity scattering in the forward direction (Barron, 2004). Furthermore, it can be shown that the specific rotation, given in Equation (3.31), is proportional to the trace of the Rayleigh magnetic dipole optical activity tensor, $G'_{xx} + G'_{yy} + G'_{zz}$, in the far-from-resonance approximation (Polavarapu, 2007). As a result, the calculation of OR is similar to that of ROA but without the need for information about the ground electronic state vibrational transitions of the molecule.

One reason to pursue the calculation of OR is the widespread availability of polarimeters for the experimental measurement of OR. If accurate calculations of OR can be carried out, then OR becomes a facile method for the determination of absolute configuration. However, several difficulties currently prevent use of this method for the routine determination of absolute configuration. The first is the relatively large range of deviations of calculated versus measured OR values for representative sets of simple chiral molecules, currently specific rotation values on the order of 50° for the highest level of calculations investigated. Second, OR is a single number or series of single numbers if more than one wavelength is considered, and there is no independent spectral measure to inform whether the calculation is a good match to the measurement. In particular, no degree of confidence calculation is available. Related to this point is the fact that there is no parent measurement that can also be calculated, such as VA for a VCD calculation, as an additional measure of whether the calculation is of sufficient quality to be reliable. Finally, for molecules that have more than one significant conformer at room temperature, the OR of each conformer must be calculated and the resulting OR spectra must be combined using a conformational distribution, but there is no simple way to know if all the important conformers have been calculated and averaged. The simplest way to solve all of these problems is to first do a VCD or an ROA calculation, and to use the resulting conformational information to perform the OR calculation. In this way, OR can serve as an additional check on the reliability of the VCD calculation, but currently for molecules with specific rotation values less that 50° such a check is of somewhat limited value.

9.4.2 Calculation of Electronic Circular Dichroism

The calculation of ECD has been carried out for many years at various levels of quantum chemistry theory, starting from semi-empirical calculations and continuing through *ab initio* Hartree–Fock theory up to the current use of TDDFT or other types of electron correlation treatments, and in some cases with vibronic sublevels included in the calculation (Autschbach, 2009). The principal difference between calculations of ECD compared with those for VCD is the need for accurate descriptions of individual excited electronic states in the molecule. For the calculation of the sign and magnitude of each ECD band in the ECD spectrum, accurate descriptions of both the *ground* and the corresponding *excited* electronic state are needed. In general, excited electronic states require a higher level of quantum theory for their accurate description compared with the ground electronic state. By contrast, only the *ground* electronic state needs to be described for VCD intensities. Hence, the calculation of a VCD spectrum is simpler and less complex than the calculation of the corresponding ECD spectrum.

Furthermore, most ECD spectral calculations provide only the integrated ECD intensity for an ECD band without calculating or simulating the actual shape or appearance of the spectrum. This is because the shape of an ECD band is governed by the underlying excited-electronic-state vibrational substructure, also called vibronic structure. The vibrational motion of excited electronic states and the vibronic coupling between excited electronic states is much more difficult to calculate than the corresponding vibrational structure of the ground electronic state needed for a VCD calculation.

Yet another point of comparison is the number of transitions or bands in an ECD spectrum compared with the corresponding VCD spectrum. For VCD, all vibrational transitions of the molecule are chromophores and in general there are tens of transitions in a typical mid-IR VCD spectrum. By contrast, a molecule must possess electronic chromophores, typically π-electrons or lone-pair electrons, in order to possess an ECD spectrum in the accessible UV–visible region, and typically there is a much smaller number of bands in a typical ECD spectrum than there is in the corresponding VCD spectrum. In addition, ECD spectra only provide stereochemical information about those parts of the molecule that are near the electronic chromophore. This is in contrast to VCD, which contains stereochemical information from all parts of the molecule that move during the normal modes of vibration covered by the spectrum. The only parts of the molecule not represented in a typical VCD spectrum are those far from sources of structural chirality, or regions that conformationally average to zero net chirality, and typically these same structural regions would also be absent in an ECD.

Given that VCD is richer in spectral content and easier to calculate than the corresponding ECD spectrum, VCD will always maintain this computational advantage even as quantum computational methods continue to improve in the future. In other words, VCD will always be the more accurate and faster calculation to carry out. On the other hand, ECD has a richness of information about the excited electronic states of a molecule, and also of the underlying vibrational substructure of these states, which will continue to be of value in understanding the chiroptical properties of molecules. Furthermore, no one chiroptical method can provide all the desired structural information about a chiral molecule, and hence the more methods, experimental and computational, that are brought to bear on the stereochemical analysis of a molecule, the more information will be obtained and the more certain one will be regarding the accuracy and validity of this information.

9.4.3 Calculation of Rayleigh Optical Activity

We conclude this section and this chapter with a brief discussion of a new proposed measure of EOA, namely Rayleigh optical activity (RayOA). The theoretical foundation of RayOA was described at the same time that the theory of ROA was first presented (Barron and Buckingham, 1971). Recently, however, the measurement and calculation of RayOA has been proposed as a new chirality measure of a molecule comparable in many ways to OR (Zuber *et al.*, 2008). It was shown computationally, particularly for backscattering DCP_I-RayOA, that the RayOA from the depolarized Rayleigh wing of light scattering in molecules possesses relatively large and experimentally accessible RayOA intensities, and in addition, that the sign of the RayOA correlates to a much higher degree with the absolute configuration of the molecule than the corresponding calculated OR. By contrast, the central polarized Rayleigh line is far more intense and possesses lower relative RayOA intensity compared with the depolarized Rayleigh wing and associated depolarized RayOA. As described in Chapter 5 and elsewhere, DCP_I-Raman, and therefore DCP_I-Rayleigh, scattering is a pure depolarized form of scattering originating only from the symmetric anisotropic invariants with no contribution from the polarized isotropic invariants (parent or OA invariant). Thus, if RayOA is to be discovered experimentally, the best strategy is to use the backscattering DCP_I form of RayOA in the search. Once reliable methods for the measurement of RayOA have been established, it is likely that the measurement and calculation of RayOA will provide a more reliable correlation between the sign of a chiroptical effect and the absolute configuration of the molecule than is now provided by OR.

References

Autschbach, J. (2009) Computing chiroptical properties with first-principles theoretical methods: Background and illustrative examples. *Chirality*, **21**, E116–E152.

Barron, L.D. (2004) *Molecular Light Scattering and Optical Activity*, 2nd edn, Cambridge University Press, Cambridge.

Barron, L.D., and Buckingham, A.D. (1971) Rayleigh and Raman scattering from optically active molecules. *Mol. Phys.*, **20**, 1111–1119.

Barron, L.D., Bogaard, M.P., and Buckingham, A.D. (1973) Raman scattering of circularly polarized light by optically active molecules. *J. Am. Chem. Soc.*, **95**, 603–605.

Barron, L.D., and Buckingham, A.D. (1975) Rayleigh and Raman optical activity. *Annu. Rev. Phys. Chem.*, **26**, 381.

Debie, E., De Gussem, E., Dukor, R.K. *et al.* (2011) A confidence level algorithm for the determination of absolute configuration using vibrational circular dichroism or Raman optical activity. *PhysChemChemPhys*. DOI: 10.1002/cphc.201100050, in press.

Devlin, F.J., Stephens, P.J., Cheeseman, J.R., and Frisch, M.J. (1997) *Ab initio* prediction of vibrational absorption and circular dichroism spectra of chiral natural products using density functional theory: alpha-pinene. *J. Phys. Chem. A*, **101**, 9912–9924.

Freedman, T.B., Shih, M.-L., Lee, E., and Nafie, L.A. (1997) Electron transition current density in molecules. 3. *Ab initio* calculations of vibrational transitions in ethylene and formaldehyde. *J. Am. Chem. Soc.*, **119**, 10620–10626.

Freedman, T.B., Lee, E., and Nafie, L.A. (2000) Vibrational current density in (*S*)-methyl lactate: visualizing the origin of the methine-stretching vibrational circular dichroism intensity. *J. Phys. Chem. A*, **104**, 3944–3951.

Galwas, P.A. (1983). Ph.D. thesis. University of Cambridge.

Haesler, J., and Hug, W. (2008) Raman optical activity: A reliable optical technique. *Chimea*, **62**, 482–488.

Holzwarth, G., and Chabay, I. (1972) Optical activity of vibrational transitions. Coupled oscillator model. *J. Chem. Phys.*, **57**, 1632–1635.

Holzwarth, G., Hsu, E.C., Mosher, H.S. *et al.* (1974) Infrared circular dichroism of carbon-hydrogen and carbon-deuterium stretching modes. Observations. *J. Am. Chem. Soc.*, **96**, 251–252.

Hug, W. (2001) Visualizing Raman and Raman optical activity generation in polyatomic molecules. *Chem. Phys.*, **264**, 53–69.

Hug, W. (2002) Raman optical activity. In: *Handbook of Vibrational Spectroscopy* (eds J.M. Chalmers, and P.R. Griffiths), John Wiley & Sons, Ltd, Chichester, pp. 745–758.

Lombardi, R.A., and Nafie, L.A. (2009) Observation and calculation of a new form of vibrational optical activity: Vibrational circular birefringence. *Chirality*, **21**, E277–E286.

Nafie, L.A. (1983) Adiabatic behavior beyond the Born-Oppenheimer approximation. Complete adiabatic wavefunctions and vibrationally induced electronic current density. *J. Chem. Phys.*, **79**, 4950–4957.

Nafie, L.A. (1992) Velocity-gauge formalism in the theory of vibrational circular dichroism and infrared absorption. *J. Chem. Phys.*, **96**, 5687–5702.

Nafie, L.A. (1997) Electron transition current density in molecules. 1. Non-Born-Oppenheimer theory of vibronic and vibrational transition. *J. Phys. Chem. A*, **101**, 7826–7833.

Nafie, L.A., and Freedman, T.B. (1983) Vibronic coupling theory of infrared vibrational intensities. *J. Chem. Phys.*, **78**, 7108–7116.

Polavarapu, P.L. (1990) *Ab initio* Raman and Raman optical activity spectra. *J. Phys. Chem.*, **94**, 8106–8112.

Polavarapu, P.L. (2007) Renaissance in chiroptical spectroscopic methods for molecular structure determination. *Chem. Rec.*, **7**, 125–136.

Schellman, J.A. (1973) Vibrational optical activity. *J. Chem. Phys.*, **58**, 2882–2886.

Stephens, P.J. (1985) Theory of vibrational circular dichroism. *J. Phys. Chem.*, **89**, 748–752.

Stephens, P.J., and Lowe, M.A. (1985) Vibrational circular dichroism. *Annu. Rev. Phys. Chem.*, **36**, 213–241.

Zerara, M. (2008) py Vib, a computer program for the analysis of infrared and Raman optical activity. *J. Comput. Chem.* **29**, 306–311.

Zuber, G., Wipf, P., and Beratan, D.N. (2008) Exploring the optical activity tensor by anisotropic Rayleigh optical activity scattering. *ChemPhysChem*, **9**, 265–271.

10

Applications of Vibrational Optical Activity

With the completion of the basic principles of VOA in terms of theory, instrumentation, measurement, and calculation, we now turn our attention to applying these concepts to practical problems where valuable new stereochemical information can be obtained. The goal of this chapter is to provide an overview of the landscape of VOA applications. As such, this is not meant to be a review article, of which many have been written over the years since the discovery of ROA and VCD. Rather, it is a description of the various classes of molecules studied using VOA, and also coverage of the different types of applications of VOA. In most cases we are seeking to provide examples of timeless or historical value that hopefully will never go out of date even though they will doubtless eventually be surpassed in quality, speed, and degree of sophistication.

10.1 Classes of Chiral Molecules

The only restriction on the molecules studied by VOA is that they must be chiral, or at least associated with a chiral molecular structure or assembly. As result, the molecules studied by VCD and ROA have ranged in size from simple molecules with only five atoms (bromochlorofluoromethane) with a single chiral center up to viruses and supramolecular protein fibril structures. Between these extremes are all manner of molecules spanning virtually all classes of known molecular species. Here we describe some the major classifications while no doubt overlooking some more specialized examples that do not fall within our current gaze of possibilities. Before doing so, we mention that there does not appear to be any fundamental distinction between the types of molecules that can be studied by VCD versus ROA. When there are differences of convenience or sensitivity, we will mention them at those points.

10.1.1 Simple Organic Molecules

The very first molecules studied by ROA and VCD were simple chiral organic molecules. The VOA spectra of these 'discovery' molecules were shown in Chapter 1. Of the various types of small organic molecules, monoterpenes are particularly favorable for VOA measurements as they have low

Vibrational Optical Activity: Principles and Applications, First Edition. Laurence A. Nafie.

molecular weights and in most cases are fairly rigid without much conformation freedom. These molecules were the focus of some of the first full-length papers on ROA (Barron, 1977a; Barron, 1977b), and they featured prominently in the first major paper on VCD (Nafie *et al.*, 1976). One molecule in particular that was of specific importance in early papers of both VCD and ROA is α-pinene. This molecule finds widespread use in the optimization and calibration of VOA instruments because it has strong VOA intensities, and because it can be sampled directly as a neat liquid. VOA spectra of α-pinene are illustrated in several figures in Chapter 8 on the principles of VOA measurements.

10.1.2 Pharmaceutical Molecules

One of the important classes of chiral molecules is that of active ingredients in pharmaceuticals. There are two subclasses of pharmaceutical molecules: one is small molecules, usually organic, and the other is biopharmaceutical molecules, which are mostly genetically engineered proteins. A decade or so ago, most pharmaceutical companies focused on small-molecule drugs while a handful of startup companies worked to develop commercial means to sell protein pharmaceuticals for various therapeutic uses. Now the distinction between these two types of companies has become blurred as small-molecule pharmaceutical companies have acquired biopharmaceutical companies and vice versa. Now most large pharmaceutical companies have research on both small-molecule pharmaceuticals and protein biopharmaceuticals. VCD and ROA can both be used as structural tools to elucidate mainly the absolute configuration of the small-molecule active ingredients and the conformational states of the protein biopharmaceuticals. A recent review highlights the applications of VOA in the pharmaceutical industry (Nafie and Dukor, 2007).

10.1.3 Natural Product Molecules

Chiral molecules isolated from plants and marine organisms represent another very important class of chiral molecules for which VCD especially has found many applications. The application is primarily to determine the absolute configuration of these molecules, many of which have natural medicinal or therapeutic applications. Numerous natural product molecules contain multiple chiral centers. Here VOA is valuable in distinguishing between mirror-image pairs of diasteriomers. Usually VCD analysis starts from a known particular diasteriomer after NMR or other techniques have been used to establish the relative configuration among multiple chiral centers. The use of VCD for stereochemical analyses of natural products has been reviewed recently (Nafie, 2008).

10.1.4 Metal Complexes

Many metal complexes have been explored using VCD and to some extent ROA. The earliest studied metal complexes were rare-earth complexes of praseodymium and europium with camphorato ligands that were featured in the first full VCD paper mentioned above (Nafie *et al.*, 1976). On the other hand, transition metal complexes have been studied more broadly over the years. Metal complexes bring two new interesting features to VOA spectra both of which arise from low-lying excited electronic states that may be present (He *et al.*, 2001; Merten *et al.*, 2010). For example, rare-earth complexes possess *f–f* electronic transitions and transition metals have *d–d* transitions. Both types of transitions are strongly magnetic-dipole allowed and hence, when present, can induce enhanced VCD or ROA by the coupling of the electronic magnetic-dipole transition moments with the smaller vibrational magnetic-dipole moments responsible for normal VCD intensity. There are also many examples of transition metal complexes where the metal simply serves as a central structural element and the surrounding ligands are stabilized or restricted in conformation, which changes their VCD spectra from that of the free ligand. These changes can then be used to understand that nature of the metal ligand binding and the degree to which the metal donates electronic charge to the ligands.

A particularly interesting transition metal complex is ferric hemoglobin azide. The azide is bound to the iron center at the distal position of the heme group. Although the azide ligand (N=N=N) is a linear, non-chiral ligand, it somehow borrows magnetic dipole intensity from the ferric center to produce a VCD intensity that is enhanced by approximately two orders of magnitude over normal VCD intensity levels (Bormett *et al.*, 1992).

10.1.5 Oligomers and Polymers

The study of large biological molecules usually begins by measuring the VOA of their smaller subunits. For example, proteins are chains of amino acids linked by peptides bonds, the conformation of which has been studied by VCD (Freedman *et al.*, 1995; Schweitzer-Stenner *et al.*, 2007). The earliest VOA studies of biological molecules involved measuring the VCD and ROA of amino acids, di-amino-acids, and simple di- and tri-peptides. Similarly, carbohydrates can be studied by starting with VOA spectra of simple sugars and disaccharides. Nucleic acids are complex but still can be studied at the level of ribose nucleosides and nucleotides. Once the VOA of oligomeric structures have been obtained, studying their homo-polymeric structures, such as poly-L-proline, poly-L-lysine, and so on, for proteins, can be extended to the study of proteins, carbohydrates, and nucleic acids (Keiderling *et al.*, 2006; Barron, 2006a; Barron, 2006b). Most VOA studies of polymeric substances are related to corresponding hetero-polymeric biological structures, but a few examples of VOA spectra of stereo-regular polymeric structures have been published.

10.1.6 Biological Molecules

More detailed descriptions of the application of VOA to biological molecules will be provided later in this chapter, but suffice it to say here that virtually all classes of biological molecules can be investigated by VCD (Keiderling *et al.*, 2006), ROA (Barron, 2006a; Barron *et al.*, 2007b; Barron *et al.*, 2007a) and VOA (Nafie and Freedman, 2000). Most VOA studies have focused on proteins because they possess such a wide variety of structural motifs and have secondary structural elements that are common across all proteins and thus allow a new form of protein conformational analysis to be carried out. Besides proteins, nucleic acids, carbohydrates, glycoproteins, bacteria, and viruses have been studied. Membranes and membrane bound proteins have not received as much attention from VOA largely due to sampling issues. Viruses have proven to be amenable to study by ROA as a means of characterizing the viral coat proteins (Blanch *et al.*, 2002). Certain bacteria have filamentous protein tails that help propel them through biological media and the VOA spectra of these tails can be easily seen by both ROA and VCD as enhanced intensities in the of the *Salmonella* bacterium (Uchiyama *et al.*, 2008).

10.1.7 Supramolecular Chiral Assemblies

Supramolecular chirality in protein amyloid fibrils has been studied by both VCD and ROA. In the case of ROA changes of native proteins can be followed into the prefibrillar state and additional changes in insulin as fibrils begin to form. The change from the native and prefibrillar state to fibrils is surprisingly dramatic in the case of VCD. Here VCD intensities can grow by up to two orders of magnitude with only comparatively minor changes in the corresponding IR spectra (Ma *et al.*, 2007). Protein amyloid fibrils are an example of a supramolecular chirality where the chirality is manifested by a level of helical chirality well beyond the driving source of chirality at the level of the chiral centers, the amino acid residues in the protein primary structure (Ma *et al.*, 2007; Kurouski *et al.*, 2010). Other examples of supramolecular chirality have been observed in VOA, primarily in VCD, for example, the supramolecular tetramer of 2,2′-dimethyl-biphenyl-6,6′-dicarboxylic acid (Urbanova *et al.*, 2005), supramolecular guanosine quartet assemblies (Stetincka *et al.*, 2006) and the association of the vitamin molecule, biotin, with silver ions (Goncharova *et al.*, 2010).

The exact reason for the heightened sensitivity of VCD, compared with that of ROA, to supramolecular chirality may due to the possible interference of scattering from larger molecular structures in ROA that is not present in VCD due to the much longer wavelengths of the interrogating radiation. Another reason is the ability of VCD to couple over distances that are larger than the coupling distances typically seen for ROA. The shorter coupling distance of ROA is related to the weaker nature of the induced dipole moments needed for Raman and ROA compared with the intrinsic dipole moments present for VA and VOA intensities.

10.2 Determination of Absolute Configuration

At present, the most powerful and important application of VOA is the determination of the absolute configuration (AC) of small- to medium-sized molecules. Most applications of AC determination have been carried out using VCD (Stephens and Devlin, 2000; Freedman *et al.*, 2003) for reasons described in the previous chapter, but ROA has been used to demonstrate the power of VOA for AC determination in molecules for which AC determination by any other means is beyond current capabilities (Polavarapu, 2002; Haesler *et al.*, 2007; Barron, 2007).

10.2.1 Importance of Absolute Configuration Determination

With the growing number of new chiral pharmaceutical molecules, there is a corresponding need for the determination of the AC of these molecules and their precursors in order to prove to regulatory agencies the unambiguous absolute stereochemistry of these molecules. The potential of VOA to provide routine assignments of AC has now been recognized worldwide by the major pharmaceutical companies. Many of these companies possess their own VCD instrumentation and carry out VCD measurements and calculations to determine ACs of tens to hundreds of compounds per year. Other such companies outsource their VCD measurements and calculations to companies such as BioTools, Inc., or the newly formed European Centre for Chirality at the University of Antwerp and Ghent University in Belgium. Simply stated, VCD and ROA offer a straightforward method for the determination of AC that bypasses the need for growing single crystals as required by anomalous X-ray diffraction, the only other established *a priori* method for determining AC without structural modification of the molecule or using rules based on empirical correlations or simple models, such as the exciton coupling model (see Appendix A).

We show in Figure 10.1 one of the first published assignments of AC for a newly synthesized iminolactone with VCD for the molecule *R*-(+)-1,4-oxazin-2-one (Solladie-Cavallo *et al.*, 2001b). In this case, the comparison of measured and calculated VA and VCD spectra was the only proof of of AC. The molecule is rigid with only one significantly populated conformer at room temperature. Visual inspection reveals an unambiguous assignment of AC, as the VCD spectrum of the opposite enantiomer, either measured or calculated, would reverse the signs relative to zero of all the VCD bands thereby destroying by reversal the agreement between calculated and measured VCD spectra. When all the major bands in the measured and calculated VCD spectra agree, and when the corresponding VA spectra show a close match, then the assignment of the AC is secure.

A molecule of intense interest for which ROA has been used to determine definitively the absolute configuration is *R*-(−)-bromochlorofluoromethane, the simplest non-isotopically substituted chiral molecule (five atoms) commonly used to illustrate the concept of chirality in molecular structure. For more than two decades this molecule was studied in order to determine its

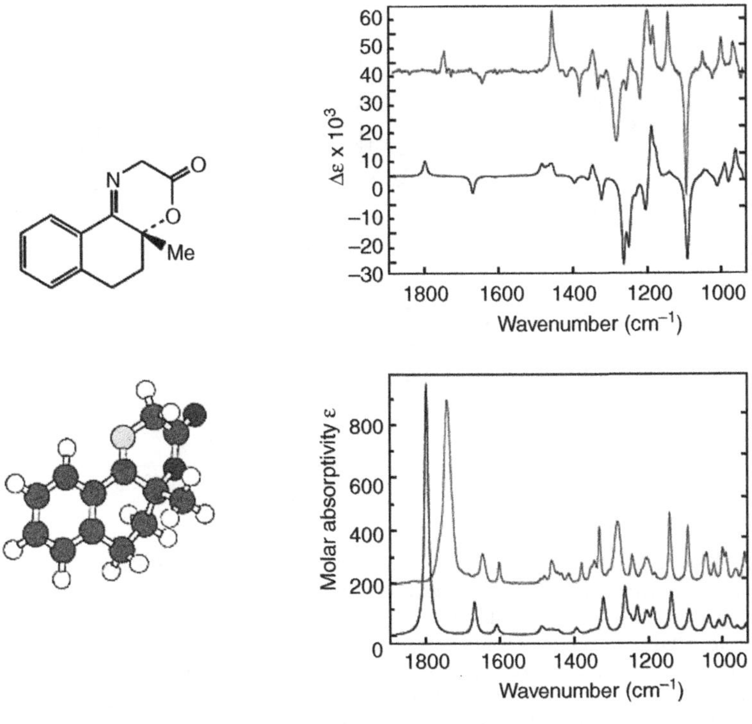

Figure 10.1 Comparison of measured and calculated VA and VCD of R-(+)-1,4-oxazin-2-one. Reproduced with permission from Elsevier Science Ltd. (Solladie-Cavallo et al., 2001b)

absolute configuration. These efforts culminated in a definitive study that combined all available experimental gas-phase data with a wide range of quantum chemistry methods for the calculations of OR and ROA that included large basis sets, origin independence and frequency dependence (Costante *et al.*, 1997; Polavarapu, 2002). In a related paper, the absolute configuration of *R*-(−)-iodochlorofluoromethane was determined by VCD measurement and *ab initio* calculation (Figure 10.2). The molecule is of interest in the search for parity violation of molecular origin (Soulard *et al.*, 2006).

10.2.2 Comparison with X-Ray Crystallography

For many years, the gold standard for AC determinations has been anomalous X-ray crystallography by the method of Bijvoet (Bijvoet *et al.*, 1951). The requirement of this method is a pure single crystal of the molecule. A typical further, but not absolute requirement is that a heavy atom (beyond C, N, O) be present in the crystal as a phase reference for the X-ray scattering. Obtaining such single-crystals in general is an art and obtaining a crystal of sufficient size and purity may take a long time, on the order of days or weeks. In some cases, a crystal can not be obtained, and of course for samples that are liquids or oils at room temperature, crystals are impossible to obtain.

The advantage of the VOA method of AC determination is that no crystals are needed. Furthermore, no additional modification of the molecule is needed as is required for the determination of AC by

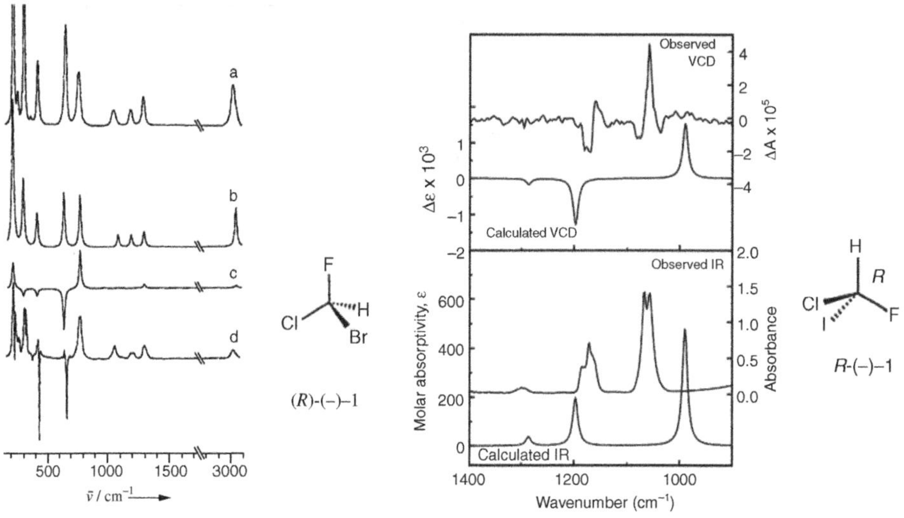

Figure 10.2 *Comparison (left panel) of measured depolarized right-angle ICP-Raman* $(I^R_z + I^L_z)$ *and -ROA* $(I^R_z - I^L_z)$*, a and d, respectively, for the neat liquid (–)-enantiomer of bromochlorofluoromethane with the corresponding ab initio* calculation of the *R*-enantiomer showing close agreement for all major bands (Costante *et al.*, 1997). *Reproduced with permission from John Wiley & Sons (Costante, J., Hecht, L., Polavarapu, P.L., Collet, A., and Barron, L.D. (1997) Absolute-Configuration of Bromochlorofluoromethane from Experimental and Ab-Initio Theoretical Vibrational Raman Optical-Activity. Angew. Chem. Int. Ed.* ***36****, 885–887. Fig. 2, p. 886. John Wiley & Sons). Comparison (right panel) of measured IR and VCD spectra in the gas phase with rotational bands shapes of the (–)-enantiomer of iodochlorofluoromethane with the corresponding DFT calculation, without rotational bandshapes of the R-enantiomer. Reproduced with permission from Elsevier Science Ltd. (Soulard et al., 2006a)*

NMR, which requires modification with an NMR shift reagent followed by further analysis. The drawback of VOA analysis is the requirement of a quantum chemistry calculation, such as DFT, to complete the AC assignment. However, DFT calculations provide additional information about the stereochemistry of the molecule by providing solution-state conformations of the molecule not provided by an X-ray analysis. One danger of X-ray analysis that can arise is incorrect assignment by virtue of the quality of the X-ray analysis or even the selection of the incorrect enantiomorphic crystal for X-ray analysis.

10.2.3 Comparison with Electronic Optical Activity

Since the 1960s electronic optical activity in the form of optical rotation (OR) and electronic circular dichroism (ECD) have been used empirically with so-called chirality rules to determine the AC of molecules. While the use of chirality rules, such as the octant rule or coupled oscillator rules, for the assignment of AC using mainly ECD have been widely used in the past, occasionally an exception to these rules appears. In some cases the exception to the rules could be understood by further conformational analysis that would lead to a correction to an earlier incorrect AC assignment. In other cases the exception could be justified by some stereochemical argument leading to a refinement of the rule, but such rationalizations offer no protection against future new exceptions.

With the advances in quantum chemistry methods for the *ab initio* calculation of OR and CD, the way has opened for these earlier methods to join VCD and ROA as *a priori* methods for AC determination (Polavarapu, 2007). However, neither OR or ECD has an internally calibrated method to be sure that the correct conformational distribution has been obtained. As a result, other sources of computational comparison rich in structural detail, such as VA and VCD intensities, are needed to make sure all important conformers have been found that are needed to calculate the OR or ECD spectrum.

Since both VCD and ROA need no specific functional groups or chromophores and sample directly the vibrational motion of all parts of the molecule, it is clear that they are preferred methods for the determination of AC in chiral molecules. On other hand, OR and ECD provide additional checks on the assignment of AC made by VOA and also as additional types of stereochemical information. No one technique contains all possible or desired stereochemical information, and thus, to obtain the maximum amount of absolute stereochemical information about a molecule, measurements and calculations of VCD, ROA, OR, and ECD need to be carried out and analyzed. Even more information and double checks on assignments can be obtained by using X-ray crystallography if available and NMR shift reagents.

10.2.4 Efficiency of VCD Determination of AC

At this time, the simplest, fastest, and most direct method of AC determination in small- to medium-sized molecules is VCD. In some pharmaceutical companies, the process of reliable determination of AC by VCD can be reduced to a 24 h service. In such cases, the IR and VCD measurements are carried out in a matter of hours. In a similar time frame, a global conformation search is carried out that produces a comprehensive set of low-energy conformers for further analysis. These conformers are then entered into an automated software program, interfaced to Gaussian 03 or 09, which computes by the following day the final conformationally-averaged VA and VCD spectra. These calculated spectra can then be compared with the measured IR and VCD spectra and, either visually or statistically with spectral similarity measures described in Chapter 9, to make the assignment of the AC. In this way tens to hundreds of determinations of AC can be completed each year in a given location.

10.2.5 Determination of Solution-State Conformation

An added bonus of the determination of AC by VOA is the determination or verification of the major conformers present in solution. This information is naturally obtained when a VCD spectrum is calculated for a molecule with more than one solution-state conformation at room temperature. The VCD spectrum of each conformer is averaged through the Boltzmann distribution using the relative free energy of the each of the conformers considered as described by Equations (9.56) and (9.57). When agreement between calculated and measured VA and VCD spectra is obtained, a confirmation of the presence of the major populated conformers present is obtained. This information in turn is useful for knowing which conformers may be present in sufficient quantity to support a specific model of drug–substrate interactions. Such modeling studies are often at the heart of strategies for understanding the mode of action of particular chiral pharmaceutical molecules in protein active sites. While such information can be predicted from quantum chemistry calculations alone, their verification under particular conditions of solvent and concentration is valuable new information in the quest to understand at the molecular level how a drug molecule interacts with its target to produce the desired therapeutic effect.

A simple early example of the determination of the AC for a molecule with more than one lowest-energy conformer is that of a new oxathiane molecule with two principal low-energy conformers (Solladie-Cavallo *et al.*, 2001a). In Figure 10.3, we show in the left panel the calculated and measured

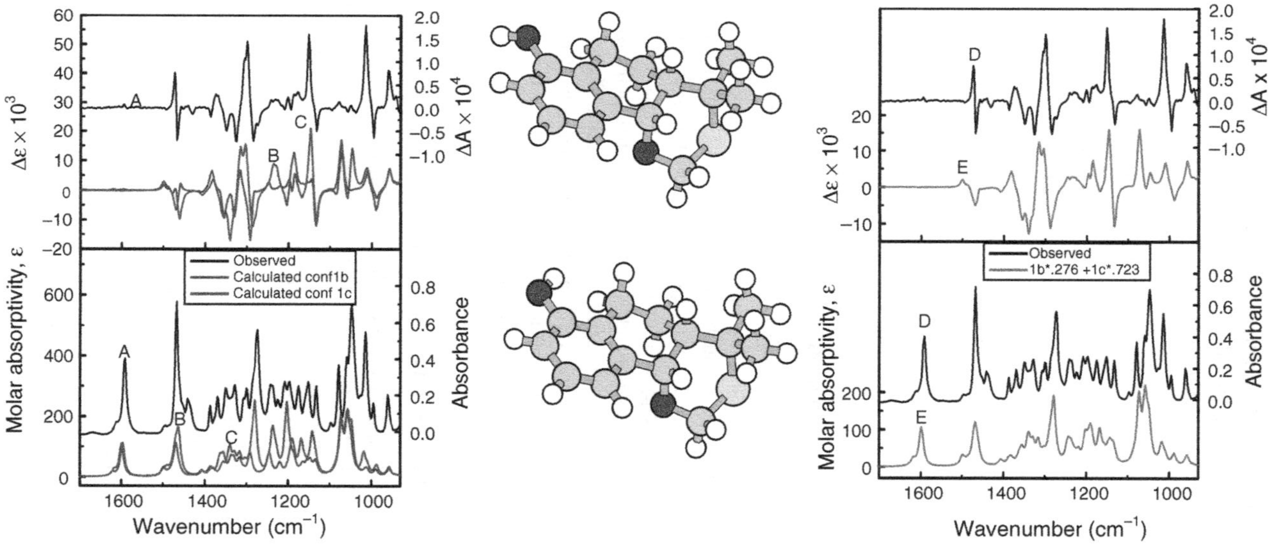

Figure 10.3 *Determination of AC of S,S-(+)-oxathiane: A, observed; B, calculated conf 1b; C, calculated conf 1c; D, observed; and E, 1b*.276+1c*.723. Reproduced with permission from the Elsevier Science Ltd. (Solladie-Cavallo et al., 2001a)*

VA and VCD spectra of these two conformers of the calculated *S,S*-configuration and the measured spectra of the (+)-enantiomer. The corresponding DFT-optimized structures of the conformers are shown in the center of the figure. The Boltzmann average of the VA and VCD spectra of the two conformers, which is 73% upper conformer and 27% lower conformer, is shown in the right panel. The two conformers differ primarily in the direction of the hydroxyl group. The upper conformer in the figure is the lowest-energy conformer, and the conformer shown below it has a calculated energy that is 0.57 kcal higher. It is clear from these spectra that the Boltzmann-averaged spectra on the right are in closer agreement with the measured spectra than are either of the conformer spectra on the left. These results not only prove the absolute configuration to be *S,S*-(+), but they also confirm that these two conformers are present in solution under the conditions of solvent and temperature. This is new information that is not available at this level of specificity from any other spectroscopic technique.

In Figure 10.4, we provide an example from ROA of conformational averaging to arrive at an AC determination (Haesler *et al.*, 2007). The molecule (*R*)-[$^2H_1$, $^2H_2$, $^2H_3$]-neopentane is the simplest chiral hydrocarbon molecule. It has no functional groups and no currently measurable optical rotation as the electronic structure of the molecule has T_d-symmetry to a high degree. The AC determination was complicated by the need to calculate all the rotomeric forms of this molecule for which there are many, as shown in the figure, and the ROA of the individual conformers varies widely in sign and

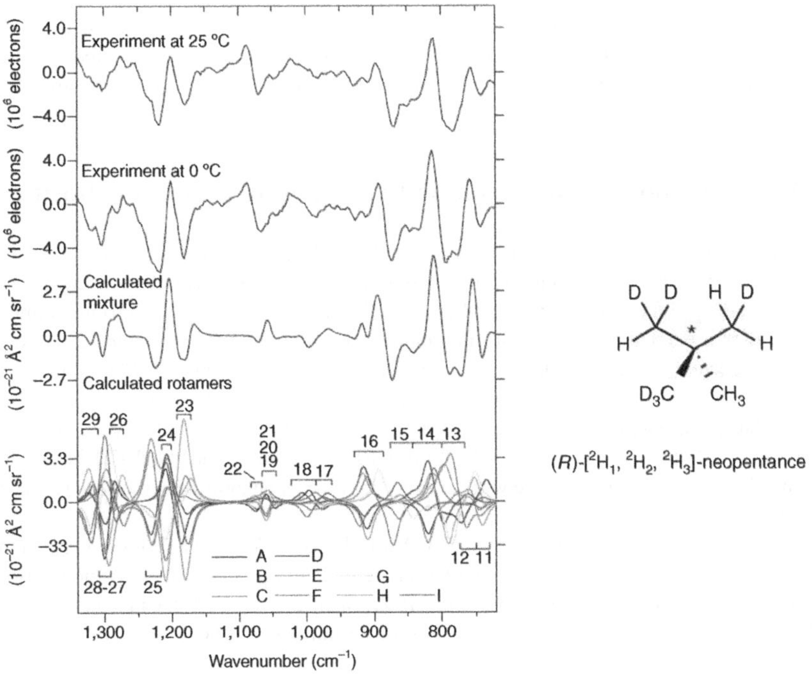

Figure 10.4 *Comparison of measured with calculated unpolarized backscattered SCP-ROA of chiral deuterium-substituted (R)-[$^2H_1$, $^2H_2$, $^2H_3$]-neopentane. Calculated ROA of individual rotomeric structures are shown in the lowest set of spectra: A, R1; B, R2; C, R3; D, R4; E, R5; F, R6; G, R7; H, R8; and I, R9. Reproduced with permission from the Nature Publishing Group (Haesler et al., 2007)*

magnitude for most of the observed bands. One simplification is that all the rotomers have the same calculated energy and as a result the Boltzmann average is a uniform weighting for all the rotomers. The final calculated ROA spectrum, both measured and calculated is a relatively small residual spectrum that remains after the largely canceling intensities of all the rotomers is summed and averaged. The remarkable degree of agreement attests to the accuracy and power of current DFT methods for calculating VOA spectra. Owing to the absence of a measurable OR the absolute configuration of this molecule could be specified by the nomenclature (*R*)-[ROA(+)800]-[$^2H_1$, $^2H_2$, $^2H_3$]-neopentane corresponding to a large positive ROA band near 800 cm^{-1}. A similar assignment has been made using VCD for the AC determination of (*R*)-(+)-[VCD(-)984]-4-ethyl-4-methyloctane, a so-called cryptochiral hydrocarbon with a quaternary chiral carbon center, the simplest chiral hydrocarbon without isotopic substitution (Kuwahara, 2010).

10.2.6 Coupled Oscillator Model AC Determination

Before leaving the topic of AC determination, we note that the empirical method of the coupled oscillator (CO) mechanism can be used not only for ECD determinations but also for VCD. The CO mechanism applied to VCD was described previously, and in Appendix A, in the context of early models of VCD that were useful in understanding VCD before full quantum chemistry methods were developed (Holzwarth and Chabay, 1972). The simple CO model describes VCD arising from a pair of degenerate, skew-oriented (twisted left or right about a line connecting the centers of the dipoles) electric-dipole moments. The predicted VCD spectrum is a conservative couplet the sense of which determines the absolute configuration of the pair of oscillators and hence the molecule by association. The model has been generalized in a number of ways to describe the VCD of polypeptides (Birke *et al.*, 1992) and also to determine the conformations of small peptides in solution (Eker *et al.*, 2004a).

10.3 Determination of Enantiomeric Excess and Reaction Monitoring

A second major area of application of VOA is the determination of the enantiomeric excess of a sample. As first described in Chapter 1, the enantiomeric excess of a sample is a measure of the excess amount of one enantiomer over the other divided by the total amount of both enantiomers. It is usually given as a percentage figure with the abbreviation *%ee* for % enantiomeric excess. For a fixed or normalized absorbance level, the magnitude of the measured VCD is linear at all wavenumber values to the *%ee* value with no constant offset if the VCD is accurately baseline corrected. Thus for a pure enantiomer, with 100*%ee*, the VCD is a maximum value. For a ratio of enantiomers of 3:1, the *%ee* is $[(3-1)/(3+1)] \times 100\%$, or 50%, and the VCD is one half its maximum value. For a racemic mixture with an exact 1:1 ratio of enantiomers, the *%ee* is 0% and the VCD is likewise zero.

One of the earliest applications of *%ee* with VOA was the use of VCD in a series of papers aimed at following the kinetics of the stereo-specific thermal degradation products of *trans*-dideuteriocyclopropane (Freedman *et al.*, 1991; Cianciosi *et al.*, 1991; Spencer *et al.*, 1990; Cianciosi *et al.*, 1989). Soon thereafter it was demonstrated that the *%ee* of a sample could be seen down to the level of approximately 1% by ROA (Hecht *et al.*, 1995).

The field of *%ee* VOA was dormant for about a decade until several papers appeared based on a simulated reaction monitoring setup that demonstrated a new level of capability of *%ee* application for VOA (Guo *et al.*, 2004; Guo *et al.*, 2005; Guo *et al.*, 2006). This involved the use of a reaction cell connected to the sample chamber of a dual-PEM FT-VCD spectrometer. As the sample in the reaction flask changed, the solution was circulated by a pump to a flow-through cell in the sampling area located between the two PEMs, as shown in Figure 10.5.

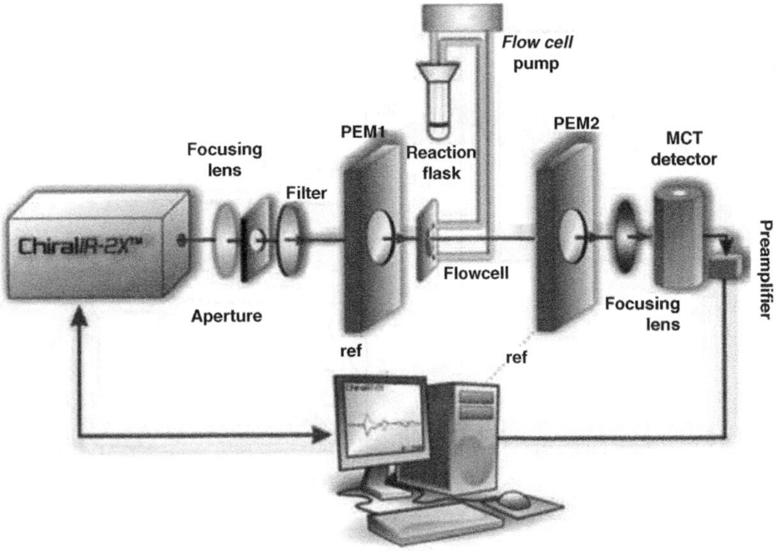

Figure 10.5 *Optical electronic layout of a dual PEM FT-VCD spectrometer showing the two PEMs and the VCD flow cell connected by tubing and a pump to a reaction vessel where changes in the composition and % ee values of the samples changed with time and could be followed quantitatively by block-scanning of the FT-VCD spectrometer*

10.3.1 Single Molecule %ee Determination

The first measurements with this setup were carried out by only changing the *%ee* of a 50% by volume solution of α-pinene in CCl_4 from 100 to 0% in 12 steps (Guo *et al.*, 2004). This was accomplished by injecting 1 or 0.5 mL solution volumes of the opposite enantiomer with a syringe without removing the flow cell from the sample area, thus giving highly stable results for the spectral measurements. At each step the IR and VCD spectra were measured, and it was found that, as expected, there were no discernable changes in the IR spectrum, while the VCD spectrum became linearly smaller as the *%ee* was systematically reduced, as shown in Figure 10.6. Although the noise level at any one spectral location was relatively large, the use of partial least squares (PLS) chemometric analysis applied across the entire spectrum gave predictions of the *%ee* for each solution with an accuracy of approximately 1%.

10.3.2 Two-Molecule Simulated Reaction Monitoring

This methodology was then extended to a two-sample solution to simulate the transformation of a reacting solution from reactants to products (Guo *et al.*, 2004). Here a solution of camphor and borneol was changed over time from pure 1*S*-(–)-camphor to pure [1*S*-endo]-(–)-borneol, as shown in Figure 10.7. The sensitivity of VCD to monitor mole fraction changes of camphor relative to borneol from 1.00 to 0.00 in 13 steps is illustrated in Figure 10.8.

As a second step the *%ee* values of the two molecules were changed in steps from 100 to 40% for camphor and from 100 to 60% for borneol, while the concentrations of camphor relative to borneol were simultaneously varied to simulate monitoring of a chemical reaction of two chiral species. The results of the chemometric analysis of this pseudo reaction are displayed in Figure 10.9.

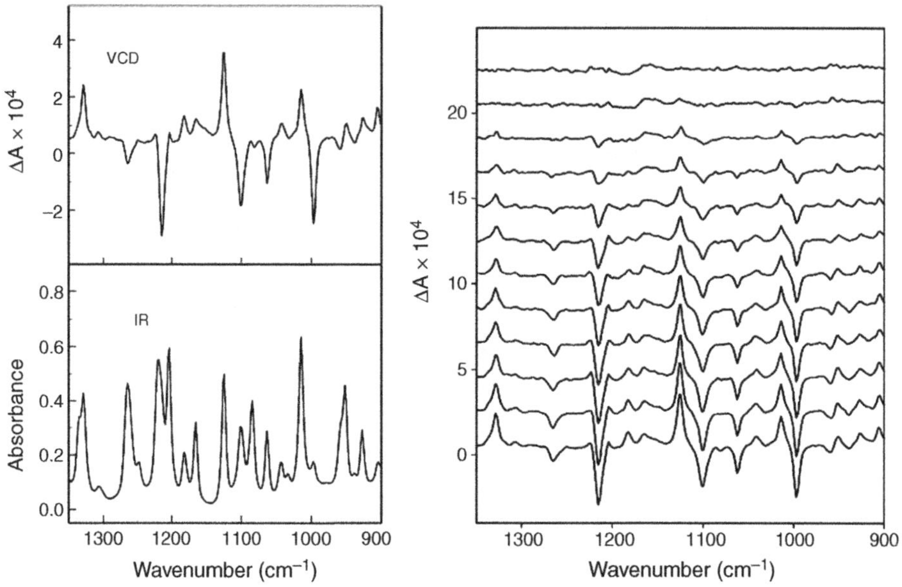

Figure 10.6 IR and VCD spectra of 3.14 M solution of (+)-α-pinene in CCl_4 (left panel) and 12 VCD spectra from the bottom to the top of (+)-α-pinene at %ee values of approximately 100, 88, 78, 66, 60, 52, 45, 33, 23, 14, 6, and 0 (right panel). Reproduced with permission from the American Chemical Society (Guo et al., 2004)

In Figure 10.9, actual concentrations of both species can be extracted from the IR concentrations (left panel). If the same analysis is carried out for the VCD spectra, a set of lower concentrations is predicted (middle panel) because the magnitude of the apparent VCD intensities is reduced by *%ee* values at each stage of the analysis. These *%ee* values can be exacted for each species, borneol and camphor in this case, by dividing the apparent VCD concentrations by the corresponding IR concentration for each block of IR and VCD spectra in the simulated kinetic run. The RMS error in the predicted versus prepared *%ee* values was approximately 2%.

10.3.3 Near-IR FT-VCD %ee and Simulated Reaction Monitoring

The same spectral measurements and analyses just presented were subsequently repeated in the near-IR region between 4000 and 5800 cm^{-1} (Guo *et al.*, 2005). Similar accuracies in the prediction of near-IR

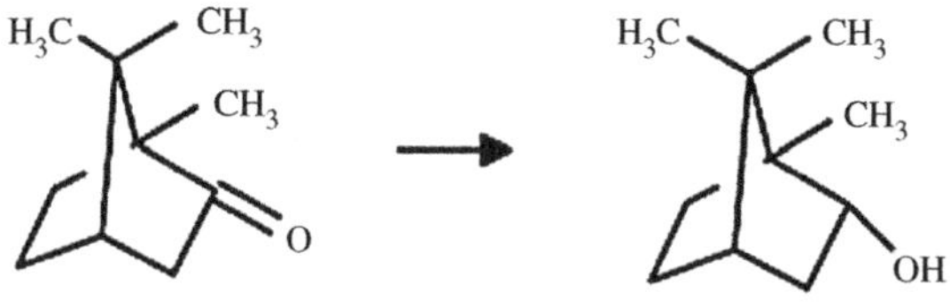

Figure 10.7 Stereo-specific diagrams of the simulated reaction from 1S-(–)-camphor to [1S-endo]-(–)-borneol as described in the text

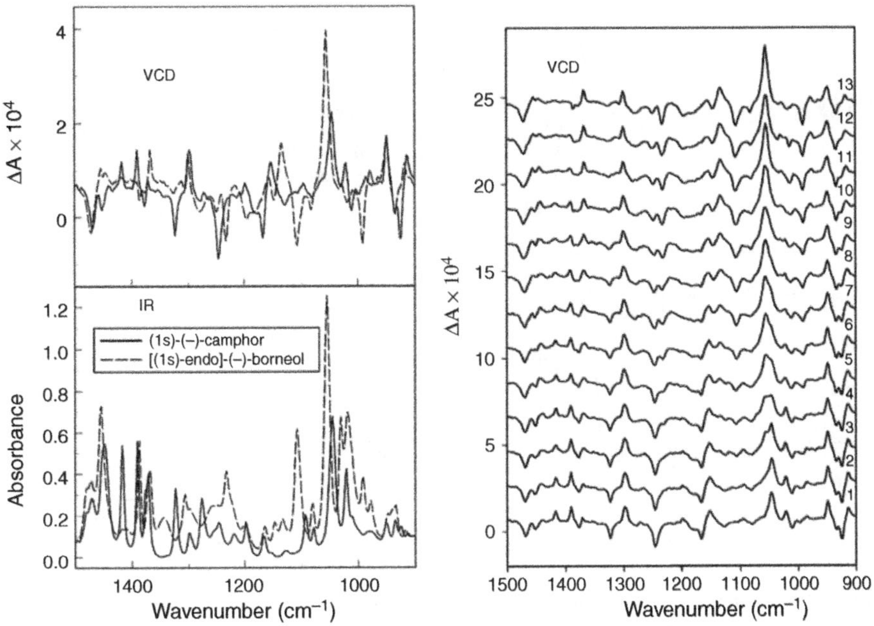

Figure 10.8 *IR and VCD spectra of 1 M CCl_4 solutions of 1S-(–)-camphor and [1S-endo]-(–)-borneol in (left panel) and 13 VCD spectra following the conversion of camphor to borneol (right panel) according to the following mole fraction of camphor with respect to borneol: 1.00, 0.91, 0.83, 0.77, 0.67, 0.59, 0.50, 0.42, 0.33, 0.23, 0.17, 0.09, 0.00. Adapted with permission from the American Chemical Society (Guo et al., 2004)*

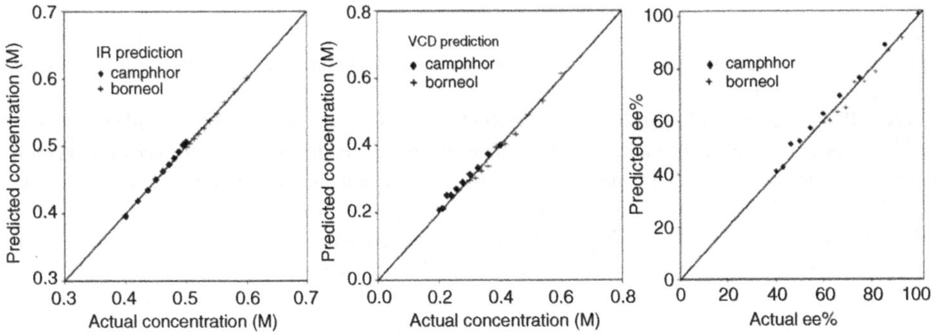

Figure 10.9 *PLS predicted versus prepared values of simultaneous changes in concentrations and %ee of CCl_4 solutions of 1S-(–)-camphor and [1S-endo]-(–)-borneol. The %ee values are obtained by dividing the apparent VCD concentrations by the corresponding IR concentrations. Reproduced with permission from the American Chemical Society (Guo et al., 2004)*

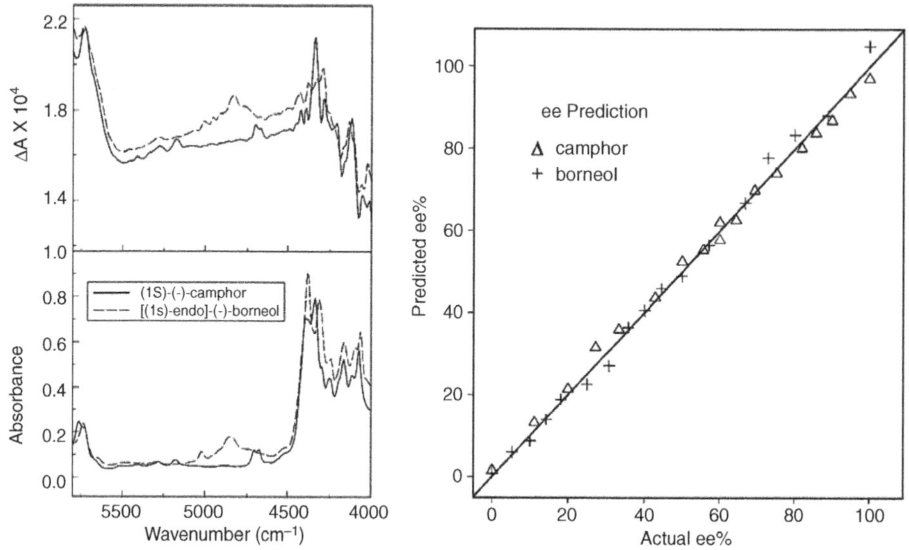

Figure 10.10 *Near-IR VA and VCD spectra of 1S-(–)-camphor and [1S-endo]-(–)-borneol (left panel). PLS predicted versus prepared values of simultaneous changes in concentrations and %ee where the %ee values are obtained by dividing the apparent VCD concentrations by the corresponding near-IR concentrations (right panel). Reproduced with permission from the Society of Applied Spectroscopy (Guo et al., 2005)*

and VCD concentrations and *%ee* were found in this spectral region as those found in the mid-IR as illustrated above. The results for the pseudo reaction between camphor and borneol are illustrated in Figure 10.10 where the pure near-IR VA and VCD spectra that formed the basis of the chemometric analysis are given in the left panel and the comparison of the simultaneous prediction of the *%ee* values is given on the right. Carrying out *%ee* determinations in the near-IR versus the mid-IR region carries all the usual advantages and disadvantages of these two regions. Most likely, for practical purposes, the near-IR region may prove more valuable for chiral reaction monitoring due to the greater use of non-chiral reaction monitoring in the near-IR region. The ability to use longer pathlengths, brighter sources, and more sensitive detectors are factors in favor of the near-IR whereas the mid-IR enjoys sharper, better understood, spectral features.

Overall, one can conclude that VCD, and also ROA, contain enough spectroscopic detail in multiplex fashion to be able to resolve simultaneously changes in both the concentrations and *%ee* of two chiral reacting species. Because VCD and ROA are the only chiroptical spectroscopic methods that are broadband multiplex spectroscopies and can measure simultaneously the parent and chiroptical spectra (VA and VCD or Raman and ROA), they are currently the *only* analytical methods that can monitor in real time in full kinetic detail reactions of chiral molecules to chiral products. With this methodology, one can in principle monitor simultaneously any number of co-reacting chiral species. The practical limit on this number is related to the signal quality and concentration of each species.

10.3.4 Near-IR Reaction Monitoring of an Epimerization Reaction

We conclude our description of *%ee* applications with a near-IR VCD reaction-monitoring study of the molecule 2,2-dimethyl-1,3-dioxolane-4-methanol (DDM). This molecule undergoes an

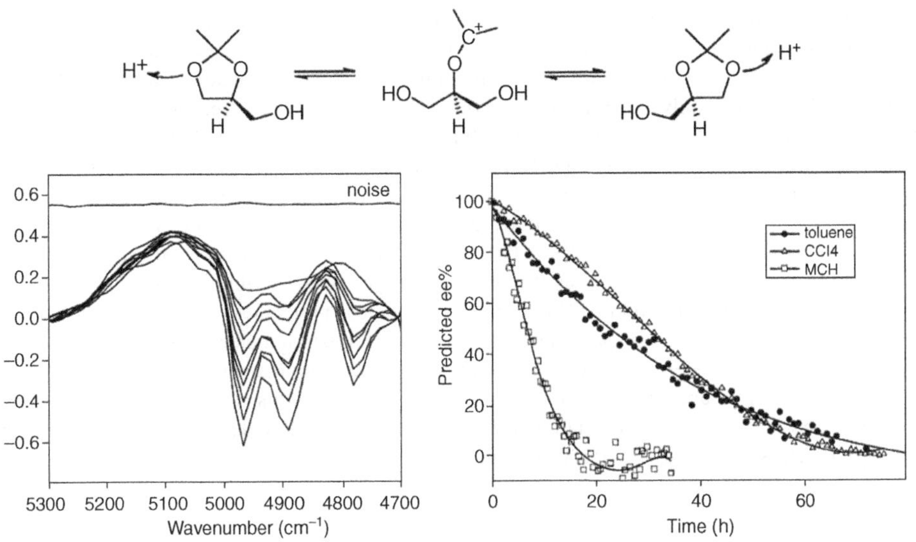

Figure 10.11 *Reaction scheme for the epimerization of 2,2-dimethyl-1,3-dioxolane-4-methanol (DDM) (top), VCD spectra of DDM undergoing epimerization in toluene in the near-IR region (left panel), and %ee kinetics for the epimerization of DDM in three different solvents (right panel). Reproduced with permission from John Wiley & Sons (Guo et al., 2006)*

acid-induced epimerization reaction which, due to the transfer of the location of the chiral center, is an internal conversion from one enantiomer into the other (Guo *et al.*, 2006). In Figure 10.11, the epimerization scheme of DDM is shown, which involves a ring opening to an intermediate species that can re-close to either of two enantiomeric structures. In the panel on the left, VCD spectra of DDM are shown that become smaller in magnitude as epimerization proceeds. To the right are shown the kinetic plots of *%ee* reduction of DDM in three different solvents, toluene, carbon tetrachloride, and methylcylohexane (MCH). The three different solvents stabilize the reaction intermediate to different degrees for which MCH is the most efficient. The VCD analysis reveals new details about the kinetics of this epimerization process, which leads to a greater understanding of this reaction and its solvent dependence.

10.4 Biological Applications of VOA

VCD and ROA have different types of sensitivities to the structure of biological molecules. As a result, each has different relative advantages over the other, and the applications of each have a different emphasis and points of view. Furthermore, there are very few publications in which both VCD and ROA are used to study a single biological problem. That being said, because of their different sensitivities, there would be much to be gained if both VCD and ROA were used in the same study in the quest for a greater understanding of a biological problem, and no doubt in future studies this strategy will be more frequently adopted. As we shall see, neither VCD nor ROA gives a sufficiently complete picture of a biological system that complementary information from the sister VOA technique would not be valuable. To emphasize this complementarity, the discussion in this section will interleaf descriptions of biological applications using VCD and ROA.

The application of VOA to biological problems is unique relative to the applications discussed so far in this chapter. This feature is that the absolute configuration is not a central issue. Nature is homochiral with respect to the absolute configuration of proteins, sugars, nucleic acids, and related structures. The only exceptions to this rule are the few cases where, for example D-amino acid residues can be found in some antimicrobial peptides of some animal species as part of a naturally-evolved survival-defense mechanism. As a result, the focus of VOA applications of biological molecules is on *conformation*, not absolute configuration. While other techniques such as NMR, ECD, Raman, and FT-IR contribute in many ways to our understanding of the conformations of biological molecules, VCD and ROA have an increasingly strong role to play in gathering this information, and, as we shall see, VOA has some unique and very powerful sensitivities to the structure and conformation of biological molecules.

The opening section of this chapter focused on the different types of molecules for which VCD and ROA have been applied. Here our focus is on the different types of applications even though the overall discussion is again grouped by the different types of molecules, in this case exclusively biological molecules and their assemblies. We start with the small biological molecules of interest and work our way up to increasingly large biological structures.

10.4.1 VCD and ROA of Amino Acids

The first and simplest biological molecule for which VOA was applied is, not surprisingly L-alanine, the simplest chiral amino acid (Diem *et al.*, 1977). The first VCD spectrum of L-alanine, and the first VCD spectrum published from Syracuse University with the third VCD instrument constructed worldwide, is presented in the left panel of Figure 10.12. The solvent was D_2O and the spectral region

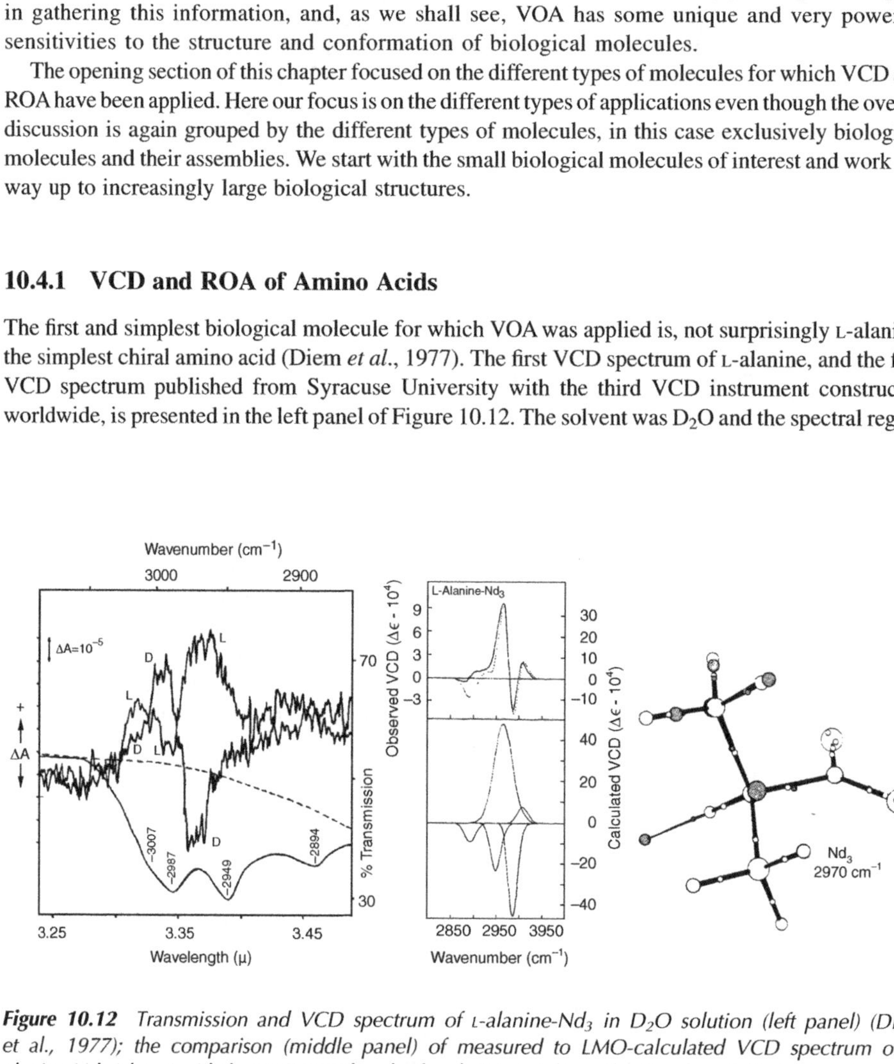

Figure 10.12 Transmission and VCD spectrum of L-alanine-Nd_3 in D_2O solution (left panel) (Diem et al., 1977); the comparison (middle panel) of measured to LMO-calculated VCD spectrum of L-alanine-Nd_3 (above) with the LMO VCD bands of each CH stretching vibrational mode (below) as described in the text; and (right) the calculated nuclear and LMO-centroid positions and displacements for the 2970 cm^{-1} C*H stretching mode with accentuated displacements for clarity (Freedman et al., 1982). Reproduced with permission from the American Chemical Society. (Diem, et al., 1977)

was the CH-stretching region. Alanine in aqueous solution is a zwitterion and the species measured was L-alanine-Nd_3. The absorption spectrum is displayed as a transmission spectrum relative to the D_2O background transmission. The two highest frequency bands are due to the near-degenerate anti-symmetric methyl stretching modes and the lowest two peaks are due the Fermi-resonance doublet of the symmetric methyl stretching mode with the first overtone of the methyl deformation mode. A large VCD appears near 2970 cm^{-1} where no absorption band appears. This VCD feature is due to the weakly absorbing methine C*H stretching mode. This strong positive VCD band forms the basis of the methine CH-stretching chirality rule for L-amino acids (Nafie *et al.*, 1983), the VCD spectra of which all have a strong positive bias due to this core vibrational feature of the CH stretching region. A number of years later, the VCD spectrum of L-alanine-Nd_3 was re-measured and compared with the results of a semi-empirical CNDO VCD calculation using the localized molecular orbital (LMO) model (see Appendix A) in the central panel of Figure 10.12 (Freedman *et al.*, 1982). The LMO model is the first quantum mechanical model of VCD (Nafie and Walnut, 1977) and was formulated as an extension of the fixed partial charge (FPC) model. In the LMO model, VA and VCD intensities arise from the motion of nuclei with positive fixed charges and negative charge at the centroids of the LMOs. The results of the LMO calculation for the C*H stretching mode of L-alanine-Nd_3 is presented on the right side of Figure 10.12 The open circles, large for nuclei and small for LMO centroids, are the equilibrium positions, whereas the dark circles are the corresponding displaced positions for this vibrational mode. The large positive VCD in the central diagram of the LMO calculations is due to this lone C*H stretching mode.

More recently, attention has again turned to this molecule where high-level quantum chemistry methods have been applied to L-alanine in H_2O solvent for both ROA and VCD over the the vibrational regions between 1000 and 1800 cm^{-1} for IR and VCD and 400 and 1800 cm^{-1} for the Raman and ROA (Jalkanen *et al.*, 2008). This work includes recently measured FT-IR and FT-VCD spectra from Syracuse University and backscattering unpolarized SCP-Raman and SCP-ROA from Glasgow University. The model for the VA and VCD calculations is DFT/B3LYP/6-31G* plus 20 water molecules, the structure for which, calculated at the level DFT/OPBE0/TZ2P + COSMO, is shown in Figure 10.13. The comparison of measured to calculated VA and VCD spectra is provided in Figure 10.14, where the calculations include four different solvent models starting with only the 20 explicit water molecules (lowest spectra) followed by three different solvent treatments in addition to the explicit water molecules. These are labeled Onsager (dielectric shell model), PCM (polarizability continuum model), and COSMO (conductor screening model). A corresponding comparison of measured and calculated Raman (left) and ROA (right) are given in Figure 10.15. Here the Raman calculations were carried out with same model chemistry as the VA and VCD, but two changes were imposed on the ROA calculations. Carrying out ROA intensity calculations with 20 explicit water molecules was computationally too demanding, but it was found that good agreement could be obtained with an improved basis set including diffuse basis functions and a continuum solvent model. The ROA calculations shown were carried out for L-alanine without explicit water molecules at the level DFT/B3LYP/cc-aug-pVDZ + COSMO for three different ROA invariants, which all give approximately the same calculated ROA spectrum.

10.4.2 VOA of Peptides and Polypeptides

The next level complexity of biological structures is that of small peptides and polypeptides. The first VCD spectra of peptides reported were the CH-stretching VCD of L-alanyl-L-alanine (L-ala-L-ala) and L-ala-L-ala-L-ala, L-ala-gly (L-alanylglycine) and Gly-L-ala, published shortly after the first amino acid spectrum (Diem *et al.*, 1978). The first VCD spectrum of a polypeptide was published a few years later (Singh and Keiderling, 1981). This paper established the VCD spectral signature of an α-helix in the NH (amide A and B) and C=O stretching (amide I and II) regions. A few years later, the mid-IR VCD spectral signatures were published of poly-L-lysine in the three classical secondary structures: α-helix,

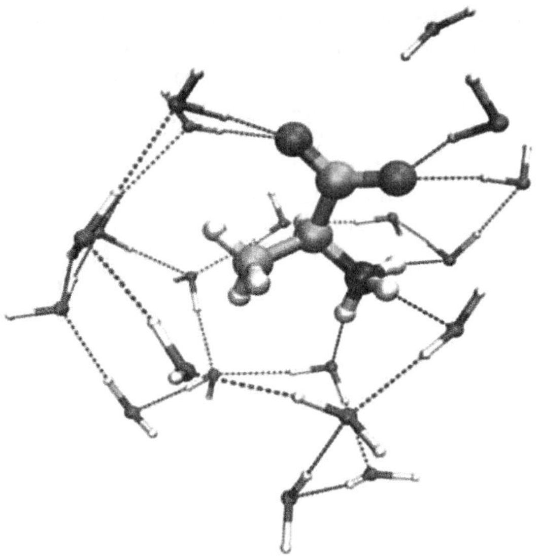

Figure 10.13 *Optimized model for L-alanine zwitterion with 20 water molecules calculated at the level DFT/OPBE0/TZ2P + COSMO. Reproduced with permission from Springer Verlag, Ltd. (Jalkanen et al., 2008)*

β-sheet, and random coil shown in Figure 10.16 (Yasui and Keiderling, 1986; Paterlini *et al.*, 1986). The paper by Paterlini *et al.* noted that by adding high concentrations of salt to a sample of poly-L-lysine in the so-called random coil conformation that a fourth secondary structural state could be reached that had no detectable VCD spectrum at the scale of sensitivity of the random coil VCD spectrum and hence might be associated with a completely unordered state. This is in turn implied that the random coil actually had a regular structure at some level of local structure. The VCD spectrum in the amide I region of the random coil was strong and opposite in sign to that of the α-helix, implying the possibility of a left-handed helical structure opposite to that of the α-helix.

A few years later this issue of a possible well-defined structure of the random coil was settled when it was proposed by Dukor and Keiderling that the so-called random coil structure of polypeptides and proteins was in fact the same VCD spectrum of polyproline II (PPII) (Dukor and Keiderling, 1991). This idea had been proposed many years earlier that the random coil was actually an extended left-handed 3_1-helix, stabilized not by internal intra-peptide hydrogen bonds but by hydrogen bonding to the solvent water molecules. The evidence provided by Dukor and Keiderling is shown in Figure 10.17 where the VCD spectrum of poly-L-glutamic acid in the random coil state is compared with that of poly-L-proline in the PPII-helix state. This work was continued where it was shown that simple unblocked oligopeptides of proline had amide I VCD spectra, which implied that they adopt a PPII conformation, contrary to conventional wisdom that small peptides had random $\varphi\psi$-angles and essentially no well-defined secondary structure (Dukor *et al.*, 1991). Further, it was shown in this work that the VCD of proline oligopeptides reach the magnitude of the full polyproline VCD intensity after only five amino acid residues.

The finding of the oligomeric secondary structure of proline systematically has been pursed more recently by Schweitzer-Stenner who analyzed the structure of an extensive series of small peptides in

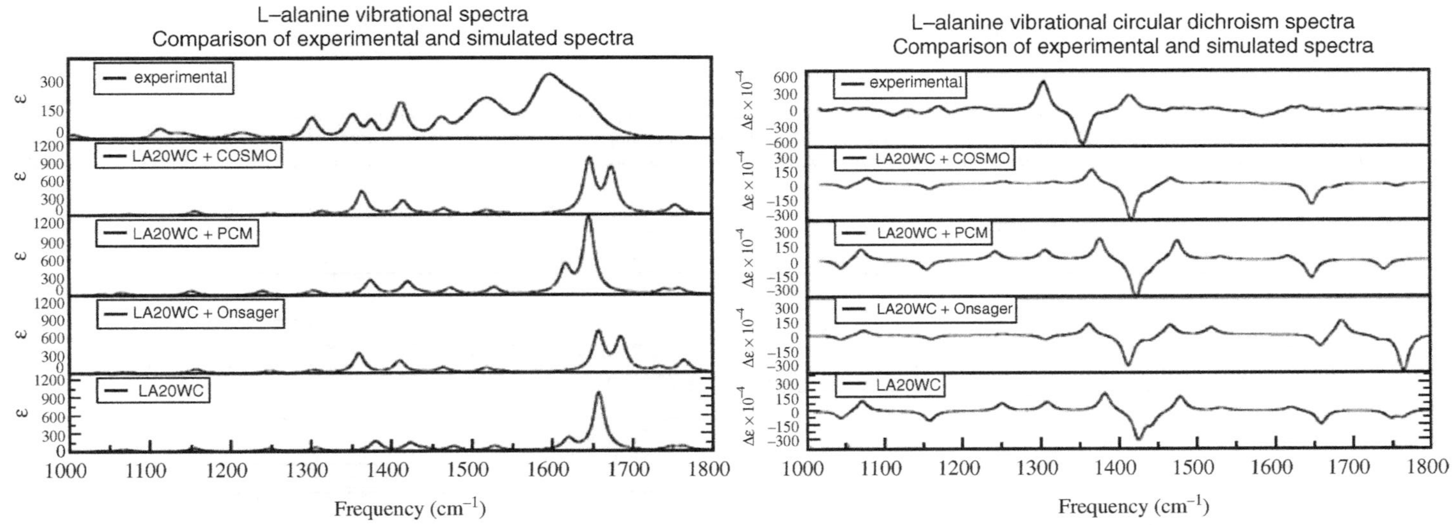

Figure 10.14 *Comparison of measured to calculated VA (left) and measured to calculated VCD (right) of L-alanine in water. Calculations were carried out with 20 water molecules as shown in Figure 10.13, at the level DFT/B3LYP/6-31G* without additional solvent modeling beyond the explicit water molecules (lowest spectra) and with three additional solvent treatments as indicated. Reproduced with permission from Springer Verlag, Ltd. (Jalkanen et al., 2008)*

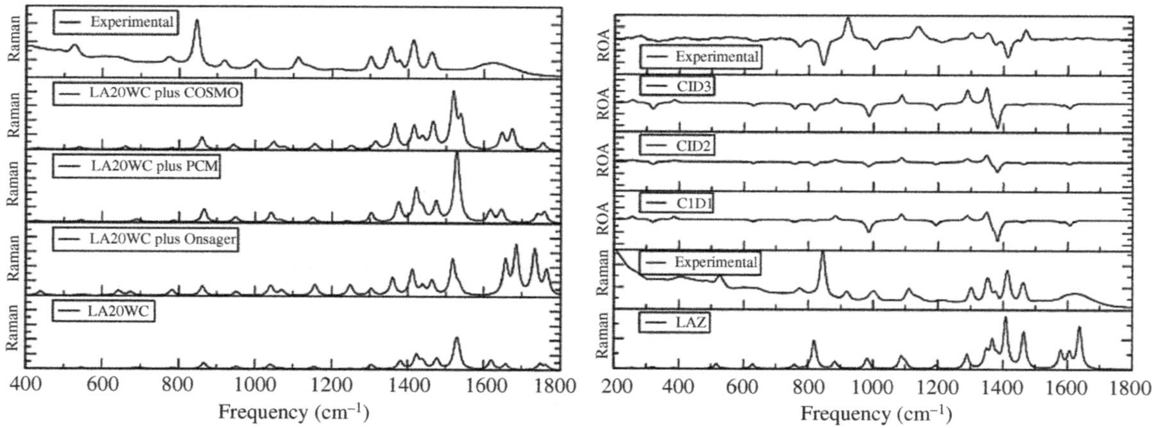

Figure 10.15 *Comparison of measured to calculated Raman (left) and measured to calculated ROA (right) of* L*-alanine in water. Calculations of Raman spectra were carried out with 20 water molecules as shown in Figure 10.13 at the level DFT/B3LYP/6-31G* with three different solvent models. ROA calculations were carried out without explicit water molecules at the level DFT/B3LYP/cc-aug-pVDZ + COSMO. Reproduced with permission from Springer Verlag, Ltd. (Jalkanen et al., 2008)*

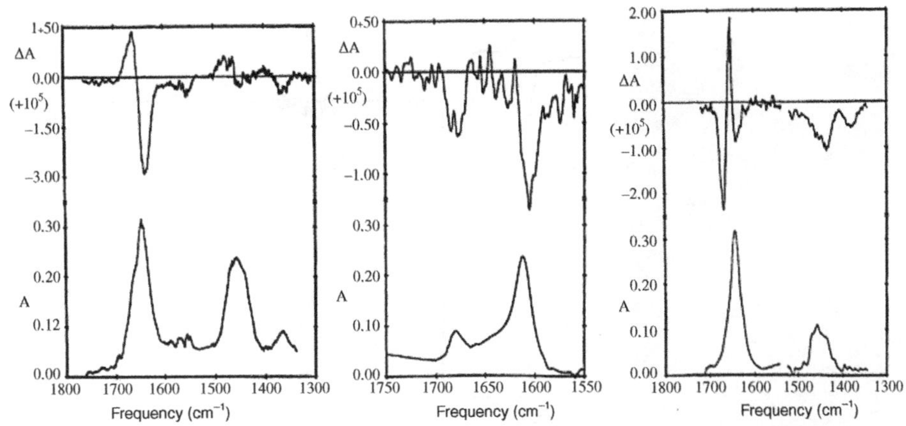

Figure 10.16 *VA and VCD spectra in the amide I and II regions of poly-L-lysine in D_2O at pD 7.3 (left) in the random coil state, at pD 11.5 after heating in the β-sheet state (middle), and as a mixture of methanol and water (96:4) in the α-helix state (right). Reproduced with permission from the American Chemical Society (Yasui and Keiderling, 1986)*

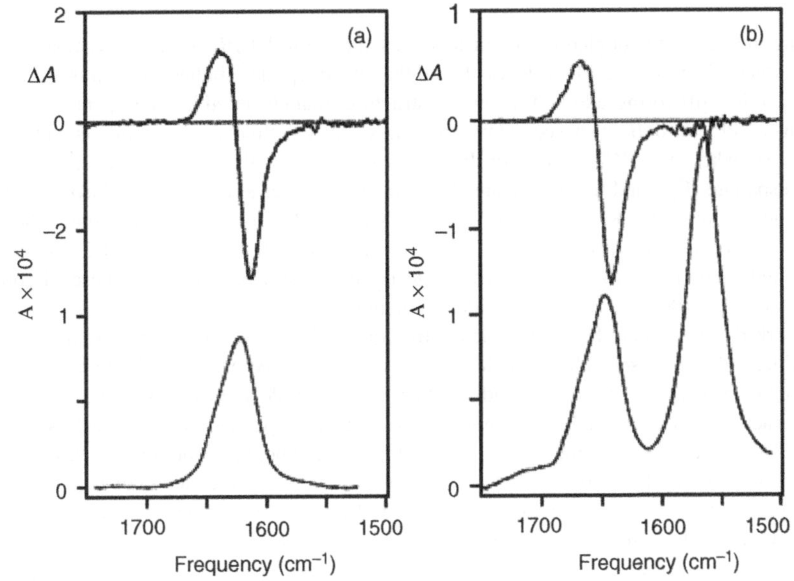

Figure 10.17 *Comparison of the VA and VCD of (b) poly-L-glutamic acid in the random coil state with (a) poly-L-proline in the PPII conformation state. Reproduced with permission from John Wiley & Sons (Dukor and Keiderling, 1991)*

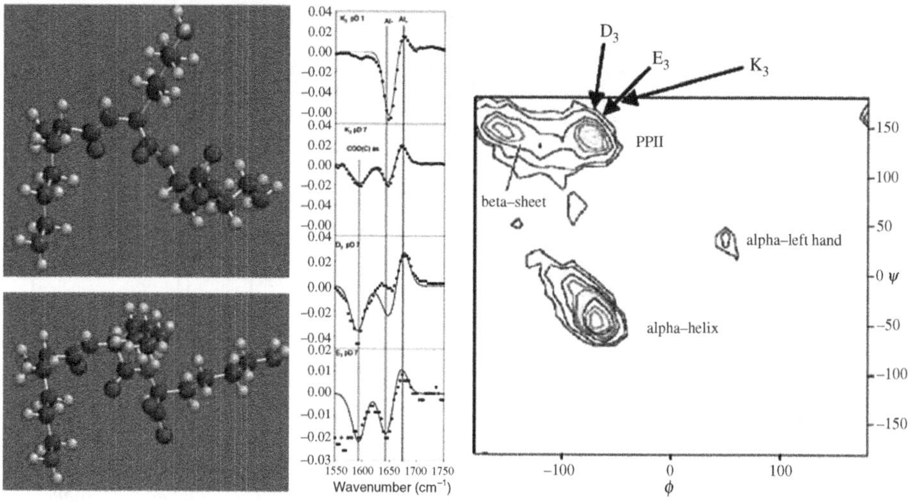

Figure 10.18 Model structure of tri-lysinate (K_3) in idealized PPII structure (left upper) and as calculated from FT-IR, VCD, and polarized Raman spectra (left lower). The middle panel compares measured (solid lines) with calculated VCD from determined $\varphi\psi$-angles from FT-IR and polarized Raman at the pD values incidated. Location of tri-lysine (K_3), tri-aspartate (E_3), and tri-glutamate (D_3) in the Ramachandran plot of $\varphi\psi$-angles near the PPII area of peptide secondary structure. Reproduced with permission from the Americal Chemical Society (Eker et al., 2004b)

D_2O using the extended coupled oscillator model as applied to FT-IR, polarized Raman, and VCD spectroscopies (Eker *et al.*, 2004a). It was found that all tripeptides studied, including zwitterionic, high-pH and low-pH forms, adopted secondary structures that clustered in the regions of $\varphi\psi$-angles belonging to PPII and β-sheet secondary structures. An example of this work is illustrated in Figure 10.18, where the VCD spectra of the three tri-peptides with ionizable side-chain, tri-lysine (K_3), tri-aspartate (E_3), and tri-gluatamate (D_3) are shown (Eker *et al.*, 2004b). Also shown is the stereo-structure found for K_3 compared with its idealized PPII structure that illustrates their close backbone conformations. The locations of the determined structures for K_3, E_3, and D_3 on a Ramanchandran plot are also shown, which clearly demonstrates how close these oligopeptide structures are to clusterings of PPII secondary structures.

The work discussed here highlights the contributions of VCD to the understanding of peptide, polypeptide, and protein secondary structure. This is accomplished primarily through a study of the amide I region where coupling between adjacent or nearby peptide units occurs by means of through-space interaction on a scale smaller and more detailed than that spanned by ECD spectroscopy in the UV near 200 nm. At the same time, parallel research in ROA of peptides and proteins has revealed an even more local type of stereo-sensitivity, one that is centered on the values of the $\varphi\psi$-angles as revealed primarily by vibrational modes in the amide III region of hydrogen bending modes.

An example of the sensitivity of ROA to the secondary structure of polypeptides is shown in Figure 10.19, were the backscattering ROA spectra of four forms of poly-L-lysine are compared and analyzed empirically (McColl *et al.*, 2003). While changes are seen across the entire spectrum in comparing the α-helix, mixed α-helix/β-sheet, β-sheet, and disordered (PPII-type) secondary structures, the most dramatic changes are seen in the extended amide III region, which involves

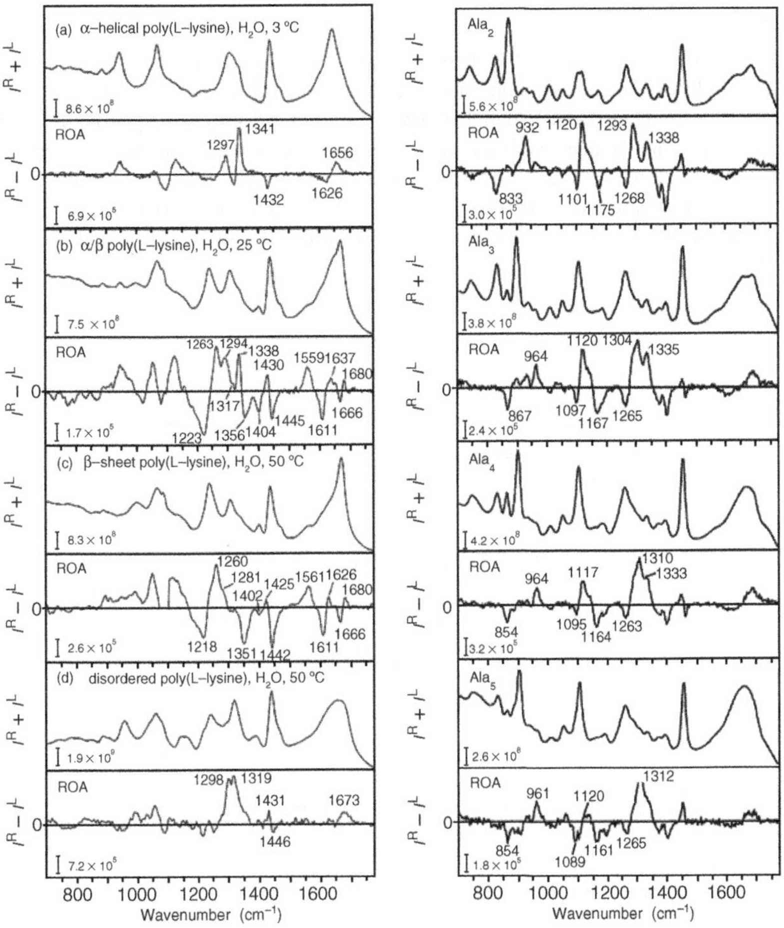

Figure 10.19 *Backscattering ICP Raman and ROA of poly-L-lysine in H_2O (left) at: (a) pH 11 at 3°C for the α-helix; (b) at 25°C for a mixed α-helix/β-sheet state; (c) at 50°C for the β-sheet state; and (d) pH 1.8 at 20°C for the disordered (PPII) state (left). Backscattering ICP-Raman and -ROA of aqueous solutions of Ala_2 to Ala_5 (right) showing the close similarity of the spectra of Ala_5 to poly-L-lysine in the disordered state (opposite on left), both PPII helical conformations. Reproduced with permission from the American Chemical Society. (McColl, et al., 2003)*

coupling between amide NH bending and the C_α–H bending modes at the adjacent chiral center of the lysine residue. A particular capability of ROA is discrimination of different types of β-sheet structures encountered in proteins. In a related paper, ROA signatures of alanine peptides in aqueous solution from Ala_2 to Ala_5, as shown in Figure 10.19, transitions from that of the dipeptides to a secondary structure corresponding to the that of the PPII helix previously assigned to poly-L-lysine and poly-L-glutamic acid (McColl *et al.*, 2004).

10.4.3 ROA of Proteins

Most of the work on applications of ROA to biological molecules in the past two decades has focused on protein secondary structure, although some work on nucleic acids and carbohydrates has also been carried out. The central theme pursued in these protein studies is first to classify proteins according to secondary structure and the degree of order in the protein. ROA spectra of proteins in water span the entire range of vibrational frequencies from 200 to 1800 cm^{-1} exhibiting an extraordinary degree of vibrational structural information. This range of spectral coverage and also the degree of changes of the ROA spectra to five proteins with different secondary structure motifs is illustrated in Figure 10.20 (Barron, 2006a).

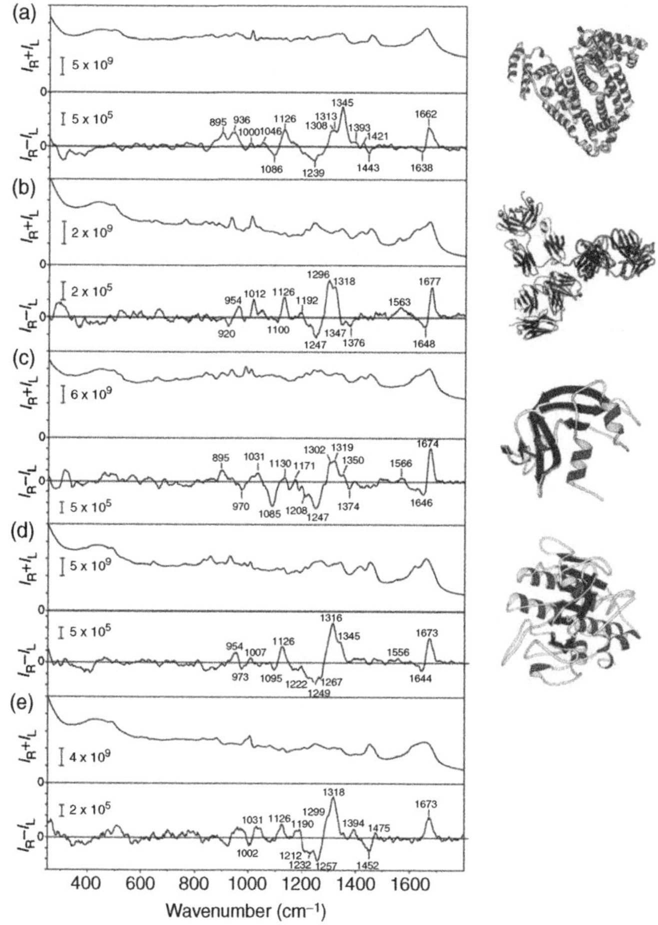

Figure 10.20 Backscattering SCP-ROA and -Raman for the five proteins from top to bottom: (a) human serum albumin; (b) human immunoglobin G; (c) bovine ribonuclease A; (d) subtilisin Carlsburg; and (e) bovine β-casein. Shown to the right of each pair of the Raman and ROA spectra are the secondary structure images for all proteins except that for bovine β-casein, absent because it is a natively unstructured protein for which such an image is not currently known or easily depicted. Reproduced with permission from Elsevier Science Ltd (Barron, 2006)

Although the most central and characteristic region of ROA sensitivity, as just mentioned, is the amide III region, the real power of ROA protein structural analysis stems from each ROA spectrum as whole, which can be used statistically as a basis for the construction of a protein-ROA database. A recent example of such a database is illustrated in Figure 10.21, where results of a multivariate analysis of 80 proteins is illustrated as a two-dimensional non-linear mapping (NLM) plot in which the secondary structure content is classified from high α-helix (right) to high β-sheet (left) along coordinate 1, and from highly-ordered (top) to highly-disordered (bottom) along coordinate 2 (Zhu *et al.*, 2006). The color scheme for locating the positions of individual protein ROA spectra corresponds to various distinct families of native protein-folding motifs and is thus valuable as a classification scheme for newly discovered proteins. Knowing the folding family is important in the design of strategies for the crystallization of the proteins, a prerequisite for X-ray structure determination. This same database can be displayed in three dimensions where an additional classification dimension adds a further degree of separation to the various clusters of secondary structure types. Also shown in this figure are averages of the ROA spectra in the groups of proteins in the plots colored for ease of association.

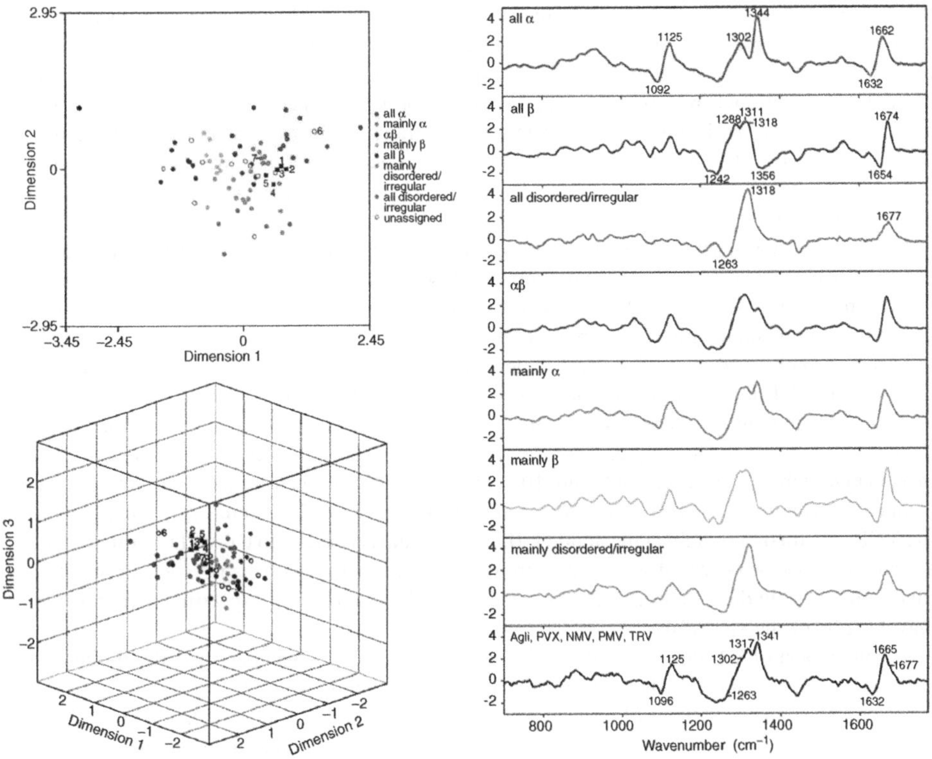

Figure 10.21 *Multivariate NLM plots in two dimensions (upper left) and three dimensions (lower left) of a database of 80 ROA spectra that cluster in secondary structure types corresponding to native protein folding families, the average ROA spectra of which are displayed to the right. Reproduced with permission from Elsevier Science Ltd (Zhu et al., 2006)*

The area of greatest potential for ROA to contribute to an understanding of proteins lies in its sensitivity to unstructured proteins for which crystal structures cannot be obtained. Such protein states are encountered not only in natively unstructured proteins that are currently coming under scrutiny for their influence on subtle biological processes, but also for partially folded or unfolded proteins that have a well-defined native structure. This of course is important as the mechanism by which proteins fold becomes better understood. Partially unfolded proteins are particularly important in describing how natively-folded proteins transition from a native state to an amyloid fibril, to be discussed further later concerning applications of VOA to supramolecular biological structures.

Before leaving this section, we briefly mention that ROA has been applied more recently to glycoproteins, which are otherwise more difficult to analyze because of their difficulty to crystallize. The ROA features from the protein and the glycosylated parts of the molecule are located primarily in different regions of the vibrational spectrum and hence their complementarity can be studied without interference from overlap or partial undesired cancellation of ROA intensity.

10.4.4 VCD of Proteins

The majority of published VCD spectra of proteins, both measured and calculated, have emerged over the years from Keiderling's laboratory at the University of Illinois, Chicago, as can seen from recent reviews (Keiderling *et al.*, 2006; Kubelka *et al.*, 2009). Most of the instrumentation used for these measurements has been based on dispersive scanning monochromators due to their stability and high signal-to-noise ratio when components are optimized for a particular spectral window, such as the amide I region in the mid-IR. On the other hand, Fourier transform instrumentation has advanced over the years since FT-VCD measurements were first carried out, leaving us today at a position where either method of VCD measurement of proteins can be used depending on which relative advantages are most important.

It was well known that the Raman scattering spectrum of water is weak and that water is a good solvent for biological samples such as proteins. Also, it is widely believed that water is a poor solvent for IR measurements due to its unusually strong absorbance spectrum. While this is generally true, if small pathlengths are used, the optimum being 6 microns, excellent IR and VCD spectra of proteins and other biological molecules can be obtained in H_2O solutions. While D_2O can be beneficial in some circumstances, there are drawbacks to using D_2O associated with the prevention of contamination by H_2O over the course of a measurement and incomplete or changing deuterium exchange in the sample. The potential to obtain high-quality IR and VCD spectra for the protein myoglobin is illustrated in Figure 10.22, where the VCD spectrum spans the entire range of frequencies from 1800 to 1000 cm^{-1} (Ma *et al.*, 2010). The capability to measure *simultaneously* the VCD spectra of a protein at the amide I, II, and III bands is unique to FT-VCD instrumentation. Kinetics can also be observed across an entire spectral range while with dispersive instrumentation this is not possible due to a time-bias associated with the scanning of a spectrum across different wavelengths of the spectrum.

A comparison of VA and VCD for a series of proteins that are changing gradually in secondary structure content from high α-helix to high β-sheet is shown in Figure 10.23. Clear trends across the frequency range from 1800 to 1400 cm^{-1} can be followed providing a solid foundation for VCD to be used as a sensitive database for the prediction of secondary structure analogous to that already established for ROA described above for Figure 10.21. VA and VCD spectra in H_2O can also be obtained in the near-IR region using FT-VCD instrumentation with appropriately changed detectors, optical filters, and PEMs as described in Chapter 6. In the near-IR, VCD has similar sensitivity to changes in secondary structure as that found in the mid-IR. Results analogous to those for the mid-IR are displayed on the right side of Figure 10.23.

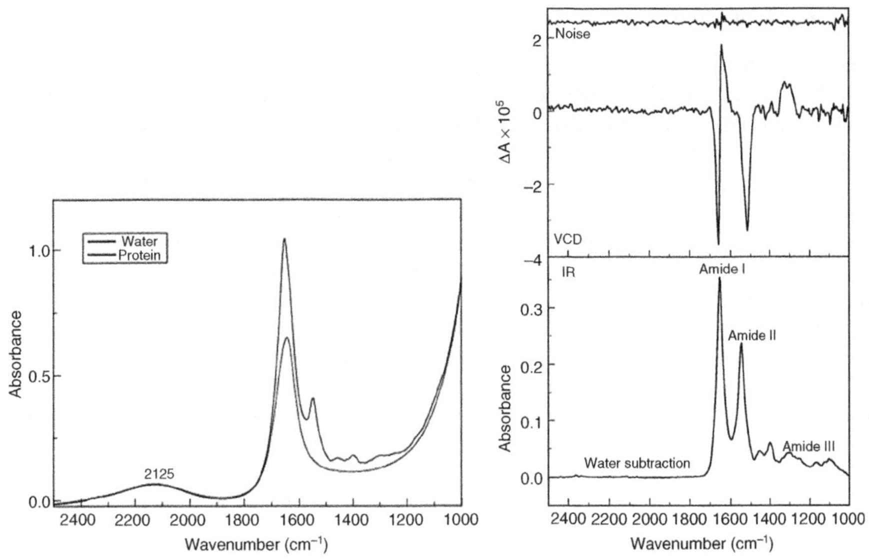

Figure 10.22 Absorbance spectrum (left) from 2500 to 1000 cm^{-1} of an aqueous solution of myoglobin superimposed on the H_2O solvent spectrum showing the exact subtraction of both spectra at 2125 cm^{-1} corresponding to a combination band of the water spectrum. The resulting solvent corrected VA and VCD spectra of myoglobin (right) over the same spectral region showing the features in the amide I, II, and III regions. Reproduced with permission from the Society for Applied Spectroscopy (Ma et al., 2010)

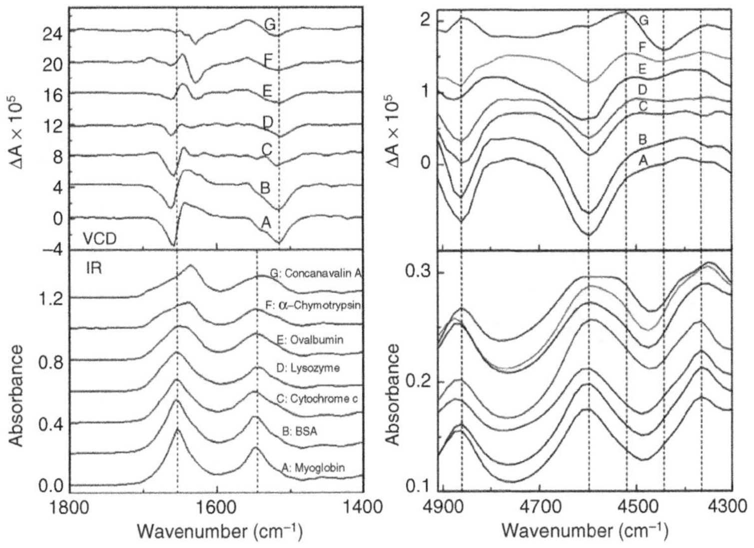

Figure 10.23 Mid-IR (left) and near-IR (right) VA and VCD spectra of seven proteins that differ in structure from high α-helix (bottom) to high β-sheet (top). Reproduced with permission from the Society for Applied Spectroscopy (Ma et al., 2010)

10.4.5 ROA of Viruses

An interesting and important application of ROA is its unique capability to obtain molecular-level structural information for viruses (Blanch *et al.*, 2002). Despite the enormous size of viruses, they are typically comprised of a capsid shell organized in a symmetrical manner by a myriad of coat proteins. ROA can characterize the secondary structure of these coat proteins and thereby obtain information specific to a particular virus. Equivalent information can only be obtained using X-ray crystallography provided that crystals of sufficient quality of the virus can be grown, typically a lengthy process. Further, the ROA of any viral coat protein can be classified against the database, such as that in Figure 10.21, to gain further information about the fold family of the coat protein.

In addition, viruses contain, as an agent of infection, one or more viral RNA molecules. These cannot be imaged by X-ray crystallography, but their ROA spectra and hence their conformational state can be obtained by subtracting the ROA spectrum of the empty capsid from the ROA spectrum of a capsid with the RNA component. An example of this analysis is given in Figure 10.24 for the cowpea mosaic virus (Barron and Buckingham, 2010). The ROA of the viral RNA-2 can be compared with

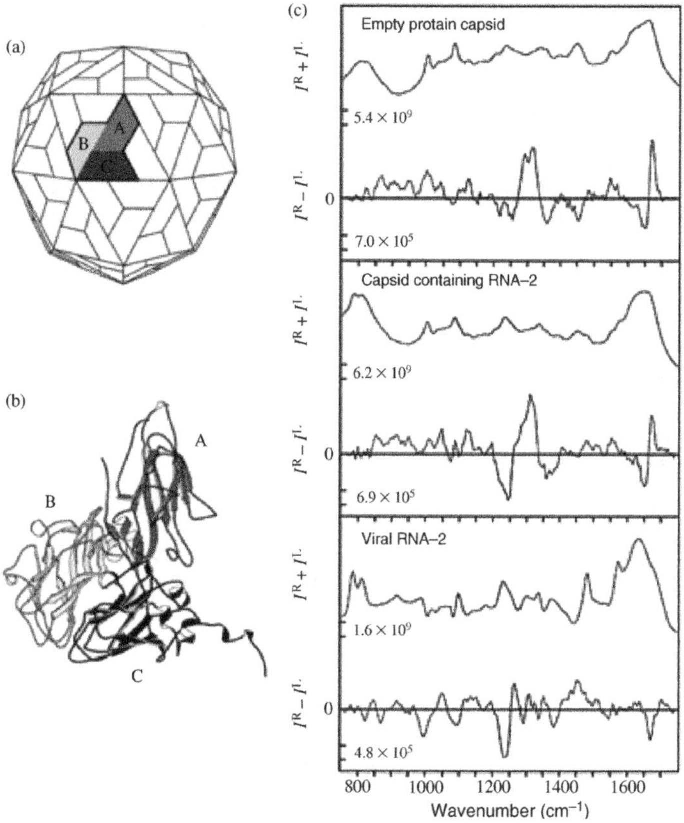

Figure 10.24 *Raman and ROA of cowpea mosaic virus, (c), of the empty capsid consisting of the coat protein (top), the capsid with viral RNA-2 (middle), and by subtraction the viral Raman and ROA spectra. The location of the coat protein on the viral shell (a) and the structure of the protein (b) are illustrated on the left. Reproduced with permission from Elsevier Science Ltd. (Barron and Buckingham, 2010)*

RNA from other sources to confirm its identity as RNA and to discern differences between free RNA and that located in the viral capsid (Blanch *et al.*, 2002).

10.4.6 VCD Calculations of Peptides

Although quantum mechanical VCD calculations for complete proteins have not yet been achieved, progress toward this goal is being made. Recently, DFT calculations have been carried out for peptides chains of up to 21 subunits by tensor transfer methods to simulate accurately the VCD in the amide I and II regions of helical secondary structures (Kubelka *et al.*, 2009). The tensor transfer method involves first calculating the VA and VCD spectra of a tripeptide or pentapeptide using a DFT model chemistry for a particular choice of secondary structure, and then extending the peptide to 20 or so units by transferring force constants, APTs and AATs needed to recalculate the vibrational modes of the longer peptide and to calculate the VA and VCD spectra for these new modes. An example of this Cartesian coordinate tensor (CCT) transfer method is given for the peptide acetyl-(alanyl)$_{20}$-NH-CH_3 in Figure 10.25. Here DFT calculated VA and VCD spectra with PCM solvent corrections are presented for the right-handed α-helix and 3_{10}-helix structures and for the extended left-handed 3_1-helix, which is the same as the PPII helix. The calculations capture the essence of all three of these helices compared with the corresponding spectra of these structures isolated experimentally. In particular, VCD can distinguish the α-helix from the 3_{10}-helix by a change in the relative magnitudes of the amide I couplet and amide II negative band, for which the amide I couplet is larger than the amide II band in the α-helix but this relative size reverses for the 3_{10}-helix. The PPII helix has an opposite signed amide I couplet due to the opposite handedness of this helix. As noted above, the PPII helix hydrogen bonded to water is the local chiral structure of the so-called random coil structure of proteins.

These calculations can be extended to include the interaction of a peptide with explicit water molecules. The results of DFT calculations of acetyl-(alanyl)$_{14}$-NH-CH_3 in an α-helix conformation without explicit water, partially-hydrated helix, and fully-hydrated helix are illustrated in Figure 10.26 for the VA and VCD spectra of the amide I′ mode (amide NH deuterated). It is clear that water has a significant effect of the frequencies and also on the vibrational coupling in the case of the partially hydrated α-helix. After full hydration the original VCD-couplet pattern returns with broadened bandshape and associated lower peak intensities.

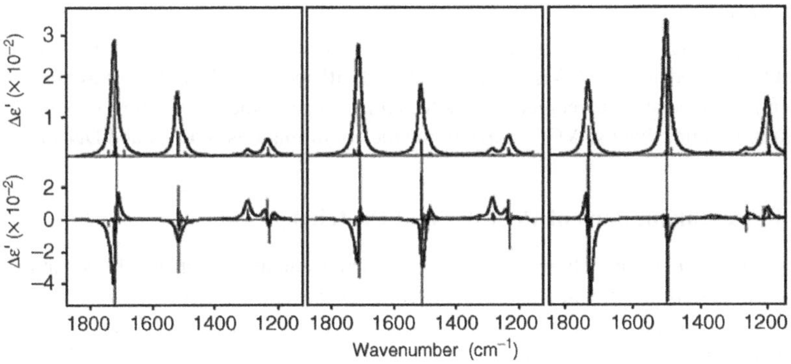

Figure 10.25 Calculated VA and VCD spectra of the peptide acetyl-(alanyl)$_{20}$-NH-CH_3 with α-helix (left), 3_{10}-helix, (middle) and 3_1-helix (PPII-type) (right) secondary structures. Reproduced with permission from IOS Press (Kubelka et al., 2009)

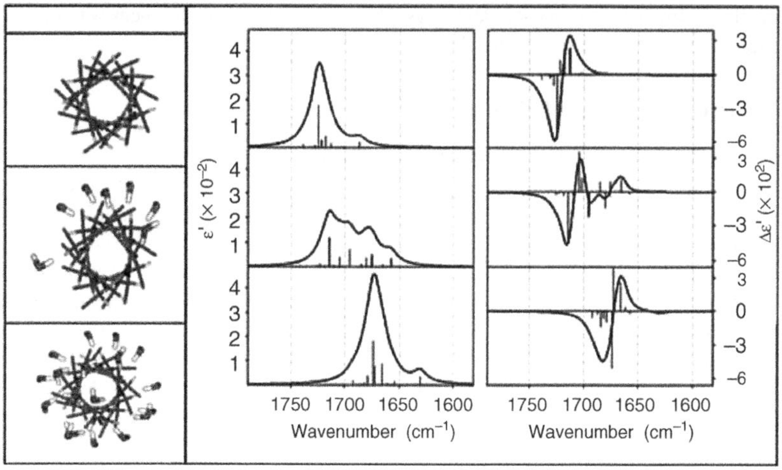

Figure 10.26 Structures with explicit waters (left) and DFT calculated VA and VCD spectra (right) of acetyl-(alanyl)$_{14}$-NH-CH$_3$ in an α-helix conformation without waters (top), partial waters (middle) and full hydration (bottom). Reproduced with permission from IOS Press (Kubelka et al., 2009)

10.4.7 VCD Calculations of Nucleic Acids

The application of VOA to the study of the solution-state conformations of nucleic acids is a promising area, despite its relative neglect compared with VOA studies of amino acids, peptides, and proteins. An example of VA and VCD spectroscopy applied to nucleic acids is a study that combines experimental VA and VCD measurements with DFT calculations (Andruschenko *et al.*, 2004). The focus of VCD studies in nucleic acids is the region of base vibrational modes above 1500 cm^{-1}. The importance of these modes is illustrated in Figure 10.27, where calculations of the AU base pair are compared with poly(rA)-poly(rU) in the base vibrational region. The effect of base planarity and hydration of the AU pair can be studied prior to VA and VCD calculations of more complex structures.

Such a complex structure is illustrated in Figure 10.28 where the calculation of VA and VCD spectra of the double stranded octomer (rA)$_8$-(rU)$_8$ is compared with the VA and VCD spectra of poly(rA)-poly (rU). A remarkable level of agreement across the entire range of the frequencies is found that bodes well for future studies that use VCD to determine the conformations of RNA and DNA molecules.

10.4.8 ROA Calculations of Peptides and Proteins

Impressive progress has been achieved in recent years for the quantum mechanical calculation of ROA, which for the first time provides detailed insight into the characteristics ROA spectral features presented above for peptides and proteins. ROA in the amide I region for model peptide structures has been analyzed using DFT calculations and a fragment approximation for both the α-helix and PPII-helix structures. Good correlation is found between the backbone conformation of these peptides and previously measured ROA spectra in this region (Choi and Cho, 2009). Model calculations of the peptide Ala$_{21}$ have been carried out for the α-helix, 3$_{10}$-helix, and PPII-helix and the results have been analyzed in terms of understanding the origin of the ROA of peptide and protein secondary structures

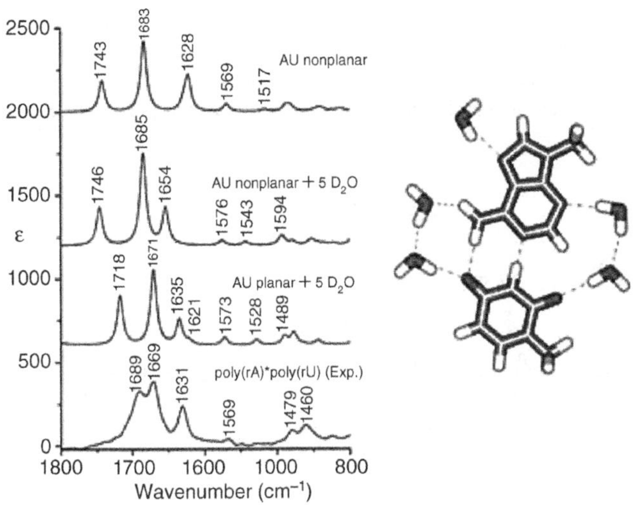

Figure 10.27 Comparison of measured VA spectra of poly(rA)-poly(rU) (bottom) with calculated VA spectra of the hydrated AU base pair as non-planar (top), non-planar hydrated with five H_2O molecules (next to top), and planar with five waters (next to bottom), with the structure shown to the right. Reproduced with permission from the American Chemical Society (Andruschenko et al., 2004)

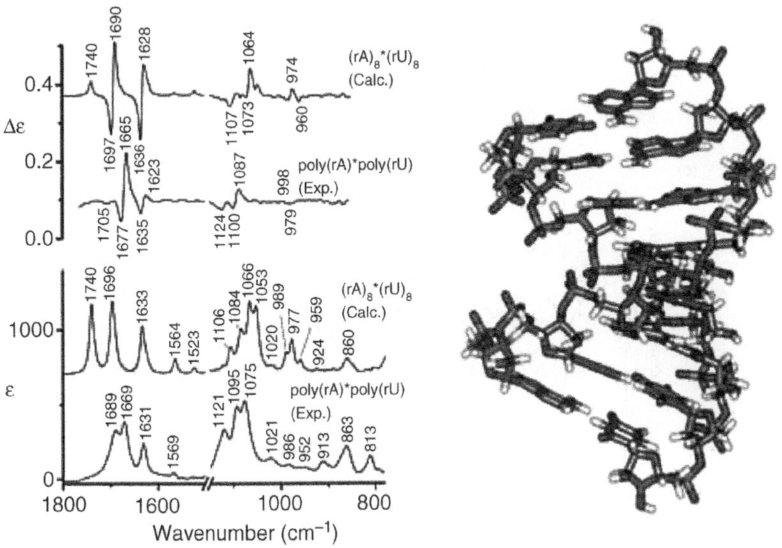

Figure 10.28 Comparison of measured VA and VCD spectra of poly(rA)-poly(rU) with DFT calculated VA and VCD spectra of the double-stranded octomer $(rA)_8$-$(rU)_8$, the structure of which is shown to the right. Reproduced with permission from the American Chemical Society (Andruschenko et al., 2004)

(Jacob *et al.*, 2009). In Figure 10.29, we show a comparison between calculated α-helical Ala_{21}, and poly-L-alanine in different solvents. It can be seen that the calculation, even without solvent corrections or explicit waters, captures the many the dominant ROA features associated with the α-helix secondary structure identified empirically for proteins and polypeptides.

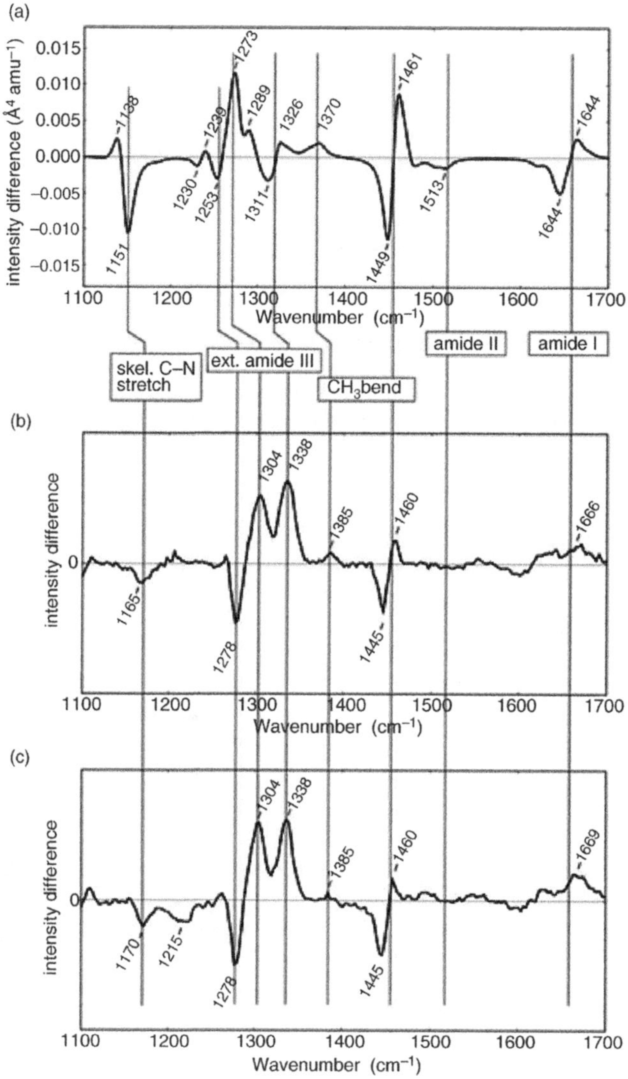

Figure 10.29 *Comparison of DFT calculated ROA for (a) the peptide Ala_{21} in the α-helix conformation with no solvent correction to the measured ROA of α-helical poly-L-alanine in (b) dichloroacetic acid (DCA) and in (c) 30% DCA/70% $CHCl_3$. Reproduced with permission from Wiley-VCH Verlag GmbH & Co. (Jacob et al., 2009)*

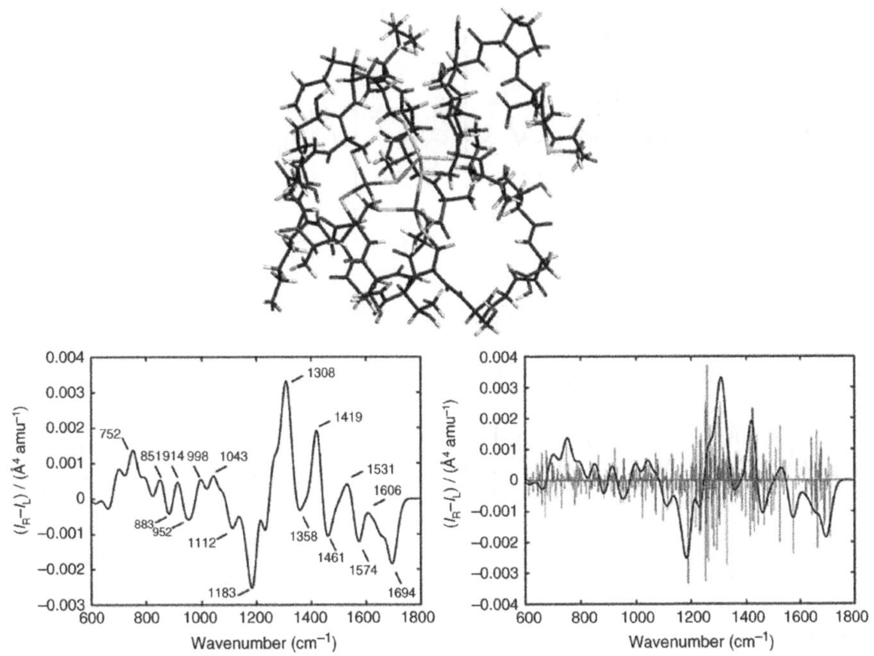

Figure 10.30 *The DFT/B86/RI/TZVP optimized structure of the β-domain of rat metallothionein (top) with the corresponding calculated ROA convolved with 20 cm⁻¹ halfwidth Gaussian bandshapes (left) and the corresponding ROA stick intensities (right) showing the contribution of each calculated normal mode of the protein. Reproduced with permission from the American Chemical Society (Luber and Reiher, 2009)*

Finally, we show the calculated VOA spectrum of a protein subunit, which is also the largest structure for which either VCD or ROA have been calculated to date with 400 atoms (Luber and Reiher, 2009). In Figure 10.30, we show the structure of the β-domain of rat metallothionein optimized at the level DFT/B86/RI/TZVP. The backbone of this structure is close to the X-ray structure of rat metallothionein after omitting the α-domain structure. Also shown in the figure is the associated calculated ROA spectrum convolved with Gaussian bandshapes of 20 cm^{-1} halfwidth and individual stick intensities for each of the normal modes of the protein. The only secondary structure units occurring in this protein are β-turns, and the calculations indicate that the 1100–1400 cm^{-1} region of the ROA is diagnostic of such turns. The experimental ROA spectrum has been published along with other proteins having native irregular folds (Smythe *et al.*, 2001). The comparison of the calculated and measured Raman and ROA of rat metallothionein is shown in Figure 10.31 where an impressive level of spectral agreement is achieved.

10.4.9 VOA of Supramolecular Biological Structures

Applications of VOA to biological structures larger than individual proteins or nucleic acids have begun to appear. VOA spectra display, under circumstances that are not yet quantitatively understood, enhanced intensities due to supramolecular assemblies of biomolecules. These applications of VOA differ from those previously discussed, even the ROA application to the coat proteins of viruses, because VOA intensities from supramolecular chiral structures are typically one to two orders of

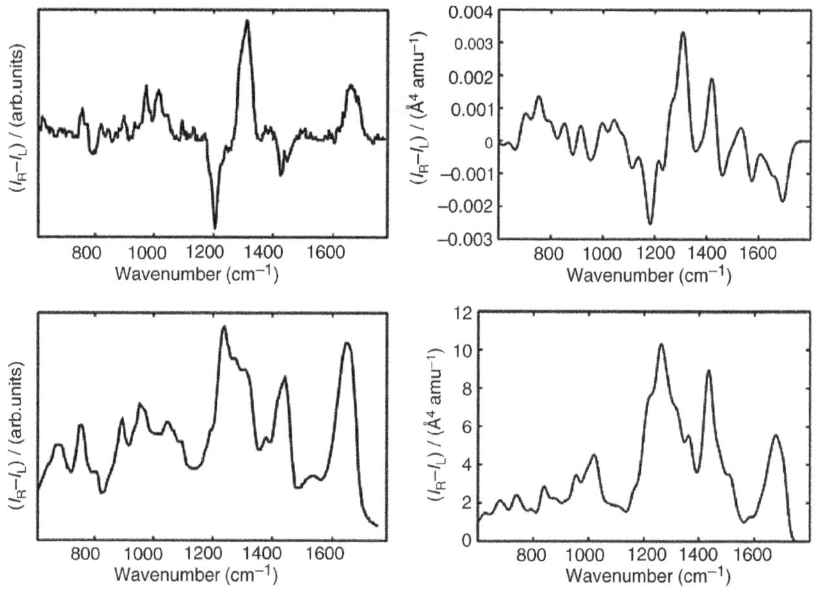

Figure 10.31 *Comparison of the aqueous solution experimental Raman and ROA (left) and DFT calculated Raman and ROA spectra (right) of the β-domain of rat metallothionein. Reproduced with permission from the American Chemical Society (Luber and Reiher, 2009)*

magnitude larger, relative to their parent VA or Raman spectrum, than ordinary VOA intensities. This enhancement is both intriguing and exciting owing to the difficulty of studying supramolecular structures at the molecular level and the relative ease with which these spectra can be measured.

10.4.9.1 VOA of Bacteria Flagella

An example of observations of supramolecular biological chirality is the ROA spectrum of the flagellar filament from *Salmonella* bacterium. The flagellar filament is an assembly of proteins that protrudes from the back of the bacterium and when rotated by an internal molecular motor propels the bacterium through its aqueous biological environment. The filaments can be isolated in three forms, L-type, Normal, and R-type (Uchihayama *et al.*, 2006). As shown in Figure 10.32, the filaments are comprised of 11 rows of proto-filaments, which in turn are comprised of individual flagellin proteins. Flagellin has a molecular weight of 51.5 kDa with 494 amino acid residues. The filaments are either straight with a left (L-type) or right (R-type) twist of the rows or the entire filament (Normal) can assume a curled morphology. When ROA spectra of these filaments are measured, the L-type shows an extraordinary level of enhancement whereby a strong ROA spectrum is obtained with only a few seconds of instrumental exposure time (Uchiyama *et al.*, 2008). The ROA spectrum of the L-type filament shown in Figure 10.32 is approximately only two orders of magnitude smaller than the parent Raman spectrum. The enhanced ROA spectrum is lost if the filament sample is heated or sonnicated resulting in a breakup of the supramolecular structure. Normal ROA spectra of flagellin can still be obtained after heating or sonnicating the filaments. Enhanced VCD has also been observed with different sample preparations for both the L-type and the R-type, but as with ROA, no enhancement of the Normal type was observed. At this stage, it is assumed that the straight type filaments, and in particular the L-type, can aggregate into some large chiral supramolecular assembly

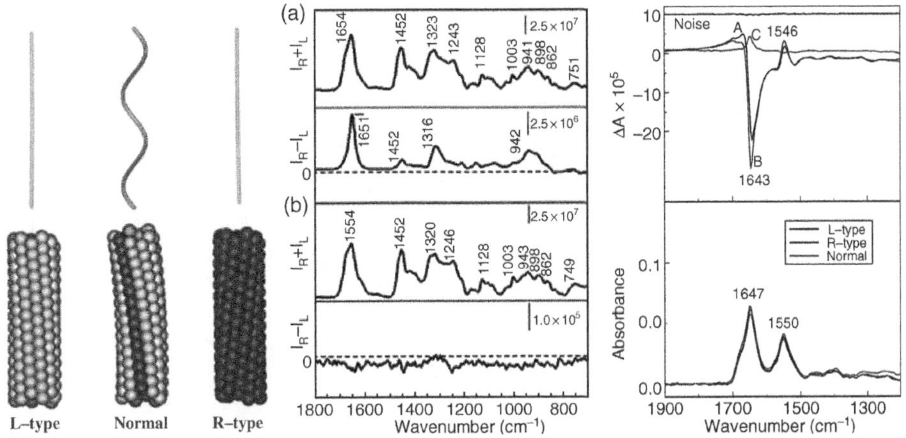

Figure 10.32 *Proposed structure of L-type, Normal, and R-type flagellar filaments. Small spheres depict individual flagellin proteins that align to form 11 rows of proto-filaments in each filament. Enhanced ROA for (a) the L-type filaments is destroyed by heating (b) with no noticeable change in the Raman spectrum. VCD of various sample preparations in a different location showed the same enhancement for both the L-type (A) and R-type filaments (B) but not of the Normal type (C) filament. Reproduced with permission from Elsevier Science Ltd (Uchiyama et al., 2008)*

of filaments, possibly a type of protein fibril that produces the enhanced ROA and VCD. Further research is needed with ROA and VCD measurements carried out for the same samples at the same time to resolve more fully the unanswered questions associated with this interesting VOA enhancement mechanism.

10.4.9.2 VCD of Protein Fibrils and Other Supramolecular Assemblies

We next consider the exciting and intriguing phenomena of enhanced VCD in protein peptides and fibrils. As background to this section, it has been observed for many years that occasionally very large VCD intensities are observed in the measurement of spectra. Initial reaction to these occurrences were to assume that something undesirable had happened to the sample, that aggregation had occurred, and that the measurements were of no value. In some cases this assumption may have been correct if the corresponding IR spectrum changed dramatically or if the sample had become opaque or cloudy. Protein samples, for example, can aggregate, cause high levels of scattering, and precipitate as solid particles from solution. However, more recently it has been noticed that under certain circumstance proteins and other biomolecules can aggregate, or assemble, into supramolecular structures that remain in solution without a significant change in the VA spectrum. The solution may become viscous or gel-like, but the sample is still viable and can be studied systematically and reproducibly. Furthermore, recently theoretical efforts to model extended supramolecular structures have shown strong indications that a detailed understanding of the phenomenon of enhanced protein fibril VCD may be achieved (Measey and Schweitzer-Stenner, 2011).

Enhanced VCD was reported for the first time for amyloid fibrils of lysozyme and insulin (Ma *et al.*, 2007). Although the VCD of these proteins in their native folded state was different, their enhanced VCD after fibrils formed showed VA and VCD bands at the same locations differing only in relative magnitude but not in sign. The spectral signature of the enhanced VCD from fibrils in

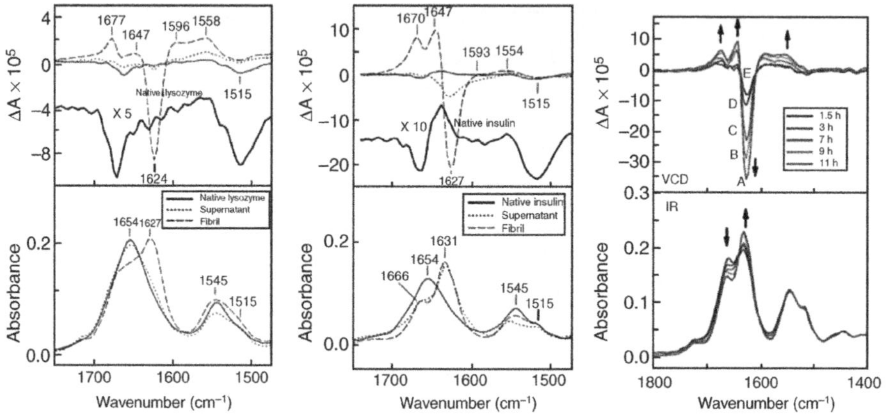

Figure 10.33 *VA and VCD spectra of lysozyme (left) and insulin (middle) as native aqueous solutions, prefibrillar supernatant solutions, and fibril gel phases. The kinetics of insulin fibril formation and development in solution without centrifugation are shown on the right: A, 1.5; B, 3; C, 7; D, 9; and E, 11 h. Reproduced with permission from the American Chemical Society (Ma et al., 2007)*

the amide I region is two positive bands to high frequency near 1675 and 1650 cm^{-1}, a strong negative VCD band near 1625 cm^{-1}, and two weaker positive bands near 1595 and 1555 cm^{-1}. Overall this pattern of five bands spans the traditional amide I and II frequency range. There are a variety of different conditions that induce fibril formation, but for these proteins, lowering the pH to near 2.0 followed by heating at 65 °C led to fibril formation and development. The VA and VCD spectra of both lysozyme and insulin in their native and fibril forms are shown in Figure 10.33. The amyloid fibrils were sampled as a gel phase after centrifugation and the VCD of the supernatant containing partially denatured prefibrillar proteins was measured for comparison to the native VCD spectrum, which, in each case, is also shown on an enlarged scale for greater clarity. Also shown in the figure is a sample of insulin that was heated at 65 °C at pH 2 for 45 min and then placed in the VCD spectrometer to monitor the growth and development of the fibrils with time. VCD is the only known technique that can follow the kinetics of fibril growth and development in solution with this level of stereo-sensitivity. Changes can also be followed by FT-IR spectroscopy, which were also obtained with VCD measurements, but the changes are relatively small and less varied by comparison.

More recently, a second remarkable finding was published showing that by careful control of the pH of the incubating solution, the sign of the supramolecular fibril VCD spectrum could be reversed as shown in Figure 10.34 (Kurouski *et al.*, 2010). Because VCD is a sensitive direct probe in the solution of the supramolecular chirality of fibrils, the reversal of sign of the VCD could only mean that the supramolecular chirality of the majority population of insulin fibrils present had a reversed helical form. It was found that fibrils incubated at below pH 2.1 exhibited reversed signed VCD compared with the normal fibrils grown at pH 2.4 and above. Reversed fibrils of insulin incubated at low pH do not show as large a VCD over time as do the normal fibrils. This led us to propose a mechanism of formation and development illustrated in Figure 10.34. Here the sense of helicity is established at the proto-filament stage depending on the pH and incubating conditions, and subsequently normal fibrils develop to a more mature stage than do reversed fibrils. Further research will be required to establish whether this proposed mechanism is correct.

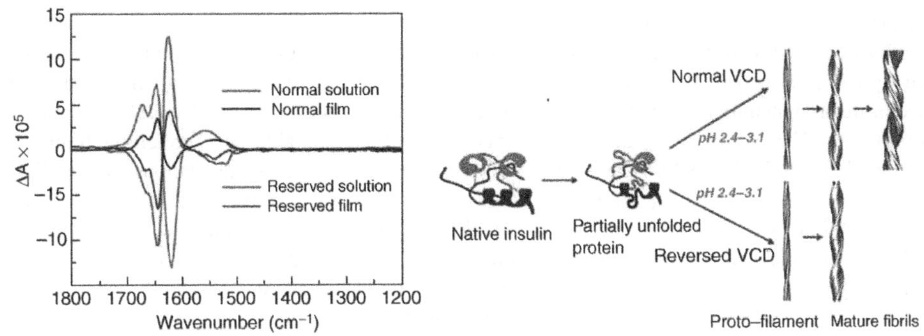

Figure 10.34 *Mirror symmetry of the VCD spectra of insulin in the solution and dry film state for fibrils prepared at pH at or above 2.4, normal, or at or below 2.1, reversed, for incubation at 65 °C for 1 h (left) and a proposed mechanism of formation and development. Reproduced with permission from The Royal Society of Chemistry (Kurouski et al., 2010)*

10.4.9.3 VCD of Spray-Dried Films

Unusually enhanced VCD has also been observed in spray-dried films of all the amino acids and some ionic forms of pharmaceutical molecules (amine salt solutions) where enhancement factors have been observed approaching 100 for some bands with little or no change in the VA spectrum relative to a standard solid-phase mull sample of the same molecule. As an example, the VA and VCD spectra for a spray-dried film of L-alanine film are shown in Chapter 8 for the measurement of solid-phase VCD, Figure 8.10 (Lombardi, 2011). Spray-dried films showing this property can be grown either from an aqueous salt solution, as shown, or from a salt-free solution with equal levels of VCD enhancement. The enhancement of the VCD in these cases is thought to arise from a chiral stacking of flat crystals that grew in two-dimensions as the spray dried on the optical surface, instead of in three dimensions from solution as is the case for normal crystalline L-alanine.

10.4.9.4 VCD of Other Biological Structures

Enhanced VCD is now emerging as a more wide-ranging phenomenon for both biological and other types of molecules. Large VCD has been observed been observed at Syracuse University in oligopeptides of gluatamine (Q) only when the number of Q-residues is beyond 30. This correlates with studies that link the presence of large poly-Q tails on certain proteins to the occurrence of fibrils in Huntington's disease (Sharma *et al.*, 2005). As mentioned earlier in this chapter, interesting examples of supramolecular chirality in biological molecules have been observed by Urbanova and co-workers, particular examples of which include supramolecular guanosine quartet assemblies (Stetincka *et al.*, 2006) and molecular associations of the of vitamin molecule, biotin, when they complexes with silver (I) ions (Goncharova *et al.*, 2010).

10.5 Future Applications of VOA

We conclude this chapter with a few thoughts regarding the future of VOA applications. In many ways the future is now. New papers on VCD and ROA are appearing at a rate that makes it difficult to keep up with the many advances that are occurring. As mentioned previously, the last decade or so has seen the emergence of VOA from two prior decades of early development in the areas of instrumentation, theory, computation, and application into a new era dominated by applications. The trigger for this emergence has been the availability of both commercial instrumentation for the measurement of VCD and ROA coupled with the availability of commercial software for VOA calculation using primarily density function theory. The real power of VCD is realized when measured and calculated VOA spectra are brought into juxtaposition for the extraction of detailed stereochemical and conformational information. The degree of agreement achievable is truly remarkable and foretells a future of continued growth in the sophistication of instrumentation, measurements, calculations, and applications of VOA. The future is bright as computers continue to become increasingly powerful, instrumentation becomes more varied and reliable, and applications are extended to varieties of biological and medical problems, many of which cannot be anticipated at this time.

The intrinsic value of VOA beyond its stereo-sensitivity, absent in non-chiral spectroscopic methods, is its capability for extracting information from solutions, solids, films, and diverse biological environments. It is likely that, based on work currently in progress, the field of VOA microscopy will emerge in the near future from the combination of VOA instrumentation with infrared and Raman microscopy. Another emerging frontier is time resolution. Here, it has been recently demonstrated that the VCD can be measured with femtosecond pulses using novel approaches for measuring the difference in IR intensities between RCP and LCP radiation states. Yet one other frontier now emerging is that of biomedical applications where ROA and VCD both show promise for shedding light on the structures of proteins related to the onset and development of neurodegenerative and other types disease. What applications beyond these will emerge in the coming years is anyone's guess. What does seem certain is that the techniques of VCD and ROA will eventually be adopted as universal fundamental techniques of great incisiveness and utility for the investigation of properties of matter at the molecular level, the so-called nanoscale regime and beyond.

References

Andruschenko, V., Wieser, H., and Bour, P. (2004) RNA Structural forms studied by vibrational circular dichroism: *Ab Initio* interpretation of the spectra. *J. Phys. Chem. B*, **108**, 3899–3911.

Barron, L.D. (1977a) Raman optical activity of simple chiral molecules; methyl and trifluoromethyl asymmetric deformations. *J. Chem. Soc. Perkin Trans.* **2**, 1790–1794.

Barron, L.D. (1977b) Raman optical activity of camphor and related molecules. *J. Chem. Soc. Perkin Trans.* **2**, 1074–1079.

Barron, L.D. (2006a) Structure and behaviour of biomolecules from Raman optical activity. *Curr. Opin. Struct. Biol.*, **16**, 638–643.

Barron, L.D. (2006b) Raman optical activity: A new light on proteins, carbohydrates and glycoproteins. *Biochemist*, **June,** 27–31.

Barron, L.D. (2007) Compliments of Lord Kelvin. *Nature*, **446**, 506.

Barron, L.D., Zhu, F., Hecht, L., and Isaacs, N.W. (2007a) Structure and behavior of proteins from Raman optical activity. In: *Methods in Protein Structure and Stability Analysis* (eds V. Uversky, and E. Permyakov), Nova Science Publishers, Inc., Hauppauge, pp. 27–68.

Barron, L.D., Zhu, F., Hecht, L. *et al.* (2007b) Raman optical activity: An incisive probe of molecular chirality and biomolecular structure. *J. Mol. Spec.*, **834-836** 7–16.

Barron, L.D., and Buckingham, A.D. (2010) Vibrational optical activity. *Chem. Phys. Lett.*, **492**, 199–213.

Bijvoet, J.M., Peerdeman, A.F., and Bommel, A. J. v. (1951) Determination of the absolute configuration of optically active compounds by means of X-rays. *Nature*, **168**, 271–272.

Birke, S.S., Agbaje, I., and Diem, M. (1992) Experimental and computational infrared CD studies of prototypical peptide conformations. *Biochemistry*, **31**, 450–455.

Blanch, E.W., Hecht, L., Syme, C.D. *et al.* (2002) Molecular structures of viruses from Raman optical activity. *J. Gen. Virol.*, **83**, 2593–2600.

Bormett, R.W., Asher, S.A., Larkin, P.J. *et al.* (1992) Selective examination of heme protein azide ligand-distal globin interactions by vibrational circular dichroism. *J. Am. Chem. Soc.*, **114**, 6864–6867.

Choi, J.-H., and Cho, M. (2009) Amide I Raman optical activity of polypoptides: Fragment approximation. *J. Chem. Phys.*, **130**, 145503.

Cianciosi, S.J., Spencer, K.M., Freedman, T.B. *et al.* (1989) Synthesis and gas-phase vibrational circular dichroism of (+)-(S,S)-cyclopropane-1,2-$^2H_2$. *J. Am. Chem. Soc.*, **111**, 1913–1915.

Cianciosi, S.J., Ragunathan, N., Freedman, T.B. *et al.* (1991) Racemization and geometrical isomerization of (2S,3S)-cyclopropane-1-^{13}C-1,2,3-d_3 at 407 °C: Kinetically competitive one-center and two-center thermal epimerizations in an isotopically substituted cyclopropane. *J. Am. Chem. Soc.*, **113**, 1864–1866.

Costante, J., Hecht, L., Polavarapu, P.L. *et al.* (1997) Absolute-configuration of bromochlorofluoromethane from experimental and *ab-initio* theoretical vibrational Raman optical-activity. *Angew. Chem. Int. Ed. Engl.*, **36**, 885–887.

Diem, M., Gotkin, P.J., Kupfer, J.M. *et al.* (1977) Vibrational circular dichroism in amino acids and peptides. 1. Alanine. *J. Am. Chem. Soc.*, **99**, 8103–8104.

Diem, M., Gotkin, P.J., Kupfer, J.M., and Nafie, L.A. (1978) Vibrational circular dichroism in amino acids and peptides. 2. Simple alanyl peptides. *J. Am. Chem. Soc.*, **100**, 5644–5650.

Dukor, R.K., and Keiderling, T.A. (1991) Reassessment of the random coil conformation. Vibrational circular dichroism study of proline oligopeptides and related polypeptides. *Biopolymers*, **31**, 1747–1761.

Dukor, R.K., Keiderling, T.A., and Gut, V. (1991) Vibrational circular dichroism spectra of unblocked proline oligomers. *Int. J. Pept. Protein Res.*, **38**, 198–203.

Eker, F., Griebenow, K., Cao, X. *et al.* (2004a) Preferred peptide backbone conformations in the unfolded state revealed by the structure analysis of alanine-based (AXA) tripeptides in aqueous solution *Proc. Nat. Acad. Sci.*, **101**, 10054–10059.

Eker, F., Griebenow, K., Cao, X. *et al.* (2004b) Tripeptides with ionizable side chains adopt a perturbed polyproline II structure in water. *Biochemistry*, **43**, 613–621.

Freedman, T.B., Diem, M., Polavarapu, P.L., and Nafie, L.A. (1982) Vibrational circular dichroism in amino acids and peptides. 6. Localized molecular orbital calculations of the carbon hydrogen stretching VCD in deuterated isotopomers of alanine. *J. Am. Chem. Soc.*, **104**, 3343–3349.

Freedman, T.B., Cianciosi, S.J., Ragunathan, N. *et al.* (1991) Optical activity arising from ^{13}C substitution: Vibrational circular dichroism study of (2S,3S)-cyclopropane-1-^{13}C,^{2}H-2, 3-$^2H_2$. *J. Am. Chem. Soc.*, **113**, 8298–8305.

Freedman, T.B., Nafie, L.A., and Keiderling, T.A. (1995) Vibrational optical-activity of oligopeptides. *Biopolymers*, **37**, 265–279.

Freedman, T.B., Cao, X., Dukor, R.K., and Nafie, L.A. (2003) Absolute configuration determination of chiral molecules in the solution state using vibrational circular dichroism. *Chirality*, **15**, 743–758.

Goncharova, I., Sykora, D., and Urbanová, M. (2010) Association of biotin with silver (I) in solution: A circular dichroism study. *Tetrahedron: Asymmetry*, **21**(15) 1916–1920.

Guo, C., Shah, R.D., Dukor, R.K. *et al.* (2004) Determination of enanitomeric excess in samples of chiral molecules using Fourier transform vibrational circular dichroism spectroscopy: Simulation of real-time reaction monitoring. *Anal. Chem.*, **76**, 6956–6966.

Guo, C., Shah, R.D., Cao, X. *et al.* (2005) Enantiomeric excess determination by Fourier transform near-infrared vibrational circular dichroism spectroscopy: Simulation of real-time process monitoring. *Appl. Spectrosc.*, **59**, 1114–1124.

Guo, C., Shah, R.D., Mills, J. *et al.* (2006) Fourier transform near-infrared vibrational circular dichroism used for on-line monitoring of the epimerization of 2,2-dimethyl-1, 3-dioxolane-4-methanol – A pseudo racemization reaction. *Chirality*, **18**, 775–782.

Haesler, J., Schindelholz, I., Riguet, E. *et al.* (2007) Absolute configuration of chirally deuterated neopentane. *Nature*, **446**, 526–529.

He, Y., Cao, X., Nafie, L.A., and Freedman, T.B. (2001) *Ab initio* VCD calculation of a transition-metal containing molecule and new intensity enhancement mechanism for VCD. *J. Am. Chem. Soc.*, **123**, 11320–11321.

Hecht, L., Phillips, A.L., and Barron, L.D. (1995) Determination of enantiomeric excess using Raman optical-activity. *J. Raman Spectrosc.*, **26**, 727–732.

Holzwarth, G., and Chabay, I. (1972) Optical activity of vibrational transitions. Coupled oscillator model. *J. Chem. Phys.*, **57**, 1632–1635.

Jacob, C.R., Luber, S., and Reiher, M. (2009) Understanding the signatures of secondary-structures elements in proteins with Raman optical activity spectroscopy. *Chem. Eur. J.*, **15**, 13491–13508.

Jalkanen, K.J., Degryarenko, I.M., Nieminen, R.M. *et al.* (2008) Role of hydration in determining the structure and vibrational spectra of L-alanine and *N*-acetyl L-alanine *N'*-methylamide in aqueous solution: A combined theoretical and experimental approach. *Theor. Chem. Acc.*, **119**, 191–210.

Keiderling, T.A., Kubelka, J., and Hilario, J. (2006) Vibrational circular dichroism of biopolymers: Summary of methods and applications. In: *Vibrational Spectroscopy of Biological and Polymeric Materials* (eds V.G. Gregoriou, and M.S. Braiman), CRC Press, Boca Raton, pp. 253–324.

Kubelka, J., Bour, P., and Keiderling, T.A. (2009) Quantum mechanical calculations of peptide vibrational force fields and spectral intensities. In: *Biological and Biomedical Infrared Spectroscopy* (eds A. Barth,and I.H. Parvez), IOS Press, Amsterdam, pp. 178–223.

Kurouski, D., Lombardi, R.A., Dukor, R.K. *et al.* (2010) Direct observation and pH control of reversed supramolecular chirality in insulin fibrils by vibrational circular dichroism. *Chem. Commun.*, **46**, 7154–7156.

Kuwahara, S., Obuta, K., Fujita, T. *et al.* (2010) (R)-(+)-[VCD(–)984]-4-ethyl-4-ethyloctane: A cryptochiral hydrocarbon with a quaternary chiral center. (2) Vibrational CD spectra of both enantiomers and absolute configuration assignment. *Eur. J. Org. Chem.* 6385–6392.

Lombardi, R.A. (2011) Ph.D. thesis, Syracuse University.

Luber, S., and Reiher, M. (2009) Theoretical Raman optical activity study of the beta domain of rat metallothionein. *J. Phys. Chem. B*, **114**, 1057–1063.

Ma, S., Cao, X., Mak, M. *et al.* (2007) Vibrational circular dichroism shows unusual sensitivity to fibril formation and development in solution. *J. Am. Chem. Soc.*, **129**, 12364–12365.

Ma, S., Freedman, T.B., Dukor, R.K., and Nafie, L.A. (2010) Near-infrared and mid-infrared Fourier transform vibrational circular dichroism of proteins in aqueous solution. *Appl. Spectrosc.*, **64**, 615–626.

McColl, I.H., Blanch, E.W., Gill, P.M.W. *et al.* (2003) A new perspective on β-sheet structures using vibrational Raman optical activity: From poly(L-lysine) to prion protein. *J. Am. Chem. Soc.*, **125**, 10019–10026.

McColl, I.H., Blanch, E.W., Hecht, L. *et al.* (2004) Vibrational Raman optical activity characterization of poly(L-proline). II Helix in alanine oligopeptides. *J. Am. Chem. Soc.*, **126**, 5076–5077.

Measey, T.J., and Schweitzer-Stenner, R. (2011) Vibrational circular dichroism as a probe of fibrillogensis: The origin of the anomolous intensity enhancement of amyloid-like fibrils. *J. Am. Chem. Soc.*, **133**, dx.doi.org/10.1021/ja1089827.

Merten, C., Li, H., Lu, X. *et al.* (2010) Observation of resonant electronic and non-resonance-enhanced vibrational natural Raman optical activity. *J. Raman Spectrosc.*, **41**, 1273–1275.

Nafie, L.A. (2008) Vibrational circular dichroism: A new tool for the solution-state determination of the structure and absolute configuration of natural product molecules. *Nat. Prod. Commun.*, **3**, 451–466.

Nafie, L.A., Keiderling, T.A., and Stephens, P.J. (1976) Vibrational circular dichroism. *J. Am. Chem. Soc.*, **98**, 2715–2723.

Nafie, L.A., and Walnut, T.H. (1977) Vibrational circular dichroism theory: A localized molecular orbital model. *Chem. Phys. Lett.*, **49**, 441–446.

Nafie, L.A., Oboodi, M.R., and Freedman, T.B. (1983) Vibrational circular dichroism in amino acids and peptides. 8. A chirality rule for the methine C*H stretching mode. *J. Am. Chem. Soc.*, **105**, 7449–7450.

Nafie, L.A., and Freedman, T.B. (2000) Biological and pharmaceutical applications of vibrational optical activity. In: *Infrared and Raman Spectroscopy of Biological Materials* (eds H.-U. Gremlich, and B. Yan), Marcel Dekker, Inc., New York, pp. 15–54.

Nafie, L.A., and Dukor, R.K. (2007) Pharamaceutical applications of vibrational optical activity. In: *Applications of Vibrational Spectroscopy in Pharmaceutical Research and Development* (eds D. Pivonka, P.R. Griffiths, and J.M. Chalmers) John Wiley & Sons, Ltd., Chichester, pp. 129–154.

Paterlini, M.G., Freedman, T.B., and Nafie, L.A. (1986) Vibrational circular dichroism spectra of three conformationally distinct states and an unordered state of poly(L-lysine) in deuterated aqueous solution. *Biopolymers*, **25**, 1751–1765.

Polavarapu, P.L. (2002) The absolute configuration of bromochlorofluoromethane. *Angew. Chem. Int. Ed.*, **41**, 4544–4546.

Polavarapu, P.L. (2007) Renaissance in chiroptical spectroscopic methods for molecular structure determination. *Chem. Rec.*, **7**, 125–136.

Schweitzer-Stenner, R., Measey, T., Kakalis, L. *et al.* (2007) Conformations of alanine-based peptides in water probed by FTIR, Raman, vibrational circular dichroism, electronic circular dichroism, and NMR spectroscopy. *Biochemistry*, **46**, 1587–1596.

Sharma, D., Shinchuk, L., Inouye, H. *et al.* (2005) Polyglutamine homopolymers having 8-45 residues form slablike beta-crystallite assemblies. *Prot. Struct. Funct. Bioinf.*, **61**, 398–411.

Singh, R.D., and Keiderling, T.A. (1981) Vibrational circular dichroism of poly-gamma-benzyl-L-glutamate. *Biopolymers*, **20**, 237–240.

Smythe, E., Syme, C.D., Blanch, E.W. *et al.* (2001) Solution structures of native proteins with irregular folds from Raman optical activity. *Biopolymers*, **52**, 138–151.

Solladie-Cavallo, A., Balaz, M., Salisova, M. *et al.* (2001a) A new chiral oxathiane: Synthesis, resolution and absolute configuration determination by vibrational circular dichroism. *Tetrahedron: Asymmmetry*, **12**, 2605–2611.

Solladie-Cavallo, A., Sedy, O., Salisova, M. *et al.* (2001b) A chiral 1,4-oxazin-2-one: Asymmetric synthesis versus resolution, structure, conformation and VCD absolute configuration. *Tetrahedron: Asymmmetry*, **12**, 2703–2707.

Soulard, P., Asselin, P., Cuisset, A. *et al.* (2006) Chlorofluoroiodomethane as a potential candidate for parity violation measurements. Spectroscopic features: supersonic beam spectroscopy and VCD in the gas phase. Preparation of its partially resolved enantiomers and enantioselective recognition by a chiral cryptophane, *PhysChemChemPhys*, **8**, 79–92.

Spencer, K.M., Cianciosi, S.J., Baldwin, J.E. *et al.* (1990) Determination of enantiomeric excess in deuterated chiral hydrocarbons by vibrational circular dichroism spectroscopy. *Appl. Spectrosc.*, **44**, 235–238.

Stephens, P.J., and Devlin, F.J. (2000) Determination of the structure of chiral molecules using *ab initio* vibrational circular dichroism spectroscopy. *Chirality*, **12**, 172–179.

Stetincka, V., Urbanova, M., Volka, K. *et al.* (2006) Investigation of guanosine quartet assemblies by vibrational and electronic circular dichroism spectroscopy: A novel approach for studying supramolecular assemblies. *Chem. Euro. J.*, **12**, 8735–8743.

Uchihayama, T., Sonoyama, M., Hamada, Y. *et al.* (2006) Raman spectroscopic study of the L-type straight flagellar filament of *Salmonella*. *Vib. Spectrosc.*, **42**, 192–194.

Uchiyama, T., Sonoyama, M., Hamada, Y. *et al.* (2008) Raman optical activity of flagellar filaments of *Salmonella:* Increase in ROA intensity upon formation of certain self-assembled protein filaments and their possible higher level structures. *Vib. Spectrosc.*, **48**, 65–68.

Urbanova, M., Stetincka, V., Devlin, F.J., and Stephens, P.J. (2005) Determination of the molecular structure in solution using vibrational circular dichroism: The supramolecular tetramer of 2,2′-dimethyl-biphenyl-6, 6′-dicarboxylic acid. *J. Am. Chem. Soc.*, **127**, 6700–6711.

Yasui, S.C., and Keiderling, T.A. (1986) Vibrational circular dichroism of polypeptides VIII. Poly lysine conformations as a function of pH in aqueous solution. *J. Am. Chem. Soc.*, **108**, 5576–5581.

Zhu, F., Tranter, G.E., Isaacs, N.W. *et al.* (2006) Delineation of protein structure classes from multivariate analysis of protein Raman optical activity data. *J. Mol. Biol.*, **363**, 19–26.

Appendix A

Models of VOA Intensity

In this first Appendix we present and describe the simplest of the models of VOA intensity as a conceptual supplement to the formal expressions of the theory of VCD and ROA presented in Chapters 4 and 5. These model descriptions appear here in an appendix rather than in the main body of the chapter because they are simply no longer used quantitatively to interpret VOA intensities, having been replaced by far more accurate quantum mechanical calculations. Nevertheless, they retain an intrinsic value in thinking about the origin of VCD intensities and where large intensities might be encountered. A comprehensive theoretical description of the various models of VCD intensities was published a number of years ago (Freedman and Nafie, 1994). A number of models of ROA have also been developed, but they have found even less use than those for VCD in the interpretation of ROA features in molecules. We will also briefly discuss the ROA models and refer the interested reader to their description in books and the literature. Before we begin a description of the models, we first present a short back-of-the-envelope calculation of the magnitude of CD intensity relative to the parent absorption intensity.

A.1 Estimate of CD Intensity Relative to Absorption Intensity

In this section we consider an extremely brief description of the ratio of CD intensities to the corresponding absorption intensities, also known as anisotropy ratio, g. By considering a simple heuristic expression for this ratio, we can understand why VCD intensities are smaller by an order of magnitude or more than both ECD and ROA intensities.

Circular dichroism intensity is given by the rotational strength, R, while the associated absorption intensity is given by the dipole strength, D. In the simplest of terms we can write

$$R = \mathrm{Im}\boldsymbol{\mu} \cdot \boldsymbol{m} \quad D = |\boldsymbol{\mu}|^2 \tag{A1}$$

where $\boldsymbol{\mu}$ is the electric dipole transition moment and $\boldsymbol{m}$ is the corresponding magnetic dipole transition moment. We can form simple expressions for these moments by writing

$$\boldsymbol{\mu} = e\boldsymbol{r} \quad \boldsymbol{m} = \frac{e}{2mc}\boldsymbol{r} \times \boldsymbol{p} \tag{A2}$$

Vibrational Optical Activity: Principles and Applications, First Edition. Laurence A. Nafie.

We now form the anisotropy ratio, given in Equation (3.53) by writing

$$g = \frac{4R}{D} = \frac{\text{Im}\,\boldsymbol{\mu}\cdot\boldsymbol{m}}{|\boldsymbol{\mu}|^2} = \frac{e^2\text{Im}(\boldsymbol{r}\cdot\boldsymbol{r}\times\boldsymbol{p})/2mc}{e^2(\boldsymbol{r}\cdot\boldsymbol{r})} \tag{A3}$$

Converting the momentum operator into a position operator using a quantum mechanical identity we have

$$\boldsymbol{p} = i\omega m\boldsymbol{r} \tag{A4}$$

$$g \approx \frac{\text{Im}(im\omega r^3\cos\theta\sin\varphi)/2mc}{r^2} \approx \frac{r\nu}{c} = \frac{r}{\lambda} \tag{A5}$$

where $\omega = 2\pi\nu$, θ is the angle in the scalar (dot) product of $\boldsymbol{\mu}$ and $\boldsymbol{m}$, and φ is the angle in vector (cross) product of $\boldsymbol{r}$ and $\boldsymbol{p}$ in the expression for $\boldsymbol{m}$ in Equation (A2). The final result is that a general estimate of the magnitude of the CD relative to the parent absorption is roughly the ratio of the spatial extent of the transition, r, divided by the wavelength, λ, of the radiation involved. One, for example, could compare the g-values for the ECD of a carbonyl $\pi - \pi^*$ transition at 200 nm with the VCD of a C=O carbonyl stretching mode near 2000 cm^{-1} or 5000 nm. We can take the spatial extent of the transition to approximately 1 nm for a carbonyl moiety and with a similar chiral environment, the g-value is 1/200, or 5×10^{-3}, for the ECD band while the g-value for the VCD is 1/5000, or 25 times smaller at 2×10^{-4}. The same type of argument applies for ROA, although here more factors need to be approximated. For excitation near 500 nm, one predicts the ratio of ROA to Raman intensities to be about an order of magnitude larger than the VCD to VA intensity ratio. It is valuable to keep in mind that optical activity magnitudes scale inversely with the wavelength, which is why terahertz and microwave CD intensities are predicted to be smaller than VCD intensities, and why there is little hope for direct detection of difference in NMR intensities for LCP and RCP radio-wave radiation.

A.2 Degenerate Coupled Oscillator Model of Circular Dichroism

The first and conceptually simplest model of VCD intensities was published in advance of the experimental discovery of VCD and gave impetus to its observational search (Holzwarth and Chabay, 1972). The model holds equally well for electronic transitions where it is known as the exciton coupling model. The model is based on two identical electric dipole transition moments, $\boldsymbol{\mu}_1$ and $\boldsymbol{\mu}_2$, which are separated from one another and twisted in orientation so they do not lie in the same plane, otherwise they do not constitute a chiral structure and their CD is zero. The separation vector, $\boldsymbol{R}_{12} = \boldsymbol{R}_2^0 - \boldsymbol{R}_1^0$, connects the two transition moments where $\boldsymbol{R}_1^0$ and $\boldsymbol{R}_2^0$ are the location of their connection points, which as we show below may be at any location along the transition moment vector. For separations that are not too distant, the two oscillators will couple energetically and oscillate coherently either in-phase (+) or out-of-phase (−) according to the expression

$$\boldsymbol{\mu}^{\pm} = \frac{1}{\sqrt{2}}(\boldsymbol{\mu}_1 \pm \boldsymbol{\mu}_2) \tag{A6}$$

The oscillators are no longer independent and result in two absorption bands that in general have intensities not equal to one another given by:

$$D^{\pm} = |\boldsymbol{\mu}^{\pm}|^2 = \left(\frac{1}{2}\right)|\boldsymbol{\mu}_1 \pm \boldsymbol{\mu}_2|^2 \tag{A7}$$

These two absorption bands are shifted in transition energy relative to one another by an amount proportional to the square of the transition moment and by their relative orientation according to:

$$E^{\pm} = \frac{\boldsymbol{\mu}_1 \cdot \boldsymbol{\mu}_2}{R_{12}^3} - \frac{3(\boldsymbol{\mu}_1 \cdot \boldsymbol{R}_{12})(\boldsymbol{\mu}_2 \cdot \boldsymbol{R}_{12})}{R_{12}^5} \tag{A8}$$

If the two oscillators are perpendicular to their separation vector, $\boldsymbol{R}_{12}$, the second term is zero, and R_{12} is the scalar length of this vector. The sum of the dipole strengths of the two coupled transitions is equal to the sum of the dipole strengths of the individual transition-dipole moments as:

$$D_T = D^+ + D^- = |\boldsymbol{\mu}_1|^2 + |\boldsymbol{\mu}_2|^2 \tag{A9}$$

Hence, the total intensity of the two oscillators does not change because of their coupling.

We next consider the coupled oscillator expression for the magnetic-dipole transition moment. Here we add a feature to generalize the model slightly by allowing a local intrinsic magnetic-dipole transition moment, $\boldsymbol{m}_1$ and $\boldsymbol{m}_2$, to be associated with each oscillator in addition to the expression involving the electric-dipole transition moments. This gives the magnetic-dipole analogue of Equation (A6)

$$\begin{aligned} \boldsymbol{m}^{\pm} &= \frac{1}{2c}\left[\frac{1}{\sqrt{2}}\left(\boldsymbol{R}_1^0 \times \dot{\boldsymbol{\mu}}_1 + 2c\boldsymbol{m}_1\right) \pm \frac{1}{\sqrt{2}}\left(\boldsymbol{R}_2^0 \times \dot{\boldsymbol{\mu}}_2 + 2c\boldsymbol{m}_2\right)\right] \\ &= \frac{1}{2c\sqrt{2}}\left(\boldsymbol{R}_1^0 \times \dot{\boldsymbol{\mu}}_1 \pm \boldsymbol{R}_2^0 \times \dot{\boldsymbol{\mu}}_2\right) + \frac{1}{\sqrt{2}}(\boldsymbol{m}_1 \pm \boldsymbol{m}_2) \end{aligned} \tag{A10}$$

The dot above the electric dipole moment signifies the velocity form of the transition dipole moment. Similar to Equation (A9), we can convert to the position form by:

$$\dot{\boldsymbol{\mu}} = i\omega\boldsymbol{\mu} \tag{A11}$$

We can now form the couple oscillator expression for the rotational strength as:

$$R^{\pm} = \mathrm{Im}\left(\boldsymbol{\mu}^{\pm} \cdot \boldsymbol{m}^{\pm}\right) \tag{A12}$$

Substituting Equations (A7) and (A10) we have

$$\begin{aligned} R_a^{\pm} &= \mathrm{Im}\left[\frac{1}{\sqrt{2}}(\boldsymbol{\mu}_1 \pm \boldsymbol{\mu}_2) \cdot \left[\frac{i\omega}{2c\sqrt{2}}\left(\boldsymbol{R}_1^0 \times \boldsymbol{\mu}_1 \pm \boldsymbol{R}_2^0 \times \boldsymbol{\mu}_2\right) + \frac{1}{\sqrt{2}}(\boldsymbol{m}_1 \pm \boldsymbol{m}_2)\right]\right] \\ &= \frac{\omega}{4c}\left(\boldsymbol{\mu}_1 \cdot \boldsymbol{R}_2^0 \times \boldsymbol{\mu}_2 \pm \boldsymbol{\mu}_2 \cdot \boldsymbol{R}_1^0 \times \boldsymbol{\mu}_1\right) + \frac{1}{2}\mathrm{Im}[(\boldsymbol{\mu}_1 \pm \boldsymbol{\mu}_2) \cdot (\boldsymbol{m}_1 \pm \boldsymbol{m}_2)] \\ &= \frac{\omega}{4c}\left(\mp\boldsymbol{R}_2^0 \cdot \boldsymbol{\mu}_1 \times \boldsymbol{\mu}_2 \pm \boldsymbol{R}_1^0 \cdot \boldsymbol{\mu}_1 \times \boldsymbol{\mu}_2\right) + \frac{1}{2}\mathrm{Im}(\boldsymbol{\mu}_1 \cdot \boldsymbol{m}_1 + \boldsymbol{\mu}_2 \cdot \boldsymbol{m}_2) \pm \frac{1}{2}\mathrm{Im}(\boldsymbol{\mu}_1 \cdot \boldsymbol{m}_2 + \boldsymbol{\mu}_2 \cdot \boldsymbol{m}_1) \end{aligned} \tag{A13}$$

The first term is the electric-dipole coupled oscillator term followed by the rotational strengths from each local oscillator, and finally rotational strength involving in- and out-of-phase of cross terms of the intrinsic magnetic-dipole moments. A much simpler expression is obtained if we combine the two electric-dipole terms and assume that the intrinsic magnetic-dipole transition moments are zero,

$$R^{\pm} = \mp\frac{\omega}{4c}\left(\boldsymbol{R}_2^0 - \boldsymbol{R}_1^0\right)\cdot\boldsymbol{\mu}_1\times\boldsymbol{\mu}_2 = \mp\frac{\omega}{4c}\boldsymbol{R}_{12}\cdot\boldsymbol{\mu}_1\times\boldsymbol{\mu}_2 \qquad \text{(A14)}$$

This equation states that in the absence of local magnetic-dipole transition moments, the rotational strengths of the in-phase and out-of-phase transitions are equal and opposite, even though their underlying parent absorption intensities are not equal. In particular the sum of the rotational strengths of the two coupled transitions is zero, namely

$$R_T = R^+ + R^- = 0 \qquad \text{(A15)}$$

This simply means that in the absence of a coupling of the two electric-dipole transition moments, they oscillate independently and have zero CD. Further, if the coupling energy is non-zero but vanishingly small, the degenerate oscillators will still couple but their transitions will overlap and effectively cancel.

These coupled oscillator equations can be generalized to more than two coupled oscillators. In the limit of very large numbers of oscillators this leads to coupled-oscillator or exciton CD descriptions of polymers or solids. A characteristic of such descriptions is that N oscillators gives rise to N coupled-oscillator modes that are distinguished by N orthogonal sets of coupling coefficients. In the case of only two oscillators considered above, the two normalized, orthogonal coupling coefficients are just $\pm 1/\sqrt{2}$, as can be seen for example in Equation (A6). Another important characteristic of and N-coupled oscillator theory is that the sum of the rotational strengths over all the transitions is zero. Such sets of rotational strengths are said to be conservative when the sum of all rotational strengths is zero.

The expressions given above apply to any type of coupled electric-dipole (and magnetic-dipole) transition moment. They can be adapted more directly to vibrational transitions following the formalism of Chapters 2 and 4 by writing the transition moments as expansions in vibrational normal mode coordinates and evaluating the coordinate $Q_a$with vibrational matrix element as

$$\frac{\boldsymbol{\mu}}{\sqrt{2}} \Rightarrow \left(\frac{\partial\boldsymbol{\mu}}{\partial Q_a}\right)\langle\phi_{g1}|Q_a|\phi_{g0}\rangle = \boldsymbol{\mu}_a\left(\frac{\hbar}{2\omega_a}\right)^{1/2} \qquad \text{(A16)}$$

This gives the following expressions for the degenerate coupled oscillator dipole and rotational strengths for coupled vibrational modes labeled a,

$$D_a^{\pm} = \left(\frac{\hbar}{2\omega_a}\right)\left|\boldsymbol{\mu}_{a,1} \pm \boldsymbol{\mu}_{a,2}\right|^2 \qquad R_a^{\pm} = \mp\frac{\hbar}{4c}\boldsymbol{R}_{12}\cdot\boldsymbol{\mu}_{a,1}\times\boldsymbol{\mu}_{a,2} \qquad \text{(A17)}$$

We will encounter analogues of these expressions when we consider the fixed partial charge model in the next section.

A.3 Fixed Partial Charge Model of VCD

The next model of VCD to be published was the fixed partial charge (FPC) model (Schellman, 1973). In this model, the role of the electrons during the vibrational motion is relegated to simply reducing the value of the nuclear charge by a fixed amount such that only a fractional charge, either positive or negative, is associated with each nucleus in the molecule. The model is therefore exactly the same form as the nuclear contribution to the full theoretical description of VCD. There is no restriction on the type of molecule or type of transition that can be described by the model.

We begin by writing first an expression for VA intensity in terms the dipole strength from Equation (4.1)

$$D^a_{r,g1,g0} = \left| \left(\frac{\partial \langle \boldsymbol{\mu} \rangle}{\partial Q_a} \right)_{Q_a=0} \left\langle \phi^a_{g1} | Q_a | \phi^a_{g0} \right\rangle \right|^2 = \left(\frac{\hbar}{2\omega_a} \right) \left| \left(\frac{\partial \langle \boldsymbol{\mu} \rangle}{\partial Q_a} \right)_{Q_a=0} \right|^2 \tag{A18}$$

We note the close resemblance of this expression to that in Equation (A17). The electric-dipole transition moment can be written in terms of S-vectors, defined in Equation (2.81), as:

$$\left(\frac{\partial \langle \mu_\beta \rangle}{\partial Q_a} \right)_{Q=0} = \sum_J \left(\frac{\partial \langle \mu_\beta \rangle}{\partial R_{J,\alpha}} \right)_{R=0} S_{J\alpha,a} = \sum_J P^J_{r,\alpha\beta} S_{J\alpha,a} = \sum_J z_J e S_{J\beta,a} = \sum_J z_J e \delta_{\alpha\beta} S_{J\alpha,a} \tag{A19}$$

Here Cartesian tensor notation is used with summation of x, y, and z for each repeated Greek subscript, $\delta_{\alpha\beta}$ is the Kroneker delta function, z_J is the fixed partial charge on nucleus J, and the S-vectors are defined as:

$$S_{J\alpha,a} = \left(\frac{\partial R_{J\alpha}}{\partial Q_a} \right)_{Q=0} = \left(\frac{\partial \dot{R}_{J\alpha}}{\partial \dot{Q}_a} \right)_{\dot{Q}=0} = \left(\frac{\partial \dot{R}_{J\alpha}}{\partial P_a} \right)_{P=0} \tag{A20}$$

The dipole strength can now be written first using Cartesian tensor notation and then vector notion as:

$$\begin{aligned} D^a_{r,g1,g0} &= \left(\frac{\hbar}{2\omega_a} \right) \left| \left(\frac{\partial \langle \boldsymbol{\mu} \rangle}{\partial Q_a} \right)_{Q_a=0} \right|^2 = \left(\frac{\hbar}{2\omega_a} \right) \sum_{J,J'} z_J z_{J'} e^2 S_{J\beta,a} S_{J'\beta,a} \\ &= \left(\frac{\hbar}{2\omega_a} \right) \sum_{J,J'} z_J z_{J'} e^2 \mathbf{S}_{Ja} \cdot \mathbf{S}_{J'a} = \left(\frac{\hbar}{2\omega_a} \right) \sum_{J,J'} \boldsymbol{\mu}_{Ja} \cdot \boldsymbol{\mu}_{J'a} \end{aligned} \tag{A21}$$

where a fixed partial charge dipole moment is defined as:

$$\boldsymbol{\mu}_{Ja} = z_J e \mathbf{S}_{Ja} \tag{A22}$$

The last expression in Equation (A21) reduces to the dipole strength expression for the coupled oscillator model given in Equation (A17) if only two nuclear motions are considered.

The corresponding rotation strength is given from Equation (4.2) by:

$$\begin{aligned} R^a_{r,g1,g0} &= \mathrm{Im}\left[\left(\frac{\partial \langle \boldsymbol{\mu} \rangle}{\partial Q_a} \right)_{Q_a=0} \cdot \left(\frac{\partial \langle \boldsymbol{m} \rangle}{\partial P_a} \right)_{P_a=0} \left\langle \phi^a_{g0} \middle| Q_a \middle| \phi^a_{g1} \right\rangle \left\langle \phi^a_{g1} \middle| P_a \middle| \phi^a_{g0} \right\rangle \right] \\ &= \left(\frac{\hbar}{2} \right) \left[\left(\frac{\partial \langle \boldsymbol{\mu} \rangle}{\partial Q_a} \right)_{Q_a=0} \cdot \left(\frac{\partial \langle \boldsymbol{m} \rangle}{\partial P_a} \right)_{P_a=0} \right] \end{aligned} \tag{A23}$$

Applying the FPC model to the magnetic-dipole transition moment we have

$$\left(\frac{\partial\langle m_\beta\rangle}{\partial P_a}\right)_{P_a=0} = \sum_J \left(\frac{\partial\langle m_\beta\rangle}{\partial \dot{R}_{J,\alpha}}\right)_{\dot{R}=0} S_{J\alpha,a} = \sum_J M^J_{\alpha\beta} S_{J\alpha,a} = \sum_J \frac{z_J e}{2c} \varepsilon_{\alpha\beta\gamma} R^0_{J\gamma} S_{J\alpha,a} \tag{A24}$$

Again this expression is the same as the nuclear contribution to the magnetic-dipole transition except for the partial nuclear charge, z_J, instead of the full nuclear charge Z_J. The rotational strength can be written as a double sum over all nuclei to give

$$R^a_{r,g1,g0} = \left(\frac{\hbar}{4c}\right) \sum_{J,J'} z_J z_{J'} e^2 \varepsilon_{\alpha\beta\gamma} R^0_{J\gamma} S_{J\alpha,a} S_{J'\beta,a} = \left(\frac{\hbar}{4c}\right) \sum_{J,J'} z_J z_{J'} e^2 \boldsymbol{R}^0_J \cdot \boldsymbol{S}_{Ja} \times \boldsymbol{S}_{J'a} \tag{A25}$$

By restricting the double summation to $J > J'$, the contribution from each pair of nuclei is considered only once. Combining terms for $J < J'$ with those for $J > J'$ by using the identity $\boldsymbol{\mu}_{Ja} \times \boldsymbol{\mu}_{J'a} = -\boldsymbol{\mu}_{J'a} \times \boldsymbol{\mu}_{Ja}$ and definition, $\boldsymbol{R}^0_{J,J'} = \boldsymbol{R}^0_J - \boldsymbol{R}^0_{J'}$, we can write Equation (A25) as:

$$R^a_{r,g1,g0} = \left(\frac{\hbar}{4c}\right) \sum_{J>J'} \boldsymbol{R}^0_{J,J'} \cdot \boldsymbol{\mu}_{Ja} \times \boldsymbol{\mu}_{J'a} \tag{A26}$$

This equation takes the same form as the coupled oscillator rotational strength in Equation (A17). This demonstrates vividly two things. Firstly, the FPC model consists of the same basic physics as the coupled oscillator model. In fact, it has the same form as the generalization of the coupled oscillator model to 3*N*–6 oscillators where *N* is the number of atoms and 3*N*–6 is the number of coupled degrees of vibrational freedom in the molecule. Secondly, because the sum of the coupled oscillator rotational strengths is zero, the same is true of the FPC model of VCD. In this case the coupling is supplied by the bonds in the molecule and is the same coupling that leads to the formation of the normal modes of vibration.

A.4 Localized Molecular Orbital Model of VCD

The next model to be developed for the description of VCD intensity, and the first quantum mechanical description of VCD intensity, is the localized molecular orbital (LMO) model (Nafie and Walnut, 1977; Walnut and Nafie, 1977). This model has two distinct derivations, although one of them is simpler and most often cited (Nafie and Walnut, 1977). The LMO model can be viewed as an extension of the FPC model. Instead of localizing, as a fixed value, the total electronic charge of an atom in a molecule at the nucleus of that atom, thereby reducing its charge to some partial charge value, z_J, the molecular orbitals of a molecule are first localized and the centroids of charge of these LMOs are taken to be the locations of charge. For inner shell LMOs, the centroids are effectively at the nuclear centers, but for the valence LMOs, the centroids correspond to bonding orbitals and lone pairs. When a molecule vibrates, the *S*-vectors describe the direction and magnitude and the direction of each nuclear displacement is a given normal mode. To apply the LMO model, a quantum chemistry program is applied to find the magnitude and direction of the corresponding displacements, or σ-vectors, of the centroids of the LMOs of the molecule. The LMO model then has two types of contribution: a full nuclear contribution of the same form as the FPC model except with full nuclear charges, Z_J, and an electronic contribution of a similar form but with charge –1 for each electron, *k*, represented by centroid of its LMO. The dipole strength is written where the first summation over *J* is for the nuclear terms and the second summation over k is for the LMO terms:

$$\begin{aligned} D^{a}_{r,g1,g0} &= \left(\frac{\hbar}{2\omega_a}\right)\left[\sum_J z_J e \boldsymbol{S}_{Ja} - \sum_k e\boldsymbol{\sigma}_{ka}\right]^2 \\ &= \left(\frac{\hbar}{2\omega_a}\right)\left[\sum_{J,J'} z_J z_{J'} e^2 \boldsymbol{S}_{Ja}\cdot\boldsymbol{S}_{J'a} + \sum_{k,k'} e^2 \boldsymbol{\sigma}_{ka}\cdot\boldsymbol{\sigma}_{k'a} - 2\sum_{J,k} z_J e^2 \boldsymbol{S}_{Ja}\cdot\boldsymbol{\sigma}_{ka}\right] \end{aligned} \tag{A27}$$

Because the dipole strength involves the square of the full transition moment, three terms are present, a pure nuclear term, and pure LMO term, and a cross term. The corresponding expression for the rotational strength is given by:

$$\begin{aligned} R^{a}_{r,g1,g0} = \left(\frac{\hbar}{4c}\right)\Bigg[&\sum_{J>J'} z_J z_{J'} e^2 \boldsymbol{R}^0_{J,J'}\cdot\boldsymbol{S}_{Ja}\times\boldsymbol{S}_{J'a} + \sum_{k>k'} e^2 \boldsymbol{r}^0_{k,k'}\boldsymbol{\sigma}_{ka}\times\boldsymbol{\sigma}_{k'a} \\ &- \sum_{J,k} z_J e^2\left(\boldsymbol{R}^0_J - \boldsymbol{r}^0_k\right)\cdot\boldsymbol{S}_{Ja}\times\boldsymbol{\sigma}_{ka}\Bigg] \end{aligned} \tag{A28}$$

The pure nuclear and the pure LMO terms take the same form as the coupled oscillator and FPC rotational strengths and hence they are a conservative theory with rotational strength contributions that sum to zero over all vibrational transitions. On the other hand, the mixed nuclear–LMO terms do not have this form and are not restricted by this sum rule. An example of the greater flexibility of the LMO mode compared with the FPC mode has been published for the CH-stretching modes of L-alanine-Nd_3 where that LMO model is able to predict the large positive intensity bias in this region (see Figure 10.20) due primarily to the methine C*H-stretching mode (Freedman *et al.*, 1982).

A.5 Ring Current Model and Other Vibrational Electronic Current Models

Models of VCD not based on a quantum mechanical calculation, such as the LMO model just described, all lead to a prediction of conservative VCD intensity over the range of vibrational modes that couple local motions. This is certainly true for the coupled oscillator model and its extensions of multiple coupled pairs of oscillators, either as coupled oscillators or as coupled fixed partial nuclear charges in the FPC model. For a spectral region, such as the hydrogen-stretching region, the CO and FPC models each predict that the sum of the VCD intensities add to zero. Gross violations of this prediction, referred to as intensity bias, obviously call for some type of extension of these models. The first, and perhaps simplest of such extensions is the ring current model first described to explain the large positive intensity bias in the CH-stretching VCD spectra of all the naturally occurring L-amino acids (Nafie *et al.*, 1983). The concept behind the ring-current model is akin to the one-electron model of electronic CD intensity, namely electronic motion in an arc or along a ring coupled with linear charge motion perpendicular to a ring is itself a source of CD or VCD intensity. These two motions can be embodied in one motion for electronic charge moving along a helical trajectory as pictured in Figure 4.1. Such CD or VCD intensity is completely biased and not balanced anywhere by an equal and opposite VCD intensity and thus is a conceptual way around the restrictions of the CO and FPC models. The ring-current model was subsequently formalized and extended to other types of molecules as a conceptual mechanism with proposed rules to allow its application to a variety of stereochemical analyses (Nafie and Freedman, 1986). As mentioned in Chapter 1, the ring current mechanism, as well as most other models of VCD intensity, fell out of use because of the emerging success of the *ab intio* VCD calculations and the occurrence of a violation of the prediction of the ring-current mechanism when compared with detailed quantum chemistry calculations (Bursi *et al.*, 1990). A summary of current density models of VCD has been published (Nafie and Freedman, 1990) that includes the so-called charge-flow models that added current-density flow along bonds to supplement the FPC model (Abbate *et al.*, 1981; Moskovits and Gohin, 1982).

A.6 Two-Group and Related Models of ROA

A variety of models of ROA intensity have been proposed, in particular the simple two-group model (Barron and Buckingham, 1975), the methyl torsion model (Barron and Buckingham, 1979), the perturbed generate oscillator model (Barron and Buckingham, 1975), the atom-dipole interaction model (Prasad and Nafie, 1979), the bond-polarizability model (Barron *et al.*, 1986), and the relationship of the bond polarizability model to the vibronic coupling theory of ROA (Escribano *et al.*, 1987). A detailed description of many of these models has been recently summarized (Barron, 2004). Of these we describe briefly the simple two-group model due to its fundamental nature. The two-group model plays a similar conceptual model for ROA akin to the degenerate couple oscillator model of VCD. Rather than lay out the entire features we describe the construction of the two-group model in terms of the polarizability $\alpha_{\alpha\beta}$ and the magnetic dipole optical activity tensor, $G'_{\alpha\beta}$. Here, if the molecule can be regarded as consisting of two groups, such as a dimer molecule, then we can write the following approximate expressions

$$\alpha_{\alpha\beta} = \alpha_{1\alpha\beta} + \alpha_{2\alpha\beta} \tag{A29}$$

$$G'_{\alpha\beta} = G'_{1\alpha\beta} + G'_{2\alpha\beta} - \frac{1}{2}\omega\varepsilon_{\beta\gamma\delta}R_{12}\alpha_{2\alpha\beta} \tag{A30}$$

We can see the similarity of the two-group model to the coupled oscillator model by comparing these equations with those for the electric dipole and magnetic dipole moments in Equations (A6) and (A10). However, the two-group model differs significantly from the coupled oscillator model in that the two groups do not need to interact to exhibit ROA intensity. Scattering provides a direct mechanism for ROA not present for VCD. The two-group model can be generalized and extended to a bond polarizability model of ROA that is somewhat akin to the FPC and LMO models of ROA. More recently, with the advent of accurate quantum chemistry calculations of ROA using TDDFT, the need for models, even conceptually, seems to have diminished to the point of near absence of use.

References

Abbate, S., Laux, L., Overend, J., and Moscowitz, A. (1981) A charge flow model for vibrational rotational strengths. *J. Chem. Phys.*, **75**, 3161–3164.

Barron, L.D. (2004) *Molecular Light Scattering and Optical Activity*, 2nd edn, Cambridge University Press, Cambridge.

Barron, L.D., and Buckingham, A.D. (1975) Rayleigh and Raman optical activity. *Ann. Rev. Phys. Chem.* **26**, 381–396.

Barron, L.D., and Buckingham, A.D. (1979) The inertial contribution to vibrational optical activity in methyl torsion modes. *J. Am. Chem. Soc.*, **101**(8), 1979–1987.

Barron, L.D., Escribano, J.R., and Torrance, J.F. (1986) Polarized Raman optical activity and the bond polarizability model. *Mol. Phys.*, **57**(3), 653–660.

Bursi, R., Devlin, F.J., and Stephens, P.J. (1990) Vibrationally induced ring currents? The vibrational circular dichroism of methyl lactate. *J. Am. Chem. Soc.*, **112**, 9430–9432.

Escribano, J.R., Freedman, T.B., and Nafie, L.A. (1987) Bond polarizability and vibronic coupling theory of Raman optical activity: The electric-dipole magnetic dipole optical activity tensor. *J. Chem. Phys.*, **87**, 3366–3374.

Freedman, T.B., Diem, M., Polavarapu, P.L., and Nafie, L.A. (1982) Vibrational circular dichroism in amino acids and peptides. 6. Localized molecular orbital calculations of the carbon hydrogen stretching VCD in deuterated isotopomers of alanine. *J. Am. Chem. Soc.*, **104**, 3343–3349.

Freedman, T.B., and Nafie, L.A. (1994) Theoretical formalism and models for vibrational circular dichroism intensity. In: *Modern Nonlinear Optics, Part 3* (eds M. Evans, and S. Kielich), John Wiley & Sons, Inc., New York, pp. 207–263.

Holzwarth, G., and Chabay, I. (1972) Optical activity of vibrational transitions. Coupled oscillator model. *J. Chem. Phys.*, **57**, 1632–1635.

Moskovits, M., and Gohin, A. (1982) Vibrational circular dichroism: Effect of charge fluxes and bond currents. *J. Phys. Chem.*, **86**, 3947–3950.

Nafie, L.A., and Freedman, T.B. (1986) The ring current mechanism of vibrational circular dichroism. *J. Phys. Chem.*, **90**, 763–767.

Nafie, L.A., and Freedman, T.B. (1990) Electronic current models of vibrational circular dichroism. *J. Mol. Struct.*, **224**, 121–132.

Nafie, L.A., Oboodi, M.R., and Freedman, T.B. (1983) Vibrational circular dichroism in amino acids and peptides. 8. A chirality rule for the methine C*H stretching mode. *J. Am. Chem. Soc.*, **105**, 7449–7450.

Nafie, L.A., and Walnut, T.H. (1977) Vibrational Circular Dichroism Theory: A Localized Molucular Orbital Model. *Chem. Phys. Lett.* **49**, 441–446.

Prasad, P.L., and Nafie, L.A. (1979) Atom dipole interaction model for Raman optical activity. Reformulation and its comparison to the general two group model. *J. Chem. Phys.*, **70**, 5582–5588.

Schellman, J.A. (1973) Vibrational optical activity. *J. Chem. Phys.*, **58**, 2882–2886.

Walnut, T.H., and Nafie, L.A. (1977) Infrared absorption and the Born–Oppenheimer approximation II. Vibrational circular dichroism. *J. Chem. Phys.*, **67**, 1501–1510.

Appendix B

Derivation of Probability and Current Densities from Multi-Electron Wavefunctions for Electronic and Vibrational Transitions

The purpose of this Appendix is to support the presentation of electron current density in molecules in Chapters 2, 4, and 9 along with a parallel and more familiar quantity is the concept of electron probability density, the absolute square of the molecular wavefunction. One goal of this Appendix is to make clear how multi-electron probability and current density is converted into density functions that depend essentially on a single electron and its unique dependence in the Cartesian coordinate space of the molecule. A second goal is to show again, more carefully, how the conservation of probability density unites at each point in the space of the molecule changes in electron probability with time with the flow of current density into or out of that point. For vibrational transitions the changes in probability density with time is a Born–Oppenheimer property, while the gradient of the vibrational transition current density is inherently a non-Born–Oppenheimer property. A more detailed description of transition current density in molecules has been published first in general (Nafie, 1997a) and then for pure electronic (Freedman *et al.*, 1998) and pure vibrational transitions (Freedman *et al.*, 1997). Subsequently applications involving vibrational transition current density maps were published (Nafie, 1997b; Freedman *et al.*, 2000a; Freedman *et al.*, 2000b).

B.1 Transition Probability Density

We begin expressing the time dependence of the exact wavefunction of a molecule in state n as an explicit function of the electron coordinates of the molecule, $\boldsymbol{r}_i$, the nuclear coordinates subsumed under a single coordinate, $\boldsymbol{R}$, and time,

$$\tilde{\Psi}_n(\boldsymbol{r}_1, \boldsymbol{r}_2, \ldots\ldots \boldsymbol{r}_N, \boldsymbol{R}, t) = \Psi_n(\boldsymbol{r}_1, \boldsymbol{r}_2, \ldots\ldots \boldsymbol{r}_N, \boldsymbol{R})\exp(-iE_n t/\hbar) \tag{B1}$$

Vibrational Optical Activity: Principles and Applications, First Edition. Laurence A. Nafie.

This wavefunction satisfies the time-dependent Schrödinger equation

$$\frac{i\hbar\partial\tilde{\Psi}_n(\boldsymbol{r}_1,\boldsymbol{r}_2,\ldots\ldots\boldsymbol{r}_N,\boldsymbol{R},t)}{\partial t}=\mathcal{H}\tilde{\Psi}_n(\boldsymbol{r}_1,\boldsymbol{r}_2,\ldots\ldots\boldsymbol{r}_N,\boldsymbol{R},t) \tag{B2}$$

Conversion to the time-independent Schrödinger equation is easily obtained by carrying out the derivative on the left side of the equation with respect to time to yield

$$i\hbar(-iE_n/\hbar)\Psi_n(\boldsymbol{r}_1,\boldsymbol{r}_2,\ldots\ldots\boldsymbol{r}_N,\boldsymbol{R})\exp(-iE_nt/\hbar)=\mathcal{H}\Psi_n(\boldsymbol{r}_1,\boldsymbol{r}_2,\ldots\ldots\boldsymbol{r}_N,\boldsymbol{R})\exp(-iE_nt/\hbar) \tag{B3}$$

Cancelling common factors within and on each side of this equation gives the starting equation of Chapter 2, Equation (2.3), namely the familiar time-independent Schrödinger equation of the molecule in state n

$$\mathcal{H}\Psi_n(\boldsymbol{r}_1,\boldsymbol{r}_2,\ldots\ldots\boldsymbol{r}_N,\boldsymbol{R})=E_n\Psi_n(\boldsymbol{r}_1,\boldsymbol{r}_2,\ldots\ldots\boldsymbol{r}_N,\boldsymbol{R}) \tag{B4}$$

For simplicity in expressions below we re-write Equation (1) using the relationship that the state frequency is related to the state energy by $\omega_n = E_n/\hbar$, which gives

$$\tilde{\Psi}_n(\boldsymbol{r}_1,\boldsymbol{r}_2,\ldots\ldots\boldsymbol{r}_N,\boldsymbol{R},t)=\Psi_n(\boldsymbol{r}_1,\boldsymbol{r}_2,\ldots\ldots\boldsymbol{r}_N,\boldsymbol{R})\exp(-i\omega_nt) \tag{B5}$$

Note that the tilda above the time-dependent wavefunction on the left designates a complex quantity and we assume that the time-independent wavefunction is in a steady state and is real.

We now form the expression for full multi-electron probability density of the molecule:

$$\begin{aligned}\tilde{\rho}_n(\boldsymbol{r}_1,\boldsymbol{r}_2,\ldots\ldots\boldsymbol{r}_N,\boldsymbol{R},t)&=\tilde{\Psi}_n^*(\boldsymbol{r}_1,\boldsymbol{r}_2,\ldots\ldots\boldsymbol{r}_N,\boldsymbol{R},t)\tilde{\Psi}_n(\boldsymbol{r}_1,\boldsymbol{r}_2,\ldots\ldots\boldsymbol{r}_N,\boldsymbol{R},t)\\&=\Psi_n^2(\boldsymbol{r}_1,\boldsymbol{r}_2,\ldots\ldots\boldsymbol{r}_N,\boldsymbol{R})=\rho_n(\boldsymbol{r}_1,\boldsymbol{r}_2,\ldots\ldots\boldsymbol{r}_N,\boldsymbol{R})\end{aligned} \tag{B6}$$

Here we note the probability density at the end of this equation is both real (no super-tilda) and time independent. This is because the complex conjugate changes the signs of all imaginary quantities thus cancelling the exponential time dependence of the two wavefunctions.

The one-electron probability density can be obtained from the full multi-electron probability density by integrating over all electron coordinates except that of electron 1,

$$\rho_n(\boldsymbol{r}_1)=\int\Psi_n^2(\boldsymbol{r}_1,\boldsymbol{r}_2,\ldots\ldots\boldsymbol{r}_N,\ \boldsymbol{R})\mathrm{d}\boldsymbol{r}_2,\ldots\ldots\mathrm{d}\boldsymbol{r}_N,\mathrm{d}\boldsymbol{R}=\Psi_n^2(\boldsymbol{r}_1) \tag{B7}$$

Integration over the nuclear coordinates is also carried out for simplicity of notation but could be retained if desired. If we now equate the position coordinate of electron 1 with the general space of the molecule without reference to a particular electron we have

$$\rho_n(\boldsymbol{r},t)=\tilde{\Psi}_n^*(\boldsymbol{r},t)\tilde{\Psi}_n(\boldsymbol{r},t)=\Psi_n^2(\boldsymbol{r})=\rho_n(\boldsymbol{r}) \tag{B8}$$

We conclude that a molecule in a single steady state, n, has a time independent probability density and, as we shall see, no current density.

We now introduce a time-dependent wavefunction that we call a transition-state wavefunction as it is a linear combination of two time-dependent wavefunctions, one for state n and one for state m,

$$\tilde{\Psi}_{nm}(\boldsymbol{r},t) = c_n\tilde{\Psi}_n(\boldsymbol{r},t) + c_m\tilde{\Psi}_m(\boldsymbol{r},t) \tag{B9}$$

This wavefunction is appropriate for a molecule under the influence of an external perturbation by for example a radiation field that leads to a transition between the states n and m, and the coefficients c_n and c_m describe the degree to which the wavefunction belongs to each of these states. We next define a transition probability density

$$\begin{aligned}\rho_{nm}(\boldsymbol{r},t) &= \tilde{\Psi}^*_{nm}(\boldsymbol{r},t)\tilde{\Psi}_{nm}(\boldsymbol{r},t)\\ &= c_n^2\Psi_n^2(\boldsymbol{r}) + c_m^2\tilde{\Psi}^2_m(\boldsymbol{r}) + c_nc_m\Psi_n(\boldsymbol{r})\Psi_m(\boldsymbol{r})\left(e^{i(\omega_n-\omega_m)t} + e^{-i(\omega_n-\omega_m)t}\right)\\ &= c_n^2\rho_n(\boldsymbol{r}) + c_m^2\rho_m(\boldsymbol{r}) + 2c_nc_m\Theta_{nm}(\boldsymbol{r})\cos\omega_{nm}t\end{aligned} \tag{B10}$$

Here, the time dependence of the wavefunctions in Equation (B5) cancels as before for the first two pure state terms but does not cancel for the two cross-terms between states. We also define the transition frequency as $\omega_{nm} = \omega_n - \omega_m$ and more importantly define the time-independent *transition* probability density (TPD) function $\Theta_{nm}(\boldsymbol{r})$ as:

$$\Theta_{nm}(\boldsymbol{r}) = \Psi_n(\boldsymbol{r})\Psi_m(\boldsymbol{r}) \tag{B11}$$

The TPD represents that part of the probability density which oscillates in time at the transition frequency for a molecule (or atom) undergoing a transition between states n and m. This time dependence is carried entirely by the cosine factor in Equation (B10).

B.2 Transition Current Density

We are now in a position to define the transition current density of a molecule. We begin, as above, by writing the quantum mechanical expression for current density of molecule in a single time-dependent steady state,

$$\begin{aligned}\boldsymbol{j}_n(\boldsymbol{r}_1,\boldsymbol{r}_2,\ldots\ldots\boldsymbol{r}_N,\boldsymbol{R},t) = \frac{\hbar}{2mi}\Big[&\tilde{\Psi}^*_n(\boldsymbol{r}_1,\boldsymbol{r}_2,\ldots\ldots\boldsymbol{r}_N,\boldsymbol{R},t)(\nabla_1+\nabla_2+\cdots+\nabla_N)\\ &\times\tilde{\Psi}_n(\boldsymbol{r}_1,\boldsymbol{r}_2,\ldots\ldots\boldsymbol{r}_N,\boldsymbol{R},t) - \tilde{\Psi}_n(\boldsymbol{r}_1,\boldsymbol{r}_2,\ldots\ldots\boldsymbol{r}_N,\boldsymbol{R},t)\\ &\times(\nabla_1+\nabla_2+\cdots+\nabla_N)\tilde{\Psi}^*_n(\boldsymbol{r}_1,\boldsymbol{r}_2,\ldots\ldots\boldsymbol{r}_N,\boldsymbol{R},t)\Big]\end{aligned} \tag{B12}$$

Here $\nabla_1 = \mathbf{i}\partial/\partial x_1 + \mathbf{j}\partial/\partial y_1 + \mathbf{k}\partial/\partial z_1$ is the vector derivative operator in Cartesian coordinates associated with the quantum mechanical momentum operator. We can reduce the multi-electron current density to a one-electron current density by integrating over all electron and nuclear coordinates except the coordinate for electron 1 as:

$$\begin{aligned}\boldsymbol{j}_n(\boldsymbol{r}_1,t) = \frac{\hbar}{2mi}\int\Big[&\tilde{\Psi}^*_n(\boldsymbol{r}_1,\boldsymbol{r}_2,\ldots\ldots\boldsymbol{r}_N,\boldsymbol{R},t)(\nabla_1+\nabla_2+\cdots+\nabla_N)\\ &\times\tilde{\Psi}_n(\boldsymbol{r}_1,\boldsymbol{r}_2,\ldots\ldots\boldsymbol{r}_N,\boldsymbol{R},t) - \tilde{\Psi}_n(\boldsymbol{r}_1,\boldsymbol{r}_2,\ldots\ldots\boldsymbol{r}_N,\boldsymbol{R},t)\\ &\times(\nabla_1+\nabla_2+\cdots+\nabla_N)\tilde{\Psi}^*_n(\boldsymbol{r}_1,\boldsymbol{r}_2,\ldots\ldots\boldsymbol{r}_N,\boldsymbol{R},t)\Big]\mathrm{d}\boldsymbol{r}_2,\ldots\ldots\mathrm{d}\boldsymbol{r}_N,\mathrm{d}\boldsymbol{R}\end{aligned} \tag{B13}$$

The time dependence cancels as it did for the probability density which, after dropping the subscript 1, gives

$$j_n(\boldsymbol{r}) = \frac{\hbar}{2mi}\left[\Psi_n(\boldsymbol{r})\nabla\Psi_n(\boldsymbol{r}) - \Psi_n(\boldsymbol{r})\nabla\Psi_n(\boldsymbol{r})\right] = 0 \tag{B14}$$

The current density vanishes as it should for a molecule in a state.

We now describe the current density for the transition-state wavefunction define in Equation (B9).

$$\begin{aligned} j_{nm}(\boldsymbol{r},t) &= \frac{\hbar}{2mi}\left[\Psi^*_{nm}(\boldsymbol{r},t)\nabla\Psi_{nm}(\boldsymbol{r},t) - \Psi_{nm}(\boldsymbol{r},t)\nabla\Psi^*_{nm}(\boldsymbol{r},t)\right] \\ &= \frac{\hbar}{2mi}\left[\Psi_n(\boldsymbol{r})\nabla\Psi_m(\boldsymbol{r}) - \Psi_m(\boldsymbol{r})\nabla\Psi_n(\boldsymbol{r})\right]\left(e^{i(\omega_n-\omega_m)t} - e^{-i(\omega_n-\omega_m)t}\right) \\ &= -2c_1c_2\boldsymbol{J}_{mn}(\boldsymbol{r})\sin\omega_{nm}t \end{aligned} \tag{B15}$$

Here the pure state current densities vanish leaving only the time-dependent cross-terms. The last line in this equation defines a real quantity that we call the *transition* current density (TCD) as:

$$\boldsymbol{J}_{nm}(\boldsymbol{r}) = \frac{\hbar}{2m}\left[\Psi_n(\boldsymbol{r})\nabla\Psi_m(\boldsymbol{r}) - \Psi_m(\boldsymbol{r})\nabla\Psi_n(\boldsymbol{r})\right] \tag{B16}$$

The TCD is a vector field over the space of a molecule that describes currents in molecules that oscillate during transitions in the molecule between states n and m. These currents describe how probability density, as a conserved fluid, flows from one part of the molecule to another during a molecule transition. For vibrational transitions, the vector field of the TCD depicts the motion of electron density that accompanies the motion of the nuclei during a normal mode of vibration. Examples of such TCD maps are shown in Figures 4.3 and 9.6.

B.3 Conservation of Transition Probability and Current Density

In this section we derive a fundamental relationship between the transition probability density (TPD), defined in Equation (B11), and the transition current density (TCD), defined in Equation (B16), and show they are related by a conservation equation. We start with the well-known conservation equation for any conserved quantity, in this case probability density.

$$-\nabla\cdot j_n(\boldsymbol{r},t) = \frac{\partial\rho_n(\boldsymbol{r},t)}{\partial t} \tag{B17}$$

The equation states that for any point in space, the change in probability with time is compensated by current density flowing across the surface surrounding that point. If we apply this equation to our transition state wavefunction we begin with

$$-\nabla\cdot j_{nm}(\boldsymbol{r},t) = \frac{\partial\rho_{nm}(\boldsymbol{r},t)}{\partial t} \tag{B18}$$

We first analyze the left-hand side of this equation, and from Equation (B15) we can write

$$-\nabla\cdot j_n(\boldsymbol{r},t) = -\nabla\cdot\boldsymbol{J}_{nm}(\boldsymbol{r})2c_1c_2\sin\omega_{nm}t \tag{B19}$$

On the other hand, from Equation (B10) we have after taking the derivative with respect to time

$$\frac{\partial\rho_{nm}(\boldsymbol{r},t)}{\partial t} = -\omega_{nm}\Theta_{nm}(\boldsymbol{r})2c_1c_2\sin\omega_{nm}t \tag{B20}$$

From these two equations, we can deduce a new relationship, taking care to note the order of the subscript states for $\boldsymbol{J}_{nm}$ and ω_{mn}, which are chosen such that the overall appearance of the equation resembles that of Equation (B18),

$$-\nabla \cdot \boldsymbol{J}_{nm}(\boldsymbol{r}) = \omega_{mn}\Theta_{nm}(\boldsymbol{r}) \tag{B21}$$

This equation relates the TCD on the left to the TPD on the right. For vibrational transition, the TCD is a property whose existence depends on relationship of the current density that is generated by nuclear velocities, a non-Born–Oppenheimer property, to the TPD, which is a function only of the nuclear positions, a Born–Oppenheimer property. This equation can be verified independently by carrying out the spatial gradient of the TCD we can write using Equation (B16),

$$\begin{aligned}-\nabla \cdot \boldsymbol{J}_{nm}(\boldsymbol{r}) = -\frac{\hbar}{2m}\Big[&\nabla\Psi_n(\boldsymbol{r},t)\nabla\Psi_m(\boldsymbol{r},t) + \Psi_n(\boldsymbol{r},t)\nabla^2\Psi_m(\boldsymbol{r},t)\\ &-\nabla\Psi_m(\boldsymbol{r},t)\nabla\Psi_n(\boldsymbol{r},t) - \Psi_m(\boldsymbol{r},t)\nabla^2\Psi_n(\boldsymbol{r},t)\Big]\end{aligned} \tag{B22}$$

The first and the third terms cancel leaving

$$-\nabla \cdot \boldsymbol{J}_{nm}(\boldsymbol{r}) = -\frac{\hbar}{2m}\left[\Psi_n(\boldsymbol{r},t)\nabla^2\Psi_m(\boldsymbol{r},t) - \Psi_m(\boldsymbol{r},t)\nabla^2\Psi_n(\boldsymbol{r},t)\right] \tag{B23}$$

It can be recognized that the del-squared operator between the two wavefunction can be replaced with the full molecular Hamiltonian

$$\mathcal{H} = -\frac{\hbar^2}{2m}\nabla^2 + V(R) \tag{B24}$$

This is because ∇^2 is essentially the kinetic-energy operator and the potential-energy terms cancel when included in both terms of Equation (B23). This allows us to write

$$-\nabla \cdot \boldsymbol{J}_{nm}(\boldsymbol{r}) = \frac{1}{\hbar}[\Psi_n(\boldsymbol{r},t)\mathcal{H}\Psi_m(\boldsymbol{r},t) - \Psi_m(\boldsymbol{r},t)\mathcal{H}\Psi_n(\boldsymbol{r},t)] \tag{B25}$$

From here, we can evaluate the action of each Hamiltonian on the wavefunction immediately to its right using Equation (B4). Thus we can finally write

$$-\nabla \cdot \boldsymbol{J}_{nm}(\boldsymbol{r}) = \frac{(E_m - E_n)}{\hbar}\Psi_n(\boldsymbol{r})\Psi_m(\boldsymbol{r}) = \omega_{mn}\Theta_{nm}(\boldsymbol{r}) \tag{B26}$$

This exactly verifies Equation (B21).

B.4 Conservation Equation for Vibrational Transitions

The theoretical expressions here apply to any transition between two quantum mechanical states of a molecule. In particular it can be applied directly to any pair of electronic states. Its extension to vibrational states within a single electronic state is more subtle due to the need to couple electron population-density changes as a function of nuclear position, a Born-Oppenheimer property, with electron current-density changes as a function of nuclear velocity, a non-Born-Oppenheimer complete adiabatic property. This formalism is described in Chapter 4, Section 4.4. For completeness we provide the generalization of the conservation equation for TPD and TCD in Equation (B21) for vibrational transitions, reproduced here from Equation (4.148):

$$-\nabla\cdot\left(\frac{\partial \boldsymbol{j}_g^{CA}(\boldsymbol{r},\boldsymbol{R},\dot{\boldsymbol{R}})}{\partial \dot{\boldsymbol{R}}_J}\right)_{\substack{\dot{R}=0\\R=0}} = \left(\frac{\partial \rho_g^A(\boldsymbol{r},\boldsymbol{R})}{\partial \boldsymbol{R}_J}\right)_{R=0} \tag{B27}$$

Here the electron probability density and current density are given, respectively, by:

$$\begin{aligned}\rho_g^A(\boldsymbol{r},\boldsymbol{R}) &= \psi_g^A(\boldsymbol{r})\psi_g^A(\boldsymbol{r}) \cong \rho_g^0(\boldsymbol{r}) + \sum_J \left(\frac{\partial \rho_g^A(\boldsymbol{r},\boldsymbol{R})}{\partial \boldsymbol{R}_J}\right)_0 \cdot \boldsymbol{R}_J \\ &= \psi_g^0(\boldsymbol{r})^2 + 2\sum_J \psi_g^0(\boldsymbol{r})\left(\frac{\partial \psi_g^A(\boldsymbol{r},\boldsymbol{R})}{\partial \boldsymbol{R}_J}\right)_0 \cdot \boldsymbol{R}_J\end{aligned} \tag{B28}$$

$$\begin{aligned}\tilde{\boldsymbol{j}}_g^{CA}(\boldsymbol{r},\boldsymbol{R},\dot{\boldsymbol{R}}) &= \frac{\hbar}{2mi}\left[\tilde{\psi}_g^{CA*}(\boldsymbol{r})\nabla\tilde{\psi}_g^{CA}(\boldsymbol{r}) - \tilde{\psi}_g^{CA}(\boldsymbol{r})\nabla\tilde{\psi}_g^{CA*}(\boldsymbol{r})\right] = \sum_J \left(\frac{\partial \boldsymbol{j}_g^{CA}(\boldsymbol{r})}{\partial \dot{\boldsymbol{R}}_J}\right)_{0,0} \cdot \dot{\boldsymbol{R}}_J \\ &= -\frac{\hbar}{m}\sum_J\left[\left(\frac{\partial \psi_g^{CA}(\boldsymbol{r})}{\partial \dot{\boldsymbol{R}}_J}\right)_{0,0}\nabla\psi_g^0(\boldsymbol{r}) - \psi_g^0(\boldsymbol{r})\nabla\left(\frac{\partial \psi_g^{CA}(\boldsymbol{r})}{\partial \dot{\boldsymbol{R}}_J}\right)_{0,0}\right]\cdot \dot{\boldsymbol{R}}_J\end{aligned} \tag{B29}$$

The vibrational version of the conservation equation in Equation (B27) can be verified by writing the probability and current densities in terms of vibronic coupling formalism by using the complete adiabatic wavefunction from Equation (2.87)

$$\begin{aligned}\tilde{\psi}_g^{CA}(\boldsymbol{r},\boldsymbol{R},\dot{\boldsymbol{R}}) &= \psi_g^A(\boldsymbol{r},\boldsymbol{R}) + i\psi_g^{CA}(\boldsymbol{r},\boldsymbol{R},\dot{\boldsymbol{R}}) \\ &= \psi_g^0(\boldsymbol{r}) + \sum_{J,e\neq g} C_{eg,\alpha}^{J,0}\psi_e^0(\boldsymbol{r})\left(R_{J\alpha} + i(\omega_{eg}^0)^{-1}\dot{R}_{J\alpha}\right)\end{aligned} \tag{B30}$$

This gives the following expressions for the vibrational derivatives of the probability and current densities as:

$$\left(\frac{\partial \rho_g^A(\boldsymbol{r})}{\partial \boldsymbol{R}_J}\right)_0 = \sum_{e\neq g}\psi_g^0(\boldsymbol{r})\boldsymbol{C}_{eg}^{J,0}\psi_e^0(\boldsymbol{r})\psi_g^0(\boldsymbol{r}) = \sum_{e\neq g}\boldsymbol{C}_{eg}^{J,0}\Theta_{ge}^0(\boldsymbol{r}) \tag{B31}$$

$$\left(\frac{\partial \boldsymbol{j}_g^{CA}(\boldsymbol{r})}{\partial \dot{\boldsymbol{R}}_J}\right)_{0,0} = \sum_{e\neq g}\frac{\boldsymbol{C}_{eg}^{J,0}}{\omega_{eg}^0}\left[\frac{\hbar}{2m}\left(\psi_g^0(\boldsymbol{r})\nabla\psi_e^0(\boldsymbol{r}) - \psi_e^0(\boldsymbol{r})\nabla\psi_g^0(\boldsymbol{r})\right)\right] = \sum_{e\neq g}\frac{\boldsymbol{C}_{eg}^{J,0}}{\omega_{eg}^0}\boldsymbol{J}_{ge}^0(\boldsymbol{r}) \tag{B32}$$

Finally, substitution of these two equations into Equation (B27) yields

$$-\nabla\cdot\sum_{e\neq g}\frac{\boldsymbol{C}_{eg}^{J,0}}{\omega_{eg}^0}\boldsymbol{J}_{ge}^0(\boldsymbol{r}) = \sum_{e\neq g}\boldsymbol{C}_{eg}^{J,0}\Theta_{ge}^0(\boldsymbol{r}) \tag{B33}$$

This equation proves the validity of Equation (B27) as for each state in the sum over excited states *e* obeys the conservation equation for transition between the pair of states, *eg*, namely,

$$-\nabla \cdot \frac{1}{\omega_{eg}^{0}} \boldsymbol{J}_{ge}^{0}(\boldsymbol{r}) = \Theta_{ge}^{0}(\boldsymbol{r}) \tag{B34}$$

This equation is identical with Equation (B21) given above for the continuity equation for electronic transitions.

References

Freedman, T.B., Gao, X., Shih, M.-L., and Nafie, L.A. (1998) Electron transition current density in molecules. 2. *Ab initio* calculations for electronic transitions in ethylene and formaldehyde. *J. Phys. Chem. A*, **102**, 3352–3357.

Freedman, T.B., Lee, E., and Nafie, L.A. (2000a) Vibrational current density in (*S*)-methyl lactate: Visualizing the origin of the methine-stretching vibrational circular dichroism intensity. *J. Phys. Chem. A*, **104**, 3944–3951.

Freedman, T.B., Lee, E., and Nafie, L.A. (2000b) Vibrational transition current density in (2*S*,3*S*)-oxirane-d_2: Visualizing electronic and nuclear contributions to IR absorption and vibrational circular dichroism intensities. *J. Mol. Struct.*, **550–551**, 123–134.

Freedman, T.B., Shih, M.-L., Lee, E., and Nafie, L.A. (1997) Electron transition current density in molecules. 3. *Ab initio* calculations of vibrational transitions in ethylene and formaldehyde. *J. Am. Chem. Soc.*, **119**, 10620–10626.

Nafie, L.A. (1997a) Electron transition current density in molecules. 1. Non-Born-Oppenheimer theory of vibronic and vibrational transition. *J. Phys. Chem. A*, **101**, 7826–7833.

Nafie, L.A. (1997b) Infrared and Raman vibrational optical activity: Theoretical and experimental aspects. *Annu. Rev. Phys. Chem.*, **48**, 357–386.

Appendix C

Theory of VCD for Molecules with Low-Lying Excited Electronic States

In this Appendix, we present the basic expression applicable to describing the vibronic wavefunction, transition matrix elements, and intensities for vibrational transitions in the ground electronic state where the usual approximation that the energies of all excited electronic states are large compared to the spacing between vibrational levels within electronic states is not made (Nafie, 2004). This occurs mainly in transition metal or rare earth metal complexes with low-lying excited electronic states that are comparable in energy to the fundamental vibrational transitions in such molecules.

C.1 Background Theoretical Expressions

A key step in the development of the quantum mechanical theory of VCD at the level of vibronic wavefunctions is the approximation that excited-state vibronic detail in the lowest-order Born–Oppenheimer correction term can be neglected as unimportant. The vibronic molecular wavefunction just before this approximation is invoked is given in Equation (4.29), which is reproduced here as Equation (C1). This wavefunction is the sum of the usual factorable Born–Oppenheimer adiabatic (A) vibronic wavefunction and the lowest-order non-Born–Oppenheimer non-adiabatic (NA) vibronic wavefunction. In Chapter 4 we referred to this wavefunction as pre-complete adiabatic (pCA) because it introduced explicitly the nuclear velocity dependence of the wavefunction, but it was still not yet factorable because we had retained vibronic detail in the energy denominator and a summation over all the excited state vibrational wavefunctions $\phi_{su}(\boldsymbol{R})$. We express here for the ground electronic-state wavefunction in a zeroth vibrational state with subscript A instead of J for the nucleus index

$$\begin{aligned}\Psi_{g0}^{pCA}(\boldsymbol{r},\boldsymbol{R},\dot{\boldsymbol{R}}) &= \Psi_{g0}^{A}(\boldsymbol{r},\boldsymbol{R})+\Psi_{g0}^{NA}(\boldsymbol{r},\boldsymbol{R},\dot{\boldsymbol{R}})\\ &= \psi_{g}^{A}(\boldsymbol{r},\boldsymbol{R})\phi_{g0}(\boldsymbol{R})+i\hbar\sum_{ev\neq g0}\sum_{A}\frac{\langle\psi_e^A\phi_{ev}|\dot{R}_{J,\alpha}(\partial/\partial R_{A,\alpha})_{elec}|\psi_e^A\phi_{g0}\rangle}{E_{ev}-E_{g0}}\psi_e(\boldsymbol{r},\boldsymbol{R})\phi_{ev}(\boldsymbol{R})\end{aligned} \quad \text{(C1)}$$

Vibrational Optical Activity: Principles and Applications, First Edition. Laurence A. Nafie.

This NA term of this wavefunction is a perturbation expansion of the adiabatic vibronic wavefunctions over all the excited states of the molecule with the nuclear kinetic energy as the perturbation operator, and in particular the perturbation with one part of the kinetic-energy operator acting on the electronic wavefunction and the other part acting on the vibrational wavefunction. We next carry out a few steps to bring the adiabatic and non-adiabatic terms closer in form to one another. For the adiabatic term, we expand the electronic part of the wavefunction to first order in nuclear-position dependence, and for the non-adiabatic term we express the electronic and vibrational parts of the perturbation matrix elements in the summation over excited states as a product of matrix elements.

$$\begin{aligned}\Psi_{g0}^{pCA}(\boldsymbol{r},\boldsymbol{R}) &= \psi_g^0(\boldsymbol{r})\phi_{g0}(\boldsymbol{R}) + \sum_{e\neq g}\sum_A \left[\langle\psi_e^0|(\partial\psi_g/\partial R_{A,\alpha})_0\rangle\psi_e^0(\boldsymbol{r})R_{A,\alpha}\right]\phi_{g0}(\boldsymbol{R}) \\ &+ i\hbar\sum_{ev\neq g0}\sum_A \frac{\langle\psi_e^0|(\partial\psi_g/\partial R_{A,\alpha})_0\rangle\langle\phi_{ev}|\dot{R}_{A,\alpha}|\phi_{g0}\rangle}{E_{ev}^0 - E_{g0}^0}\psi_e^0(\boldsymbol{r})\phi_{ev}(\boldsymbol{R})\end{aligned} \tag{C2}$$

This wavefunction is still not factorable because the summation over excited states depends on the vibronic detail, in the numerator and the denominator of the NA part of the wavefunction. We shall return to consider this wavefunction further, but first, for reference, we go on to invoke the critical approximation that leads to the complete adiabatic (CA) wavefunction.

To achieve a factorable wavefunction and separation of the electronic and vibrational parts of the wavefunction characteristic of the CA approximation, the vibronic detail in the denominator is removed by $E_{ev}^0 - E_{g0}^0 \approx E_e^0 - E_g^0$, which permits the summation over excited-state vibrational wavefunctions to be carried out to closure as $\sum_v|\phi_{ev}\rangle\langle\phi_{ev} = 1|$ in Equation (C2) to yield

$$\begin{aligned}\Psi_{g0}^{CA}(\boldsymbol{r},\boldsymbol{R},\dot{\boldsymbol{R}}) = \Bigg\{\psi_g^0(\boldsymbol{r}) + \sum_{e\neq g}\sum_A \Bigg[&\langle\psi_e^0|(\partial\psi_g/\partial R_{A,\alpha})_0\rangle R_{A,\alpha} \\ &+ i\hbar\frac{\langle\psi_e^0|(\partial\psi_g/\partial R_{A,\alpha})_0\rangle}{E_e^0 - E_g^0}\dot{R}_{A,\alpha}\Bigg]\psi_e^0(\boldsymbol{r})\Bigg\}\phi_{g0}(\boldsymbol{R})\end{aligned} \tag{C3}$$

The electronic part of the wavefunction between the curled braces can be consolidated by writing

$$\Psi_{g0}^{CA}(\boldsymbol{r},\boldsymbol{R},\dot{\boldsymbol{R}}) = \left\{\psi_g^0(\boldsymbol{r}) + \sum_{e\neq g}\sum_A \langle\psi_e^0|(\partial\psi_g/\partial R_{A,\alpha})_0\rangle\psi_e^0(\boldsymbol{r})\left[R_{A,\alpha} + \frac{i\hbar\dot{R}_{A,\alpha}}{E_s^0 - E_e^0}\right]\right\}\phi_{g0}(\boldsymbol{R}) \tag{C4}$$

Using the CA wavefunction the rotational strength and the dipole strengths can be written in either the position or the velocity forms for a vibrational transition between the zeroth and first vibrational levels of normal mode a in the ground electronic state as:

$$D_{g1,g0}^a = \left|\left\langle\Psi_{g1}^a\middle|\mu_\beta\middle|\Psi_{g0}^a\right\rangle\right|^2 \quad D_{g1,g0}^a = \omega_a^{-2}\left|\left\langle\Psi_{g1}^a\middle|\dot{\mu}_\beta\middle|\Psi_{g0}^a\right\rangle\right|^2 \tag{C5}$$

$$R_{g1,g0}^a = \mathrm{Im}\left[\left\langle\Psi_{g0}^a\middle|\mu_\beta\middle|\Psi_{g1}^a\right\rangle\cdot\left\langle\Psi_{g1}^a\middle|m_\beta\middle|\Psi_{g0}^a\right\rangle\right] \tag{C6}$$

$$R_{g1,g0}^a = \omega_a^{-1}\mathrm{Re}\left[\left\langle\Psi_{g0}^a\middle|\dot{\mu}_\beta\middle|\Psi_{g1}^a\right\rangle\cdot\left\langle\Psi_{g1}^a\middle|m_\beta\middle|\Psi_{g0}^a\right\rangle\right] \tag{C7}$$

The electric and magnetic dipole transition matrix elements that appear in these expressions can be written as in Chapters 2 and 4 as:

$$\langle\Psi^a_{g1}|\mu^E_\beta|\Psi^a_{g0}\rangle = 2\sum_{e\neq g}\sum_A \langle\psi^0_g|\mu^E_\beta|\psi^0_e\rangle\langle\psi^0_e|(\partial\psi_g/\partial R_{A,\alpha})_0\rangle S_{A\alpha,a}\langle\phi^a_{g1}|Q_a|\phi^a_{g0}\rangle \tag{C8}$$

$$\langle\Psi^a_{g1}|\dot{\mu}^E_\beta|\Psi^a_{g0}\rangle = 2i\hbar\sum_{e\neq g}\sum_A \frac{\langle\psi^0_g|\dot{\mu}^E_\beta|\psi^0_e\rangle\langle\psi^0_e|(\partial\psi_g/\partial R_{A,\alpha})_0\rangle S_{A\alpha,a}\langle\phi^a_{g1}|P_a|\phi^a_{g0}\rangle}{E^0_e - E^0_g} \tag{C9}$$

$$\langle\Psi^a_{g1}|m^E_\beta|\Psi^a_{g0}\rangle = 2i\hbar\sum_{e\neq g}\sum_A \frac{\langle\psi^0_g|m^E_\beta|\psi^0_e\rangle\langle\psi^0_e|(\partial\psi_g/\partial R_{A,\alpha})_0\rangle S_{A\alpha,a}\langle\phi^a_{g1}|P_a|\phi^a_{g0}\rangle}{E^0_e - E^0_g} \tag{C10}$$

C.2 Lowest-Order Vibronic Theory Including Low-Lying Electronic States

We now return to the wavefunction given in Equation (C2) where excited-state vibronic detail is still retained. This wavefunction can be used to obtain more general expressions for the transition moments in Equations (C8),(C9) and (C10). The non-factorable wavefunctions using normal nuclear coordinates for mode a and the transformation S-vectors, $S_{A\alpha,a} = (\partial R_{A\alpha}/\partial Q_a)_{Q=0} = (\partial \dot{R}_{A\alpha}/\partial P_a)_{Q=0}$ are given by:

$$\Psi^a_{g0} = \psi^0_g\phi^a_{g0} + \sum_{e\neq g}\sum_A \langle\psi^0_e|(\partial\psi_g/\partial R_{A,\alpha})_0\rangle Q_a S_{A\alpha,a}\psi^0_e\phi^a_{g0}$$
$$+ i\hbar\sum_{ev\neq g0}\sum_A \frac{\langle\psi^0_e|(\partial\psi_g/\partial R_{A,\alpha})_0\rangle\langle\phi_{ev}|P_a|\phi^a_{g0}\rangle S_{A\alpha.a}}{E^0_{ev} - E^0_{g0}}\psi^0_e\phi_{ev} \tag{C11}$$

$$\Psi^{a*}_{g1} = \psi^0_g\phi^a_{g1} + \sum_{e\neq g}\sum_A \langle\psi^0_e|(\partial\psi_g/\partial R_{A,\alpha})_0\rangle Q_a S_{A\alpha,a}\psi^0_e\phi^a_{g1}$$
$$+ i\hbar\sum_{ev\neq g1}\sum_A \frac{\langle\psi^0_g|(\partial\psi_e/\partial R_{A,\alpha})_0\rangle\langle\phi^a_{g1}|P_a|\phi_{ev}\rangle S_{A\alpha.a}}{E^0_{ev} - E^0_{g1}}\psi^0_e\phi_{ev} \tag{C12}$$

For complex conjugate wavefunction, Ψ^{a*}_{g1}, we have used the relationship $\langle\phi_{ev}|P_a|\phi^a_{g1}\rangle = -\langle\phi^a_{g1}|P_a|\phi_{ev}\rangle$ to restore a positive sign to the imaginary term.

Insertion of these wavefunctions into electronic contributions to Equation (C8) for the position form of the electric-dipole transition moment yields

$$\langle\Psi^a_{g1}|\mu^E_\beta|\Psi^a_{g0}\rangle = \left[2\sum_{e\neq g}\sum_A \langle\psi^0_g|\mu^E_\beta|\psi^0_e\rangle\langle\psi^0_e|(\partial\psi_g/\partial R_{A,\alpha})_0\rangle\langle\phi^a_{g1}|Q_a|\phi^a_{g0}\rangle\right.$$
$$+ i\hbar\sum_{ev\neq g0}\sum_A \frac{\langle\psi^0_g|\mu^E_\beta|\psi^0_e\rangle\langle\psi^0_e|(\partial\psi_g/\partial R_{A,\alpha})_0\rangle\langle\phi^a_{g1}|\phi_{ev}\rangle\langle\phi_{ev}|P_a|\phi^a_{g0}\rangle}{E^0_{ev} - E^0_{g0}}$$
$$\left.+ i\hbar\sum_{ev\neq g0}\sum_A \frac{\langle\psi^0_g|(\partial\psi_e/\partial R_{A,\alpha})_0\rangle\langle\psi^0_e|\mu^E_\beta|\psi^0_g\rangle\langle\phi^a_{g1}|P_a|\phi_{ev}\rangle\langle\phi_{ev}|\phi^a_{g0}\rangle}{E^0_{ev} - E^0_{g1}}\right]S_{A\alpha,a} \tag{C13}$$

In contrast to Equation (C8), this equation contains three terms instead of one. The first arises from within the BO approximation and the next two are due to non-BO contributions that normally do not appear at the lowest level of the position form of the electric-dipole transition moment. For the two non-BO terms, the first arises from substitution of non-BO part of Equation (C11) for the initial vibronic state, $g0$, whereas the second term represents substitution of the non-BO part of Equation (C12) for the final state, $g1$. In the absence of low-lying electronic states (LLESs) where the separation between excited electronic states is large compared with the vibronic sublevel energies, Equation (C13) reduces to the standard expression given in Equation (C8). This is accomplished by removing vibronic detail from the denominators in Equation (C13) and summing over excited vibronic states to closure. The two non-BO terms, when the electronic matrix elements are put in the same form as carried out below, have opposite signs and cancel, and make a zero non-BO contribution.

The corresponding expressions for the velocity form of the electric dipole transition moment starting from the electronic part of Equation (C9) are given by;

$$\langle \Psi^a_{g1} | \dot{\mu}^E_\beta | \Psi^a_{g0} \rangle = i\hbar \sum_{ev \neq g0} \sum_A \left[\frac{\langle \psi^0_g | \dot{\mu}^E_\beta | \psi^0_e \rangle \langle \psi^0_e | (\partial \psi_g / \partial R_{A,\alpha})_0 \rangle \langle \phi^a_{g1} | \phi_{ev} \rangle \langle \phi_{ev} | P_a | \phi^a_{g0} \rangle}{E^0_{ev} - E^0_{g0}} + \frac{\langle \psi^0_g | (\partial \psi_e / \partial R_{A,\alpha})_0 \rangle \langle \psi^0_e | \dot{\mu}^E_\beta | \psi^0_g \rangle \langle \phi^a_{g1} | P_a | \phi_{ev} \rangle \langle \phi_{ev} | \phi^a_{g0} \rangle}{E^0_{ev} - E^0_{g1}} \right] S_{A\alpha,a} \tag{C14}$$

Here the two non-BO terms in this equation do not cancel when the vibronic detail is removed and the electronic matrix elements are brought into the same form. Both represent the entire corrected contribution to the transition moment. The corresponding two terms in Equation (C13) serve only to correct the main BO term.

Similarly, substitution of Equations (C11) and (C12) into the electronic contribution to Equation (C10) for the magnetic-dipole transition moments yields,

$$\langle \Psi^a_{g1} | m^E_\beta | \Psi^a_{g0} \rangle = i\hbar \sum_{e(A3.2.3) \neq g0} \sum_A \left[\frac{\langle \psi^0_g | m^E_\beta | \psi^0_e \rangle \langle \psi^0_e | (\partial \psi_g / \partial R_{A,\alpha})_0 \rangle \langle \phi^a_{g1} | \phi_{ev} \rangle \langle \phi_{ev} | P_a | \phi^a_{g0} \rangle}{E^0_{ev} - E^0_{g0}} + \frac{\langle \psi^0_g | (\partial \psi_e / \partial R_{A,\alpha})_0 \rangle \langle \psi^0_e | m^E_\beta | \psi^0_g \rangle \langle \phi^a_{g1} | P_a | \phi_{ev} \rangle \langle \phi_{ev} | \phi^a_{g0} \rangle}{E^0_{ev} - E^0_{g1}} \right] S_{A\alpha,a} \tag{C15}$$

This equation closely follows the velocity form of the electric-dipole transition moment in Equation (C14). Equations (C14) and (C15) reduce to the standard CA expressions given in Equations (C9) and (C10), respectively, by eliminating the vibronic detail in the denominators, as was described for the reduction of Equation (C13) to Equation (C8).

C.3 Vibronic Energy Approximation

To simplify these equations that allow LLESs, the following approximation is used. It can be argued an LLES is not likely to have potential-energy surfaces that differ significantly from those of the ground state. This holds in particular for metal-centered *d–d* or *f–f* transitions that have little effect on the vibrational motion of the associated ligands. A possible exception to this assumption may arise for complexes with sufficiently high symmetry that the ground electronic state or the LLESs are degenerate. In such cases, the molecule will distort its geometry in the degenerate state in a way that splits the degeneracy in the so-called the Jahn–Teller effect. Under the assumption that any Jahn–Teller effect is not large compared with the splitting imposed by the coordination geometry of the complex, the following approximation can be invoked:

$$\phi_{ev} = \phi_{ev}^{a} = \phi_{gv}^{a} \tag{C16}$$

This approximation may still work well for many vibrational modes in molecules with Jahn–Teller distortion, but only comparisons between theory and experiment can address this point. Using the approximation of Equation (C16) in Equation (C13) and invoking harmonic oscillator selection rules allows evaluation of the summation over excited vibrational states, v in the non-adiabatic terms

$$\begin{aligned}
\langle \Psi_{g1}^{a} | \mu_{\beta}^{E} | \Psi_{g0}^{a} \rangle = \Bigg[& 2\sum_{e \neq g}\sum_{A} \langle \psi_{g}^{0} | \mu_{\beta}^{E} | \psi_{e}^{0} \rangle \langle \psi_{e}^{0} | (\partial \psi_{g} / \partial R_{A,\alpha})_{0} \rangle \langle \phi_{g1}^{a} | Q_{a} | \phi_{g0}^{a} \rangle \\
& + i\hbar \sum_{e \neq g}\sum_{A} \frac{\langle \psi_{g}^{0} | \mu_{\beta}^{E} | \psi_{e}^{0} \rangle \langle \psi_{e}^{0} | (\partial \psi_{g} / \partial R_{A,\alpha})_{0} \rangle \langle \phi_{g1}^{a} | \phi_{g1}^{a} \rangle \langle \phi_{g1}^{a} | P_{a} | \phi_{g0}^{a} \rangle}{E_{e}^{0} - E_{g}^{0} + \hbar\omega_{a}} \\
& + i\hbar \sum_{e \neq g}\sum_{A} \frac{\langle \psi_{g}^{0} | (\partial \psi_{e} / \partial R_{A,\alpha})_{0} \rangle \langle \psi_{e}^{0} | \mu_{\beta}^{E} | \psi_{g}^{0} \rangle \langle \phi_{g1}^{a} | P_{a} | \phi_{g0}^{a} \rangle \langle \phi_{g0}^{a} | \phi_{g0}^{a} \rangle}{E_{e}^{0} - E_{g}^{0} - \hbar\omega_{a}} \Bigg] S_{A\alpha,a}
\end{aligned} \tag{C17}$$

Here we can explicitly evaluate the vibronic energy difference in the denominator, which differs from the CA approximation of ignoring the excited state detail by a quantum of vibrational energy. This expression can be simplified further by converting the electronic matrix elements into the same form by interchanging wavefunctions, with a change of sign of the nuclear-derivative matrix element but not the dipole-moment matrix element, and taking into account of the normalization of the vibrational wavefunctions,

$$\begin{aligned}
\langle \Psi_{g1}^{a} | \mu_{\beta}^{E} | \Psi_{g0}^{a} \rangle = \Bigg[& 2\sum_{e \neq g}\sum_{A} \langle \psi_{g}^{0} | \mu_{\beta}^{E} | \psi_{e}^{0} \rangle \langle \psi_{e}^{0} | (\partial \psi_{g} / \partial R_{A,\alpha})_{0} \rangle \langle \phi_{g1}^{a} | Q_{a} | \phi_{g0}^{a} \rangle \\
& + i\hbar \sum_{e \neq g}\sum_{A} \langle \psi_{g}^{0} | \mu_{\beta}^{E} | \psi_{e}^{0} \rangle \langle \psi_{e}^{0} | (\partial \psi_{g} / \partial R_{A,\alpha})_{0} \rangle \\
& \times \left(\frac{1}{E_{eg}^{0} + \hbar\omega_{a}} - \frac{1}{E_{eg}^{0} - \hbar\omega_{a}} \right) \langle \phi_{g1}^{a} | P_{a} | \phi_{g0}^{a} \rangle \Bigg] S_{A\alpha,a}
\end{aligned} \tag{C18}$$

It is now even easier to see that this equation reduces to Equation (C8) when the energy difference between excited and ground electronic states is large compared with vibrational energy spacing. The non-BO term simply vanishes. Despite the simplifying nature of the approximation leading to Equation (C18), the combination of excited-state energies and particular vibrational normal-mode frequencies mixes the contributions of the electrons and nuclei to the transition moment in a non-separable way, in keeping with the non-BO nature of including correction terms for LLESs. The BO and non-BO terms in Equation (C18) can be brought into closer form in the following way. The momentum (velocity) vibrational matrix element in the non-BO term can be converted into the position form of this matrix element by using $\langle\phi^a_{g1}|P_a|\phi^a_{g0}\rangle = i\omega_a\langle\phi^a_{g1}|Q_a|\phi^a_{g0}\rangle$. After combining the energy terms over a common denominator, one obtains

$$\langle\Psi^a_{g1}|\mu^E_\beta|\Psi^a_{g0}\rangle = 2\sum_{e\neq g}\sum_A \langle\psi^0_g|\mu^E_\beta|\psi^0_e\rangle\langle\psi^0_e|(\partial\psi_g/\partial R_{A,\alpha})_0\rangle S_{A\alpha,a}\left(1+\frac{\omega_a^2}{(\omega^0_{eg})^2-\omega_a^2}\right)\langle\phi^a_{g1}|Q_a|\phi^a_{g0}\rangle \tag{C19}$$

One additional algebraic simplification gives

$$\langle\Psi^a_{g1}|\mu^E_\beta|\Psi^a_{g0}\rangle = 2\sum_{e\neq g}\sum_A \langle\psi^0_g|\mu^E_\beta|\psi^0_e\rangle\langle\psi^0_e|(\partial\psi_g/\partial R_{A,\alpha})_0\rangle S_{A\alpha,a}\left(\frac{(\omega^0_{eg})^2}{(\omega^0_{eg})^2-\omega_a^2}\right)\langle\phi^a_{g1}|Q_a|\phi^a_{g0}\rangle \tag{C20}$$

In Equation (C19), the first term is the BO contribution and the second is the non-BO correction. In the limit where the electronic energy spacing is large relative to the vibrational energy spacing, the correction term vanishes and this expression again reduces to the standard CA expression given by Equation (C8). Equation (C20) is the generalized expression for the position form of the electric dipole transition moment taking into account the possible close approach of an excited electronic state to the energy region of vibrational transitions. Equations (C19) and (C20) represent two equivalent algebraic ways of expressing the frequency-dependent correction associated with the presence of an LLES: one is the standard term plus a correction term and the other is a modified standard term in a more compact representation.

From Equation (C20), it is possible to write a generalization of the electronic atomic polar tensor (ATP) defined in Equation (2.74) as:

$$E^A_{r,\alpha\beta}(\omega_a) = 2\sum_{e\neq g}\langle\psi^0_g|\mu^E_\beta|\psi^0_e\rangle\langle\psi^0_e|(\partial\psi_g/\partial R_{A,\alpha})_0\rangle\left(\frac{(\omega^0_{eg})^2}{(\omega^0_{eg})^2-\omega_a^2}\right) \tag{C21}$$

Here the ATP is no longer independent of the normal modes of the molecule, as indicated by its parametric dependence on the vibrational frequency of the ath normal mode. The, inclusion of vibronic detail in BO correction terms introduces an interdependence of the vibrational motion on the electronic response of the molecule. Thus, in the presence of LLESs, one must envision a *set* of electronic ATPs for each atom A in the molecule, one for each normal mode a. The factor in parentheses in Equation (C21) depends on both the frequencies of the excited electronic states and the individual vibrational modes in a non-separable way.

The velocity form of the electric dipole transition moment can be developed in a similar way. Here both the matrix elements of the dipole velocity and nuclear derivative change sign upon interchange of the electronic wavefunctions needed to bring the two non-BO terms of Equation (C14) into the same form,

$$\langle\Psi^a_{g1}|\dot{\mu}^E_\beta|\Psi^a_{g0}\rangle = i\hbar\sum_{e\neq g}\sum_A \langle\psi^0_g|\dot{\mu}^E_\beta|\psi^0_e\rangle\langle\psi^0_e|(\partial\psi_g/\partial R_{A,\alpha})_0\rangle S_{A\alpha,a}$$
$$\times\left(\frac{1}{E^0_{eg}+\hbar\omega_a}+\frac{1}{E^0_{eg}-\hbar\omega_a}\right)\langle\phi^a_{g1}|P_a|\phi^a_{g0}\rangle \tag{C22}$$

Combining the energy denominators and factoring out the pure electronic energy difference yields

$$\langle \Psi_{g1}^{a} | \dot{\mu}_{\beta}^{E} | \Psi_{g0}^{a} \rangle = 2i\hbar \sum_{e \neq g} \sum_{A} \frac{\langle \psi_{g}^{0} | \dot{\mu}_{\beta}^{E} | \psi_{e}^{0} \rangle \langle \psi_{e}^{0} | (\partial \psi_{g} / \partial R_{A,\alpha})_{0} \rangle S_{A\alpha,a}}{E_{eg}^{0}} \left(\frac{(\omega_{eg}^{0})^{2}}{(\omega_{eg}^{0})^{2} - \omega_{a}^{2}} \right) \langle \phi_{g1}^{a} | P_{a} | \phi_{g0}^{a} \rangle \tag{C23}$$

The same correction factor appears here that appears in Equation (C21) where the reduction to the standard expression in Equation (C14) follows in the same way. The corresponding expression for the velocity form of the electronic ATP given is:

$$E_{v,\alpha\beta}^{A}(\omega_{a}) = 2i\hbar \sum_{e \neq g} \frac{\langle \psi_{g}^{0} | \dot{\mu}_{\beta}^{E} | \psi_{e}^{0} \rangle \langle \psi_{e}^{0} | (\partial \psi_{g} / \partial R_{A,\alpha})_{0} \rangle}{E_{eg}^{0}} \left(\frac{(\omega_{eg}^{0})^{2}}{(\omega_{eg}^{0})^{2} - \omega_{a}^{2}} \right) \tag{C24}$$

The hypervirial equation given in Chapter 2 in Equation (2.24) exactly converts the new generalized expressions for the velocity form of the APT in Equation (C24) into the corresponding position form in Equation (C21), and vice versa. Neglecting the vibrational frequency term in the denominators of these expressions reduces them to their standard forms for high-energy excited electronic states.

The LLES expressions for the electronic contribution to the magnetic dipole transition moment follow closely the corresponding expressions for the velocity form of the electric dipole transition moment.

$$\left\langle \Psi_{g1}^{a} \middle| m_{\beta}^{E} \middle| \Psi_{g0}^{a} \right\rangle = \left[i\hbar \sum_{e \neq g} \sum_{A} \left\langle \psi_{g}^{0} \middle| m_{\beta}^{E} \middle| \psi_{e}^{0} \right\rangle \left\langle \psi_{e}^{0} \middle| (\partial \psi_{g} / \partial R_{A,\alpha})_{0} \right\rangle \right.$$
$$\left. \times \left(\frac{1}{E_{eg}^{0} + \hbar\omega_{a}} + \frac{1}{E_{eg}^{0} - \hbar\omega_{a}} \right) \left\langle \phi_{g1}^{a} \middle| P_{a} \middle| \phi_{g0}^{a} \right\rangle \right] S_{A\alpha,a} \tag{C25}$$

The two energy terms can be combined over a common denominator and the pure electronic energy difference factored out to yield

$$\langle \Psi_{g1}^{a} | m_{\beta}^{E} | \Psi_{g0}^{a} \rangle = 2i\hbar \sum_{e \neq g} \sum_{A,\alpha} \frac{\langle \psi_{g}^{0} | m_{\beta}^{E} | \psi_{e}^{0} \rangle \langle \psi_{e}^{0} | (\partial \psi_{g} / \partial R_{A,\alpha})_{0} \rangle S_{A\alpha,a}}{E_{eg}^{0}} \left(\frac{(\omega_{eg}^{0})^{2}}{(\omega_{eg}^{0})^{2} - \omega_{a}^{2}} \right) \langle \phi_{g1}^{a} | P_{a} | \phi_{g0}^{a} \rangle \tag{C26}$$

This expression reduces to the traditional CA form given in Equation (C10) upon neglect of the vibrational frequency term in the denominator of Equation (C26). Finally, the general form of the electronic contribution to the AAT is given by:

$$I_{\alpha\beta}^{A}(\omega_{a}) = 2i\hbar \sum_{e \neq g} \frac{\langle \psi_{g}^{0} | m_{\beta}^{E} | \psi_{e}^{0} \rangle \langle \psi_{e}^{0} | (\partial \psi_{g} / \partial R_{A,\alpha})_{0} \rangle}{E_{eg}^{0}} \left(\frac{(\omega_{eg}^{0})^{2}}{(\omega_{eg}^{0})^{2} - \omega_{a}^{2}} \right) \tag{C27}$$

C.4 Low-Lying Magnetic-Dipole-Allowed Excited Electronic States

The expressions above do not depend on the nature of the excited electronic states. A common case is when an LLES is approximately electric-dipole forbidden and magnetic-dipole allowed. This occurs for the *d–d* transitions in transition metals with unfilled *d*-levels, and similarly for *f–f* transition in rare earth elements. In the limit of pure magnetic-dipole character, there is no effect of these LLES transitions on the electric-dipole transition moments, ATPs, or IR vibrational absorption intensities. On the other hand, the vibrational magnetic-dipole transition moments will be significantly affected and the more general expressions developed above must be utilized to calculate the corresponding atomic axial tensors and VCD intensities.

The generalized expression for the electronic contribution to the AAT given in Equation (C27) written as a sum of the conventional, frequency-independent term, $I^A_{\alpha\beta}$, plus a frequency-dependent correction term is given by:

$$I^A_{\alpha\beta}(\omega_a) = 2i\hbar \sum_{e \neq g} \frac{\langle \psi^0_g | m^E_\beta | \psi^0_e \rangle \langle \psi^0_e | (\partial \psi_g / \partial R_{A,\alpha})_0 \rangle}{E^0_{eg}} \left(1 + \frac{\omega^2_a}{(\omega^0_{eg})^2 - \omega^2_a} \right) \tag{C28}$$

The correction term in Equation (C28) is less than a 1% correction for electronic states when the transition energy for state *e* is an order of magnitude, or more, greater than the vibrational energy of the *a*th normal mode of the molecule. As a result, as a reasonable approximation, it is necessary to include the correction term only for LLESs. The standard term and the correction term are written in terms of AAT symbols in Equation (C29) where the primed term is the frequency-dependent correction term,

$$I^A_{\alpha\beta}(\omega_a) = I^A_{\alpha\beta} + I'^A_{\alpha\beta}(\omega_a) \tag{C29}$$

An approximation for the correction term is given in Equation (C30) where only LLESs labeled e' are included in the summation,

$$I'^A_{\alpha\beta}(\omega_a) \cong 2i\hbar \sum_{e'} \frac{\langle \psi^0_g | m^E_\beta | \psi^0_{e'} \rangle \langle \psi^0_{e'} | (\partial \psi_g / \partial R_{A,\alpha})_0 \rangle}{E^0_{e'g}} \left(\frac{\omega^2_a}{(\omega^0_{e'g})^2 - \omega^2_a} \right) \tag{C30}$$

A further division of these equations that is useful to consider is to separate the standard AAT tensor, $I^A_{\alpha\beta}$, into terms involving electronic states that are far from vibrational energies from the state or states that are low-lying and needed for the correction term in Equation (C30). This expression for the AAT is given by:

$$I^A_{\alpha\beta}(\omega_a) = I^A_{\alpha\beta}(e \neq e') + I^A_{\alpha\beta}(e') + I'^A_{\alpha\beta}(\omega_a) \tag{C31}$$

Here the first two terms depend only on the electronic energies in the usual way, and the last term is the same as that in Equation (C30). For the case of two molecules that possess identical bonding properties but differ in the presence or absence of LLESs, such as a transition metal complex with these properties, the first term in Equation (C31) applies to the molecule with no LLESs, whereas two correction terms apply to an otherwise identical molecule with LLESs. The first correction term can be calculated using standard VCD algorithms, whereas the second correction term must be added for each LLES and each vibrational mode under consideration.

If only the magnetic dipole transition moment needs correction for an LLES, this separation of correction carries directly forward to the rotational strength in a straightforward, linear manner. Thus, using the same notation as used for the atomic axial tensor, the rotational strength can be written as:

$$R^{a}_{g1,g0}(\omega_a) = R^{a}_{g1,g0}(e \neq e') + R^{a}_{g1,g0}(e') + R'^{a}_{g1,g0}(\omega_a) \qquad \text{(C32)}$$

and no corrections are needed for the ATPs and the dipole strength.

Reference

Nafie, L.A. (2004) Theory of vibrational circular dichroism and infrared absorption: Extension to molecules with low-lying excited electronic states. *J. Phys. Chem. A*, **108**, 7222–7231.

Appendix D

Magnetic VCD in Molecules with Non-Degenerate States

We present in this Appendix the perturbation theory for magnetic vibrational circular dichroism (MVCD) applicable to molecules without degenerate states. In general, magnetic circular dichroism (MCD) possesses three intensity mechanisms termed the A-, B-, and C-term mechanisms. A-term MCD arises from degenerate excited final states that are split in spectral frequency due to their opposite angular momentum properties. Such split states have opposite MCD intensity, and the splitting leads to a couplet intensity pattern that grows with magnetic field strength until the bandshapes of the split states no longer overlap, a rare occurrence. The B-term arises from magnetic field couplings between excited states. The C-term arises only for molecules that originate in degenerate ground states that are split by the magnetic field and populated to different degrees. The C-term of MCD is therefore temperature dependent and can be separated from the A-term by this dependence. Degenerate states only occur in molecules with threefold or higher symmetry, and as mentioned in Chapter 3, there is no requirement of chirality to observe MCD. On the other hand chiral molecules can exhibit MCD with the CD and MCD effects independent of each other to at least the first order in interaction of light with matter. For molecules of lower symmetry that possess no degenerate states, the only source of MCD intensity is the B-term mechanism.

Before beginning the development of B-term MVCD formalism, we note that the nuclei, unlike for natural VCD, do not make a contribution to MVCD. This is because there is no detectable response of the nuclei to the applied magnetic field. In particular, electronic currents arise due to the magnetic field but no such currents within nuclei contribute to MVCD intensities.

D.1 General Theory

The observed MVCD for a vibrational transition $g0$ to $g1$ in a low-symmetry molecule with no degenerate states is given by:

$$\Delta\varepsilon = \varepsilon_L - \varepsilon_R = Yh\nu B_{g0-g1}\beta H \tag{D1}$$

Vibrational Optical Activity: Principles and Applications, First Edition. Laurence A. Nafie.

where Y is a constant, $h\nu$ is the photon energy, β is the Bohr magneton, H is the applied magnetic field in the direction of the propagation of the light beam, and B_{g0-g1} is the MVCD B-term given by:

$$B_{g0,g1} = -\frac{2}{\beta}\sum_{ev}\text{Im}\left\{\langle\Psi_{g0}\big|\boldsymbol{\mu}\big|\Psi_{g1}\rangle \cdot \left[\frac{\langle\Psi_{g1}|\boldsymbol{\mu}|\Psi_{ev}\rangle\times\langle\Psi_{ev}|\boldsymbol{m}|\Psi_{g0}\rangle}{E_{ev}-E_{g0}} - \frac{\langle\Psi_{g1}|\boldsymbol{m}|\Psi_{ev}\rangle\times\langle\Psi_{ev}|\boldsymbol{\mu}|\Psi_{g0}\rangle}{E_{ev}-E_{g1}}\right]\right\} \tag{D2}$$

The corresponding ordinary infrared absorption for this transition is given by the dipole strength $D_{g1,g0}$

$$D_{g1,g0} = \langle\Psi_{g0}|\boldsymbol{\mu}|\Psi_{g1}\rangle\cdot\langle\Psi_{g1}|\boldsymbol{\mu}|\Psi_{g0}\rangle \tag{D3}$$

Using this same notation, natural VCD arising from molecular chirality in the absence of a magnetic field is given by the rotational strength

$$R_{g1,g0} = \text{Im}\langle\Psi_{g0}|\boldsymbol{\mu}|\Psi_{g1}\rangle\cdot\langle\Psi_{g1}|\boldsymbol{m}|\Psi_{g0}\rangle \tag{D4}$$

By analogy, the B-term of MVCD can be written as:

$$B_{g1,g0} = K\,\text{Im}\langle\Psi_{g0}|\boldsymbol{\mu}|\Psi_{g1}\rangle\cdot\langle\Psi_{g1}|\boldsymbol{m}_{mag}|\Psi_{g0}\rangle \tag{D5}$$

Here magnetic-field *induced* magnetic dipole transition moment can be written as

$$\langle\Psi_{g1}|\boldsymbol{m}_{mag}|\Psi_{g0}\rangle = \sum_{ev}\left[\frac{\langle\Psi_{g1}|\boldsymbol{\mu}|\Psi_{ev}\rangle\times\langle\Psi_{ev}|\boldsymbol{m}|\Psi_{g0}\rangle}{E_{ev}-E_{g0}} - \frac{\langle\Psi_{g1}|\boldsymbol{m}|\Psi_{ev}\rangle\times\langle\Psi_{ev}|\boldsymbol{\mu}|\Psi_{g0}\rangle}{E_{ev}-E_{g1}}\right] \tag{D6}$$

D.2 Combined Complete Adiabatic and Magnetic-Field Perturbation Formalism

In this section we provide from Chapter 4 the basic formalism of the CA and MFP wavefunctions that are needed to reduce Equation (D6) to a more familiar form amenable to computation evaluation in terms of perturbed electronic wavefunctions. First we consider the electronic Hamiltonian as sum of the usual Born–Oppenheimer adiabatic term, the nuclear velocity perturbation term of the CA wavefunction, and finally the magnetic field perturbation term of the MFP formalism. Because the magnetic-dipole moment operator, $\boldsymbol{m}^E$, is imaginary, electronic wavefunctions with nuclear velocity and/or magnetic-field dependence are complex, denoted by an overscript tilda. From Equation (4.86) we have

$$\tilde{H}_E^{HCA}(\boldsymbol{R},\dot{\boldsymbol{R}},\boldsymbol{H}) = H_E^A(\boldsymbol{R}) - i\hbar\left(\frac{\partial}{\partial\boldsymbol{R}}\right)_E\cdot\dot{\boldsymbol{R}} - \boldsymbol{m}^E\cdot\boldsymbol{H} \tag{D7}$$

The resulting complex CA-MFP wavefunction, separable into electronic and vibrational parts is given by:

$$\tilde{\Psi}_{gv}^{HCA}(\boldsymbol{r},\boldsymbol{R},\dot{\boldsymbol{R}},\boldsymbol{H}) = \tilde{\psi}_g^{HCA}(\boldsymbol{r},\boldsymbol{R},\dot{\boldsymbol{R}},\boldsymbol{H})\varphi_{gv}(\boldsymbol{R}) \tag{D8}$$

The two imaginary perturbations result, to first order, in two addition imaginary terms in the electronic wavefunction written as:

$$\tilde{\psi}_g^{HCA}(\boldsymbol{r},\boldsymbol{R},\dot{\boldsymbol{R}},\boldsymbol{H}) = \psi_g^A(\boldsymbol{r},\boldsymbol{R}) + i\psi_g^{CA}(\boldsymbol{r},\boldsymbol{R}_0,\dot{\boldsymbol{R}}) + i\psi_g^H(\boldsymbol{r},\boldsymbol{R}_0,\boldsymbol{H}) \tag{D9}$$

More explicitly, from using first-order perturbation theory developed in Chapters 2 and 4, we have

$$\tilde{\psi}_g^{HCA}(\boldsymbol{R},\dot{\boldsymbol{R}},\boldsymbol{H}) = \psi_g^0 + \sum_A\sum_{e\neq g}\left[\langle\psi_e^0|(\partial\psi_g/\partial\boldsymbol{R}_A)_0\rangle\psi_e^0(\boldsymbol{R}_A + i\hbar\dot{\boldsymbol{R}}_A/E_{eg}^0) + \left\langle\psi_e^0|\boldsymbol{m}^E\middle|\psi_g^0\right\rangle\psi_e^0\cdot\boldsymbol{H}/E_{eg}^0\right] \tag{D10}$$

As a result, we can write derivatives of the wavefunction with respect to the normal coordinate momentum, or velocity because $P_a = \dot{Q}_a$, and magnetic field, H, respectively, as:

$$\left(\frac{\partial\tilde{\psi}_g^{CA}}{\partial P_a}\right)_{P=0} = i\left(\frac{\partial\psi_g^{CA}}{\partial P_a}\right)_{P=0} = \sum_{e\neq g}\frac{i\hbar\left\langle\psi_e^0|(\partial\psi_g/\partial Q_a)_0\right\rangle}{E_e^0 - E_g^0}\psi_e^0 \tag{D11}$$

$$\left(\frac{\partial\tilde{\psi}_g^{HCA}}{\partial H_\beta}\right)_{H=0} = i\left(\frac{\partial\psi_g^{H}}{\partial H_\beta}\right)_{H=0} = \sum_{e\neq g}\frac{\langle\psi_e^0|m_\beta^E|\psi_g^0\rangle}{E_e^0 - E_g^0}\psi_e^0 = \sum_{e\neq g}\frac{i\hbar\langle\psi_e^0|e\varepsilon_{\beta\gamma\delta}r_\gamma\dot{r}\delta/2c|\psi_g^0\rangle}{E_e^0 - E_g^0}\psi_e^0 \tag{D12}$$

The last expression in Equation (D12) shows explicitly the imaginary character of the magnetic-dipole moment operator and also the similarity to the nuclear velocity perturbation derivative in Equation (D11).

D.3 Vibronic Coupling B-Term Derivation

Using Cartesian tensor notation, Equation (D6) can be written in terms of the β-component of the magnetic-field induced moment operator, $m_{\beta,mag}$, as:

$$\langle\Psi_{g1}|m_{\beta,mag}|\Psi_{g0}\rangle = \varepsilon_{\beta\gamma\delta}\sum_{ev}\left[\frac{\langle\Psi_{g1}|\mu_\gamma|\Psi_{ev}\rangle\langle\Psi_{ev}|m_\delta|\Psi_{g0}\rangle}{E_{ev} - E_{g0}} - \frac{\langle\Psi_{g1}|m_\gamma|\Psi_{ev}\rangle\langle\Psi_{ev}|\mu_\delta|\Psi_{g0}\rangle}{E_{ev} - E_{g1}}\right] \tag{D13}$$

This equation can be viewed as the application of Rayleigh–Schrodinger perturbation on the initial, and final vibronic wavefunctions of the electric-dipole transition moment with an external magnetic field. Reversing the application of the perturbation, as is done in the derivation of the MFP expression for VCD described in Chapter 4, we can write

$$\langle \Psi_{g1} | m_{\beta,mag} | \Psi_{g0} \rangle = \varepsilon_{\beta\gamma\delta} \left[\left\langle \Psi_{g1} | \mu_\gamma \middle| \left(\frac{\partial \tilde{\Psi}_{g0}}{\partial H_\delta} \right)_{H=0} \right\rangle - \left\langle \left(\frac{\partial \tilde{\Psi}_{g1}}{\partial H_\gamma} \right)_{H=0} \middle| \mu_\delta | \Psi_{g0} \right\rangle \right] \tag{D14}$$

Here it can be seen that the magnetic field perturbation is acting in the first term on the initial state, $\tilde{\Psi}_{g0}$, the ground vibrational state of the molecule, and in the second term the final state, $\tilde{\Psi}_{g1}$, is perturbed. Factoring out the vibrational wavefunctions gives

$$\langle \Psi_{g1} | m_{\beta,mag} | \Psi_{g0} \rangle = \varepsilon_{\beta\gamma\delta} \langle \varphi_{g1} | \left[\left\langle \tilde{\psi}_g | \mu_\gamma \middle| \left(\frac{\partial \tilde{\psi}_g}{\partial H_\delta} \right)_{H=0} \right\rangle - \left\langle \left(\frac{\partial \tilde{\psi}_g}{\partial H_\gamma} \right)_{H=0} \middle| \mu_\delta | \tilde{\psi}_g \right\rangle \right] | \varphi_{g0} \rangle \tag{D15}$$

The electronic wavefunctions of the ground state, $\tilde{\psi}_g = \psi_g + i\psi'_g$, are written here as complex wavefunctions due to either the nuclear velocity or magnetic field perturbation terms. For simplicity we do not explicitly write the source of the perturbation as a superscript as this is clear from the context. The real part of $\tilde{\psi}_g$, namely, ψ_g, vanishes in this expression because in this case the second term can be written as:

$$\varepsilon_{\beta\gamma\delta} \left\langle \left(\frac{\partial \tilde{\psi}_g}{\partial H_\gamma} \right)_{H=0} \middle| \mu_\delta | \psi_g \right\rangle = -\varepsilon_{\beta\gamma\delta} \left\langle \left(\frac{\partial \tilde{\psi}_g}{\partial H_\delta} \right)_{H=0} \middle| \mu_\gamma | \psi_g \right\rangle = \varepsilon_{\beta\gamma\delta} \left\langle \psi_g | \mu_\gamma \middle| \left(\frac{\partial \tilde{\psi}_g}{\partial H_\delta} \right)_{H=0} \right\rangle \tag{D16}$$

The first equality occurs with a change of sign due to the interchange of Cartesian subscripts and the form of the alternating tensor, $\varepsilon_{\beta\gamma\delta}$. A positive sign is restored when the wavefunctions are interchanged bearing in mind that the wavefunction derivative with respect to magnetic field is pure imaginary, whereas no sign change occurs for ψ_g because it is real. It is clear that the first and second terms in Equation (D15) are equal for real ψ_g and therefore cancel.

If, on the other hand, the imaginary part of $\tilde{\psi}_g$ is retained, and in particular the part that carries the nuclear velocity dependence in the CA electronic wavefunction, we have

$$\langle \tilde{\Psi}^a_{g1} | m_{\beta,mag} | \Psi^a_{g0} \rangle = \varepsilon_{\beta\gamma\delta} \langle \varphi^a_{g1} | \left[\langle i\psi'_g | \mu_\gamma \middle| \left(\frac{\partial \tilde{\psi}_g}{\partial H_\delta} \right)_{H=0} \right\rangle - \left\langle \left(\frac{\partial \tilde{\psi}_g}{\partial H_\gamma} \right)_{H=0} \middle| \mu_\delta | i\psi' \rangle \right] | \varphi^a_{g0} \rangle$$

$$= 2\varepsilon_{\beta\gamma\delta} \langle \varphi^a_{g1} | \langle i\psi'_g | \mu_\gamma \middle| \left(\frac{\partial \tilde{\psi}_g}{\partial H_\delta} \right)_{H=0} \right\rangle | \varphi^a_{g0} \rangle \tag{D17}$$

In the last step, we have combined equal terms as there are three sign changes in converting the second term into the first for the imaginary perturbed wavefunction, $i\psi'_g$.

We next need to create an explicit normal coordinate dependence to keep the vibrational matrix element non-zero. The derivative with respect to Q_a cannot be taken because to first order the imaginary part of the complex electronic wavefunction has no dependence on Q_a. We therefore must take the derivative of the complex wavefunction with respect to P_a, which gives us the final result,

$$\langle \tilde{\Psi}^a_{g1} | m_{\beta,mag} | \Psi^a_{g0} \rangle = 2\varepsilon_{\beta\gamma\delta} \left\langle \left(\frac{\partial \tilde{\psi}_g}{\partial P_a} \right)_{P=0} \middle| \mu_\gamma \middle| \left(\frac{\partial \tilde{\psi}_g}{\partial H_\delta} \right)_{H=0} \right\rangle \langle \varphi^a_{g1} | P_a | \varphi^a_{g0} \rangle \tag{D18}$$

This expression for the B-term induced magnetic dipole transition moment is a mix of the two perturbation expressions for the magnetic-dipole moment associated with natural VCD, namely the NVP formalism adapted from Equation (4.84) as:

$$\langle \tilde{\Psi}^a_{g1} | m_\beta | \Psi^a_{g0} \rangle = 2 \left\langle \tilde{\psi}_g | m_\beta \left| \left(\frac{\partial \tilde{\psi}_g}{\partial P_a} \right)_{P=0} \right\rangle \right. \left\langle \varphi^a_{g1} | P_a | \varphi^a_{g0} \right\rangle \tag{D19}$$

and the MFP formalism adapted from Equation (4.85) given by:

$$\langle \tilde{\Psi}^a_{g1} | m_\beta | \Psi^a_{g0} \rangle = 2i\hbar \left\langle \left(\frac{\partial \psi_g}{\partial Q_a} \right)_{P=0} \middle| \left(\frac{\partial \tilde{\psi}_g}{\partial H_\beta} \right)_{H=0} \right\rangle \left\langle \varphi^a_{g1} | P_a | \varphi^a_{g0} \right\rangle \tag{D20}$$

From these expressions, in particular Equation (D18), it is clear that given formalism for the NVP electronic wavefunction, the induced magnetic-dipole transition moment of MVCD can be calculated from computation pieces used for VCD calculations without the need to carry out an explicit sum over all excited electronic states, which appears in the opening definition of MVCD B-term intensity Equation (D2).

The potential value of the measurement of B-term MVCD is its use as a third vibrational intensity expression, supplementing the VA and VCD spectra, for the analysis of the vibration mode structure and electron population and current densities modulated by the vibrational motion of the molecule. For instance, MVCD could be used as a further point of spectral reference in making assignments of absolute configuration using VCD and VA spectra. MVCD spectra for typical magnetic fields are weaker than VCD intensities, but given improvements in FT-VCD instrumentation and access in the future to higher superconducting magnetic fields, MVCD may some day prove a value additional spectroscopic probe of molecules for comparison with quantum chemistry calculations.

D.4 MCD from Transition Metal Complexes with Low-Lying Electronic States

As a first example of B-term MVCD, we present here the VA, VCD, ECD, MVCD, and MECD spectra in the hydrogen stretching region of five transition metal complexes of the chiral ligand (–)-sparteine (Ma, 2007). The structure of the (–)-sparteine ligand and its mode of complexation to transition metals, with two chloride ligands are shown in Figure D.1. Also shown are the associated vibrational and electronic spectra in this spectral region.

The VA spectra of all five complexes are nearly indistinguishable. Any low-lying excited electronic states (LLES) are primarily magnetic-dipole allowed and exhibit no discernable electronic absorption spectra. The VCD spectra for Zn(II) and Cu(II), which have no LLES, are the same and show VCD of normal magnitude. The complexes of Fe(II), Co(II), and Ni(II) show broad ECD and narrow CH-stretching VCD that are all different and enhanced by the LLES. A similar pattern occurs for the MCD spectra. The Zn(II) and Cu(II) complexes show very small if any MVCD spectra, and the Fe(II), Co(II), and Ni(II) complexes show enhanced MVCD spectra in the CH-stretching region superimposed on broad MECD spectra from the LLES. The magnetic field strength from a permanent magnet was 1.4 T, a modest field compared with superconducting magnets that are typically in the 7 T or higher range. The MCD were separated from the natural CD by reversing the field direction of the magnet relative to the IR beam from the FT-VCD spectrometer. These MCD spectra are the first measurements of MVCD for a chiral molecule

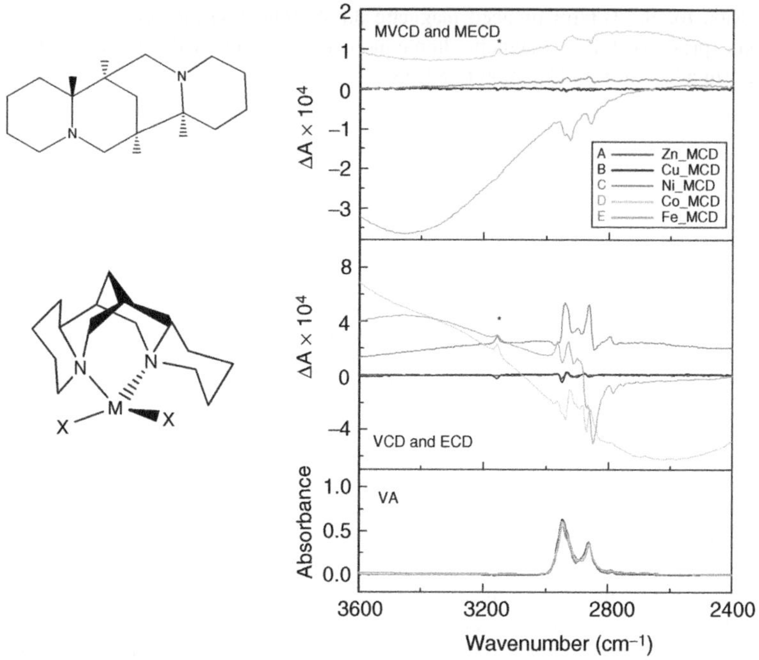

Figure D.1 *VA (lower), VCD and ECD (middle), and MVCD and MECD (upper) spectra (right) of (–)-sparteine transition metal complexes for Fe(II), Co(II), Ni(II), Cu(II), and Zn(II). The structure of (–)-sparteine ligand and its transition metal complexes are shown on the left*

and the first for which the B-term mechanism was the only source of MCD intensity. Some of the natural VCD spectra of these complexes were published previously (He *et al.*, 2001).

References

He, Y., Cao, X., Nafie, L.A., and Freedman, T.B. (2001) *Ab initio* VCD calculation of a transition-metal containing molecule and a new intensity enhancement mechanism for VCD. *J. Am. Chem. Soc.*, **123**, 11320–11321.

Ma, S. (2007). Mid-infrared and near-infrared vibrational circular dichroism: New methodologies for biological and pharmaceutical applications. Ph.D. thesis, Syracuse University.

Index

Vibrational Optical Activity: Principles and Applications, First Edition. Laurence A. Nafie.

Printed and bound by CPI Group (UK) Ltd, Croydon, CR0 4YY
16/04/2025
14658472-0003